Second Editio[n]

AUSTRALIAN *Bird* NAMES

ORIGINS AND MEANINGS

Ian Fraser and Jeannie Gray

PUBLISHING

A catalogue record for this book is available from the National Library of Australia.

ISBN: 9781486311637 (pbk.)
ISBN: 9781486311644 (epdf)
ISBN: 9781486311651 (epub)

Published by:

CSIRO Publishing
Locked Bag 10
Clayton South VIC 3169
Australia

Telephone: +61 3 9545 8400
Email: publishing.sales@csiro.au
Website: www.publish.csiro.au

Cover illustrations from the *Companion to Gould's Handbook; or Synopsis of the Birds of Australia* by Silvester Diggles

Edited by Peter Storer
Cover design by Andrew Weatherill
Typeset by Thomson Digital
Printed in China by Asia Pacific Offset Ltd

The paper this book is printed on is in accordance with the standards of the Forest Stewardship Council®. The FSC® promotes environmentally responsible, socially beneficial and economically viable management of the world's forests.

May19_01

Contents

Acknowledgments

As we wrote this book, we were more and more amazed and impressed by the naturalists of old: their extraordinary devotion to the collection and dissemination of knowledge, and the sheer volume of work many of them produced. And we were impressed by and grateful for the collaboration of libraries across the world, in making ancient and historic texts so readily available. In this we would particularly like to pay tribute to the National Library of Australia and the national Trove project, both operating under considerable financial constraints.

We would like to say how much gratification we have had from researching and writing *Australian Bird Names: Origins and Meanings* – it has been constantly interesting to us, as lovers of words and birds, and usually great fun too. Our collaboration has been smoother than we could have imagined – the times when a difference of opinion took more than one or two email exchanges to resolve could be counted on the toes of a cockatoo's foot. We both want to express too our great appreciation to our partners, Lou and David, for their love, patience and support.

We are grateful to CSIRO Publishing for taking on the project, and in particular to John Manger and Tracey Millen (now Tracey Kudis) for overseeing the production.

All of this remains true for this Second Edition, for which we are also greatly indebted to Briana Melideo, Publisher at CSIRO, for supporting our suggestion that an entire revision was warranted for this edition, and for encouraging the rewriting process. Tracey Kudis again oversaw the production, later assisted by Eloise Moir-Ford, and Lauren Webb negotiated the cover. Peter Storer was a most assiduous editor.

Illustrations

Acknowledgements are also due to the extraordinary Biodiversity Heritage Library, a consortium of natural history and botanical libraries from around the world that makes available digital versions of many marvels of writing and illustration. Special thanks go to the Smithsonian Institution Libraries who contributed to BHL the volume from which we have taken the illustrations for the current book. These are from the *Companion to Gould's Handbook; or, Synopsis of the Birds of Australia* by Silvester Diggles (1817–1880), published in Brisbane in 1877 by Thorne and Greenwell. Diggles was a musician, artist and natural historian who took it upon himself to provide for Australians an affordable guide to their country's birds, at a time when Gould's excellent but very costly *Birds of Australia* was the only book available. Only three volumes of Diggles' original large-scale work, *Ornithology of Australia*, were published but he subsequently reissued parts of it as the *Companion to Gould's Handbook*. All the illustrations were done by Diggles himself, though the signature of his lithographer Henry Eaton may be seen on some of them. Those we have chosen show the beauty, accuracy of observation and quirkiness (though not always all in the one illustration!) that characterise Diggles' work.

Introduction to the First Edition

Australian Bird Names: A Complete Guide was written to fill a gap we perceived in the literature, and has been a focus for our own thirsts for knowledge about words and names. We examined bird names in as much breadth and depth as possible, to give an accurate picture of their origins and, where possible, of the intentions of the name-givers.

As well as being a book about names this is also a book about the history of ever-developing understandings about birds, about the people who contributed and, most of all, about the birds themselves. Ultimately names are just human conceits; they can never be more important or more interesting than the organisms onto which we've foisted them.

This introduction to the main text gives a background to the names, common and scientific, then offers a brief guide to the order of proceedings and our suggested pronunciations. By taking a few minutes to read this introduction, we think you will find the answers to most of the questions that might arise as you read further into the book.

Introduction to the Second Edition

It is only 6 years since we wrote the first edition of this book, and it may seem surprising that a new edition is already warranted, but much has happened in that time, both to Australia's bird fauna and to their taxonomy. The former might seem a rash claim, but the fact is that some 55 species have been added to the national list in those years in the form of vagrants – mostly to the seas and islands off the north-west of Australia. There is a small industry now based on enabling determined and skilful birders to head out each year to an area focused on Christmas Island, Cocos (Keeling) Islands and Ashmore Reef to record errant seabirds and Asian migrants that have been blown off course or have simply continued flying south and overshot their intended destinations to the north of Australia. Most of the species we have added are those accepted by BARC (the Birdlife Australia Rarities Committee) whose deliberations can readily be found online via the Birdlife Australia website. In addition we have included a small number of reported vagrants on which BARC at the time of writing had not yet made determinations, but which Menkhorst *et al.* (2017) felt confident enough about to warrant inclusion in their recent comprehensive field guide.

This is a book about names. It is not intended as a definitive list of Australian birds, so, if a couple of species dealt with here are not formally on the national list, nothing is lost – in the interest of interpreting the names we preferred to err on the generous side.

Even more has happened on the taxonomic front. Ever more powerful and precise tools have become available to add to the analytical arsenal, to the extent that sequencing of complete mitochondrial DNA is pretty much standard practice, and the term most recent common ancestor (MRCA) is widespread in the literature. This is effectively a measure of the relatedness of any two species or higher taxa. Evidence of the confidence these techniques have inspired can be seen in the increasing convergence of opinions on formerly vexed and divisive taxonomic questions. Scientists are ultimately human, so, of course, there is not unanimity, but for the most part disagreements are now relatively minor compared with times not long past. This means that one's choice of which taxonomy to use is no longer as significant a decision as it once was, and the International Ornithologists' Union is working actively to achieve an agreed global checklist of birds.

When we wrote the first edition of *Australian Bird Names* we opted to base it on Christidis and Boles (2008): an internationally respected Australian taxonomy, not least because Birds Australia (now Birdlife Australia, or BA) was then guided by it. Even by the time we went to print, however, BA's commitment to it was wavering, in large part because by then it was known that the authors did not intend to update it. We have opted to base this edition on the IOC (International Ornithological Committee, now International Ornithologists' Union) list Version 8.2 (27 June 2018). This is a widely used and respected taxonomy, but we pass no judgement on its merits or those of 'competing' taxonomies – that is not our interest and we are not qualified to do so. Among its foremost merits, however, is that it is very readily available online and it is updated every 6 months. Moreover, the most recent Australian bird guide (Menkhorst *et al.* 2017) has based its scientific names (though not its species or family order, or some common names) on the IOC list.

There is a good case for arguing that, as Australians, we ought to have used BirdLife Australia's *Working List of Australian Birds (WLAB)*, based on *Handbook of the Birds of the World* and Birdlife International's *Illustrated Checklist*, but at the time of writing this is still in the hands of a committee.

Perhaps more importantly though, we reiterate that our book is **not** intended as an authoritative checklist, but an attempt to interpret the meanings and origins of Australia's bird names. As such it cannot, and does not, clash with the work of those preparing WLAB. A few scientific names may differ between lists, but we believe these instances will be few. With regard to common names, if our IOC name is not that preferred by WLAB, we are confident we will have picked up that name and discussed it under 'Other names'. In this overall context, our choice of taxonomy is simply a frame on which to drape our work, and another would have done just as well.

Taxonomy (including the use of IOC, though the same would have been said of other world lists) has demanded significant changes in this edition in several ways. First, the entire disposition of the book has been reorganised to reflect current understandings of the relationships of families and orders, (we didn't mention orders in the first edition, but have included them here, spurred in part by a suggestion from a reviewer), as well as of those within families and genera, all of which have improved greatly in just a few years.

Second, in addition to the 55 new vagrant species on the Australian list, 21 species have been added to our previous list by the taxonomic splitting of resident species; notable in this category is the expansion of the number of quail-thrushes from four to seven. Mostly, however, these new species result from one population of a mostly widespread species being split off, as the new tools reveal hitherto unrecognised divergences. Examples of such split species are Rainbow Lorikeet, Variegated Fairywren, and western races of Golden Whistler and White-naped Honeyeater. In nearly all cases we may be surprised to find that such splits are not really novel, because these 'new' species have generally been recognised at times in the past, and most were assigned scientific names in the 19th century.

There have been a few new families recognised (e.g. for boatbills and for bellbirds) and some genus name changes. Notable here was the recent splitting of the honeyeater genus *Lichenostomus* into seven genera. It is worth reiterating that most of these are recognised by the other major taxonomies too.

Finally, with regard to common names, we have received information since the first edition from reviewers and correspondents on local usages with which we were previously unfamiliar – and to those people of course we are very grateful. Of great significance too has been the continued expansion of the National Library of Australia's magnificent *Trove* project: the digitisation for ready public access of its vast collection of newspapers, pamphlets, illustrations and older books (*inter alia*), as well as those of libraries, museums and other collections around the country. Renewed searches here have enabled us to push back the oldest recorded first instances of some names by decades. Where IOC-recommended names are at odds with standard Australian usage (the usage of Myzomela and Flyrobin are jarring examples, though there are quite a few involving the usage or otherwise of hyphens), we have made this discrepancy clear.

How common names of Australian birds have been derived

The non-scientific names of Australian birds – the names by which they are known outside the scientific literature – are a mix of words imported from Britain and applied to apparently similar birds, Indigenous names (sadly not many), names that spontaneously arose among the people who encountered the birds and names carefully coined by influential writers. Since the early 20th century, formal comprehensive checklists of birds have been prepared, forming the basis of recommended names, both scientific and common; inevitably the authors of such lists have at times been required to coin their own names.

British names

Many older groups of birds, especially the non-passerines, are spread throughout much of the world and so had acquired British group names long before Australian ones were named: duck, gull, owl and finch are just a few. However, many of the namers were not ornithologists – and indeed some of the early ornithologists did not recognise the same relationships as we do – so British names became associated with utterly unrelated groups too: magpie, chough, wren and treecreeper are some examples. In other cases it was realised that the birds in question were new and, with a lack of knowledge of or interest in Indigenous names, people were driven to coin strange chimeric names based on unrelated bird groups with which the Australian birds were supposed to share some characteristics. In really bizarre cases this involved merging names of birds from different orders – cuckoo-shrike, for instance – though in other cases they were just from different families (from each other, as well as from the bird being named). Shrikes, not found naturally in Australia, were popular in such names, hence shrike-thrush, shrike-tit, crow-shrike and robin-shrike (the latter two not often used these days).

Of course, adjectives were then needed, coined informally or formally, to identify the different species.

For pre-Australian British names, the *Oxford English Dictionary* (and other reputable dictionaries) is an excellent source of information, as is the sometimes contentious *Oxford Dictionary of British Bird Names* (Lockwood 1984).

Indigenous names

Of course all Australian birds had been named by humans before Europeans arrived in the country, but very few of those names were adopted by our forebears. We see this as a sad situation. In some cases (in particular from the Sydney area language), a couple of bird names are among the relatively few words surviving. Boobook is apparently one such name that was adopted early. Gang-gang is another example of an Indigenous name adopted early in European settlement, but we don't even know with any certainty from which language it came. Others – kookaburra and currawong, for instance – were presumably known to the newcomers (kookaburra certainly was) but not accepted into general English usage until the early 20th century.

Other languages

Some Australian birds have relatives in non-English speaking countries, from where the group's local name (or an Anglicised version of it) entered English. Cockatoo, cassowary, koel and jacana are examples of this group of names.

Names 'of the Colonists'

This is a phrase that recurs in the writings of John Gould, who recorded such folk names (and Indigenous names), though he seldom adopted them himself. These are names that apparently arose spontaneously among the British settlers and that were perpetuated in books, newspaper articles and, perhaps most significantly, in oral tradition. Many did not survive or did not enter the later birding literature, because they were too vague, were limited to one colony or were simply supplanted by later names that proved more popular. Some survived however, and have even become first-choice names. Brush-turkey, Stubble Quail, Cape Barren Goose, White-necked Heron, lyrebird and catbird are examples of modern names that apparently arose spontaneously in the early days of the colonies.

Names coined by ornithologists

Many, perhaps most, of our currently used names can be traced to a small number of hugely influential publications by professional ornithologists. Intuitively it might be supposed that the derivation of common names is not the domain of these professionals, but this has not always been the case. Unlike scientific names, there are no rules for the setting and adoption of common names, so successive publications have sought to supplant much of what went before; the adoption of such recommendations is optional, but it is more likely to succeed if the proposing body also publishes a significant journal over which it holds editorial sway.

The next section describes some of the most significant authors who stamped their mark on the landscape of Australian bird names. In former times, it was clearly an ornithological advantage to be called John!

John Latham

John Latham, who died in 1837 at the age of 96, was a successful English doctor whose passion was natural history and especially birds. He is sometimes described as the greatest ornithologist of his time; he was probably not that, though he was assiduous and dedicated and, most importantly, lived at just the right time to make the first significant contribution to Australian ornithology. He was at the height of his career when Captain James Cook was making his voyages of discovery and the first specimens and illustrations were being sent back from the fledgling colony. Joseph Banks lent Latham drawings from all the Cook expeditions, which he then copied! Latham also had access to works by early colonial artists including Thomas Watling, via the collection of botanist Aylmer Lambert. Many of his species descriptions were based on these drawings, with some odd results.

Latham was elected a Fellow of the Royal Society in 1775, even before his major publications, and in 1788 was a moving force behind the formation of the influential Linnean Society. The latter was a bit ironic, because his reluctance to embrace the Linnaean system of binomial nomenclature represented the major weakness in his earlier works. In Latham's great *General Synopsis of Birds* (1781–85), which he also illustrated, he opted for the old non-system of arbitrary common names; he basically disapproved of Latin as foreign and couldn't see the point of the new aim of consistency in naming species. As a result the honour of formally describing the species he introduced in the *Synopsis* fell to later workers, who used Latham's work but assigned proper scientific names.

In Latham's later works, *Index Ornithologicus* (1790 plus a later supplement) and *A General History of Birds* (1821–28), he did fall into line with modern requirements and formally described ~60 Australian species, including some of the best-known ones. His long reliance on vernacular names meant that he contributed many of those to the language, though few now survive. The reason for this is that by modern standards Latham's insights into bird relationships were pretty clouded – lots of unrelated birds were referred to as 'creepers', 'honey-eaters', 'warblers' or 'manakins'. Further, much of his work contains confusions – he was dealing with vast amounts of new material, some just in the form of notes or sketches – and in several identifiable cases he described the same bird more than once under different names. He even did that within the same book: Buff-rumped Thornbill is an example. Some of the duplications were due to sexual dimorphism or juvenile plumage.

It was Latham who first applied the word 'honey-eater' to Australian birds, though he used it for everything from bee-eaters to robins to whistlers to bowerbirds to whipbirds (and some honeyeaters). His White-eared Thrush survives as White-eared Honeyeater.

John White

John White was a naval officer who was chief surgeon to the 1788 settlement expedition to New South Wales. He did an excellent job of maintaining health standards in the colony, and indulged his passion for natural history during expeditions with Governor Arthur Phillip. His *Journal of a Voyage to New South Wales*, published in 1790, was very successful and included sketches and accounts of Australian birds. Some of the accounts were actually written by George Shaw, from specimens sent by White, so attributions vary; we can be fairly sure that names that appeared in his 1788 diary were White's, unless he altered them later to fit the book. One of his best-known (though not very inspired) names was New Holland Creeper, now known as New Holland Honeyeater.

George Shaw was a British polymath-naturalist who took a great interest in the flood of Australian material, including some birds, though he is best known for his work on mammals. His eight-volume *General Zoology* (1800–12) was highly influential, and included many Australian species.

John Lewin

John Lewin (1770–1819) was a naturalist, talented artist and engraver who came to Sydney in 1800 to collect specimens for English patrons, bringing etching equipment that included a printing press, copper plates and paper. His father William had written the seven-volume *Birds of Britain*. Lewin's first plan, backed by influential patrons, was to produce a book on the colony's insects but his interest broadened to include the birds. He produced two books in the colony: *Prodromus Entomology* in 1805 and *Birds of New Holland* in 1808 for six English subscribers (plus of course the King!). His 67 Australian subscribers did not receive their copies for reasons still unknown, though it is suggested that the books were lost at sea en route back from the English printers. Only eight copies of this book remain. A Sydney edition, called *Birds of New South Wales*, was accordingly produced in 1813 for the Australian subscribers, using proof impressions that Lewin had taken before he sent the plates to England (Lewin 1813). This was the first illustrated such book to be printed in Australia and there are only 13 known copies. Although Lewin drew on Latham's and White's names, he was willing to provide his own as well: Variegated Fairy-Wren (from his Variegated Warbler) is one of Lewin's, but many of his names were 'near misses' (e.g. Scarlet-backed Warbler for our Red-backed Fairy-wren).

John Gould

Books could be – and have been – written about this remarkable man. He has occasionally been dismissed as an entrepreneur (which he was) and as a charlatan who got credit for others' artistic work (credit that he never claimed, though he did many of the sketches that others, notably the equally remarkable Elizabeth Gould, filled in). In truth, John Gould was an exceptional ornithologist who changed the landscape of 19th century Australian bird taxonomy: by our calculation, he is the author of some 175 bird names (and 39 mammal names) **still in use**. He also assigned another 150 or so names that have been superseded. Gould's contribution to vernacular names was equally immense, but fewer of those survived because of his habit of using scrupulous translations (often his own) of Latin names, in particular using genus names as group names. Swift-flying Hemipode, Little Chthonicola, Black-throated Psophodes or Ground Grauculus were never going to catch the public attention.

John Leach

John Leach, a Victorian country schoolteacher, became in the early decades of the 20th century one of the most influential bird people in Australia, introducing natural history into the school curriculum and training its teachers, founding the Gould League of Bird Lovers, becoming president of the Royal Australasian Ornithologists Union (RAOU) and acting as editor of *Emu* for 10 years. He wrote the first Australian bird field guide, *An Australian Bird Book*, in 1911, in which he introduced some names not generally used previously. Most significantly, he chaired the committee that produced the RAOU's second *Official Checklist of the Birds of Australia* in 1926, and was doubtless influential in having some of his names adopted in that checklist. Whistler, White-backed Swallow and Rainbow Lorikeet are some of the names that are apparently due to Leach's influence in that forum.

Gregory Mathews

Mathews' story is surely one of the strangest in ornithology. A field birder and egg collector in his youth in late 19th century Australia, he became wealthy through inheritance, share trading and marriage, then went to England to lose himself in the horsey set. However, in a bizarre damascene moment during a visit to the British Museum he was inspired to write a monumental work on Australian bird taxonomy. This massive work – *The Birds of Australia* – was published in 12 volumes between 1910 and 1927. Although all the work was conducted from England, Mathews couldn't be ignored in Australia. He was a compulsive splitter (he recognised five genera of Australian woodswallows) and constantly changed his mind after publishing, and there were strong suggestions that some of his taxonomy was influenced by a desire to create names that honoured his friends and family. Nonetheless, Mathews was one of the first to recognise the importance of subspecies. He coined many vernacular names for these, as well as for the many 'new' species that were later deemed to exist only in Mathews' mind. Most of those names disappeared or were never widely adopted, but Mathews had quite an influence on the first *Official Checklist of the Birds of Australia* (RAOU 1913).

Royal Australasian Ornithologists Union 1913

The need for an agreed checklist of Australian bird species and a uniform set of names (scientific and vernacular) was strongly felt by the start of the 20th century. Several ornithologists were working towards it, with considerable acrimonious interstate and interpersonal rivalries complicating the process. A preliminary list was drawn up under the auspices of the Australasian Association for the Advancement of Science in 1898 and was promptly rejected by the South Australian Ornithological Association, supported by some eastern state individual ornithologists. The original purpose of a national ornithological body, which did not then exist, was seen as coordinating a national checklist; although the Australian Ornithologists Union did not consider this its chief role when formed, setting up a checklist committee in 1903 was nonetheless one of its first acts.

The logistics of creating a checklist must have been horrendous, given the lack of communication technology that we would regard as essential – and bear in mind that Mathews was communicating from Britain! Moreover two members of the RAOU, Archibald Campbell and Alfred North, detested each other.

Nonetheless, a list – the first *Official Checklist of the Birds of Australia* – was presented and endorsed in 1912 (RAOU 1913). It was a list firmly rooted in the 19th century, drawing heavily on Gould's work and on Mathews' already profligate outpourings. There seems to have been little attempt to examine the material critically, but the situation was very fluid with new material appearing all the time. It appears that the initial task was seen primarily as getting all the data into one document to enable further refinement. The list introduced few, if any, new names, but its significance – and the reason we discuss it in some detail here – is that it formed the basis of the next document, a most influential source of names indeed. For a fuller discussion of the sometimes tortuous process, see Robin (2001).

Royal Australasian Ornithologists Union 1926

Following the production of the first checklist, the work continued unabated, not least because large numbers of points of divergence from Mathews' ideas had to be dealt with one by one. The resultant new checklist (with the same name) was a revelation, bringing Australian bird nomenclature (scientific and vernacular) decisively into the 20th century. It is invidious to single out individuals from a committee but there is no doubt that John Leach, who was convening the group by the time the list was published, and Alec Chisholm, one of the great naturalist-communicators, had immense influence. Leach's introduction to the list discussed the replacement of names that were 'indefinite' (e.g. Ground-bird, though we might debate whether Quail-thrush was a big improvement), 'inelegant' (e.g. Black-vented) or 'long formal' names (e.g. Great Brown Kingfisher). It was this list that gave us Kookaburra, Galah, thornbill, triller and currawong, for which we should be grateful. In many cases there is little evidence of prior usage of these names, so this was a most creative committee, as well as being diligent and scholarly.

Royal Australasian Ornithologists Union 1978

The task was seen to have been done – or perhaps everyone was thoroughly fed up with it all by then – because 50 years elapsed before an updated checklist was produced, though a checklist committee operated for at least part of the intervening period. However, the committee that was set up in 1975 had a narrower task than the overall checklist committee – its goal was the production of a report on *Recommended English Names for Australian Birds* (RAOU 1978). One of the main aims was to 'internationalise' Australian bird names, to remove any confusions with names in use elsewhere.

One such manifestation was dealing with the 'wren/warbler' syndrome, which saw many Australian bird groups named for unrelated Northern Hemisphere groups. To this end the committee introduced, or formalised, such group names as fairywren, gerygone, monarch, boatbill, hobby and jaeger. At a species level it brought us, for example, Spinifex Pigeon, Gould's Petrel, Black-faced Cormorant and Sulphur-crested Cockatoo. Inevitably there was controversy, especially with the imposition of such 'foreign' group names as baza, thick-knee, needletail and lapwing. (A few of these, including thick-knee, were subsequently rescinded following a 'referendum' of members.) To the credit of the authors, for the first time the rationale for each decision was explained in a lengthy addition to the report.

Les Christidis and Walter Boles 1994 and 2008

The most recent completed 'definitive lists' of Australian birds were not prepared under the direct auspices of the RAOU (which from 1996 was known as Birds Australia), but were compiled by Les Christidis and Walter Boles, both then of the Australian Museum in Sydney. The first was published by the RAOU as a monograph in 1994; the second, totally revised version was published externally in 2008. The recommendations of both were adopted by the RAOU/Birds Australia and subsequently by field guides (mostly). Although vernacular names were not a key part of their work, Christidis and Boles nonetheless made recommendations about them, in some cases 'overruling' the earlier RAOU report. The more recent book, like the first, was remarkably detailed, including all Australian island territories and oceanic waters. Both publications, quite properly, took a conservative approach in that new positions were not adopted until all the evidence was in, though likely future changes were flagged. Where new species or new combinations of species were recognised the authors recommended names accordingly (e.g. Kimberley Honeyeater, Tasman Parakeet, Eastern Koel).

We followed Christidis and Boles (2008) in the first edition of this book, as the most recent authoritative account available. We made no judgement on their taxonomy – we are not qualified to do so – but noted where changes seemed likely or where controversy was significant.

Working List of Australian Birds (ongoing as at 2018)

At the time of writing this (October 2018) Birdlife Australia is in the process of preparing a *Working List of Australian Birds (WLAB)*, based on *Handbook of the Birds of the World* and Birdlife International's *Illustrated Checklist*. It is anticipated that this will become the touchstone for future works on Australian birds, and will be updated as appropriate.

International lists

Finally, there are other worldwide lists maintained by reputable international authorities and organisations. The *Clements Checklist of the Birds of the World*, for instance, is maintained by the prestigious Cornell Ornithology Laboratory; the name commemorates the list's founder, the late Dr Jim Clements of the USA. Originally published in book form, it is now available online; it is the formal list for the American Birding Association.

The monumental *Handbook of the Birds of the World* (published in 16 volumes from 1992 to 2011 by Lynx Editions in Spain) inevitably suffers from the changes in understanding over the 20-year period, but it has now been fully digitised and is constantly updated online.

Perhaps the most widely used list, however, is that of the International Ornithologists' Union, formerly known as the International Ornithological Committee (also sometimes referred to as the International Ornithological Congress, though that is really the name of its 4-yearly conference). The list is maintained and regularly updated online as the *IOC World Bird List*. This is the authority used by Birds Australia Rarities Committee (Palliser *pers. comm.*) as well as the most recent Australian bird guide (Menkhorst *et al.* 2017), and its most recent version at the time of writing (8.2) has been used by us for this edition of *Australian Bird Names*.

Hyphens

It might seem more than a tad pedantic to devote an entire paragraph to this topic, but it is pertinent. Many group names of birds have been developed over the years by combining adjective and noun separated by a hyphen: familiar ones are Black-Cockatoo, Storm-Petrel, Fairy-wren and Button-quail. The convention has been to capitalise the second word if the bird really is one of that group (e.g. Cockatoo, Petrel in these examples) but lower case if the bird is not a 'true' wren or quail for instance. However, the trend by some authorities now is to do away with hyphens, in favour of either two separate words, so now IOC (and thus we) use Black Cockatoo, Storm Petrel, or a single unhyphenated one, hence Fairywren and Buttonquail. However, does retain the hyphen if two bird names are combined, so Quail-thrush and Emu-wren, though Cuckooshrike mysteriously loses its hyphen, seemingly because of a contradictory rule relating to compound names. Moreover, despite the IOC's claim to have 'finally rejected the practice' (of hyphenating group names) this rejection is not applied consistently, so Stone-curlew, Painted-snipe still appear. And elsewhere there is no consensus, with *Handbook of the Birds of the World* for instance retaining Black-cockatoo (albeit with lower case 'cockatoo'!) so unanimity on this question would seem to be a distant dream.

How scientific names of birds are derived

1 January 1758 is a date to remember – at least it is if you're an animal taxonomist. That was the publication date of a truly seminal work in the history of biology, the 10th edition of Carl Linnaeus' *Systema Naturae*, and the correct name of any animal is the first one published since that date.

But who says so? The rules of zoological nomenclature are strict and labyrinthine, administered by the International Union of Biological Sciences, which maintains the International Code of Zoological Nomenclature. It is an evolving process, begun in the 1840s by the British Association for the Advancement of Science, and the rules laid out then were subsequently adopted internationally in later decades.

The Code is vast and does not make for light reading, but there a few basic rules that are readily understood.

Systematics

First, it is a given that the system of naming reflects our understanding of the relationships of the bird (or any other animal, but we'll use birds as examples for obvious reasons). The basic unit is the **species**, which is a population, or groups of populations, whose members readily interbreed with others of that species but not with those of other species. (Yes, hybrids do occur, but they reflect the fact that evolution is an ongoing process. Magpie races, such as Black-backed and White-backed, readily interbreed because they have not been separated long enough to form full species. Crimson and Eastern Rosellas are regarded as full species but have not been apart for so long that they are totally reproductively isolated, and they occasionally hybridise, though such hybrids are nearly always sterile.) With this proviso, the species is a self-defining unit, but 'higher' units of taxonomy are subjectively determined by taxonomists.

All rosellas belong to one **genus** (*Platycercus*); all members of *Platycercus* are more closely related to each other than they are to any other species of parrot. 'Relatedness' is a measure of the time since they last had a common ancestor. There is no limit to the number of species in a genus: it is purely a question of how many species there are with that degree of relatedness. For example, the Budgerigar is the only member of the genus *Melopsittacus*, while *Platycercus* has six species. The plural of genus is genera; all parrot genera belong to the **Family** Psittacidae. The family is often an intuitively recognised group; cockatoos form another family, as do ducks–geese–swans, and honeyeaters. In Australia, Psittacidae and Cacatuidae (the cockatoo family) together form the **Order** Psittaciformes. A very important Order of birds is the vast Passeriformes, loosely referred to as songbirds or perching birds. This huge order comprises some two-thirds of all living birds; this is why the rest tend to be lumped as non-passerines, defined only by what they aren't. Thirty-nine orders of birds (according to IOC, though as recently as our First Edition 27–29 was the generally agreed number) together comprise the **Class** Aves.

The process of naming a bird

Having determined to which genus the bird belongs, it is necessary to propose a species name. (In circumstances that are rare today, but that were not at all rare in Australia in the 19th century, a new genus name might also be required.) The genus name must be unique among animals (though doubling up with a plant name is OK); the species name need only be unique in that genus. For example, once *Dromaius* was ascribed as the Emu genus, no beetle, whale or starfish genus can also be called *Dromaius*. On the other hand, any number of other species may be – and are – called *novaehollandiae*. There is no requirement for the author to explain the meaning of the name; many do so now, but the purposes of some many older names are clouded in frustrating uncertainty.

The name must be in Latin (or in a Latinised form if the name comes from another language, traditionally often Greek, or is based on a personal or place name). The genus name is a noun and must be capitalised. The species name is an adjective or adjectival in intent and must always be lower case; in earlier times, species names based on proper nouns were capitalised, but that is no longer done. If the species name is an accepted Latin adjective it must agree with the genus name in ending. In a formal setting such as a checklist, though not generally a field guide, the name is accompanied by the surname of the author (with initials if more than one author shares the surname) and the year of publication of the name. Hence, the Mallee Fowl is properly known as *Leipoa ocellata* Gould, 1840, indicating that it was named by John Gould in an 1840 publication.

A family name is based on the type genus – that is, the one that includes the type species (the first one described, see next paragraph). It basically involves adding '-idae' to the stem of the genus, with minor adjustments for euphony permitted. For instance, the type genus of honeyeaters is *Meliphaga*, stem is Meliphag- and the Family is Meliphagidae; for ducks the type genus is *Anas* and the Family is Anatidae.

The publication of the name must be accompanied by a description (formerly in Latin, but not now) of a specimen that is designated the **type specimen**. This specimen must be deposited in a recognised collection, usually a museum, and be available for later researchers to compare with other specimens. The type specimen is the ultimate arbiter as to whether later specimens also belong to that species. (Duplicate specimens, known as **paratypes**, are sometimes used to help describe variations in the species, and as back-ups in case calamity befalls the type specimen. **Syntypes** refer to a group of specimens from which an author made a composite description of the species; this is discouraged nowadays.) These days the description must include a statement of where the type is lodged but in the past that was not required, which has led to confusion. Note that, in the early days of taxonomy, authors such as John Latham often published based only on descriptions or drawings.

Publication is not strictly defined, but is understood to mean reproducing multiple copies of the description and making them freely available. In general this means a peer-reviewed journal, but publishing in books was a common practice in the 19th century and earlier. Publication in newspapers, while frowned upon, was not unknown where it helped to beat a rival to the punch!

The question of electronic publication is a source of debate today.

Once published, a species name cannot be changed by anyone, including the author, be it misspelt or infelicitous, or even if the description conveyed is misleading or just plain wrong. The Green Rosella, endemic to Tasmania, was called *Platycercus caledonicus*, based on Johann Friedrich Gmelin's misconception that it came from New Caledonia: whoops, but too late!

Why names change

Amateur botanists are much more likely to use scientific names than are birdwatchers, and most complaints from laypeople about name changes come from them. However, changes occur as often in animal names as in plant names, which might seem surprising given the previous comment about inflexibility with regard to amending published names.

There are two basic reasons for such a change. The first is to do with priority. It is a fundamental tenet of taxonomy that, where two names have been applied to one species, the first-published name must prevail. It is easy to see how doubling-up could have happened in the early 19th century, with specimens flooding into Europe from all over the world and communications between taxonomists being limited by the available technology. There were also many complications from specimens of dimorphic species, where males and females were quite different, or of young birds. Birds in Australia were described as new species, the author not realising that they had already been described from specimens from Indonesia or elsewhere.

Where there is a close finish between publishers, the actual date of publication (not the date the description was written) is the determinant. Many times, later researchers have found an old name that pre-dated an existing one, or found evidence to settle a disputed situation. In such cases the old name must always replace the more recent one.

The other basic reason for name changes is our changing understandings of the birds. The Musk Duck's name appears as *Biziura lobata* (Shaw, 1796). The brackets mean that John Shaw named it in 1796, but although the species name is his, he didn't use *Biziura* as a genus. In fact he called it *Anas lobata*, putting it in with the 'conventional' ducks. A later worker, James Stephens, recognised that it was not closely related to them and erected a new genus to accommodate it. The genus is thus Stephens', but the species name remains Shaw's.

New methods of studying taxonomy – including biochemical, DNA and RNA analyses – have enabled more precise measurements of the similarities and differences between populations. In some cases, different populations have proved to represent separate species, which requires the erection of a new name (to be applied to the population that did **not** supply the type specimen, because of the requirements of priority). A recent example is the Kimberley population of the White-lined Honeyeater *Meliphaga albilineata*, which was recognised as a separate species from the Top End birds and thus became *M. fordiana* in 1989.

Races and subspecies

Although subspecies names do not feature heavily in this book (in fact, they only appear where they have attracted their own common names), we use the terms often and so should clarify them.

A race of birds is a population which differs enough from other populations of the species to be recognisably different (e.g. we talk of three races of magpie, each with differently coloured backs), or is just isolated (e.g. the Kangaroo Island race of Crescent Honeyeater). If taxonomists decide that the race is sufficiently different from the rest of the species, it may be assigned its own subspecies name. If the subspecies is of the same race as the type species, it automatically takes the species name – the Pied Currawong of the mid-east coast of Australia, from where the type specimen originated, is *Strepera graculina graculina*. All other populations (one or many) must also then be given their own subspecific name: for example, Cape York Peninsula Pied Currawongs are *S. g. magnirostris*.

Procedures and pronunciations

How we proceeded

The bird names are arranged in the order in which they appear in the IOC World Bird List 8.2. Put simplistically, the order of families in modern systematic lists seeks to reproduce, as accurately as knowledge allows, the order in which we believe those families evolved; likewise, the order of genera within families, and species within those genera.

Our choice of Christidis and Boles (2008) as the template for our first edition reflected the fact that it was then widely used in Australia, from forming the basis of Birds Australia's publications guidelines, to the authority by which many Australian twitchers prepared comparable national lists. Since then much has changed, as explained in the Introduction to this edition. The Christidis and Boles list is no longer being updated and the IOC list is now very widely adopted by professionals and individual birders, and is continually updated on line. (It is formally updated twice a year, but pending changes are flagged before that.) For this edition, we began with our First Edition, comprising the Christidis and Boles list, plus those species subsequently accepted onto the Australian list by the Birds Australia Rarities Committee (BARC), up to March 2012. To that we have added another 55 vagrants since accepted by BARC plus a few records hitherto undetermined by BARC but incorporated by Menkhorst *et al.* (2017) based on the evidence, in addition to 21 species arising from taxonomic decisions (see 'Introduction' for more information).

We start each family with the family name, only explaining it if the genus it is named for does not appear in the Australian list (see 'How scientific names of birds are derived' for more information). The namers of families are really identifiers of families; for example, Dasyornithidae Schodde 1975 – Schodde did not invent the name but he identified the family, recognising *Dasyornis* (named by Vigors and Horsfield in 1827) as the type genus, (i.e. the first in the family to be named, and basing the family name on that). (Note that all family names end with 'idae' from the Greek suffix '*-idēs*', meaning 'like' or 'child of', so Dasyornithidae is a family like, or stemming from, the genus *Dasyornis*.) Under the 'family name' heading, we also explain any common name that has been given generally to birds in this family.

A list of the genera that belong in that family follows, and we give a translation of the name, a guide to pronunciation, a derivation, an explanation of why that name was chosen (if we can), and other information we hope will be of interest. We then list every species in the family, with its often many-and-varied common names (we have attempted, doubtless in vain, to track down every name used in English, at least in print, for each species), followed by its full scientific name, with translation, pronunciation, and so on, following the process used for the genera. To establish exact meanings of the names, we have consulted original sources wherever possible. Many of these are not in English, and all translations are ours.

Translations, pronunciation guide and derivations

As far as our translations of scientific names are concerned, our aim has been to render a graspable concept, with a view to helping comprehension and memory (and to having a bit of fun!) while retaining, where possible, the original sense and spirit of the namer's intentions. We have therefore used free, and sometimes creative, translations, especially where the author's thinking was not made clear. Where we have speculated, we have been careful to say so. In our derivations, we used the basic form of nouns; for adjectives we used the Latin masculine form *us* (as opposed to feminine *a* or neuter *um*) and for verbs in both Greek and Latin we generally used the first person singular. Note that the species name often takes the form of an adjective (with suffix *-atus*, *-ctus*, etc., often -ed in English). Here Greek words were needed, we used the so-called classical transliteration of Greek (e.g. 'kh' for χ, 'nk' for γκ, 'ē' for η and 'ō' for ω).

Scientific names are universal but their pronunciation unfortunately is not! It varies according to the speaker's own language and their level of desire for correctness. We have observed that even the most knowledgeable, expert and meticulous bird people will have idiosyncratic elements in their pronunciation of these names. This is because there is **no** definitive right way to say them. Our pronunciation guide is therefore only a guide! These are Latin words, and we have tried to stick to correct Latin pronunciation and intonation (e.g. as explained by Covington (2010) and using the pronunciation he recommends for scientific use) but with the occasional nod to the Greek (e.g. using a long vowel where a Greek origin has one).

Sometimes, though, we have ignored those rules completely in favour of common usage. For example, all the family names end in -ae, which we have normally pronounced as ay as in play, and written as 'eh'), but in this case we have followed common usage and put 'ee' as in speed. We have done the same in words beginning with *haema,* where common English usage also has ee in the majority of words, such as haemophilia, haematology, though not haemorrhage! Another example is in Greek Y and u (upsilon) and Latin y where it exists, which are really to be pronounced like French u or German ü, but we have gone with a less strict version, putting them as 'i' or 'ee'. With *chryso-*, for example, we are saying 'kri-so', because most English words from that source are pronounced that way (e.g. chrysanthemum, chrysalis), but we fully understand that many people will prefer to say krie-so. Conversely, we put *hy-* as 'hie' in *hylacola*, because there is a different usage imperative for English words beginning in that way (hydrangea, hymen).

Our general principles are as follows.

- Vowels. If they are short, they appear as a e i o, (as in fan, fen, fin, fon) but u is, **not** like fun, unless you come from the northern parts of England! So, if we write stroo-thee-O-ni-dee, the O is as in dog, the ni as in nip – always! This includes syllables like no or to where the inclination might be to pronounce them differently.

  If vowels are long, then ah (as in father), ee, eh (as in may), ie (as in spider), oh (as in goat). Long u is oo; for short u (as in push rather than pooch), we are using the symbol ʊ.

  Double vowels or diphthongs: *au* = ow as in clown; *ae* = ie (but ee in family names); *oe* = oy; *eu* is like feud in English, and we shall present it as yoo, except in *leuko*, where we have it as loo-ko because most of English words containing it are so pronounced.

- Consonants. We consider them to be generally pretty much the same as in English. Note that c before i and e, is soft, so we write it as s, otherwise it appears as k; g is always hard (though we have made an exception due to universal usage in '*gerygone*' and called it dj); v is v (i.e. not following what you might have learned during Latin classes at school, which would have been w).
- As far as personal names are concerned, we have tried as far as possible to keep them as their owners might have pronounced them, so we have PAR-kin-son-ee and MATH-yooz-ee, keeping the stress as it is in the name; KOW-pee for Kaup, lehr-meen-YEHR-ee for L'Herminier, to approximate the German and French pronunciations.

So! Read the name aloud exactly as written in our guide, with stress on the capitalised syllable. You can't be wrong, in any case – you can only be idiosyncratic like the rest of us.

NON-PASSERINES

STRUTHIONIDAE (STRUTHIONIFORMES): ostrich

Struthionidae Vigors, 1825 [stroo-thee-O-ni-dee]: the Ostrich family, see genus name *Struthio*.

The genus

Struthio Linnaeus, 1758 [STROO-thee-oh]: 'bird' from Greek *strouthos*, often given the meaning 'sparrow', though the Ancient Greeks actually used *strouthos* (often unqualified) for both sparrow and ostrich, so the context is all-important. For example, we have Herodotus (c. 435 BC) referring in his *Histories* to shields made of the skins of '*strouthoi*' (not sparrows?), and in Homer's *Iliad* (c. 700 BCb), eight young '*strouthoi*' nestle with their mother in the top of a tree (not ostriches!). (See Arnott 2007 for a detailed discussion.)

The species

Ostrich (introduced breeding resident, though possibly only one small population survives.)

Via stages, notably old French, from Latin *Avis struthio* – literally, 'ostrich bird'! Native of Africa, known to the Romans and indeed the Greeks.

Struthio camelus Linnaeus, 1758 [STROO-thee-oh ka-MEH-lʊs]: 'camel-bird', see genus name, and from Greek *kamelos*, a camel or dromedary. Xenophon, writing in ~380 BC, tells of Ancient Greeks in Arabia coming upon a great crowd of '*strouthoi*'. He praises the speed of these birds, describing them as uncatchable when running at full speed, 'hoisting their wings and using them like a sail' (not sparrows!). The two Greek words are combined in Latin as *struthiocamelus* (translated as ostrich), as in Pliny's (77–79 AD) *Naturalis Historia*, which gives a detailed and sometimes imaginative description of the bird and its habits. Overall not a bad name really, if you think of the ostrich and camel both being large, somewhat ungainly looking but very fast moving.

CASUARIIDAE (CASUARIIFORMES): emus, cassowaries

Casuariidae Kaup, 1847 [kaz-yoo-a-REE-i-dee]: the Cassowary family, see genus name *Casuarius*.

The genera

Casuarius Brisson, 1760 [kaz-yoo-AH-ri-ʊs]: 'cassowary' from *kasuari* or *kasavari* in Malay, rendered in French as *casoar* and thence to Modern Latin.

Dromaius Vieillot, 1816 [dro-MEH-ʊs]: 'racer', from Greek *dromaios*, running at full speed (as in velodrome, for example, from *dromos* a race or race-course).

The species

Southern Cassowary (breeding resident)

From Malay *kasuari*, but it would have come to English via Dutch, due to the Dutch influence in the East Indies; there was already an unfortunate individual bird in an English menagerie in the early 17th century. In earlier times 'cassowary' and 'emu' were used interchangeably. In addition to this species, there are two others in New Guinea and associated islands; 'Southern' is in relation to these, and the name was only adopted after the RAOU's third official list (RAOU 1978).

Other names: Australian Cassowary, as used by Gould (1869) and still used into the 20th century (e.g. Mathews 1913, but also Pizzey 1980); Cassowary (as this is the only Australian species) through much of the 20th century, at least until CSIRO (1969) and Slater (1970); Double-wattled Cassowary, to distinguish from the New Guinea Northern or Single-wattled Cassowary, *appendiculatus* – not much used in Australia.

Casuarius casuarius (Linnaeus 1758) [kaz-yoo-AH-ri-ʊs kaz-yoo-AH-ri-ʊs]: 'cassowary-cassowary', see genus name.

Emu (breeding resident)

Of Portuguese origin, from '*ema*'. (There are suggestions, not well substantiated, that it might have had origins further back in Arabic.) In the 17th century the word primarily referred to a crane but in more recent times it has been used to refer to virtually all the ratites, including rheas and cassowaries. The sailors who bestowed the name in Australia were apparently not Portuguese but Dutch – at the time Portuguese was the *lingua franca* among Dutch sailors in the East Indies, probably because the early maps were in Portuguese. Gould (1848) formally referred to it as The Emu.

For race (formerly species) *minor* (formerly *ater*): King Island, Dwarf or Emu; a small dark race endemic to King Island, which was driven to extinction almost as soon as it was named.

For race (formerly species) *baudinianus*: Kangaroo Island or Dwarf Emu, another small island race, extinguished early after European settlement.

Other names: New Holland Cassowary (White 1790); Southern Cassowary (Shaw 1819); Van Diemen's Land Cassowary (Latham 1823); Emeu, a common form throughout the 19th century; Spotted or Bartlett's Emu, for a supposedly more speckled form from western Australia, described by Edward Bartlett (an English ornithologist with no particular knowledge of Australian ornithology) as *D. irroratus* in 1859 – it is not now recognised at any level.

Dromaius novaehollandiae (Latham, 1790) [dro-MEH-ʊs no-veh-hol-LAN-di-eh]: 'New Holland racer', see genus name, and from Modern Latin *Nova Hollandia*, New Holland, the old name for Australia.

MEGAPODIIDAE (GALLIFORMES): mound builders

Megapodiidae Lesson, 1831 [me-ga-po-DEE-i-dee]: the Bigfoot family, see genus name *Megapodius*.

The genera

Alectura Latham, 1824 [a-lek-TOO-ruh]: 'cock-tail' from Greek *alektor*, cock, plus *oura*, tail. As in alectryomancy, divination using a cock or hen pecking at scattered grain, or alectryomachy, cock-fighting.

Leipoa Gould, 1840 [leh-POH-uh]: 'egg-leaver' from Greek *leipo*, leave or abandon, and ōon, an egg.

Megapodius Gaimard, 1823 [me-ga-POH-dee-ʊs]: 'big-foot', from Greek *megas*, great or large, and Greek *pous/podos*, the foot.

The species

Australian Brushturkey (breeding resident)

Poultry names abound among Australian birds: familiarity was more important than systematics. However, megapodes are indeed primitive members of the same order as fowls, turkey and quails. 'Brush', meaning a thicket, was used in Australia to refer to rainforest, particularly in NSW. Gould (1848) records that Brush Turkey was already a name 'of the Colonists'; it began to appear in newspapers in 1840. 'Australian' distinguishes this species from five other brush-turkeys in related genera in New Guinea and associated islands.

Other names: Brush Turkey or Brush-turkey more common usages; New Holland Vulture, the somewhat surprising name used by Latham (1823), presumably in acknowledgment of the bare head and neck; Wattled Talegalla, the name coined by Gould (1848) – he very often used a genus name as a group English name, and the bird was at that time ascribed to genus *Talegalla* (of which there are still three species recognised, in Indonesia and New Guinea); Pouched Talegallus, cited by HANZAB (1996–2006, vol. 2) but not otherwise evident; Scrub Turkey (see Orange-footed Scrubfowl). Barnard's or Purple-wattled Brush-turkey, for race *purpureicollis* of Cape York Peninsula (which has a pale mauve wattle), described as a full species by Dudley Le Souef in 1898; Henry Greensill Barnard was an eminent and assiduous natural history collector, and Le Souef (1899) cites Barnard's field notes – we may assume that Barnard collected the type specimen. Yellow-pouched or Yellow-wattled Brush-turkey, to distinguish southern birds from the northern subspecies, but little used.

Alectura lathami Gray JE, 1831 [a-lek-TOO-ruh LEH-thuh-mee]: 'Latham's cock-tail', see genus name, and after John Latham (1740–1837), English naturalist and ornithologist. The tail of the Brush-turkey does indeed resemble that of the domestic cock in being vertical and rudder-like, with feathers one above the other rather than horizontal and side-by-side (like the tail of a shark rather than a whale).

Malleefowl (breeding resident)

See comments under Australian Brushturkey about poultry names. 'Mallee', of Aboriginal origin, refers to the bird's habitat – a large swathe across semi-arid southern Australia of often dense multi-stemmed eucalypt shrubland. The name seems to have arisen spontaneously – Gould (1848) does not mention it, but it was in general use by the end of the century (Morris 1898), see 'Other names' and was the first-choice name of Gregory Mathews by 1913. The earliest reference we can find in newspapers is to the form 'mallee-hen' in 1861.

Other names: Ocellated Leipoa, a direct 'translation' by Gould of his species name – he regularly thus achieved some blink-inducing results; Native Pheasant, recorded by Gould (1848) as being used by the 'Colonists of Western Australia' – Morris (1898) also records this name as being used in South Australia; Mallee Hen, Mallee Bird and Scrub Turkey, all recorded by Morris (1898) in his *Dictionary of Austral English*; Lowan, from *Lawan*, a word from the Wembawemba language of western Victoria – it is only in relatively recent times that this name has, perhaps unfortunately, dropped from general use; Gnow (in south-western Australia), from *Ngaw* in Nyungar.

Leipoa ocellata Gould, 1840 [leh-POH-uh o-se-LAH-tuh]: 'eyelet egg-leaver', see genus name plus Latin *ocellus*, meaning a small eye, eyelet or buttonhole – referring to the bird's barred markings but also coincidentally a term of endearment. This bird certainly leaves its eggs but it hardly abandons them – the male's constant monitoring and adjustment of the temperature of the mound in which they are laid allows them to incubate and hatch successfully. Indeed, in German, it is known as the *Thermometerhuhn* – thermometer hen. But it must surely be a coincidence that there is also a Finnish word *leipoa* which means to bake?

Orange-footed Scrubfowl (breeding resident)

See comments under Australian Brushturkey about poultry names. 'Scrub' means low dense vegetation; in reference to rainforest, it was probably used in a derogatory sense. It is unclear why the adjective is 'orange-footed' rather than 'orange-legged'.

Other names: Jungle Fowl, which more properly refers to the Indian wild ancestor of domestic hens but which was recorded by Gould (1848) as being the name used by the 'Colonists of Port Essington'; Mound-raising Megapode, by Gould (1848), as his wont, a more or less direct rendering of his species name, which was *Megapodius tumulus* (he was unaware of Dumont's name); Orange-footed Megapode; Mound-bird, recorded by Morris (1898).

Megapodius reinwardt Dumont, 1823 [me-ga-POH-dee-ʊs REHN-vaht]: 'Reinwardt's big-foot', see genus name, and for Caspar Georg Carl Reinwardt (1773–1854), Dutch ornithologist, botanist and general polymath, keeper of Napoleon's Amsterdam menagerie and first-hand expert on Java. It is likely that he collected the type specimen, which is from Lombok.

NUMIDIDAE (GALLIFORMES): guineafowl

Numididae de Sélys Longchamps, 1842 [nyoo-MI-di-dee]: the Numidian family, see genus name Numida.

The genus

Numida Linnaeus 1764 [NYOO-mi-duh]: 'Numidian', from the name of the ancient kingdom in North Africa where the guineafowl originated. Related to the Greek word *nomadikos*, which was used both to describe a herdsman's way of life, and as a proper noun meaning from Numidia.

The species

Helmeted Guineafowl (introduced breeding resident)

Although the Romans introduced Helmeted Guineafowl from North Africa (they are now found naturally only south of the Sahara) they disappeared from Europe with the Roman Empire and were rediscovered and introduced from West Africa (including Guinea) in the 15th century. The 'helmet', or casque, is unique among the six guineafowl species.

Numida meleagris (Linnaeus, 1758) [NYOO-mi-duh me-le-AH-gris]: 'Numidian guineafowl', see genus name, and from Greek *meleagris*. The word is from the name of the warrior Meleager, whose sisters, the Meleagrides of Greek legend, were so grief-stricken at his death that they were turned into birds (usually described as guineafowl) by the goddess Artemis. She did this out of pity for their misery: an interesting change of heart given that she had maliciously engineered Meleager's death in the first place. The white spots on the birds' plumage are said in some versions of the story to be the indelible mark of the tears shed by the sisters. However, Zeuner (1963) claimed that the Greeks had originally called the bird *melanargis* (black and white or silver), which then became corrupted into *meleagris* and was secondarily associated with Greek myth. Arnott (2007) suggested (based on Aelian) that *meleagris* may just have 'echoed the female bird's staccato call 'melag''.

ODONTOPHORIDAE (GALLIFORMES): New World quail

Odontophoridae Gould 1844 [o-don-to-FO-ri-dee]: the Tooth-bearing family, from Greek *odous*, tooth, and *phoros*, bearing or carrying. These are the New World quails and, unlike the Old World quails, they have a serrated lower mandible (rather than teeth).

'Quail' was originally onomatopoeic, from an old Germanic base resembling 'kwak', thence via Old French *quaille* into English.

The genus

Callipepla Wagler, 1832 [kal-li-PEP-luh]: 'beautiful robe', from the Greek *kalli-*, beautiful, and *peplos*, a rich outer robe or shawl worn by women in Ancient Greece, hanging in folds and sometimes drawn over the head; refers to the handsome plumage (and head-gear) of the bird.

The species

California Quail (introduced breeding resident, King and Norfolk Islands)

The species does occur in California but it is also found naturally all along the west coast of North America.

Callipepla californica (Shaw, 1798) [kal-li-PEP-luh ka-li-FOR-ni-kuh]: 'beautifully robed Californian', see genus name, and for the place name.

PHASIANIDAE (GALLIFORMES): quails, domestic hen relatives

Phasianidae Horsfield, 1821[faz-i-A-ni-dee]: the Pheasant family, see genus name *Phasianus*. 'Quail' was originally onomatopoeic, from an old Germanic base resembling 'kwak', thence via Old French 'quaille' into English. (In Greek it was *ortux* and the Ancient Greeks appear to have had an interesting relationship with quails – as well as having words for quail-catcher, quail-keeper, and so on, we find *ortugomania*, for madness after quails, and *ortugokopia*, the game of quail-striking!)

The genera

Meleagris Linnaeus, 1758 [me-le-AH-gris]: 'guineafowl', from Greek *meleagris*, guineafowl (see Numididae). Linnaeus (1758) uses the name for a grab-bag of birds, including the Wild Turkey (see species profile), a guan and a trapogan.

Coturnix Bonnaterre, 1791 [ko-TER-niks]: 'quail', from Latin *coturnix*, quail, a word also used in Latin as a term of endearment – a bit like 'duckie' or 'you old coot', perhaps, but with a better ring to it.

Excalfactoria Bonaparte, 1856 [ex-kal-fak-TOR-i-uh]: 'warmer' from Latin *excalfactorius* (warming, heating). The name refers to the alleged extraordinary warmth of these birds which, it seems, were used by the Chinese as hand-warmers in cold weather. Jardine (1834) tells us this was because the quails' bodies 'are thought to contain a large proportion of animal heat, from the pugnacious disposition of their tempers'. Perhaps the users were mainly the poverty-stricken Chinese though, who couldn't afford to own any of the beautifully crafted hand-warmers now prized by antique collectors.

Gallus Brisson, 1760 [GAL-lʊs]: 'cock', from Latin *gallus*, cock. This refers to the common-or-garden barnyard cock, which used to be called in English the Dunghill Cock.

Phasianus Linnaeus, 1758 [fa-zi-AH-nʊs]: 'pheasant', from Greek *phasianos*, pheasant, for the bird's supposed origin on the Phasis River, now known as the Rion, in modern Georgia.

Pavo Linnaeus, 1758 [PAH-voh]: 'peacock' from Latin *pavo* (or occasionally *pavus*), the peacock. As far as the origins of Latin *pavo* are concerned, is it a coincidence that the word *pavor* means quaking or panic – and that peacocks quake their tails in spectacular fashion in their display? And on the subject of that tail, its eyes are, according to Greek myth, those of the hundred-eyed giant Argos. He died in the service of the goddess Hera, whereupon she placed his eyes in the tail of her favourite bird, the peacock (*taōs*).

The species

Wild Turkey (introduced breeding resident, Bass Strait and Norfolk Islands)

African Helmeted Guineafowl were taken to Turkey, whence they arrived in Europe apparently in the 16th century. (As noted earlier, the Romans had previously introduced them but they vanished with the Empire.) Here they were known as Turkey-cock and -hen, and later simply as Turkey (or Turky).

When the birds we now know as turkeys arrived later in the same century from the Americas, there was confusion between the two species; they were even regarded as variants of the same species. In time, as the distinctions between them were recognised, for reasons that aren't entirely clear, the word turkey came to be applied to the current species rather than to guineafowl (the 'real' turkey)! Note the analogy with Red Junglefowl – the Australian feral population derives from long-domesticated, not wild stock.

Other names: Feral Turkey.

Meleagris gallopavo Linnaeus, 1758 [me-le-AH-gris gal-loh-PAH-voh]: 'cock-peacock guinea-fowl' – a hedging of bets and muddying of waters, and seemingly not much to do with turkeys! Even Linnaeus was confused! See genus name plus Latin *gallus*, cock and *pavo*, peacock.

Stubble Quail (breeding resident)

Stubble Quail are essentially birds of grassland and low shrubland, generally avoiding forests. They are not particularly more likely than Brown Quail to use stubble, but presumably were noted when they moved into croplands formed by forest clearing. Stubble Quail was recorded by Gould (1848) as being the name 'of the Colonists of Van Diemen's Land' and it apparently spread spontaneously, because it was recorded as first-choice name for the species by Morris in 1898.

Other names: Pectoral Quail, the name used (inevitably) by Gould (1848) from his specific name; Grey Quail, presumably in contrast to Brown Quail – it seems to be mostly used by caged bird enthusiasts, though the Jardine brothers noted 'grey quail' (not otherwise identified) in north Queensland (Jardine and Jardine 1867).

Coturnix pectoralis Gould, 1837 [ko-TER-niks pek-to-RAH-lis]: 'quail with a breast[-plate]', see genus name, and from Latin *pectus/pectoris*, the breast. It also evokes the Old French word *pectorale* meaning a breast-plate. The male bird has the equivalent in the shape of a rich buff bib, which is diagnostic of the species.

Brown Quail (breeding resident)

Although this is one of the least colourful of its genus (apart from the southern New Guinea race *dogwa*, which is almost black), 'brown' is not a particularly helpful adjective for quails! Nonetheless, Gould (1848) reported it as being used by 'the Colonists' of both Swan River and Van Diemen's Land and it was in general use by the end of the 19th century.

Other names: New Holland Quail, Latham (1801); Australian Partridge, first choice of Gould (1848); Swamp Quail or Swamp Partridge (in Tasmania), perhaps in habitat contrast to Stubble Quail; Tasmanian Quail or Tasmanian Brown Quail, from a time when the Tasmanian race was believed to be a separate species – Gould described it as *Synoïcus diemensis* ('from Van Diemen's Land') and called it Van Diemen's Land Partridge (1848) and Tasmanian Swamp-Quail (1865); Tasmanian Brown Quail for the species was still in use into the 20th century (RAOU 1913); Greater Brown Quail, reported by Gould to be the name 'of the Colonists' (of Van Diemen's Land); Sombre or Sordid Quail, for another of Gould's species, *S. sordidus* of South Australia, also no longer recognised.

Coturnix ypsilophora Bosc, 1792 [ko-TER-niks ip-sil-O-for-uh]: 'Y-bearing quail', see genus name, and because the breast markings are similar to this letter. From the Greek letter *upsilon* (= Y, u), and Greek *phoros*, bearing or carrying. Related to the word ypsiliform, Y-shaped.

King Quail (breeding resident)

The origin of this name is a mystery; Gould was apparently unaware of it, though a *Sydney Morning Herald* article (21 February 1842, widely republished thereafter) on edible Australian wildlife referred to the 'minum 'king' quail'. By 1899 Hall was using Chestnut-bellied Quail, but with King Quail as an alternative; by 1907 he was using King Quail as the first-choice name, though Lucas and Le Soeuf were still favouring Chestnut-bellied in 1911.

Other names: Chinese Quail, used by Latham (1821) and Gould (1848); Chestnut-bellied Quail (Hall 1899); Blue-breasted Quail, a descriptive name that applies only to the male and is the name used through much of its Asian range; Dwarf or Least Quail, in reference to its diminutive size relative to the two Australian *Coturnix* quails.

Excalfactoria chinensis (Linnaeus, 1766) [ex-kal-fak-TOR-i-uh chi-NEN-sis]: 'Chinese warmer', see genus name, and Modern Latin, China plus the suffix *-ensis*, usually indicating the place of origin of the type specimen.

Red Junglefowl (introduced breeding resident, tropical and Norfolk Islands)

There would be grounds for suggesting that this name is a touch pretentious; realistically, the few Australian island populations are of feral chooks! Although it is probable that the Red Junglefowl of South-East Asia and western Indonesia is the ancestor of the modern domestic fowl, it is some 5000 years since domestication took place. The Red Junglefowl male has long red-gold neck feathers.

Other names: Feral Fowl or Chicken.

Gallus gallus (Linnaeus, 1758) [GAL-lʊs GAL-lʊs]: 'cock-cock', from the Latin word *gallus*, meaning (you guessed it) cock.

Green Junglefowl (introduced breeding resident, Cocos (Keeling) Islands)

Unlike Australian wild Red Junglefowl, the Green Junglefowl population of Cocos (Keeling) Islands is probably derived directly from wild birds, which live naturally in relatively nearby southern Indonesian islands. The male plumage features iridescent green-bronze hues.

Gallus varius (Shaw, 1798) [GAL-lʊs VAH-ri-ʊs]: 'variegated cock', see genus name, and referring to the colours in the bird's beautiful plumage.

Common Pheasant (introduced breeding resident, Rottnest and Bass Strait Islands)

A native of middle Northern Hemisphere latitudes right across Asia, where it is known as the Ring-necked Pheasant. Elsewhere – notably Europe (where it was probably brought in by the Romans), and North America – there are no native pheasants and it is referred to as Common Pheasant or just Pheasant. The word comes to us, via French, from the Latin *phasianus* (see genus name).

Other names: Ring-necked, Chinese or Mongolian Pheasant.

Phasianus colchicus Linnaeus, 1758 [fa-zi-AH-nʊs kol-CHEE-kʊs]: 'Colchian pheasant', see genus name, plus *colchicus* for Colchis, ancient name of the region east of the Black Sea through which the Phasis River flowed (see genus name) and where the Golden Fleece sought by Jason in Greek myth was said to be found.

Indian Peafowl (introduced breeding resident, Rottnest and Bass Strait Islands)

Pea appeared in Old English, derived from the Latin *pavo* (see genus name). The old forms gave rise to peacock and peahen – the more general 'peafowl' is relatively recent. The species is found naturally throughout the subcontinent.

Other names: Common Peafowl; Peacock, Peahen.

Pavo cristatus Linnaeus, 1758 [PAH-voh kri-STAH-tʊs]: 'crested peacock', see genus name, and *cristatus*, meaning tufted or crested.

ANSERANATIDAE (ANSERIFORMES): Magpie Goose

Anseranatidae Sclater, 1880 [an-ser-a-NA-ti-dee]: the Goose-duck family, see genus name Anseranas.

The genus

Anseranas Lesson, 1828 [an-ser-AN-as]: 'goose-duck', from Latin *anser*, a goose and *anas*, a duck.

The species

Magpie Goose (breeding resident)

Magpie is in reference to the black and white colouring; the 'goose' is a bit more problematic, in that this bird is not truly either duck or goose, belonging to a separate, though related, family. It is unclear when the word was first used, though the *Dictionary of Austral English* (Morris 1898) gives it as 'a common name for the Australian goose', so presumably it arose spontaneously – the *Australian National Dictionary* gives the first known usage as 1861 and we can't better that. However, the first 'official' use of it as first-choice name is in the *Index of Australian Bird Names* (CSIRO 1969).

Other names: Black and White Goose, from Latham (1801); Semipalmated Goose, used by Latham (1823) directly from his specific name (or perhaps vice versa – he was easygoing on such things); Pied Goose, the name by which it was most widely known in ornithological circles until 1969; Swan Goose, recorded by Morris (1898) but hard to find otherwise in print, so presumably mostly a folk name.

Anseranas semipalmata (Latham, 1798) [an-ser-AN-as se-mi-pal-MAH-tuh]: 'half-webbed goose-duck', see genus name plus Latin *semi*, half, and *palma*, palm, *palmatus*, webbed, an accurate description of their feet.

ANATIDAE (ANSERIFORMES): geese, ducks, swans

Anatidae Leach, 1820 [a-NA-ti-dee]: the Duck family, see genus name *Anas*.

An old group of birds, with members on every unfrozen continent, so the basic names pre-date European Australia.

'Duck' is from an Old English word *duce*, meaning to dive.

'Whistling-Duck' because the loud whistles and twitterings of the three Australian *Dendrocygna* are very diagnostic.

'Shelduck' is from the now obsolete English word *sheld*, meaning pied (the European species of *Tadorna* is primarily black and white). The older form was Sheldrake.

'Goose' is also from Old English *gos* (pronounced like 'goats' without the 't').

'Swan' comes to us unchanged in spelling, also from Old English. The word appears to have the same origin as a word meaning 'sound' or 'call', perhaps linking with the concept of a 'swan song' (although the implication of that, from the ancient Greeks, was that swans sang only when about to die).

'Pygmy Goose' (or 'Pygmy-goose', as mostly used in Australia) is an odd name for three small ducks in the genus *Nettapus* (the two Australian species, which both also have ranges to the north of Australia, and an African one), with no goose associations. However, they were originally believed to be geese and the name has stuck; in 1896 Alfred Newton in his *Dictionary of Birds* noted somewhat peevishily 'systematists will have it that they are Geese, which the formation of their trachea shows they are not'. Current thinking places them at least temporarily in the primarily Gondwanan loose grouping of perching ducks (along with the Australian Wood Duck), though the grouping is largely one of convenience.

'Teal' was applied to the Common Teal *Anas crecca* of England (and most of the Northern Hemisphere), a very colourful and distinctive little duck; forms of the word date back to at least the 14th century, when it appeared as *teles*, presumably originating in a now lost Old English word. It is said to be onomatopoeic. It developed through a variety of forms until the current spelling was recorded in the 17th century. The word subsequently became applied to small members of *Anas* throughout the world.

'Wigeon': Lockwood (1984) traces 'wigeon' in various forms back to 'weygons' in 1513, and thence via French to Latin *vipionem*, which he explains was onomatopoeic for the drake's piercing whistle.

The genera

Dendrocygna Swainson, 1837 [den-dro-SIG-nuh]: 'tree-swan' from Greek *dendron*, tree and *kuknos*, Latin *cycnus/cygnus*, swan.

Cereopsis Latham, 1801 [se-ri-OP-sis]: 'wax-face' from Greek *kerinos*, waxen/wax-coloured (Latin *cera*, wax) and *opsis*, appearance or face. This refers to the bird's extensive yellow/green cere.

Branta Scopoli, 1769 [BRAN-tuh]: 'burnt goose' from Old Norse *brandgas* and Late Middle English *brant* or *brent*, in full *brent-goose*, thought to be *Branta bernicla*. The *brent* part of this is assumed to be the same as *brent*, meaning burned, and to refer to the largely black plumage characteristic of these geese.

Cygnus Garsault, 1764 [SIG-nus]: 'swan' from Greek *kuknos*, Latin *cycnus/cygnus*.

Stictonetta Reichenbach, 1853 [stik-to-NET-tuh]: 'dappled duck' from Greek *sticto*, punctured or dappled, and *netta*, duck.

Radjah Reichenbach, 1853 [RAH-juh]: 'rajah'. The name comes from Lesson (1828) as a species name (*Anas radjah*), and is the French spelling of English rajah, from Hindi rājā, originally from Sanskrit.

Tadorna Boie, 1822 [ta-DOR-nuh]: 'shelduck'. *Tadorna* appears to have somewhat uncertain origins. It is usually said to be from the French word for shelduck, *tadorne*, which comes from … well, at least one French dictionary claims it comes from Greek meaning 'wandering' and another claims it is from the Latin *anas tadorna*. However, the Latin was Linnaeus' term, first published in 1758, and the French word *tadorne/tadourne* is far older than that. It was used by Rabelais in 1534 (dozens of them are mentioned as a very small part of a feast in *Gargantua*) and by Belon (1555) in his description of the bird (cited by Ray 1713, who in turn is cited by Linnaeus, so going from French to Latin, not the other way). There is a separate, and very respectable, claim by English ornithologist Janet Kear that it was Celtic for 'pied waterbird' (Kear 2005), but we have been unable to verify this. Its exact origins may remain a mystery.

Malacorhynchus Swainson, 1831 [ma-la-ko-RIN-kʊs]: 'soft-bill' from Greek, *malacos*, soft and *rhunkhos*, bill, for the unusual bill. Swainson (1831a) described it thus: 'The edge of the upper mandible, instead of being smooth, as in the European species, is furnished with a thin membranaceous skin, which projects considerably, and hangs down somewhat like a wattle on each side.'

Chenonetta von Brandt, 1836 [ke-no-NET-tuh]: 'goose-duck' (hedging bets again), from Greek *khen*, goose and *netta*, duck.

Nettapus von Brandt, 1836 [NET-ta-pʊs]: 'duck-foot', from Greek *netta*, duck and *pous*, foot. Why were duck feet chosen as the highlighted feature to name the genus? Ducks differ from geese in having shorter legs set much further back on the body (this is why they swim better than they walk). So though these birds were originally believed to be geese (see Pygmy-goose), it did not go unnoticed that they had duck-like characteristics.

Spatula Boie, 1822 [SPA-tʊ-luh]: 'spoon' from Modern Latin *spatula*, a spoon. (In Classical Latin the word is a diminutive of *spatha*, which indicated a variety of things, including a flat piece of wood for stirring, a palm frond and a broadsword.) Boie (1822) used the German *Löffelente*, making it clear that 'spoon' is the meaning he wanted to convey.

Mareca Stephens, 1824 [MAR-eh-kuh]: 'duck' from the Brazilian Portuguese word *marreco*, a drake or duck. Stephens wrote that because the name *Penelope* had already been used for another genus, 'I have adopted the appellation given to some of the species by the illustrious Ray'. John Ray himself pointed out that he had the word from Marcgrave who had already used it in his *Historiae rerum naturalium Brasiliae*.

Anas Linnaeus, 1758 [AN-as]: 'duck' from Latin *anas, anatis*, a duck.

Aythya Boie, 1822 [AY-thee-uh]: 'seabird' from Greek *aithuia*, an unidentified seabird. The word *aithuia* (as well as being one of the surnames of Athena, apparently indicating that she had special knowledge of ships and navigation) was used by Homer (c. 700 BCa) as a simile to describe Ino/Leucothea, a minor marine goddess, as she rises from the waves to save Odysseus from drowning and then dives back into the water. Although it has been variously translated in this context as diver-bird, sea-crow, sea-gull and sea-mew, it now refers to a genus of diving ducks that includes scaup and some pochard.

Oxyura Bonaparte, 1828 [ok-see-YOO-ruh]: 'sharp-tail' from Greek *oxus*, sharp or acute, and *oura*, tail.

Biziura Stephens, 1824 [bi-zi-OO-ruh]: 'straw-tail' perhaps from Greek *bizeai*, straws, and certainly from *oura*, tail. Stephens simply says that the tail is 'acute', and later 'somewhat pointed' (Stephens and Shaw 1824).

The species

Note that the generic 'duck' has been used for many species; this is sometimes cited as an alternative name for some common Australian duck species, but we've not followed that path, believing it to be a reflection simply of non-discrimination rather than the application of a specifying name. The same comment could be made with regard to 'parrot' for instance.

Spotted Whistling-Duck (breeding resident)

See family introduction. Liberally white-spotted beneath.

Other names: Spotted Tree-Duck.

Dendrocygna guttata Schlegel, 1866 [den-dro-SIG-nuh gʊt-TAH-tuh]: 'spotted tree-swan', see genus name, and from Latin *gutta*, a droplet.

Plumed Whistling Duck (breeding resident)

See family introduction. The long wing plumes of this species are diagnostic; on the other hand, the Wandering Whistling Duck also has plumes! The name was well established by the end of the 19th century (Hall 1899).

Other names: Plumed Whistling-Duck, the form mostly used in Australia; Eyton's Duck (from the species name), used by Gould (1848); Grass Whistling-Duck, widely used, refers to the more terrestrial habits of this species compared with to the Wandering Whistling-Duck (which is also known as Water Whistling-Duck) – it was offered as first-choice name by CSIRO (1969) but was firmly suppressed by RAOU (1978). 'Tree-Duck', influenced by the genus name though not particularly descriptive, is used for this genus throughout its mostly Gondwanan range; hence Eyton's, Plumed, Red-legged and Whistling Tree-Duck (the last being notably unhelpful!). 'Whistler', usually with one of the previous descriptors, has also been used confusingly. Monkey Duck, something of a mystery (see also Freckled Duck), but perhaps related to the Tree-Duck concept.

Dendrocygna eytoni (Eyton, 1838) [den-dro-SIG-nuh eh-TOH-nee]: 'Eyton's tree-swan', see genus name, and for Thomas Campbell Eyton, English ornithologist (1809–1880). It is definitely not done to name a species for yourself, but Gould actually did the naming in an unpublished manuscript as 'a just tribute of respect to T.C. Eyton, Esq., of Donnerville, a gentleman ardently attached to the science of ornithology and well known for his valuable 'Monograph of the Anatidae''. However, it seems that Gould was too tardy in publishing for Eyton's liking, so in the end Eyton published the manuscript himself!

Wandering Whistling Duck (breeding resident)

See family introduction. Wandering is a less than useful name, since the Plumed Whistling Duck wanders much more into southern Australia than this species does; nonetheless RAOU (1978) favoured it as 'the name most widely used from Malaysia to Papuasia'.

Other names: Wandering Whistling-Duck, the form mostly used in Australia; Whistling Duck, used by Gould (1848), taking his cue from 'the Colonists'; perhaps oddly this was used alone throughout the 19th century and into the 20th (RAOU 1913). See Plumed Whistling Duck for Tree-Duck; hence Wandering, Black-legged and Whistling (!) Tree-Duck – the last was favoured by RAOU (1926); Red and Wandering Whistler. Water Whistling-Duck, was widely used in the 20th century until RAOU (1978).

Dendrocygna arcuata (Horsfield, 1824) [den-dro-SIG-nuh ark-yoo-AH-tuh]: 'bowed tree-swan', see genus name, and from Latin *arcuatus*, bow or rainbow. Horsfield (1824) writes 'The name of *arcuata* corresponds with its external marks.' He mentions 'semilunar marks' on breast, neck and upper back, as well as 'the plumes of the back [which] are bordered at the extremity, in the form of an arch or semicircle, by a narrow chestnut band'. (Strangely, the rich chestnut colour of the bird also evokes *morbus arcuatus*, the rainbow-coloured disease, or jaundice, so in that case it would be 'jaundiced tree-swan'!)

Cape Barren Goose (breeding resident)

See family introduction. Apparently a true goose, though an aberrant one. Cape Barren Island is in Bass Strait, just south of Flinders Island off north-eastern Tasmania. It was George Bass and Matthew Flinders in 1798 who reported the bird from Bass Strait (the *d'Entrecasteaux* expedition had previously found it in south-western Australia), but Bass referred to it as a 'Brent or Barnacle Goose'. The current name may have been applied by the sealers who followed the explorers or it may have already been in place from 1797, given that survivors of the wreck of the *Sydney Cove* on nearby Preservation Island would certainly have encountered the bird – doubtless with some relish! Gould (1848) reported that it was the name 'of the Colonists'.

Other names: New Holland Cereopsis, used by Latham (1801) – his species name thus seems to have followed it, but presumably was in the process of being published by then; Cereopsis Goose, by Gould (1848); Pig Goose, apparently for its grunting calls; Pigeon Goose, perhaps for the flesh but perhaps too an association with Pig Goose – either way it is not easy to find examples of its use.

Cereopsis novaehollandiae Latham, 1801 [se-ri-OP-sis no-veh-hol-LAN-di-eh]: 'New Holland wax-face', see genus name, and from Modern Latin *Nova Hollandia*, New Holland, the old name for Australia.

Canada Goose (vagrant)

See family introduction. Found naturally throughout North America, though the breeding grounds are in Canada and Alaska.

Branta canadensis (Linnaeus, 1758) [BRAN-tuh ca-na-DEN-sis]: 'Canadian burnt goose', see genus name, plus Modern Latin Canada plus the suffix *-ensis*, usually indicating the place of origin of the type specimen.

Black Swan (breeding resident)

See family introduction. A true swan, and the name is inevitable because the bird is the only all-black swan of the world's seven species. (It was also a serious challenge for Northern Hemisphere traditionalists, for whom 'white swan' was a tautology and 'black swan' an oxymoron!) It can be traced back to the earliest days of the colony (White 1790) but doubtless pre-dates that.

Other names: Notable in having had, apparently, no other names.

Cygnus atratus (Latham, 1790) [SIG-nʊs at-RAH-tʊs]: 'swan in mourning', see genus name, and Latin *atratus*, clothed in black for mourning, a reference to the bird's feathers rather than its emotional state.

Mute Swan (introduced breeding resident, only in Northam, WA)

See family introduction. Although not loquacious, the Mute Swan does have a voice; Soothill and Whitehead (1978) referred to 'various grunting, snoring and hissing notes'. It does tend to keep its own counsel, however, which is always likely to be mistaken for muteness by humans.

Other names: White Swan.

Cygnus olor (Gmelin JF, 1789) [SIG-nus OH-lor]: 'swan-swan', see genus name, and Latin *olor*, also meaning swan. Clearly the quintessential swan, perhaps rivalled only by the Trumpeter Swan from Europe, *Cygnus cygnus*.

Freckled Duck (breeding resident)

See family introduction. Descriptive, in the sense of covered in small spots; in the field this is helpful only in good light. Gould (1848) used the name, noting that it was used by the 'Colonists of Western Australia' – his species name apparently followed.

Other names: Oatmeal and Speckled Duck have the same descriptive origin as the primary name; Diamantina Duck, reflecting the fact that the Freckled Duck is essentially a bird of the south-east inland, though more especially of the Murray–Darling Basin than the Lake Eyre Basin (of which the Diamantina is a river). Monkey Duck, said to be an old shooters' name (Gould doesn't mention it in 1848, but it appears in a newspaper column in 1897 (Campbell 1897) and Leach offers it as an alternative in 1911), but the origin has been lost; a possible clue is the observation that in open water the bird will roost hunched on top of emergent stumps (Frith 1967); another is the observation that 'monkey' was slang for sheep, but there is no suggestion that the bird is mutton-flavoured.

Stictonetta naevosa (Gould, 1841) [stik-to-NET-tuh neh-VOH-suh]: 'spotty dappled duck', see genus name plus Latin *naevus*, a spot or mole (as in small growth on the body, not the furry animal, though the colours of the latter are coincidentally not dissimilar to those of the Freckled Duck).

Radjah Shelduck (breeding resident)

See family introduction. 'Radjah' was apparently borrowed directly from the rather grandiose species name; it was used by Gould in the old form Radjah Shieldrake. Thereafter it sank from popularity until being reinstated by the RAOU's third official list of recommended names (RAOU 1978) on the grounds of its use overseas.

Other names: Burdekin Duck, widely used, presumably for the Burdekin District around Ayr, south of Townsville, where Edmund Kennedy reported large numbers in 1870; although this is south of its main range now, it was originally well within its normal distribution. To people coming up the east coast from the south it would have been a new species; it would have been already known from the Top End, but there it was probably known as Radjah, the name used outside Australia (New Guinea and the Moluccas). White-headed Shelduck or Sheldrake or Shieldrake, the latter being the preferred name of the RAOU's first *Official Checklist* (RAOU 1913). As White-headed Shelduck, it was retained by the second checklist (RAOU 1926), with Burdekin Duck as an alternative.

Radjah radjah (Garnot & Lesson, 1828) [RAH-juh RAH-juh]: 'Rajah rajah', see genus name. Lesson does not explain the species name, saying only that he and Garnier 'procured for ourselves a pretty species of duck, which we called *anas radjah*' (Lesson 1839a).

Australian Shelduck (breeding resident)

See family introduction. Of the two Australian *Tadorna* species, only this one is endemic, justifying the 'Australian'. It did not appear, however, until coined as a recommended name by the RAOU (1978).

Other names: New Holland Sheldrake, used by Jardine and Selby (1826–35, vol. 2); Chestnut-coloured Shelduck or Sheldrake or Shieldrake (the latter used by Gould 1848 and RAOU 1913, Sheldrake used by RAOU 1926); also Chestnut or Chestnut-breasted Shelduck, and so on; Mountain Duck, widely used, including as an alternative name by RAOU (1926), reported by Gould (1848) as used by the 'Colonists of Swan River' – it supposedly implies 'over the mountains' (also used in this way in the name of the lizard *Moloch horridus*, known as Mountain Devil); Grunter, for the calls.

Tadorna tadornoides (Jardine & Selby, 1828) [ta-DOR-nuh ta-dor-NOY-dehz]: 'shelduck-like shelduck', see genus name, and from Latin *-oides*, resembling.

Paradise Shelduck (vagrant)

See family introduction. A New Zealand endemic, recorded from Lord Howe Island and the south coast of NSW. A striking bird, but the name is perhaps a little fulsome.

Tadorna variegata (Gmelin JF, 1789) [ta-DOR-nuh va-ri-e-GAH-tuh]: 'variegated shelduck', see genus name, and Latin *variegatus*. Although both sexes are very handsome birds, 'variegated' probably describes the female with her white head slightly better than the male, who appears more uniform.

Pink-eared Duck (breeding resident)

See family introduction. A classic case of a bird named from a skin (or from a shooter's bag) – the lovely pink 'ears' are almost impossible to see in the field without a telescope. It seems to have arisen spontaneously in the second half of the 19th century, because Gould doesn't mention it (unless his Pink-eyed Duck,

Pink-eared Duck *Malacorhynchus membranaceus*

see 'Other names', was indeed intended to be Pink-eared), but a record of a donation to the Australian Museum in 1855 does so. In 1898 Morris used it as first-choice name in his *Dictionary of Austral English.*

Other names: New Holland Duck (Latham 1823); Membranaceous Duck (it's true!) used by Gould (1848), from the species name; Zebra Duck or Teal, in reference to the stripes, which are one of its most conspicuous field characters; Pink-ear; Pink-eyed Duck, for no obvious reason (except that the pink patch could be mistaken for an eye from a distance, if the light was good enough to see it!) – however, Gould records this as the name used by the 'Colonists of Swan River', and perhaps this was a mistranscription, though if so it was repeated at times in the names of some 19th century paintings and photos; Whistler or Whistling Teal, for the distinctive musical twittering of the flocks; Widgeon, the old spelling of the Northern Hemisphere Wigeon, *Mareca penelope*, with which there is no obvious resemblance except for the whistling call – nonetheless, it was obviously widespread in the 19th century, because Morris (1898) cites it as the only alternative name.

Malacorhynchus membranaceus (Latham, 1801) [ma-la-ko-RIN-cʊs mem-bra-NAY-see-ʊs]: 'membrane-y soft-bill', from Latin *membrana*, a thin skin, for the strange flaps on the outer corners of the bird's bill, see also genus name.

Maned Duck (breeding resident)

See family introduction. The name (with Maned Goose) was widely used as first choice from at least the time of Gould (who used Maned Goose) until recently (e.g. Maned Duck in HANZAB 1990–2006, vol. 1b). It was taken from the species name but, while a somewhat aberrant duck due to its grazing habits, it is certainly not a goose.

Maned Duck *Chenonetta jubata*

Other names: Australian Wood Duck (usually the first-choice name in Australia today) or just Wood Duck; although this species is often placed in the same subtribe as the American Wood Duck (genus *Aix*), the relationship is not particularly close and the physical resemblance only passing. It seems more likely that the Australian bird was named independently for its habit of nesting and perching loudly and conspicuously in trees away from water. Certainly Gould (1848) reported that Wood Duck was already used by 'the Colonists of New South Wales and Swan River', most of whom would have been unaware of the American bird. Blue Duck, cited by HANZAB (1990–2006, vol. 1b) but not evident elsewhere.

Chenonetta jubata (Latham, 1802) [ke-no-NET-tuh yoo-BAH-tuh]: 'crested or maned goose-duck', see genus name, and Latin *iubatus*, crested or maned for the little crest on the male's nape.

Cotton Pygmy Goose (breeding resident)

See family introduction. This name has been widely used for this species in eastern and southern Asia (presumably for the snowy white face and throat) and has only recently become the first-choice name in Australia, following the recommendations of RAOU (1978).

Other names: Cotton Pygmy-goose, the usual name in Australia; Pygmy Goose, used by Gould (1848); Cotton Teal (teal is a term more properly applied to small members of the genus *Anas* – Goose-teal has also been widely used in apparent acknowledgment of this); White Pygmy-goose; White-quilled Pygmy-goose or Dwarf-goose, for the broad black-tipped white band across the tips of the primaries, visible in flight; Australian Pygmy Goose, which is not helpful given its wide range through Asia and the presence of another Australian species. The form Pigmy, known from at least the 14th century, has also been widely used, including as late as RAOU (1926).

Nettapus coromandelianus (Gmelin JF, 1789) [NET-ta-pʊs co-ro-man-de-lee-AH-nʊs]: 'Coromandel duck-foot'. The Coromandel Coast referred to is in south-eastern India (not the region in New Zealand) and is named after the ancient kingdom of Cholamandalam.

Green Pygmy Goose (breeding resident)

See family introduction. The green is in obvious contradistinction to the white neck of the Cotton Pygmy-goose – a distinctive feature on the water– though both have a green back.

Other names: Green Pygmy-goose, the usual name in Australia; Beautiful Pygmy Goose, by Gould (1848) from his species name – he was quite smitten by it, using both 'beauty' and 'beautiful' in his paragraph describing it; Little Goose, reported by Gould to be used by 'residents at Port Essington' (north of Darwin); Green Goose or Teal; Goose-teal (or Green Goose-teal); Green Dwarf-goose.

Nettapus pulchellus Gould, 1842 [NET-ta-pʊs pʊl-KEL-lʊs]: 'beautiful little duck-foot', see genus name, with Latin *pulchellus*, a diminutive of beautiful, usually translated as 'pretty' or 'very pretty', which the bird, with its iridescent green back and delicately marked flanks, certainly is.

Garganey (vagrant or rare migrant)

From an Italian diminutive of the name used locally for the bird around Bellinzona, Switzerland. It began to replace older English names during the 17th century.

Spatula querquedula Linnaeus, 1758 [SPA-tʊ-luh kwehr-KWEH-dʊ-luh]: 'spoon duck', see genus name, and from Latin *querquedula* (some kind of duck) and the old genus name for ducks including the Garganey and the Eurasian Teal. The Greek words *kerkēdēs* or *kerkēris* indicated waterbirds, probably teal, and the word *querquedula* may well come from this. Both versions may simply have been onomatopoeic (as in 'quack, quack').

Australasian Shoveler (breeding resident)

The Australian and New Zealand analogue of the pan-Northern Hemisphere Northern Shoveler (there are also southern African and South American species). The name refers to the unusually large bill, though the shovelers are filter feeders, like other *Anas*.

Other names: New Holland Shoveller (Latham 1801); Australian Shoveller, used by Gould (1848) – the oddly inexplicable dropping of an 'l' to make 'shoveler' didn't take place until the 20th century (RAOU 1926); Shovel-nosed Duck, reported by Gould (1848) to be a name 'of the Colonists'; Shovelbill; Blue-wing, Blue-winged Shoveler; Spoonbill Duck, a curious coincidence given that the name shoveller (or shoveler) in England was applied to the Spoonbill until at least the 16th century; Stinker – as Frith (1967) commented politely, it is 'not considered a good table bird'.

Spatula rhynchotis Latham, 1802 [SPA-tʊ-luh rin-KOH-tis]: 'beaked spoon', see genus name, and from Greek *rhunkhos*, bill.

Northern Shoveler (vagrant)

Found across the Northern Hemisphere (and migrating south, including into South-East Asia).

Spatula clypeata Linnaeus, 1758 [SPA-tʊ-luh kli-pe-AH-tuh]: 'spoon with a shield', see genus name, and from Latin *clipeatus*, armed with a shield (*clipius* or *clupius*, a round shield), the huge bill, which Linnaeus (1758) described as having a dilated round end.

Eurasian Wigeon (vagrant)

See family introduction. This species occurs right across northern Eurasia, and there are other wigeons in North and South America.

Mareca penelope Linnaeus, 1758 [MAR-eh-kuh pe-NE-lo-pe]: 'Penelope duck', see genus name, and from Penelope, wife of Ulysses, who is said to have been thrown into the sea by her parents Icarius and Periboea, rescued by these ducks and then named for them. The use of the name for ducks/wigeons dates back to antiquity. Linnaeus himself cited Gesner, Aldrovandus, Ray and Willughby as using the name.

Pacific Black Duck (breeding resident)

See family introduction. A source of some confusion, since it is not very black at all (unlike for instance the closely related African Black Duck), nor is it particularly similar to the American Black Duck. Perhaps the comparison was with the other very common Australian duck, the much paler Grey Teal. It arose spontaneously – Gould (1848) reported that Black Duck was used by 'Colonists of New South Wales and Van Diemen's Land'. 'Pacific' is in reference to its presence in New Zealand and some quite remote offshore islands (as well as New Guinea).

Other names: Supercilious Duck, by Latham 1801; Australian Wild Duck, used by Gould (1848), an acknowledgment of its ubiquity; Brown or Grey Duck, equally valid and equally unhelpful – Gould reported that Grey Duck was a product of the creativity of the 'Colonists of Swan River'; Blackie.

Anas superciliosa Gmelin JF, 1789 [AN-as soo-pehr-si-li-OH-suh]: 'eyebrowed duck' rather than 'supercilious duck', though of course the word supercilious indeed involves what you do with your eyebrows when you're feeling it. The word describes the very strong and unmistakable facial markings of this duck. See genus name, and from Latin *supercilium*, above the eyelid (i.e. eyebrow).

Mallard (introduced breeding resident)

It was not until the 20th century that this name came into general use for the species. Although the word had been in the language since the 14th century, from Old French *malard*, it was used as the French had done, to designate the drake. The species as a whole was simply referred to in England as Wild Duck.

Anas platyrhynchos Linnaeus, 1758 [AN-as pla-ti-RIN-kos]: 'broad-billed duck', see genus name, and from Greek *platus*, broad and *rhunkhos*, bill, so-called because the bill is uniformly broad, with no tapering towards the end.

Eaton's Pintail (vagrant to the Australian Antarctic Territory)

The male in breeding plumage has long thin tail feathers, and for the Rev. Alfred Edmond Eaton (1845–1929), English clergyman and entomologist, who sailed as naturalist on one of the two British Transit of Venus expeditions to the Kerguelen Islands in 1874. While there, he collected the specimen which Bowdler Sharpe described the following year.

Other names: Kerguelen Pintail, as it is endemic to the southern Indian Ocean Islands of Kerguelen and Crozet.

Anas eatoni (Sharpe, 1875) [AN-as EE-ton-ee]: 'Eaton's duck', see genus name and common name.

Northern Pintail (vagrant)

See previous species for pintail; found across the Northern Hemisphere (and migrating into southern Asia).

Anas acuta Linnaeus, 1758 [AN-as a-KOO-tuh]: 'pointy duck', see genus name, and from Latin *acutus*, sharp or pointed, an apt description of the look of the tail.

Eurasian Teal (vagrant to Cocos (Keeling) Islands)

See family introduction. Found throughout most of Europe and Asia.

Other names: Common Teal, from when North American Green-winged Teal *A. carolinensis* was included in the species.

Anas crecca Linnaeus, 1758 [AN-as KREK-kuh]: 'quacking duck', from a Swedish dialect word *kricka*, or *kräcka*, presumably onomatopoeic. The word enters standard Swedish c. 1806 as *kricka*, the current word for teal (Svenska Akademiens ordbok). Linnaeus (1746), in his *Fauna svecica*, puts the Swedish as År*ta*; but in 1758 he used *crecca* as his species name.

Grey Teal (breeding resident)

See family introduction. The bird is no more grey than it is brown, though doubtless its general lack of obvious physical characteristics challenged early labellers; perhaps more relevantly it was used as a contrast to Chestnut Teal, widely known as just Teal in the 19th century (e.g. Morris 1898).

Other names: Slender Teal, perhaps a name applied by settlers eating the bird (while the difference is not obvious in the field, Grey Teal is around 25% lighter than Chestnut Teal) but probably just from the species name; Wood Teal, perhaps reflecting the fact that it nests in tree hollows much more regularly than the often ground-nesting Chestnut Teal; Oceanic Teal – while it does use sheltered bays, it does so less than the next species; Mountain Teal, though also applied to the next species.

Oddly, Gould seems to have overlooked this most abundant of ducks – the clue to the mystery lies in his comment relating to Chestnut Teal that 'It is very rare that the male is killed in nuptial dress'; he was assuming that Grey Teal were female or non-breeding male Chestnuts. As a result it was named surprisingly late.

Anas gracilis Buller, 1869 [AN-as GRA-si-lis]: 'slender duck', see genus name, and from Latin *gracilis*, slender.

Chestnut Teal (breeding resident)

See family introduction. The 'Chestnut' makes this bird one of those named only for the male, because only he has the rich chestnut body.

Other names: Chestnut-breasted Duck, used by Gould (1848); Teal, reported by Gould to be used by the 'Colonists of Swan River'; Brown (oh dear!), Black or Red Teal; Mountain Teal; Green-headed Teal.

Anas castanea (Eyton, 1838) [AN-as kas-TA-ne-uh]: 'chestnut duck', see genus name, and from Latin *castanea*, the chestnut or chestnut-tree, hence chestnut-coloured.

Hardhead (breeding resident)

A shooters' name. Although there is no evidence that its skull is particularly solid, Frith (1967) commented that 'owing to a very dense plumage and apparently great stamina, [it] is hard to kill'. It presumably arose spontaneously – by 1898 Morris was using Hard-head as an alternative to White-eyed Duck.

Other names: White-eyed Duck (or White-eye), the name used by Gould (1848) and the one by which it was long known, which may well be thought a more useful one because the white eyes are very conspicuous in the field, though this applies only to the male; Brownhead, Copperhead or Coppertop are all self-evident; Bar-wing, White-wing or White-winged Duck (the name 'of the Colonists' according to Gould) for the considerable amount of wing white shown in flight, emphasised against the dark plumage.

Aythya australis (Eyton, 1838) [AY-thee-a ost-RAH-lis]: 'southern seabird', see genus name, and Latin *australis*, southern.

Blue-billed Duck (breeding resident)

See family introduction. An inevitable and most appropriate name. Gould (1848) used it himself and reported that the colonists were already doing so.

Other names: Blue-bill; Spinetail or Stifftail, Spiny-tailed Duck or Stiff-tailed Duck, descriptive, but Stifftail is also used as the group name of the eight mostly Gondwanan (particularly South American) species to which this species belongs (traditionally the Musk Duck was also considered to belong, but no longer); Diver or Diving Duck are also descriptive, but not unique; Little Musk Duck, particularly in reference to the female.

Oxyura australis Gould, 1837 [ok-see-OO-ruh ost-RAH-lis]: 'southern sharp-tail', see genus name plus Latin *australis*, southern.

Musk Duck (breeding resident)

See family introduction. In 1791 Archibald Menzies, with Captain George Vancouver in HMS *Discovery*, first collected this species at the site of the modern Albany; he found the name compelling when the

Musk Duck ***Biziura lobata***

bird's musky aroma proceeded to dominate the ship! Only the male smells thus, apparently exuding the chemical with the preening oil from his uropygial gland; it is particularly intense during courtship. Gould (1848) adopted the name in the form Musk-Duck.

Other names: Diver or Diving Duck; Steamer, for splashing courtship display (for the same reason the South American steamer ducks were named); Mould Goose, perhaps also due to the scent; Lobed Duck, from the species name, but very rarely used.

Biziura lobata (Shaw, 1796) [bi-zi-OO-ruh lo-BAH-tuh]: 'lobed straw-tail', see genus name plus Latin *lobatus*, lobed. This refers to the large expandable flap under the bill; both this and the stiff tail are involved in the dramatic display of the courting male.

PODARGIDAE (CAPRIMULGIFORMES): frogmouths

Podargidae Bonaparte, 1838 [po-DAR-gi-dee]: the Gouty family, see genus name *Podargus*. 'Frogmouth' refers to the broad flattened bill, in combination with a huge gape, which are adaptations to taking prey from the ground. The name can be traced back to Gould's genus *Batrachostomus*, a South and East Asian genus of frogmouths; Thomas Jerdon anglicised it in his 1863 *Birds of India* as Frog-mouth. It was probably restricted to Anglo-Indian birders for a while after that, because the *Oxford Dictionary*'s first citation is from 1888. By 1899 Hall was using frogmouth in Australia, but we can find no usage down under before that.

The genus

Podargus Vieillot, 1818 [po-DAR-gʊs]: 'gouty', from Greek *podagra*, gout, itself derived from *agra*, a trap in which a *pous/podos* (foot) might be caught – an apt use of words to anyone familiar with the exquisite pain of gout. The feet of the frogmouths are rather weak, hence their use of the bill for catching prey. Not to be confused (even though the word looks a more likely source) with Greek *podargus*, swift-footed – not in any case a description we immediately associate with the frogmouth.

The species

Marbled Frogmouth (breeding resident)

See family introduction. The pale marbled effect of the wing and underside feathers is indeed notable and beautiful, but not obviously more so than in the other species; based on Gould's name *P. marmoratus* for his Cape York 'species', now a race.

Other names: Plumed Frogmouth, especially applied to the southern race *plumiferus* (which has sometimes been treated as a separate species), for the striking feathery plumes above the bill; Plumed Podargus, used by Gould (1848); Marbled Podargus, Gould (1869); Little Papuan Frogmouth, in contradistinction to the Great Papuan Frogmouth, as both are found throughout New Guinea.

Podargus ocellatus Quoy & Gaimard, 1830 [po-DAR-gʊs o-sel-LAH-tʊs]: 'gouty eyelet bird', see genus name, and from Latin *ocellus*, small eye or eyelet, referring to the bird's markings.

Papuan Frogmouth (breeding resident)

See family introduction. Essentially a New Guinea species, which spills into far north Queensland; named from the species name.

Other names: Papuan Podargus, used by Gould (1869); Large or Great Papuan Frogmouth, because it is the largest of its genus; Plumed Frogmouth, obscure, because its small above-bill plumes are no more conspicuous than those of the Tawny Frogmouth, but it could be from confusion with the Marbled (also known as Plumed) Frogmouth where they overlap at the top of Cape York Peninsula.

Podargus papuensis Quoy & Gaimard, 1830 [po-DAR-gʊs pah-PWEN-sis]: 'gouty Papuan bird', see genus name, and from Papua plus the suffix *-ensis*, indicating the place of origin of the type specimen.

Tawny Frogmouth (breeding resident)

See family introduction. Most frogmouth species (and all three Australian ones) are very variable in colour; although most Tawny Frogmouths are mottled tawny-grey, there is a rufous morph (females of the widespread race *phalaenoides*), and darker birds of the other two species, especially males, are similar in colour to most Tawnies. The name probably arose from Gould's Tawny-shouldered, see 'Other names'.

Tawny Frogmouth ***Podargus strigoides***

Other names: Cold River Goatsucker, enigmatically used by Latham; see Caprimulgus in the next family for goatsucker, but Cold River? – even the estimable *Australian Gazetteer* is unable to assist. Frogmouth Owl, in the good Australian tradition of 'five bob each way'; Podargus, an old name, for the genus; Tawny-shouldered Frogmouth or Podargus, the latter was the name Gould (1848) used, not very obvious, but Gould was using small samples; Short-winged Podargus, not useful, but taken from an old species name of Gould's for Western Australia birds, *brachypterus* – still recognised by RAOU (1913); Mopoke or Morepork, for the call (reported by Gould to be used by 'the Colonists') but not descriptive of it (it is a monotonously thrumming oo-oo-oo-oo) and probably due to confusion with Southern Boobook calls – the names are applied to it too; Mope-hawk, Moreport, Mawpork, much less commonly encountered variants of the previous; Night Hawk (see White-throated Nightjar); Cuvier's Podargus, for the no longer recognised Tasmanian species *P. cuvieri*; Tasmanian Frogmouth for the same former species, still used by RAOU (1913); Moth-plumaged Podargus, for Gould's species *P. phalaenoides* from Port Essington –no longer recognised as a species, but race *phalaenoides* is found right across Australia except for the east and south-east coast; also Freckled Frogmouth, for the same former species.

Podargus strigoides (Latham, 1801) [po-DAR-gʊs stri-GOY-dehz]: 'gouty bird like an owl', see genus name, and from Latin *strix/strigis*, an owl, and *-oides*, resembling.

CAPRIMULGIDAE (CAPRIMULGIFORMES): nightjars

Caprimulgidae Vigors, 1825 [kap-ri-MUL-gi-dee]: the Goat-sucker family (Mark 1 – Latin; see Aegothelidae for Mark 2 – Greek). See genus name *Caprimulgus*.

Although not immediately obvious, 'nightjar' is (at least in part) another onomatopoeic name, from the distinctive extended nocturnal churring courtship call of the male European Nightjar *Caprimulgus europaeus* – in fact an old alternative name is nightchurr.

The genera

Eurostopodus Gould, 1838 [yoo-ro-STO-po-dʊs]: 'strong-foot', from Greek *eurōstos*, strong, and *pous/podos*, foot. Gould (1838b) described the legs as 'stout' and the toes as 'short, thick and fleshy'.

Caprimulgus Linnaeus, 1758 [kap-ri-MʊL-gʊs]: 'goat-sucker' from Latin *capra*, a female goat, and *mulgere*, to milk. This comes from an ancient European folk belief, reported by classical naturalists (e.g. Pliny 77–79 AD), that the bird sucks the milk from goats at night, somehow blinding them in the process.

The species

Spotted Nightjar (breeding resident)

See family introduction. The big white wing spots are very evident when the bird is flushed; based on the species name, and used by Gould, see 'Other names'.

Other names: Goatsucker, the name 'of the Colonists', reported by Gould (1848) – see also under *Caprimulgus*; Spotted Goatsucker, used by Gould, from the then name *guttatus*; Laughing Owl, Night Hawk.

Eurostopodus argus (Hartert, 1892) [yoo-ro-STO-po-dʊs AR-gʊs]: 'spotted strong-foot', see genus name, and from Argos, the hundred-eyed giant described under Indian peafowl. The name was first used by Rosenberg in 1867, unfortunately with no description of the species, and was taken up by Hartert (1892). Hartert's only mention of spots involved the large white ones on the wings, but it seems to us that the bird is covered in other spots, at least the hundred that would justify the choice of name. Is it possible that Hartert did not understand Rosenberg's classical reference? Could he not count? (Though on the other hand, whatever made us think there has to be logic in bird naming?)

White-throated Nightjar (breeding resident)

See family introduction. It does indeed have a white throat, but so does the Spotted Nightjar, albeit less conspicuously. We certainly acknowledge the difficulty in assigning unambiguous names to such a similar-looking group of birds – maybe their calls would have offered more choices! Having said that, it comes directly from the species name *Caprimulgus albogularis* Vigors & Horsfield 1827, which had been gazumped by Temminck.

Other names: White-throated Goat-sucker, from Gould (1848) – see Caprimulgidae family introduction; Fern Owl, not clear – the bird is associated more with open dry forest habitats than ferny ones (the 'owl' is simply a reminder that those who first recorded Australian birds were not generally taxonomists); Laughing Owl, for the wonderful bubbling call; Moth-hawk or Night-hawk, surely for the image of the large bird fluttering silently through a patch of light – note, however, that 'nighthawk' is also formally used for a group of American nightjars.

Eurostopodus mystacalis (Temminck, 1826) [yoo-ro-STO-po-dʊs mis-ta-CAH-lis]: 'moustachioed strong-foot', see genus name, and from Greek *mustax*, the upper lip or moustache, Modern Latin *mystacalis*, moustached. This refers to the white cheek marks, which are common to the three species of nightjar in Australia. Indeed the other two seem, if anything, to be slightly better endowed in this respect than *mystacalis*.

Grey Nightjar (vagrant)

See family introduction. Perhaps a bit greyer than some other nightjars.

Caprimulgus jotaka Temminck & Schlegel, 1845 [kap-ri-MʊL-gʊs djo-tuh-kuh]: 'jotaka goat-sucker', see genus name, and from 'the name it bears in Japan' as the authors wrote (Siebold *et al.* 1850).

Large-tailed Nightjar (breeding resident)

See family introduction. This is one of those names that are distinguished by being of no conceivable usefulness; if you were standing over the bird without disturbing it, you might note that the wings cover

Large-tailed Nightjar *Caprimulgus macrurus*

a fraction less of the tail than in other species. It is taken directly from the scientific name but that is no excuse! It was formalised by RAOU (1978) as being 'traditional' in Australia.

Other names: Large-tailed Goatsucker, used by Gould (1848), directly from the species name; White-tailed Nightjar, the name in general use until quite recently, when it was discarded presumably for being descriptive and helpful in distinguishing it from other Australian nightjars (though RAOU 1978 pointed out that the name is also used for an African species); Carpenter-bird, Axe-bird, Woodcutter, Mallet-bird, Joiner-bird, Hammer-bird, Tok-tok, all evocative of the echoing repetitive chok-chok-chok call; Long-tailed Nightjar.

Caprimulgus macrurus Horsfield, 1821 [kap-ri-MʊL-gʊs mak-ROO-rʊs]: 'large-tailed goat-sucker', see genus name, and from Greek *macros*, large, and *oura*, tail. Horsfield (1821) said: 'This species is strikingly distinguished by the length of the tail' (see common name).

Savanna Nightjar (vagrant to Christmas Island)

See family introduction. An inhabitant of grassy habitats in southern and eastern Asia (northern birds migrate as far as Java after breeding).

Caprimulgus affinis Horsfield, 1821 [kap-ri-MʊL-gʊs af-FEE-nis]: 'related goat-sucker', see genus name, and Latin *affinis*, related. Horsfield (1821) considered it allied to the Indian Nightjar, *Caprimulgus asiaticus*.

AEGOTHELIDAE (CAPRIMULGIFORMES): owlet-nightjars

Aegothelidae Bonaparte, 1853 [eh-go-THEH-li-dee]: the Goat-sucker family (Mark 2 – Greek; see Caprimulgidae for Mark 1 – Latin). See *Aegotheles*.

Although there are seven other species in the genus (and family) in New Guinea, the Moluccas and New Caledonia, the Australian one was described from Sydney long before them, so the name owlet-nightjar almost certainly arose in Australia. Various newspaper articles in 1841 reported Gould's usage of it in the first volume of *Birds of Australia* (as Owlet Nightjar) – though in the 1848 7-volume final edition it opens Volume 2. (Shaw, in White 1790, had used the wonderful Crested Goatsucker, see Australian Owlet-nightjar). The allusion is to its small size, nocturnal habits and superficially nightjar-like form and behaviour; note that owlet-nightjars are not now regarded as particularly closely related to true nightjars, with swifts being apparently their closest relatives.

The genus

Aegotheles Vigors & Horsfield, 1827 [eh-go-THEH-lehz]: 'goat-sucker', from Greek *aigothēlas*, a nightjar (from *aix/aigo-*, goat and *thelazo*, to suckle). The folk belief in this behaviour ascribed to the bird is said to have arisen in Ancient Greece, but may well pre-date that civilisation. See also Caprimulgidae.

The species

Australian Owlet-nightjar

See family introduction. The only Australian species.

Other names: Crested Goatsucker (see family introduction, also used by Governor Phillip), from the specific name and the wonderful old name for nightjars (see genus name and *Caprimulgus*), or Crested Owlet-nightjar; New Holland or Bristled Goatsucker, both used by Latham (1801) on different pages; Banded Goatsucker, for the barred back and tail (Latham 1823); Little Nightjar, Fairy Owl, Moth Owl, all for its small size (and, it must be said, for its undeniable cuteness!); Rufous Nightjar, for formerly recognised species *A. rufa* from north-western Australia, described in 1901 and recognised by RAOU (1913), but not RAOU (1926).

Aegotheles cristatus (Shaw, 1790) [eh-go-THEH-lehz kri-STAH-tʊs]: 'bristly goat-sucker', see genus name, and Latin *cristatus*, meaning crested. This refers to the rictal feathers – bristly feathers above the beak, although all owlet-nightjar species have them. White (1790), who edited the work in which Shaw's illustration appeared, described the bristles as 'standing up on each side, in the manner of a crest'.

Australian Owlet-nightjar ***Aegotheles cristatus***

APODIDAE (CAPRIMULGIFORMES): swifts

Apodidae Olphe-Galliard, 1887 [a-PO-di-dee]: the Footless family, see genus name *Apus*.

'Swifts' are perhaps more familiar in Europe than they are in Australia, though there was – and is – much confusion with swallows and martins. The name goes back to the 17th century in England, but as an alternative to martin. The reference is obvious to anyone who has thrilled at these spectacular aerialists.

'Swiftlet' is applied to a (taxonomically confused) group of small swifts.

'Needletail' is applied to the four species of the genus *Hirundapus*, for the stiff feather shafts which extend beyond the end of the tail; it is an arbitrary application, because other genera in the same subgroup of the family have the same characteristic.

The genera

Collocalia Gray GR, 1840 [ko-lo-KAH-li-uh]: 'glue-nester' from Greek *kolla*, glue, and *kalia*, nest.

Aerodramus Oberholser, 1906 [eh-ro-DRAH-mʊs]: 'air-racer' from Greek *aer-*, air, and *dromos*, racing (cf. *dramema* or *dromema*, a race, running).

Hirundapus Hodgson, 1837 [hi-rʊn-DAH-pʊs]: 'swallow-swift' from the genera *Hirundus*, swallows, and *Apus*, swifts, because the bird seemed to Hodgson to make the link between the two.

Mearnsia Ridgway, 1911 [MERN-zi-uh]: 'Mearns' after Edgar Mearns (1858–1916), a US army surgeon and dedicated naturalist, who extensively studied birds in Africa and the Philippines. (He also 'discovered' the familiar Australian Black Wattle *Acacia mearnsii* in Kenya, many decades after it had first been invalidly described from Australia, and scored its name in a taxonomic farce, but that's a different story.)

Apus Scopoli, 1777 [AH-pʊs]: 'footless [bird]', from Greek *apous*, (*a-*, without, and *pous/podos*, foot). Aristotle's (c. 330 BC) version of this was: 'Some birds have feet of little power, and are therefore called *Apodes*.' The footlessness has been taken literally at some stages in history – curious readers should look carefully at the illustrations in field guides and books such as HANZAB.

The species

Glossy Swiftlet (vagrant)

See family introduction. Glossier above than most other swiftlets, though the taxonomy is changing.

Other names: Edible-nest Swiftlet, from confusion with *Aerodramus fuciphagus*, the cave-dwelling species whose nests are harvested to make soup (by those who are fond of eating bird spittle).

Collocalia esculenta (Linnaeus, 1758) [ko-lo-KAH-li-uh es-kʊ-LEN-tuh]: 'edible glue-nester' (see 'Other names' for the soup, but don't make it from this inedible one!), see genus name, plus Latin *esca* food, and *esculens*, edible (untrue!).

Christmas Island Swiftlet (breeds on Christmas Island)

See family introduction and for its breeding ground.

Other names: Linchi Swiftlet from a previous misidentification of this species for *C. linchi*.

Collocalia natalis Lister, 1889 [ko-lo-KAH-li-uh na-TAH-lis]: 'Christmas glue-nester', see genus and from Latin *natalis*, birthday, for the birth of Christ, and hence for the place-name.

Australian Swiftlet (breeding resident)

See family introduction. The only swiftlet breeding on the Australian mainland.

Other names: Grey Swiftlet, only of use in Australia, where there is nothing generally to confuse it with – otherwise it's the equivalent of 'Black Raven'; Grey-rumped Swiftlet, probably in contrast to the vagrant white-rumped Glossy Swiftlet (see species profile); White-rumped Swiftlet, used until quite recently (Pizzey and Knight 2003), due to past confusions with closely related species.

Aerodramus terrareginae (Ramsay EP, 1875) [eh-ro-DRAH-mʊs te-rah-reh-GEE-nie]: 'Queensland air-racer', see genus name, and from Latin *terra*, land, and *regina*, a queen (in this case Queen Victoria), and for its range.

Uniform Swiftlet (vagrant)

See family introduction. It's a bit hard to see how a species with 15 different subspecies can be regarded as very uniform, though it is pretty much uniformly coloured top and bottom. Most of the *Aerodramus* swiftlets are dauntingly similar.

Aerodramus vanikorensis Quoy & Gaimard, 1830 [eh-ro-DRAH-mus va-ni-ko-REN-sis]: 'Vanikoro air-racer', see genus name, plus Vanikoro, the type locality, an island in the Santa Cruz group (now under Solomon Islands administration) plus the suffix *-ensis*, indicating the place of origin of the type specimen.

White-throated Needletail (non-breeding summer migrant)

See family introduction. The only *Hirundapus* with a clear white throat.

Other names: New Holland and Needle-tailed Swallow (Latham 1801), Pin-tailed Swallow (Latham 1823) – it wasn't until much later in the 19th century that they were separated from swallows; Australian

Spine-tailed Swallow, used by Gould (1848); Spine-tailed Swift, the name in primary use in Australia until recently and a pertinent one, in contrast with Fork-tailed, the only other regularly occurring large swift in Australia; Needle-tailed Swift.

Hirundapus caudacutus (Latham, 1801) [hi-rʊn-DAH-pʊs kow-da-KOO-tʊs]: 'sharp-tailed swallow-swift', see genus name, plus Latin *cauda*, tail, and *acutus*, sharpened or pointed, referring to the 'needles' at the end of the tail (see family introduction).

Papuan Spine-tailed Swift (vagrant to Torres Strait)

See family introduction. The only spine-tailed swift breeding in New Guinea.

Mearnsia novaeguineae (D'Albertis & Salvadori, 1879) [MERN-zi-uh no-veh-GI-ni-eh]: 'New Guinea Mearns-bird', see genus name, and from Modern Latin version of place name.

Pacific Swift (non-breeding summer migrant)

See family introduction, and from the species name.

Other names: Fork-tailed Swift, the name used in Australia, where it is unambiguous (though not so elsewhere in its range); Australian Swift, used by Gould (1848) though quite inaccurately, because it migrates from north-eastern Asia – however, he thought it was indigenous and called it *Cypselus australis*; White-rumped Swift, a useful distinguisher from White-throated Needletail.

Apus pacificus (Latham, 1802) [AH-pʊs pa-SI-fi-kʊs]: 'Pacific footless bird', see genus name, and from Latin *pacificus*, peaceful, not for its temperament but for the Pacific Ocean (the type locality), which it frequents during migration.

Little Swift (vagrant)

See family introduction. One of the smallest members of the extensive genus *Apus*.

Other names: House Swift, commonly breeds in wall hollows throughout its wide African-Asian range – a widely used name.

Apus affinis (Gray JE, 1830) [AH-pʊs af-FEE-nis]: 'related footless bird', see genus name, and from Latin *affinis*, related. Unfortunately, Gray (1830), who named it *Cypselus affinis*, did not vouchsafe what it was supposed to be related to, unless it be the preceding bird in the same plate, Latham's Balassian Swift, *Cypselus palmarum*.

OTIDIDAE (OTIDIFORMES): bustards

Otididae Rafinesque, 1815 [oh-TI-di-dee]: the Bustard family, from genus name *Otis*, not represented in Australia (see genus name *Ardeotis*).

'Bustard' has a somewhat tangled etymology, though its route into English from Latin seems clear enough. The Romans, when they encountered the Great Bustard *Otis tarda* in Spain, called it *Avis tarda*. The logical linguistic conclusion would be that this meant 'slow bird' but that is not at all a logical ornithological conclusion. Pliny explained that these are 'the birds which in Spain they call the 'tarda', and in Greece the 'ōtis'', so 'tarda bird' implied just 'bustard bird'. It appears in French as *ostarde* by the 14th century, evolving into *outarde* (the word for it in French today); later it apparently re-entered France from Italy as *vistarde*. The suggestion is that both came to England (where the Great Bustard occurred until the early 19th century) and blended to give 'bustard' (or 'bustarde') by the 15th century!

The genus

Ardeotis Le Maout, 1853 [ar-de-OH-tis]: 'heron-bustard', from Latin ardea, heron, and Greek ōtis, used by Xenophon (c. 380 BC) in his description of the Greek expedition to seize the Persian throne (see also Ostrich). He was probably referring to the Great Bustard *Otis tarda*. Xenophon wrote, 'The bustards … can be caught if one is quick in starting them up, for they fly only a short distance, like partridges, and soon tyre; and their flesh was delicious.' But why the heron part of the name? Emmanuel Le Maout (1853), using the late Isidore Geoffroy St Hilaire's lectures and pictures to complete the latter's *Histoire Naturelle des Oiseaux*, explained that the name is made up of Ardea and Otis. The only feature of the new genus he described is the bill, which he said is long and pointed, so presumably the bill seemed to St Hilaire a little more heron-like than that of the Otis bustards.

The species

Australian Bustard

The only Australian species, endemic to Australia and southern New Guinea.

Other names: Plain, Bush, Native or Wild Turkey – probably all reflecting an interest in the bird as a culinary item, which certainly had an impact on bird populations, especially in southern Australia.

Ardeotis australis (Gray JE, 1829) [ar-de-OH-tis ost-RAH-lis]: 'southern heron-bustard', see genus name, and from Latin *australis*, southern.

CUCULIDAE (CUCULIFORMES): cuckoos

Cuculidae Leach, 1820 [kʊ-KOO-li-dee]: the Cuckoo family, see genus name *Cuculus*.

'Cuckoo' is clearly onomatopoeic – but only if you're European. In Australia the word which pre-dated Chaucer in England, having crossed the Channel with the Normans, was adopted for closely related birds to the European one, even though the only bird in Australia to declaim 'cuckoo' was an owl!

'Bronze Cuckoo'; 'Bronze-cuckoo' is first mentioned by Gould (1848) with regard to our Shining Bronze-cuckoo; though he used Shining Cuckoo, he noted that it is the 'Golden or Bronze Cuckoo of the Colonists'. The reference is to the iridescent greenish wing speculum. The hyphenated form is the one generally used in Australia.

'Coucal' is apparently of a derivation far too bizarre to have been invented. It is said to have derived in Africa (where several species are prominent) from the French combination *coucou-alouette* 'cuckoo-lark' – couc-al, get it? We are presumably not supposed to ask what the wonderful big shambling creature has in common with a lark. Coucals were previously placed in a separate family, but are now regarded as 'mainline' cuckoos.

'Koel' is very clearly from the ringing two-note call of the male, deriving from Hindi, whose speakers obviously heard their closely related species in much the same way as we do ours. The name seems to have been first used in Australia by Gould (1869).

The 'Hawk-Cuckoos' are a group of Asian cuckoos in genus *Hierococcyx* (often still included in *Cuculus*), named for their quite striking similarity, in colour and flight pattern, to the Shikra Goshawk *Accipiter badius*; it seems to be a case of defensive mimicry.

The genera

Centropus Illiger, 1811 [SEN-tro-pus]: 'spur-foot', from Greek *kentron*, a goad or spur, and *pous*, foot, first named for the Senegal Coucal *C. senegalensis*, though the long inner hind-claw is common to many coucals.

Eudynamys Vigors & Horsfield, 1827 [yoo-DEE-na-mis]: 'very powerful [bird]', from Greek *eu-*, well or very, and *dunamis*, power or strength. The bill and feet are considered strong by comparison with those of other cuckoos.

Urodynamys Salvadori, 1880 [oo-ro-DEE-na-mis]: 'tailed powerful [bird]', from Greek *oura*, tail, and genus name *Eudynamis*, in which this bird was formerly placed.

Scythrops Latham, 1790 [SKEE-throps]: 'angry-faced [bird]', from Greek *skuthros*, angry or sullen, *skuthropos*, angry-looking, and ōps, the face or eye. The red ring round the bird's eye could well look angry, and the massive size of the bill could contribute to this impression.

Chrysococcyx Boie, 1826 [kri-so-KOK-siks]: 'golden cuckoo', from Greek *khrusos/khruso-*, gold/golden, and *kokkux*, cuckoo.

Cacomantis Müller S, 1843 [ka-ko-MAN-tis]: 'prophet of evil', from Greek *kakos*, evil, and *mantis*, a seer or prophet. Müller (1839–44) described the birds as 'prophets of misfortune' and 'misfortune cuckoos' but didn't explain further. This is thought to allude to the traditional belief in the association of the birds with bad weather (see Pallid Cuckoo), though Jobling (2010) asserted that it might derive from superstitious Javanese who heard the birds calling at night from burial grounds.

Hierococcyx Müller S, 1845 [hi-e-ro-KOK-siks]: 'hawk-cuckoo', from Greek *hierax*, hawk or falcon, and *kokkux*, cuckoo.

Cuculus Linnaeus, 1758 [KʊU-kʊ-lʊs]: 'cuckoo', from Latin *cuculus*, a cuckoo (Greek *kokkux*, and it says *kokku*!). Both Latin and Greek also use the word as a term of reproach, for foolish, dilatory or unprincipled people.

The species

Lesser Coucal (vagrant)

See family introduction. One of the smallest coucals (it is tiny relative to the Pheasant Coucal, though the two are probably not compared by the name, because their normal ranges do not overlap except in Timor).

Centropus bengalensis (Gmelin JF, 1788) [SEN-tro-pus ben-ga-LEN-sis]: 'Bengal spur-foot', see genus name, and for the type locality, plus the suffix *-ensis*, usually indicating a place of origin of the type specimen.

Pheasant Coucal (breeding resident)

See family introduction. Latham (1801) commented that it was already known in New South Wales as Pheasant (no mention of Coucal, though he makes it clear that he knows it's a 'cuckow') and explained that this was related to the patterning of its back, though doubtless its terrestrial habits and long tail contributed; note that Lyrebirds were also 'pheasants'.

Other names: Coucal, Common Coucal, New Holland Coucal (Latham 1823), Variegated Coucal (also Latham 1823, based on Vigors and Horsfield's now invalid species *variegatus*); Pheasant Cuckoo, used by Gould (1848), taken from Latham (1801); Swamp Coucal or Pheasant, for its habitat; Cane Pheasant, likewise – they are common in sugarcane; North-west Pheasant, from Gould who considered those birds to be a separate species (as he also did Top End populations).

Centropus phasianinus (Latham, 1801) [SEN-tro-pus fa-zi-a-NEE-nʊs]: 'pheasant spur-foot', see genus name, and from Latin *phasianus*, pheasant, with suffix *-inus*, of or belonging to. See common name.

Asian Koel (Christmas Island vagrant)

See family introduction. The taxonomy of Koels is convoluted and until recently only one variable species was recognised from southern Asia to Australia; Christidis and Boles (2008) accepted three species, naming this one for its range, right across southern Asia to the Philippines.

Other names: Common or Indian Koel, for the whole complex.

Eudynamys scolopaceus (Linnaeus, 1758) [yoo-DEE-na-mis sko-lo-PAH-se-ʊs]: 'very powerful bird like a snipe', see genus name, and from Latin *scolopax*, the snipe or woodcock, whose markings were apparently thought to be echoed in the female *Eudynamys* birds.

Pacific Koel (breeding resident)

See family introduction and the discussion of Asian Koel. This name was suggested by Ian Mason, who proposed the three species as per the earlier comments, in Schodde and Mason's monumental *Zoological Catalogue of Australia* (1997), though the original species name was applied by Linnaeus.

Other names: Eastern Koel from the species name; Common or Indian Koel, for the overall complex; Black Cuckoo; Cooee Bird, a good interpretation of the call; Rainbird or Stormbird or Corn-planter, all for its summer arrival, which coincides with the wet season (see also Channel-billed Cuckoo); Flinders' Cuckoo for the great navigator, scientific explorer and cartographer Matthew Flinders, applied by Vigors and Horsfield in the form *E. flindersi* (though Linnaeus had already named the bird, from a Moluccas specimen) and retained by Gould (1848); Blue-headed Cuckoo (Latham 1801) – the basis of his names is not always evident.

Eudynamys orientalis (Linnaeus, 1766) [yoo-DEE-na-mis o-ri-en-TAH-lis]: 'very powerful bird from the East', see genus name, and from Latin *orientalis*, eastern, for its range to eastern Indonesia.

Long-tailed Cuckoo (non-breeding migrant to Lord Howe and Norfolk islands)

See family introduction. The name used in New Zealand, its breeding home; all koels have long tails (until recently, and currently in some quarters, this species was included in *Eudynamys*; either way it is very close to the koels).

Other names: Long-tailed Koel.

Urodynamys taitensis (Sparrman, 1787) [oo-ro-DEE-na-mis tah-ee-TEN-sis]: 'tailed powerful bird from Tahiti', see genus name, and from the type locality Tahiti plus the suffix *-ensis*, usually indicating a place of origin of the type specimen; after breeding it scatters right across the south-west Pacific.

Channel-billed Cuckoo (breeding resident)

See family introduction. For the grooved upper mandible – though we might think that the overall bill might have attracted more attention! Latham (1801) focused on the fact that the bill is 'furrowed or

Channel-billed Cuckoo ***Scythrops novaehollandiae***

channelled on the sides', calling it a Channel-Bill. (He also complained that it couldn't 'easily be tamed', because a wounded, and presumably terrified, one insisted on biting its captor.)

Other names: Channel-bill (Latham 1801; Gould 1848), Australian Channel-bill (Shaw 1819); Channel-bill Cuckoo; Giant Cuckoo or Bird; Hornbill or Anomalous Hornbill (White 1790), an understandable misconception; Rainbird, Stormbird, Floodbird, see also Eastern Koel, though Latham (1801) also claimed that 'the natives … consider its presence as an indication of wind and blowing weather'; Fig Hawk, for its diet; Thornbill, as used in a list of donations of specimens to the Australian Museum (supported by the scientific name) reported in the *Sydney Morning Herald* of 11 June 1856 – however, we do wonder if this was a typo for Hornbill.

Scythrops novaehollandiae Latham, 1790 [SKEE-throps no-veh-hol-LAN-di-eh]: 'angry-faced New Holland bird', see genus name, and from Modern Latin *Nova Hollandia*, New Holland, the old name for Australia.

Horsfield's Bronze Cuckoo (breeding resident)

See family introduction. For Dr Thomas Horsfield, a US doctor and naturalist who worked for the British East India Company in Java and later at its London museum. He named the species from a specimen from Java after he had returned to London, but it is unclear whether he collected it himself while he was there. In Australia it was introduced to general use by RAOU (1926).

Other names: Horsfield's Bronze-Cuckoo, the usual form in Australia – see family introduction; Narrow-billed Cuckoo or Bronze-Cuckoo, an odd name, but used into the 20th century (Lucas and Le Soeuf 1911; RAOU 1913) to distinguish from Broad-billed Bronze-Cuckoo (i.e. Shining Bronze-Cuckoo), though we might have thought there were more obvious distinguishing characters; Rufous-tailed Bronze-Cuckoo, for the bases of the outer tail feathers, a reliable way of distinguishing from Shining Bronze-Cuckoo – in fact this character is one that betrayed Gould (1848), in that he illustrated this species with his description of Shining Bronze-Cuckoo! Rainbird, see also Eastern Koel.

Chrysococcyx basalis (Horsfield, 1821) [kri-so-KOK-siks ba-SAH-lis]: 'basal golden cuckoo', see genus name, and from Modern Latin *basalis* (from Greek and Latin *basis*, a base or pedestal). For the rufous bases of the outer tail feathers, described under the common name.

Black-eared Cuckoo (breeding resident)

See family introduction. For the obvious broad black eye-ear stripe on a plain face. Gould (1848) stated that it was the name used by the Swan River colonists, and this was one of the few occasions that he adopted such a name.

Chrysococcyx osculans (Gould, 1847) [kri-so-KOK-siks OS-koo-lanz]: 'kissing golden cuckoo', see genus name, and from Latin *osculor*, to kiss. Banish those thoughts of snogging birds – Gould (1847) named it so because he believed it linked the characteristics of two genera, *Cuculus* and *Chalcites* (as this genus was formerly known).

Shining Bronze Cuckoo (breeding resident)

See family introduction. It does shine, in the right light; straight from the species name, applied by Gould in 1848.

Other names: Shining Bronze-Cuckoo, the usual form in Australia – see family introduction; Broad-billed Bronze-Cuckoo, see Horsfield's Bronze-Cuckoo; Golden or Bronze Cuckoo 'of the Colonists' (Gould 1848); Golden Bronze-Cuckoo, specifically for the Australian race *plagosus*, which is darker on the back than the New Zealand race *lucidus* (which also appears sometimes on migration) – these races were formerly seen as two species; Greenback.

Chrysococcyx lucidus (Gmelin JF, 1788) [kri-so-KOK-siks LOO-si-dʊs]: 'bright golden cuckoo', see genus name, and from Latin *lucidus*, clear or bright, for the shining plumage.

Little Bronze Cuckoo (breeding resident)

See family introduction. The smallest bronze cuckoo (albeit marginally) – see the species name.

Other names: Little Bronze-Cuckoo, the usual form in Australia – see family introduction; Red-eyed Bronze-Cuckoo, for a distinctive character, albeit male-only; Rufous-breasted, Rufous-throated, Gould's Bronze-Cuckoo, all for race *russatus*, named by John Gould and formerly regarded as a separate species.

Chrysococcyx minutillus Gould, 1859 [kri-so-KOK-siks mi-NOO-til-lʊs]: 'very tiny golden cuckoo', see genus name, and from Latin *minutulus*, a diminutive of minutus, tiny, so very small indeed.

Pallid Cuckoo (breeding resident)

See family introduction. Directly from the species name; by 1865 Gould was using this name.

Other names: Unadorned Cuckoo, as used by Gould (1848), from the old species name *Cuculus inornatus* (applied by Vigors and Horsfield after Latham had already named it); Greater Cuckoo, reported by

Gould to be used by 'the Colonists', in contrast to the equally common Fan-tailed Cuckoo; Rainbird, Stormbird, Weatherbird, Wetbird, Harbinger-of-Spring, see Eastern Koel for explanation of similar names; Brainfever Bird, Semitone Bird, Scale Bird, for the ascending stepped whistle, which can be repeated for hours on end; Mosquito or Grasshopper Hawk, for its active insect hunting.

Cacomantis pallidus (Latham, 1801) [ka-ko-MAN-tis PAL-li-dʊs]: 'pale prophet of evil', see genus name, and from Latin *pallidus*, pale, colourless.

Chestnut-breasted Cuckoo (breeding resident)

See family introduction. For its chestnut breast, richer than that of the Fan-tailed Cuckoo; Gould (1869) used it, being unwilling to use the coarse 'belly' even though it was the meaning of his species name.

Cacomantis castaneiventris (Gould, 1867) [ka-ko-MAN-tis kas-ta-ne-i-VEN-tris]: 'chestnut-bellied prophet of evil', see genus name, and from Latin *castanea*, chestnut or chestnut tree, hence chestnut-coloured, and *venter/ventris*, belly, see common name description, though the chestnut extends all the way from chin to undertail.

Fan-tailed Cuckoo (breeding resident)

See family introduction. Straight from the somewhat mystifying species name – best explained by the fact that John Latham never saw a live one.

Other names: Bar-tailed or Barred-tailed Cuckoo (Latham 1823) – hardly an original or helpful name; Ash-coloured Cuckoo, by Gould for the species name *Cuculus cineraceus* (for the grey back), again a Vigors and Horsfield name, overlooking Latham's earlier work; Lesser Cuckoo 'of the Colonists' (Gould 1848) – see Pallid Cuckoo. (It seems strange that such a common and vocal bird should have attracted virtually no alternative names.)

Cacomantis flabelliformis (Latham, 1801) [ka-ko-MAN-tis fla-bel-li-FOR-mis]: 'small-fan-shaped prophet of evil', see genus name, and from Latin *flabellum*, a small fan or fly-whisk, and *forma*, shape, for the bird's tail.

Brush Cuckoo (breeding resident)

See family introduction. Gould (1848) reported, in describing the bird and introducing the name, that he encountered it 'while traversing the cedar brushes of the Liverpool Ranges', brush being a somewhat derogatory word for dense forests, especially rainforests.

Other names: Square-tailed or Brush Square-tailed Cuckoo, for no evident reason, though Mathews (1913) and RAOU (1913) both used it; Brush-tailed Cuckoo, if used, surely a misprint based on the previous name (or a Fan-tailed/Brush Cuckoo hybrid ...); Grey-breasted Brush-Cuckoo, perhaps to distinguish from the previous two species.

Cacomantis variolosus (Vigors & Horsfield, 1827) [ka-ko-MAN-tis va-ri-o-LOH-sʊs]: 'pock-marked prophet of evil', see genus name, and from Mediaeval Latin *variola* (6th century), meaning 'infectious illness', diminutive of *varius* (600 AD), having smallpox (influenced also by Latin *varius*, varied or mottled). This describes the plumage of the juvenile only, and in fact Vigors and Horsfield (1827) commented that the specimen 'has much the appearance of a young bird'.

Large Hawk-Cuckoo (vagrant to Christmas Island)

See family introduction. This species is close to the largest of the hawk-cuckoos.

Hierococcyx sparverioides (Vigors, 1832) [hi-e-ro-KOK-siks spar-ve-ri-OY-dehz]: 'sparrowhawk-like hawk-cuckoo', see genus name, and from Mediaeval Latin *sparverius*, sparrowhawk, with Greek *-oeidēs* (Latin *-oides*), resembling. Just in case you were in any doubt that it looks like a hawk.

Hodgson's Hawk-Cuckoo (vagrant)

See family introduction. Brian Hodgson represented the East India Company and the British Government from 1825 to 1859 in Nepal and Darjeeling, amassing a huge collection of bird specimens – presumably including this one – and describing many of them.

Hierococcyx nisicolor (Blyth, 1843) [hi-e-ro-KOK-siks NI-si-co-lor]: 'sparrowhawk-coloured hawk-cuckoo', see genus name and from Mediaeval Latin *nisus* (sparrowhawk) and Latin *color*, colour. In his description of the bird, Blyth mentions the 'very dark pure ash-colour' of the upper parts, and the hints of 'bright rufous' on wings and breast.

Indian Cuckoo (vagrant)

See family introduction. From the type locality, though it ranges south-west to Java and north-east to eastern Russia.

Cuculus micropterus (Gould, 1838) [KƱ-kʊ-lʊs mik-ro-TE-rʊs]: 'small-winged cuckoo', see genus name, and from Greek *mikros*, small, and *pteros*, winged.

Oriental Cuckoo (non-breeding migrant)

See family introduction. This bird belongs to a species complex (until recently regarded as a single species) and is found throughout far eastern Asia, though it was described by Gould in Australia before it was described from its north Asian breeding grounds. RAOU (1926) introduced the name into Australia for general use.

Other names: Australian Cuckoo, applied by Gould (1848); this may seem a strange name now, for one of our 'least Australian' cuckoos (not even breeding), but he was in raptures over it as being the local representative of the European Cuckoo and 'the harbinger of spring, and the index of the revivifying of nature' – he even predicted that it would prove to say 'cuckoo'! Surprisingly, the name was still used by RAOU (1913). Other names used cannot safely be referred to any one of the three species that are now recognised from the former *C. saturatus*, and in any case are not relevant to Australia.

Cuculus optatus Gould, 1845 [KƱ-kʊ-lʊs op-TAH-tʊs]: 'welcomed cuckoo', see genus name, and from Latin *optatus*, longed-for or welcomed. Indeed a harbinger of spring, at least in its breeding grounds in northern Russia, Japan, and so on. Not the one English people write to *The Times* about: that's the Common Cuckoo, *Cuculus canorus*.

COLUMBIDAE (COLUMBIFORMES): pigeons and doves

Columbidae Illiger, 1811 [co-LUM-bi-dee]: the Pigeon/Dove family, see genus name *Columba*. Taxonomically, 'pigeon' and 'dove' are interchangeable, though we tend to use dove for smaller long-tailed members of the family. (However, the name Rock Dove in a genus of species generally called pigeons reminds us that this is arbitrary.)

'Pigeon' came from the Old French *pijon*, written variously and first recorded in its current form in England in the late 15th century. Originally it apparently referred to a young dove, though it was also used for any young bird. It did not become the generally first-choice term until the early 19th century, gradually replacing terms such as queece, culver and cushat (some of which referred specifically to the Wood Pigeon) and in part replacing the older 'dove'.

'Dove' was of Old English origin, from an apparently onomatopoeic word since lost, but presumed to resemble *dufe*.

'Cuckoo-Dove' is a somewhat awkward and obscure term used for all members of the long-tailed genus *Macropygia*.

'Bronzewing' (or 'Bronze-wing') was used from the earliest days of British colonisation; the *Australian National Dictionary* reported that Arthur Phillip in his *Voyage to Botany Bay* in 1789 talked of Bronze-winged Pigeons; Latham (1801) and later Gould (1848) consolidated the name. The reference is of course to the iridescent bronzey-gold secondary wing coverts. The dropping of 'pigeon' from the name is a curious piece of linguistic evolution; Cayley's *What Bird is That?* was still using it in 1931, but Crosbie Morrison had dropped it in his 1958 revision of John Leach's 1918 *An Australian Bird Book* (Leach himself had used 'pigeon'). Presumably it was simply a function of familiarity.

'Rock Pigeon' (nearly always hyphenated in Australia) is a direct reference to the very close affinity of the two *Petrophassa* pigeons for tropical sandstone escarpments; it comes directly from the genus name.

'Imperial Pigeon' (nearly always hyphenated in Australia) is the name used for the genus *Ducula*, ~40 species of large colourful fruit pigeons from the South-East Asia/south-west Pacific region. It seems to have arisen in the 20th century, perhaps influenced by the genus name, though if so it was misconstrued; there is no evidence of the bird's desire for an empire.

The genera

Columba Linnaeus, 1758 [co-LUM-buh]: 'pigeon/dove', from Latin *columba*, pigeon or dove.

Streptopelia Bonaparte, 1855 [strep-to-PEH-li-uh]: 'collared dove' from Greek *streptos*, a collar of twisted or linked metal, and *peleia*, rock-pigeon (see also Rock Dove), named for the collared doves – '*tourterelles à collier*' – of which Bonaparte (1855) said there were nine species known at the time.

Spilopelia Sundevall, 1873 [spi-lo-PEH-li-uh]: 'spotted dove', from Greek *spilos*, spot and *peleia*, rock-pigeon.

Macropygia Swainson, 1837 [mak-ro-PI-gi-uh]: 'big-bum' from Greek *macros*, large, and *pugē*, rump, for its large rear end. Swainson (1837) said: 'Tail long, graduated; the feathers very broad and obtuse. The rump feathers very thick set.'

Chalcophaps Gould, 1843 [KAL-ko-faps]: 'bronze-pigeon' from Greek *khalkos*, bronze or copper, and *phaps*, wild pigeon. The name doesn't quite capture those brilliant wings.

Phaps Selby, 1835 [faps]: 'wild pigeon' from Greek *phaps*.

Ocyphaps Gray GR, 1842 [OH-kee-faps]: 'swift pigeon' from Greek *phaps* for 'wild pigeon', with Greek ōkus, quick or swift. Gould (1842) asserted that the bird has rather short wings. In 1848 he repeated that, but added that they have long pointed wings (sic!) and 'can pass at pleasure' over the 'vast expanse' of the Australian interior. In 1865 he stuck with the long pointed version, dropped the 'pass at pleasure' and added that the bird has 'a flight so rapid as to be unequalled by those of any group to which it belongs'. See *Geophaps* for Gray *v.* Gould.

Geophaps Gray GR, 1842 [GEH-oh-faps]: 'ground pigeon' from Greek *geo*, earth, for its preference for being on the ground, plus *phaps*, wild pigeon.

[A slightly odd story attaches to this and the preceding genus, *Ocyphaps*: Gray included these names in the 1842 Appendix to his *List of the Genera of Birds* (Gray and Wetmore 1841), putting them as 'OCYPHAPS, Gould (1842)' and 'GEOPHAPS, Gould (1842)', presumably referring to the fact that Gould had presented *Ocyphaps* and *Geophaps* as new genera to a meeting of the Zoological Society on 11 January 1842. Gray's formal publication of the names trumped Gould's detailed oral description, as the rules of nomenclature dictate. So Gray was recorded as the author of the two genera by the skin of his teeth, perhaps to his own surprise.]

Leucosarcia Gould, 1843 [loo-ko-SAR-si-uh]: 'white-flesh', from Greek *leukos*, white, and *sarx/sarkos*, flesh – good eating? We wouldn't know.

Petrophassa Gould, 1841 [pe-tro-FAS-suh]: 'rock-pigeon' from Greek *petros*, rock, for its preferred habitat, plus *phassa*, another Greek word for wild pigeon.

Geopelia Swainson, 1837 [ge-o-PEH-li-uh]: 'ground-dove' from Greek *geo*, earth, and *peleia*, rock-pigeon, for its preferred habitat.

Caloenas Gray GR, 1840 [ka-LO-e-nas]: 'beautiful wine-coloured pigeon' from Greek *kalos*, beautiful, and *oinas*, a wild pigeon with vinous colouring.

Alopecoenas Sharpe, 1899 [a-lo-pe-CO-e-nas]: 'foxy wine-coloured pigeon' from Greek *alōpēx* (fox or bat – see Arnott 2007 for a detailed discussion of this) and *oinas*, a wild pigeon with vinous colouring.

Ptilinopus Swainson, 1825 [ti-LI-no-pʊs]: 'feather foot' from Greek, feather, and *pous*, foot. Birds of this genus generally do have feathered legs (if not feet, though *pous* can include the leg), some species more so than others. The genus was named for the Rose-crowned Fruit-Dove, *Ptilinopus regina*, whose legs Swainson (1825) described as 'covered with soft and thick-set feathers'.

Ducula Hodgson, 1836 [doo-KOO-luh]: 'dukul'. This seems to come from *dukul*, the common name (very likely onomatopoeic) for the Mountain Imperial-Pigeon, *D. badia*, found throughout much of South and South-East Asia. Hodgson (1836) described the bird as 'dukul of the Nipalese'. So not from Latin *dux/ducis*, a leader with diminutive *-ulus*, hence little duke, as has been asserted. (See Imperial-Pigeon in family introduction.)

Lopholaimus Gould, 1841 [lo-fo-LIE-mʊs]: 'crested-throat' from Greek *lophos*, tuft or crest and *laimos*, throat. Refers to the spiky throat feathers.

Hemiphaga Bonaparte, 1854 [heh-mi-FAH-guh]: 'half an imperial-pigeon' from Greek *hēmi*, half, plus the latter part of the old genus name for imperial pigeons, *Carpophaga*, itself from Greek *karpos*, fruit, and *phagos*, eating. Bonaparte (1854) considered that this new genus fell midway between the first five new genera he includes under '*Carpophagiens*', and genus *Megaloprepia*, which we now know as *Ptilinopus*.

The species

Rock Dove (introduced breeding resident)

See family introduction. The ancestral wild populations, from the Mediterranean to southern Asia, bred largely on sea-cliffs; in Britain the name was used to distinguish it from the closely related and similar Wood Pigeon and Stock Dove ('stock' referring to a tree trunk, because the species nests in hollows). The name in Australia is somewhat analogous to Red Junglefowl, in that all Australian birds are derived from feral domestic birds, very many generations removed from their Rock Dove ancestors.

Other names: Feral Pigeon, arguably a more relevant name; Pigeon, an unconscious acknowledgment of their ubiquity; Domestic Pigeon, Homing Pigeon.

Columba livia Gmelin JF, 1789 [co-LUM-buh LEE-vi-uh]: 'dark-coloured/bluish-grey pigeon', see genus name plus the word *livia* (Mediaeval Latin), which was used by Gaza (1513) to translate Aristotle's *peleia*, dove (Greek). These words are both linked to others meaning dark-coloured, blue-black or bluish-grey (Greek *pelios/pellos* and Latin *livens/lividus*). So it sounds as if an ancient pigeon was by definition the colour of a Rock Dove – maybe we should just call it the 'pigeon-coloured pigeon'.

Metallic Pigeon (Lord Howe Island, extinct)

See family introduction, and for the shining dark plumage. The Lord Howe population was a subspecies, the parent species being found from the Philippines and Indonesia to the south-west Pacific.

Other names: White-throated Pigeon, for the extensive white chin and throat; Lord Howe Pigeon.

Columba vitiensis Quoy & Gaimard, 1830 [co-LUM-buh vi-ti-EN-sis]: 'Fiji pigeon', see genus name, and from *Viti*, the Fijian word for that country and the suffix *-ensis*, usually indicating a place of origin of the type specimen.

White-headed Pigeon (breeding resident)

See family introduction. An unusually logical and unambiguous name, used by Gould (1848) as White-headed Fruit Pigeon (perhaps he was using it in contrast to seed-eating pigeons).

Other names: Norfolk Pigeon, used by Latham (1801) in the mistaken belief that it came from Norfolk Island; Baldy, Baldy Pigeon, a reminder that 'bald' originally meant white, especially in reference to animals' heads ('baldy' is also used in Australia to refer to Hereford cattle); Bally Pigeon, presumably from the same origin; Cook Pigeon, uncertain, but we suspect it's onomatopoeic.

Columba leucomela Temminck, 1821 [co-LUM-buh loo-ko-ME-luh]: 'white and black pigeon'. See genus name, plus from Greek *leukos*, white, and *melas* black. A fair description, though at close quarters the 'black' parts have some beautiful green and purple highlights.

Oriental Turtle Dove (vagrant)

Turtle, from Latin 'turtur', clearly onomatopoeic; hence the apparently bizarre 'the voice of the turtle is heard in our land' from the biblical Song of Solomon. Turtle-Dove was in wide use until recently, but for obscure reasons has fallen out of favour in Australia at least. This one is found widely from central to eastern and southern Asia.

Streptopelia orientalis (Latham, 1790) [strep-to-PEH-li-uh o-ri-en-TAH-lis]: 'Eastern collared dove', see genus name, and from Latin *orientalis*, eastern.

African Collared Dove (scarce introduced breeding resident)

See family introduction. Most of this genus (and the next) have black half-collars, though only some of their names reflect this. At present both 'collared' and 'turtle' (see 'Other names') are being randomly added to and subtracted from names of these birds.

Other names: Barbary Dove, the more familiar name, and that given to a long-domesticated (perhaps for 3000 years) form of the African Collared Dove (though its taxonomy is still being debated). Barbary (or Barbary Coast) was the term for the land of the Berbers, which corresponded to north Africa west of the Nile. Ring Dove, Ringed Turtle-Dove (or Turtle-dove or Turtledove), for the dark half-collar.

Streptopelia roseogrisea (Linnaeus, 1758) [strep-to-PEH-li-uh roh-ze-o-GREE-se-uh]: 'rosy-grey collared dove', see genus name, and Latin *roseus*, rosy pink, and Mediaeval Latin *grisus-griseus*, grey. The bird is grey, with a very faint hint of pink.

Red Turtle Dove (vagrant to Christmas Island)

See family introduction and Oriental Turtle Dove for 'turtle'; this one, especially ssp. *humilis*, is richly rusty.

Other names: Red Turtle-dove; Red Collared Dove; Red Ring-dove; Dwarf Turtle-dove, because it is the smallest *Streptopelia*.

Streptopelia tranquebarica (Hermann, 1804) [strep-to-PEH-li-uh tran-ke-BAH-ri-kuh]: 'Tranquebar collared dove', see genus name, and from the name Tranquebar, more correctly called Tharangambadi, on the east coast of India.

Spotted Dove (introduced breeding resident)

See family introduction. From the adult's extensive white-spotted black neck cape.

Other names: Spotted Turtle-dove or Turtle Dove (see African Collared Dove), the name by which it is probably most widely known in Australia; Laceneck or Necklace Dove for the same reason; Indian, Chinese, Burmese Dove or Spotted Dove or Turtle-dove, for its origins in southern Asia.

Spilopelia chinensis Scopoli, 1786 [spi-lo-PEH-li-uh chi-NEN-sis]: 'Chinese collared dove', see genus name, and from the type locality China (Canton) with *-ensis*, indicating place of origin.

Laughing Dove (introduced breeding resident)

See family introduction. From the attractive (for an introduced species!) bubbly chuckling call.

Other names: Laughing Turtle-dove (see African Collared Dove); Senegal Dove or Turtle-dove, from the origin of the first specimen described by Linnaeus, though it occurs over most of Africa and western Asia.

Spilopelia senegalensis (Linnaeus, 1766) [spi-lo-PEH-li-uh se-ne-ga-LEN-sis]: 'Senegal spotted dove', see genus name, and from the name Senegal plus the suffix *-ensis*, indicating the place of origin.

Brown Cuckoo-Dove (breeding resident)

See family introduction. Originally this was described as a solely Australian species, then until recently it was lumped into a species found through New Guinea and eastern Indonesia; the recent splitting up of this super-species simply represents a return in Australia to the original situation. Most of the ~10 species of *Macropygia* are substantially brown, but the name is a hangover from when the name applied only to Australian birds – and we're back there again!

Other names: Brown Pigeon, at the time a surprisingly unconfusing name; Brownie. Large-tailed Pigeon (the long broad tail is conspicuous), also Pheasant Pigeon and the hyperbolic Pheasant-tailed Pigeon for the original name used by Gould 1848, and now resurrected; it was still in use in RAOU (1913) but by 1926 the RAOU had amended it to Brown (Pheasant) Pigeon.

Macropygia phasianella (Temminck, 1821) [mac-ro-PI-gi-uh fa-zi-a-NEL-luh] 'big-bum like a little pheasant', see genus name and from Latin *phasianus* (Greek *phasianos*), pheasant, with diminutive ending 'ella'.

Common Emerald Dove (Christmas Island breeding resident)

See family introduction. 'Emerald' refers to the bright green wings, the rest of the bird being pale brown; it was formally adopted in Australia in 1978 by the RAOU on the grounds that it was then the 'established international name'. Until recently there was only one widespread emerald dove recognised, from Australia to India and China, but now the Australian birds have been separated out (see next species), hence the addition of Common for this one, still with a huge range.

Other names: Emerald Dove, from before the split; Grey-capped Emerald Dove, used by the authoritative HBW in contradistinction to their Brown-capped Emerald Dove for the next species.

Chalcophaps indica (Linnaeus, 1758) [KAL-ko-faps in-DEE-kuh]: 'Indian bronze-pigeon', see genus name, and from Latin *indicus*, Indian, for the type locality (East Indies).

Pacific Emerald Dove (breeding resident)

See previous species; this one is found in Australia, New Guinea and east to Vanuatu and New Caledonia.

Other names: Emerald Dove and Pigeon; Little Green Pigeon, used by Gould (1848) – although it is not clear with what large green pigeon he was comparing it, we assume it was his *C. longirostris*, his Long-billed Green-Pigeon from the Top End (which name has now been resurrected for the whole species); similarly Little Green-Pigeon, used by the RAOU (1913); Green Pigeon or Dove; Green-winged Pigeon or Dove, the next choice of the RAOU (1926); Lilac-mantled Pigeon, not at all an obvious name in Australia, and probably referable to the previous species; Brown-capped Emerald Dove, see previous species.

Chalcophaps longirostris Gould, 1848 [KAL-ko-faps lon-gi-ROS-tris]: see genus name, and from Latin *longus*, long, and *rostrum*, bill.

Common Bronzewing (breeding resident)

See family introduction. 'Common' is relative to the more restricted-range Brush Bronzewing, though it is, and doubtless always was, a common bird in absolute terms.

Other names: Bronze Pigeon, reported by Gould to be used by the 'Colonists of Swan River'; Common Bronzewing or Bronzewing Pigeon; Forest or Scrub Bronzewing (Pigeon), in reference to its generally more open forest habitat than the Brush Bronzewing, though in practice their habitats overlap to a degree, and 'scrub' was never precisely defined.

Phaps chalcoptera (Latham, 1790) [faps kal-ko-TE-ruh]: 'bronze-winged wild pigeon', see genus name, and from Greek *khalkos*, bronze or copper, and *pteron*, wing.

Brush Bronzewing (breeding resident)

See family introduction. Although 'brush' in Australia is mostly associated with rainforest (especially in a somewhat derogatory way) its English use was more general, referring to dense shrubby understorey of a forest or wood. Hence it could be equally applied to dense coastal heaths, with which the species is associated – think of Brush Wattlebird, another heathland specialist – but in truth the description is an excellent one of its affinity for a heavy understorey under various canopies. Gould made reference to this in 1848, commenting that in contrast with the Common Bronzewing, this species 'affects the most scrubby localities'; he called it Brush Bronze-winged Pigeon.

Other names: Box-poison Pigeon, a reference to various Western Australian peas, especially of the genera *Gastrolobium* and *Oxylobium*, which contain the poison sodium fluoroacetate ('1080') – although there is limited information on its diet, it would be surprising if it did not eat pea seeds, like Common Bronzewing, and was immune to poison peas, like other Western Australian wildlife; Little Bronze Pigeon, because it is notably smaller than the Common Bronzewing – the name was reported by Gould (1848) as used by the 'Colonists of Swan River'.

Phaps elegans (Temminck, 1809) [faps EH-le-ganz]: 'elegant wild pigeon', see genus name, and from Latin *elegans*, fine, elegant, handsome. Indeed it is, but maybe not more so than its fellows!

Flock Bronzewing (breeding resident)

See family introduction. This inland nomad is famously flocky; until the early 20th century vast flocks were recorded in good seasons. The name seems to have arisen spontaneously; Morris (1898) reported that the Harlequin-Pigeon was 'commonly called in the interior the 'flock' pigeon' and by 1913 the name was first choice for the RAOU's first *Official Checklist*.

Other names: Flock Pigeon; Harlequin Pigeon (used by Gould 1848) or Bronzewing, for the mask-like facial pattern (see also species name).

Phaps histrionica (Gould, 1841) [faps hist-ree-O-ni-kuh]: 'theatrical wild pigeon', see genus name, and from Latin *histrio*, a harlequin or actor or pantomime performer – a reference to the bird's looks, not its behaviour.

Crested Pigeon (breeding resident)

See family introduction. Unequivocally appropriate, with only the Spinifex Pigeon able to mount a challenge; Gould was using the name in 1848 and reported that Charles Sturt had already used it, but so had Edward Eyre in an 1838 newspaper account of his cattle drive along the Murray in South Australia.

Other names: Crested Dove; Topknot Pigeon, widely used and also descriptive, but the name has been appropriated – see species profile; Crested Bronzewing, for its bronzewing-like wing patches; Whistle-winged, Whistling-winged Pigeon, Wirewing, for its characteristic signalling whirring, caused by the alignment of a single flight feather; Saddleback Pigeon, origin unclear – this is also the name of a breed of racing pigeon but there is no obvious connection.

Ocyphaps lophotes (Temminck, 1822) [OH-kee-faps lo-FOH-tehz]: 'crested swift pigeon', see genus name, and *lophotos*, crested.

Spinifex Pigeon (breeding resident)

See family introduction. An appropriate name, because it is more strongly associated with *Triodia* (Spinifex) hummock grasslands than any other pigeon. It was only recently formally adopted – the first authoritative usage seems to have been by RAOU (1978), followed closely by Pizzey (1980) – but there are newspaper references to it going back to 1900.

Other names: Plumed Partridge-bronzewing, Gould's name in 1848; Plumed Pigeon or Dove, the name of choice in relatively recent times, quite appropriate; Ground Dove, apt, but not very helpful; White-bellied Bronzewing, by Gould (1869) for his Victoria River species *leucogaster* (i.e. 'white-bellied'), now included in race *plumifera*; Red- and White-bellied Pigeon, Dove or Plumed-Pigeon refer to races *ferruginea* and *leucogaster*, respectively – as late as 1913 these were regarded as separate species, in addition to the current one, which was Plumed-Pigeon. (In 1926 Plumed-Pigeon and Red-bellied Plumed-Pigeon were still recognised by RAOU.) Plumed Rust-coloured Bronzewing, used by Gould for his north-western species, now race *ferruginea*.

Geophaps plumifera Gould, 1842 [GEH-oh-faps ploo-MI-fe-ruh]: 'plumed ground pigeon', see genus name, and Latin *pluma*, plume, and *ferus*, bearing (see common name).

Squatter Pigeon (breeding resident)

See family introduction and discussion of next species. Probably from its habit of landing and freezing, or possibly from its reputed affinity for cattle camps and homesteads. It seems to have arisen spontaneously, and appears in the Sydney *Empire* in 1862 in a notice of donations to the public aviary in the botanic gardens; thereafter it appears fairly regularly, including in an 1865 list of birds protected by the *NSW Game Act*. Nonetheless, Morris (1898) didn't seem to know the name and it was first used officially by RAOU (1926).

Other names: Partridge Bronzewing, Gould's name (see next species) – it was still being used for this species into the 20th century (RAOU 1913); Partridge Bronzewing Pigeon.

Geophaps scripta (Temminck, 1821) [GEH-oh-faps SKRIP-tuh]: 'ground pigeon with writing', see genus name, and from Latin *scribo/scriptus*, write/written, for the distinctive markings on the face and neck. Temminck (1822) called the bird the *Colombe marquetée* (French) and *Columba scripta* (Latin), and described how each patch of white on the face and throat is framed in black, producing an effect like 'a sort of marquetry'.

Partridge Pigeon (breeding resident)

See family introduction. Gould, who introduced the name, was referring more to 'jizz' than physical resemblance. In the following description he was actually talking about the previous species, Squatter Pigeon, which he came upon first, but he regarded the current species as very similar, calling it Smith's Partridge Bronze-wing: 'While on the ground it has so much the carriage and actions of a partridge that it might readily be mistaken for one' (Gould 1848). He was referring to a single bird, but we might also think of a flock running jinking along the ground then bursting into the air in all directions. It is not clear how this species came to be promoted to 'first Partridge Pigeon' but the current allocation of names happened, without explanation, in the RAOU's second *Official Checklist* (RAOU 1926).

Other names: Smith's Partridge Bronze-wing, Gould's name, from the species name; Bare-eyed, Red-eyed, Naked-eyed Pigeon or Partridge Pigeon (used by RAOU as the formal name in 1913), from the large bare skin patch round the eye (though it is yellow in the western race).

Geophaps smithii (Jardine & Selby, 1830) [GEH-oh-faps SMITH-i-ee]: 'Smith's ground pigeon', see genus name, and for Smith – though which Smith has been a contested question, with the candidates being Sir James Smith (1759–1828), eminent botanist, and Sir Andrew Smith (1797–1872), Scottish zoologist. However, the authors make it clear that it was Sir James, a wealthy man who purchased the invaluable Linnaean collection after the death of Linnaeus the younger and later founded the Linnean Society around the collection. Moreover they explain that it was a sort of posthumous apology for having 'robbed' Smith of having his name attached to the Green Catbird, by showing that the bird had already been described.

Wonga Pigeon (breeding resident)

See family introduction. Undoubtedly from an Indigenous language but as usual we don't know which one, although we might reasonably suppose it was from one of the Sydney area language groups; Gould (who adopted the name Wonga-wonga) said only that it was used by 'Aborigines of New South Wales'. Especially given this apparently original duplicated name, we suggest it was probably onomatopoeic for the truly incessant call. Governor Lachlan Macquarie in 1821 referred to Wanga Wanga Pigeon. The 'one-Wonga' name was formally adopted by RAOU (1926).

Other names: Wonga Wonga (with or without Pigeon appended), used as recently as 1913 by the RAOU; Pied Pigeon (Latham 1801); White-faced Pigeon (Latham 1823); White-fleshed Pigeon, attributed by Gould (1848) to the 'Colonists of New South Wales'.

Leucosarcia picata (Latham, 1801) [loo-ko-SAR-si-uh pi-KAH-tuh]: 'pied white-flesh', see genus name, and from Latin *picatus* from *pica*, a jay or magpie (i.e. black and white, or in this case, grey and white and reputedly good eating). Indeed, this is almost the first feature of the bird that Gould (1848) remarked on: 'it is a great delicacy for the table; its large size and the whiteness of its flesh rendering it in this respect second to no other member of its family'.

Chestnut-quilled Rock-Pigeon (breeding resident)

See family introduction. The two *Petrophassa* species differ most obviously in the colour of the primaries, especially in flight, and are named for them with commendable clarity.

Other names: Red-quilled Rock-Pigeon.

Petrophassa rufipennis Collett, 1898 [pe-tro-FAS-suh roo-fi-PEN-nis]: 'red-winged rock-pigeon', see genus name, and from Latin *rufus*, red, and *penna*, feather, hence *pennis*, winged, for the large red patch on the outspread wing.

White-quilled Rock-Pigeon (breeding resident)

See family introduction and Chestnut-quilled Rock-Pigeon.

Other names: It is perhaps indicative of the appropriateness of the given name that no alternatives have been recorded since Gould's White-quilled Rock Dove.

Petrophassa albipennis Gould, 1841 [pe-tro-FAS-suh al-bi-PEN-nis]: 'white-winged rock-pigeon', see genus name, and from Latin *albus*, dull white, and *penna*, feather, hence *pennis*, -winged, for the large white patch on the outspread wing.

Diamond Dove (breeding resident)

See family introduction. A reference to the spangles of somewhat elongated white wing spots. There exists a series of 1887 newspaper advertisements for 'fancy birds' including 'diamond doves', but most of the species on the list are introduced, so their identity is uncertain. It doesn't seem to have appeared in print in unequivocal reference to the current species until the 20th century – the first apparent use is in Leach (1911), but surely he collected it as a spoken name; shortly afterwards it was adopted by the RAOU (1913).

Diamond Dove *Geopelia cuneata* and Peaceful Dove *Geopelia placida*

Other names: Graceful Ground-Dove, used by Gould (1848); Turtle-dove, a reference to its similarity to the Old World *Streptopelia* doves and according to Gould the name used by the 'Colonists of Swan River' – there are certainly ecological similarities, but they are not particularly closely related; Little Dove or Little Turtle-dove – because the species is, with Peaceful Dove, Australia's smallest dove – reported by Morris (1898); Red-eyed Dove, for the conspicuous red skin eye surrounds.

Geopelia cuneata (Latham, 1801) [ge-o-PEH-li-uh koo-neh-AH-tuh]: 'wedge-shaped ground-dove', see genus name, and from Latin *cuneus*, wedge, for the shape of the tail, which is long and strongly tapered.

Peaceful Dove (breeding resident)

See family introduction. A blatantly anthropomorphic name; although perhaps less overtly aggressive to other species than some of the turtle doves for instance, it is not clear that they are inherently more pacific than their confamilials. It is tempting to think that Gould (1848), the author of the name, was beguiled by the unquestionably restful call of the dove from a perch on a hot afternoon. He named it at different times both *G. tranquila* and *G. placida* but didn't share his reasons. The name Ground Dove was later used for some time until the RAOU resurrected Gould's name Peaceful Dove in 1926.

Other names: Zebra Dove, for the finely striped upper breast – this name is now applied to *G. striata* from western Indonesia and the Malay Peninsula; Ground Dove, descriptive but far from uniquely so – nonetheless it was the name of choice of the RAOU as late as 1913; Doodle-doo, Doo-doo, Four o'Clock Dove, all valid renditions of the repetitive call.

Geopelia placida Gould, 1844 [ge-o-PEH-li-uh PLA-si-duh]: 'peaceful ground-dove', see genus name, and Latin *placidus*, meaning gentle or peaceful.

Bar-shouldered Dove (breeding resident)

See family introduction. Refers to the conspicuous dark scallops on the coppery neck; first formal use was by RAOU (1926).

Other names: Barred-shouldered Dove, Gould's name (1848), still in use into the 20th century (RAOU 1913); Bronze-necked or Copper-necked Dove; Mangrove Dove or Pigeon (reported by Gould to be used by a 'resident at Port Essington'), Pandanus Pigeon, River Pigeon, Scrub Dove, for common (but not obligatory) habitats, especially in the tropics; Kook-a-wuk, a lovely onomatopoeia.

Geopelia humeralis (Temminck, 1821) [ge-o-PEH-li-uh hoo-me-RAH-lis]: 'shouldered ground-dove', see genus name, and from Latin *umerus*, the shoulder – though it's really the neck more than the shoulder that is distinctive.

Nicobar Pigeon (vagrant)

See family introduction. From the origin of the type specimen, though it is found from north of Sumatra to the Solomons, mostly on small islands.

Caloenas nicobarica (Linnaeus, 1758) [ka-LO-e-nas ni-co-BA-ri-cuh]: 'beautiful wine-coloured pigeon from Nicobar' – though the vinous colour is not so obvious.

Norfolk Ground Dove (Norfolk Island endemic, extinct)

There are Ground Doves found in many genera. This species and the Emerald Dove were the only pigeons found on Norfolk Island.

Alopecoenas norfolkensis Forshaw 2015 [a-lo-pe-CO-e-nas nor-foe-KEN-sis]: 'foxy wine-coloured pigeon from Norfolk'. Whereby lies a curious tale. Latham originally named the bird *Columba norfolciensis* – but two other pigeons also bore that name and it is not clear to which of them Latham was referring. Moreover the only record of the species' existence is a watercolour painting. In 2010 the International Commission on Zoological Nomenclature suppressed the name (which by then was *Gallicolumba norfolciensis)* and in 2015 Joseph Forshaw redescribed it from the painting, changing the spelling slightly to effectively retain the original name.

Black-Banded Fruit Dove (breeding resident)

See family introduction. *Ptilinopus* is a large genus of some 50 species in South-East Asia and the Pacific, all called Fruit Dove (or Fruit-Dove, as generally used in Australia); they are essentially all rainforest fruit-eaters but are far from unique in this among pigeons. In this species, a broad black breastband separates the pure white throat from a grey belly – it is especially obvious in flight.

Other names: Banded Fruit-Dove, the previous name before this species was split from one in Bali and the Sundas; Banded Pigeon; Black-banded Pigeon or Fruit-Pigeon; Grey-rumped Fruit-dove, used by the authoritative HBW and which does distinguish the species from others in the genus more than the current name does, but not in Australia.

Ptilinopus alligator (Collett, 1898) [ti-LI-no-pʊs AL-li-geh-tor]: 'alligator feather-foot', see genus name, and from the South Alligator River, where the type-specimen was found.

Wompoo Fruit Dove (breeding resident)

See family introduction and Black-Banded Fruit Dove. A beautifully onomatopoeic name for the superbly resonating 'ONK-OOO' call. The *Australian National Dictionary* cites the explorer Edmund Kennedy as referring to the Whompoa Pigeon in north Queensland in 1870, so it seems that it was of Indigenous origin. It was not adopted as first-choice name for another 56 years (RAOU 1926).

Other names: Wompoo Pigeon or Fruit-Pigeon or Fruit-Dove; Purple-breasted Pigeon or Fruit-Pigeon, the name of choice until 1926; Whampoo, Bubbly Mary, Bubbly Jock, all for the call – it was a common phenomenon to name familiar or popular birds with human names, such as Robin, Martin and Willie Wagtail; Green Pigeon or Greeny, Parrot Pigeon, Painted Pigeon (also recorded by Kennedy in 1870), King Pigeon, Magnificent Fruit-Dove or Fruit-Pigeon (referring to the specific name, as used by Gould 1848) – this is a truly spectacular bird, and all these names seek to capture that. Allied Fruit Pigeon, applied by Gould to his Cape York species *assimilis*, not now recognised, though it was still accepted into the 20th century (RAOU 1913) – his point was that it was close to the Wompoo (closer than he thought, in fact). Lesser Purple-breasted Fruit-Pigeon is a puzzling name, given that this is the largest species in the genus and the Australian race is among the largest ones! We suspect this more properly belongs to one of the other subspecies, which are found in and around New Guinea.

Ptilinopus magnificus (Temminck, 1821) [ti-LI-no-pʊs mag-NI-fi-kʊs]: 'magnificent feather-foot', see genus name, plus Latin *magnificus*.

Orange-fronted Fruit Dove (vagrant)

See family introduction and Black-Banded Fruit Dove. From the apposite species name.

Other names. *Ptilinopus aurantiifrons* (Gray GR, 1858), [ti-LI-no-pʊs o-RAN-tii-fronz]: 'orange-fronted feather-foot', see genus name, and from Scientific Latin post 15th century *aurantius*, orange-coloured (from Classical Latin *aurum*, gold), and Latin *frons*, forehead.

Superb Fruit Dove (breeding resident)

See family introduction and Black-Banded Fruit Dove. An unhelpful name – there is scarcely a *Ptilinopus* that is not superb! It reflects the species name, but of course this just begs the question. It was used by Gould (1848) and reinstated by RAOU (1978) on the grounds that it is known by this name elsewhere in its range.

Other names: Superb Fruit-Pigeon or Fruit-Dove; Purple-crowned Fruit-Pigeon or Fruit-Dove, the name by which it was known until recently in Australia and arguably a more useful one, providing comparison with the next species.

Ptilinopus superbus (Temminck, 1809) [ti-LI-no-pʊs soo-PEHR-bʊs]: 'superb feather-foot', see genus name, plus Latin *superbus*, which as well as having a positive meaning can also mean haughty or arrogant, but we don't think this was Temminck's intention.

Rose-crowned Fruit Dove (breeding resident)

See family introduction and Black-Banded Fruit Dove. At least in Australia (it is also found in Timor and Indonesian islands between New Guinea and Sulawesi) this is an unequivocally useful descriptive name, though Gould didn't use it (he called it Swainson's Fruit Pigeon).

Other names: Rose-crowned Pigeon or Fruit-Dove or Fruit-Pigeon; Red-crowned Pigeon or Fruit-Pigeon; Pink-headed or Pink-capped Dove or Fruit-Dove, Pink Cap, all variations on the theme. Ewing's Fruit-Pigeon, for race *ewingii* of the Top End and Kimberley – the Reverend Thomas Ewing was a Tasmanian amateur naturalist honoured with the name by Gould (who thought it was a separate species). Swainson's Fruit-Pigeon, for William Swainson, zoological artist and self-styled naturalist, who described the species in 1825 from a NSW bird – Gould, perhaps somehow forgetting that Swainson had already named both genus and species, later called it *P. swainsonii* for him, though it seems a curious coincidence. 'I have therefore named it after Mr. Swainson, the author of the genus to which it belongs, as a slight testimony of the respect I entertain for the talents of one who has done so much towards the advancement of ornithology, at once the most interesting and popular branch of the science of natural history.' Not everyone shared this view of Swainson; see Superb Parrot for more on him.

Ptilinopus regina Swainson, 1825 [ti-LI-no-pʊs re-GEE-nuh]: 'queen feather-foot', see genus name, and from Latin *regina*, queen. Swainson (1825) gave no clues but we may guess that it was not for a specific queen, but rather playing on the double meaning of its rosy pink crown. Its French name is *ptilope à diadème* – dove with a tiara.

Orange-bellied Fruit Dove (vagrant to Torres Strait islands)

See family introduction and Black-Banded Fruit Dove. The orange belly stands out on the little dark-green bird – if you can see the bird.

Ptilinopus iozonus (Gray GR, 1858) [ti-LI-no-pʊs i-o-ZOH-nʊs]: 'rusty-belted feather-foot', see genus name, and probably from Greek *ios*, rust, and *zōnē*, belt. Gray (1858) described the 'middle of the abdomen deep orange'. Another suggested interpretation, which we consider less convincing, is that it is from Greek *ion*, violet, rather than *ios*, rust. Gray did describe the 'bend of the wings' as being 'greyish-violet mixed with green', presumably referring to the smallish lilac and purple band on the shoulder.

Elegant Imperial Pigeon (vagrant)

See family introduction. Overall a comprehensively unhelpful name; in addition to the 'imperial', we defy anyone to explain why this undoubtedly elegant bird is more so than any other, scientific name notwithstanding.

Ducula concinna (Wallace, 1865) [doo-KOO-luh kon-SIN-nuh]: 'elegant dukul', see genus name, and Latin *concinna*, elegant or neat.

Christmas Imperial Pigeon (Christmas Island endemic)

See family introduction. The only imperial pigeon likely to be found on Christmas Island (though Pied turns up occasionally).

Other names: Christmas Imperial-Pigeon; Christmas Island Pigeon, Imperial-Pigeon or Imperial Pigeon; Black or Dusky Imperial-Pigeon, both appropriate names.

Ducula whartoni (Sharpe, 1887) [doo-KOO-luh WOR-ton-ee]: 'Wharton's dukul', see genus name, and for Sir William Wharton (1843–1905), a British admiral and naval surveyor who was later appointed official Hydrographer of the Navy. His relationship, if any, to Richard Sharpe, the British Museum zoologist who named the bird, is not known to us, though he had been made a Fellow of the Royal Society the previous year, so this may have played a part in Sharpe's thinking; Sharpe himself does not assist us.

Collared Imperial Pigeon (vagrant to Torres Strait islands)

See family introduction. The snappy broad black collar round the white face/throat is undoubtedly its most obvious characteristic.

Other names: Black-collared Pigeon or Imperial-Pigeon; Collared Imperial Fruit-Pigeon, which seems a little excessive; Pink-capped Imperial-Pigeon, also descriptive; Müller's or Mueller's Fruit Pigeon or Imperial-Pigeon, from the species name.

Ducula mulleri (Temminck, 1835) [doo-KOO-luh MIL-ler-ee]: 'Müller's dukul', see genus name, and for Dr Saloman Müller, a German naturalist, taxidermist and collector in New Guinea, Timor, Java and Sumatra in the 1820s and 1830s.

Zoe's Imperial Pigeon (vagrant)

See family introduction. As with the species name, for Zoé Lesson, first wife of French zoologist René Lesson; she died aged just 19; Anselme Desmarest named the bird for her 7 years later. He and Lesson were working in Paris at the same time and doubtless knew each other.

Ducula zoeae (Desmarest, 1826) [doo-KOO-luh ZO-i-eh]: 'Zoe's dukul', see genus name and common name.

Torresian Imperial Pigeon (breeding resident)

See family introduction and because it appears annually in northern Australia on migration across the strait from New Guinea. Gould (1848), who used Torres Strait Fruit Pigeon, noted that it was 'so abundant there that few voyagers pass the straits during its breeding-season without encountering it'.

Other names: Torresian Imperial-Pigeon, the more usual Australian form; Pied Imperial Pigeon or Imperial-Pigeon, widely used – pied originally meant black and white, as in a (European) magpie, and this is almost the only white and black imperial-pigeon; Imperial or Torres Strait Pigeon or Fruit Pigeon. Nutmeg or White Nutmeg Pigeon – it is suggested (Goodwin 1970) that this refers to a preference for native nutmegs (e.g. *Myristica insipida* and *M. globosa*) but while HANZAB (1990–2006, vol. 3) confirms that this fruit is eaten, there is no suggestion that it is especially significant; in the absence of an opportunity to sniff or taste the pigeon, however, we can't offer an alternative! (Although untold thousands were shot for food in earlier times, we are unaware of any record of the bird having a spicy flavour.) Gould used the name in 1865; it was still first-choice name by the RAOU 38 years later (RAOU 1913) though by 1926 it had gone back to Torres Strait Pigeon; Spice Pigeon, which presumably follows from Nutmeg.

Ducula spilorrhoa (Gray GR, 1858) [doo-KOO-luh spi-lo-ROH-uh]: 'spotted-rump dukul', see genus name, and from Greek *spilos*, spot, and *orrhos*, rump (though, as Gray pointed out, the 'spots' are on the undertail-coverts rather than the rump).

Topknot Pigeon (breeding resident)

See family introduction. Although the name hardly does justice to the extraordinary front-and-back double crest, this is certainly the reference. It arose among and was used by 'the Colonists of New South Wales' (Gould 1848) and Gould himself atypically adopted their name. The earliest reference we can find is in a 'wanted ad' of 1846 for live birds in the *Sydney Morning Herald*, with genus specified.

Other names: Flock Pigeon, for its default setting of travelling and feeding in huge flocks – but see Flock Pigeon profile; Quook Quook, surely onomatopoeic, but curious in that the bird is mostly silent; Blue Pigeon, from the wanted ad already mentioned.

Lopholaimus antarcticus (Shaw, 1793) [lo-fo-LIE-mʊs ant-ARK-ti-kʊs]: 'southern crested-throat', see genus name, with Latin *antarcticus*, southern, as opposed to *arcticus* – Greek *arktikos* – pertaining to the constellation of the Bear (Greek *arktos*, Latin *ursa*) Ursa Major, hence northern. It is something of a puzzle that the feature that the genus name refers to (the spiky throat) is much less striking than the splendid flat-cap crest.

New Zealand Pigeon (Norfolk Island, extinct)

See family introduction. The nominate race is found throughout most of New Zealand; whether the Norfolk Island population was a subspecies or full species is still being debated.

Other names: Norfolk Island Pigeon; it is probable that this was its name on Norfolk, at least until it tragically disappeared, largely into saucepans, by the end of the 19th century.

Hemiphaga novaeseelandiae (Gmelin JF, 1789) [he-mi-FAH-guh no-veh-zeh-LAN-di-eh]: 'New Zealand half-imperial-pigeon', see genus name, and Modern Latin name for New Zealand, the type locality.

RALLIDAE (GRUIFORMES): crakes and rails

Rallidae Rafinesque, 1815 [RAL-li-dee]: the Rail family, from genus name *Rallus* (which is not present in Australia). *Rallus* is a Latinised form of European names for the rail: English rail, French *râle* or German *Ralle*. There is a possible echo of classical Latin too, either via *rallus*, thin (hence 'thin as a rail'), or *rallum*, a scraper for cleaning a ploughshare (i.e. the accompanying noise).

As with Megapodiidae, this is a family where poultry names abound, especially 'hen' in various combining forms. The two key base names (crake and rail) go well back to their British relations and did not arise in Australia. As might be expected for such cryptic, though widespread, birds, both names are onomatopoeic.

'Rail' came with the Normans, like so many other currently used names (many of which gradually replaced older English words). It arrived as *raale*, and meant a rattle.

'Crake' comes from the name of a particular European dryland species, Corn Crake (*Crex crex*); both 'crake' and 'crex' refer to the call, an insect-like rasp repeated in short bursts. It tends to be used for smaller members of the family, but 'crake' and 'rail' are interchangeable. The recurring use of Water, Swamp or Marsh Crake seems superfluous but was used to contrast with the Corn Crake's waterless habitats.

The genera

Rallina Gray GR, 1846 [ral-LEE-nuh]: '[bird] belonging to the rails', from genus name *Rallus* (see family introduction) with *-ina* (indicating relationship or origin, or sometimes a diminutive).

Gallirallus Lafresnaye, 1841 [gal-li-RAL-lʊs]: 'cock-rail', from Latin *gallus*, cock, and Modern Latin *rallus*, rail, via those genus names.

Lewinia Gray GR, 1855 [loo-WI-ni-uh]: 'Lewin's [bird]', for John Lewin, English naturalist and engraver who came to Australia with etching paraphernalia in 1800 to collect specimens (initially insects but later birds) to produce books for English patrons. His 1813 *Birds of New South Wales* (based on his earlier *Birds of New Holland* for six English subscribers) was the first illustrated natural history book to be printed in Australia. It's not clear that he had any connection with the bird – it doesn't appear in his published paintings.

Crex Bechstein, 1803 [kreks]: 'crake', from Greek *krex*, though the identity of this bird as described by the Ancient Greeks is actually very uncertain (according to Arnott, it may have been the Corn Crake *Crex crex*, the Ruff *Calidris pugnax* or even the Black-winged Stilt *Himantopus himantopus*). However, the word was chosen in the belief that it stood for a crake and for its similarity to the bird's call – the related word *krekō* means to strike a stringed instrument or to make a sharp noise. Sounds like a Corn Crake to us!

Amaurornis Reichenbach, 1853 [a-mow-ROR-nis]: 'dark and shadowy bird', from Greek *amauros*, dark, shadowy, dim, and *ornis*, bird.

Porzana Vieillot, 1816 [por-ZAH-nuh]: 'crake', apparently from a Venetian dialect word *porzana* or *sporzana*, used for smaller crakes.

Eulabeornis Gould, 1844 [yoo-la-be-OR-nis]: 'cautious or discreet bird', from Greek *eulabēs*, cautious or discreet, and *ornis*, bird.

Gallicrex Blyth, 1852 [GAL-li-kreks]: 'cock-crake', from Latin *gallus*, cock, and Greek, *krex*, crake.

Porphyrio Brisson, 1760 [por-FI-ri-o]: 'water-hen', from Linnaeus' 1758 species name for the bird. From Greek *porphuriōn* or Latin *porphyrio*, both meaning the Purple Gallinule (in Australia, the Purple Swamphen), a bird with many subspecies in Europe, Africa, the Indian subcontinent and Australasia. It seems to have been quite common in antiquity across southern Europe but is now rare there. Attempts were made in both Ancient Greece and Rome to domesticate it (Arnott 2007). The name must ultimately have been from the Greek word *porphuroeis/porphur-*, meaning purple.

Gallinula Brisson, 1760 [gal-LIN-yoo-luh]: 'little hen', from Latin *gallina*, hen, and diminutive ending *-ulus*. Gallinule, derived thus, was used in the 16th century for Common Moorhen *Gallinula chloropus*; the genus name was formed by translating this back into Latin.

Tribonyx Du Bus de Gisignies, 1840 [tri-BO-niks]: 'worn-claw', from Greek *tribō*, to rub or wear away, and *onux*, talon or nail, apparently in reference to the short hind claw.

Fulica Linnaeus, 1758 [Fʊ-li-kuh]: 'coot', from Latin *fulica*, coot. Called *phalaris* by the Ancient Greeks, close to *phalaros*, having a patch of white, and indeed not being far from *phalacros*, bald.

The species

Red-necked Crake (breeding resident)

The lovely rufous head and neck are unique among Australian crakes (but occur in the vagrant and closely related Red-legged Crake).

Other names: Red-necked Rail, used by Gould (1869) and still in use into the 20th century (RAOU 1926).

Rallina tricolor Gray GR, 1858 [ral-LEE-nuh TRI-ko-lor]: 'three-coloured rail', see genus name. The three colours of Latin name *tricolor* are the rusty red head and neck, the olive-brown back and the spectacular white patterning on the underwing.

Red-legged Crake (vagrant)

Strikingly red legs indeed.

Rallina fasciata (Raffles, 1822) [ral-LEE-nuh fas-see-AH-tuh]: 'banded rail', see genus name, plus common name.

Lord Howe Woodhen (Lord Howe Island endemic)

It might seem that Woodhen comes from the scientific name, but, given that it was formally described as late as 1869, it was more likely to have been the other way round.

Other names: Woodhen, Lord Howe Island Woodhen.

Gallirallus sylvestris (Sclater PL, 1869) [gal-li-RAL-lʊs sil-VES-tris]: 'wood cock-rail', see genus name, plus Latin *silvestris*, of the woods, woody (from *silva*, forest or woodland).

Buff-banded Rail (breeding resident)

For the distinctive buffy breastband against zebra-striped underparts. The name seems to have been coined by Tom Iredale (1956) and formalised (somewhat grudgingly) by RAOU (1978) – 'perhaps the best of a bad lot'.

Other names: Landrail, Buff-banded Landrail, Banded Rail or Landrail, for its tendency to wander away from waterbodies; Pectoral Rail, the name used by Gould (1848) from a very old species name *pectoralis* by the great Frenchman Cuvier, but this could have been a confusion with Lewin's Rail – nonetheless it was still recommended by RAOU (1913); Painted or Striped Rail, for its patterning.

Gallirallus philippensis (Linnaeus, 1766) [gal-li-RAL-lʊs fi-li-PEN-sis]: 'Philippines cock-rail', see genus name, plus the Philippines plus the suffix *-ensis*, usually indicating the place of origin of the type specimen.

Slaty-breasted Rail (vagrant)

See family introduction. For the clear grey breast, despite the species name focused elsewhere.

Gallirallus striatus (Linnaeus, 1766) [gal-li-RAL-lʊs stree-AH-tʊs]: 'streaked cock-rail, see genus name, and from Latin *striatus*, streaked or furrowed. The breast is indeed grey, but the streaks are on most other parts.

Lewin's Rail (breeding resident)

For John Lewin, see genus name, though the name had been associated with him long before 1855. Swainson named the bird twice (both times after Temminck had already done so), one of those being *Rallus lewinii.* It was formalised as first-choice name by RAOU (1926).

Other names: Lewin's Water Rail or Water Rail; Pectoral Rail (Gould 1848), from the scientific name; Slate-breasted Rail, a characteristic like that of many other rails, but it was the preferred name of RAOU (1913). Short-toed Rail, from an old species name, *Rallus brachypus*, by William Swainson (see Rose-crowned Fruit Dove), but this just begs the question of why; Gould (1848) supplied the answer, in that Swainson was referring to the short claws on some island specimens, which Gould maintained were atypical and due to their stony habitat.

Lewinia pectoralis (Temminck, 1831) [loo-WI-ni-uh pek-to-RAH-lis]: 'Lewin's [bird] with a chest', see genus name, plus Latin *pectus/pectoris.* Although the breast is the least striking feature of the bird, being a palish grey-olive and really quite plain, it does stand out as different when compared with the bold striping and barring of the belly, back and folded wings.

Corn Crake (vagrant)

See family introduction, 'crakes'.

Crex crex (Linnaeus, 1758) [kreks kreks]: 'crake crake', see genus name.

Pale-vented Bush-hen (breeding resident)

Bush-hen is the group name used for the majority of this (somewhat fluidly defined) mostly South-East Asian genus. Pale-vented was only coined in the late 1990s in an attempt to find a uniquely descriptive name for the recently separated species from Australia plus New Guinea and associated islands; formerly it was included in *A. olivaceus*, now regarded as a Philippines endemic. It's not a great field character, though.

Other names: Bush-hen (widely in Australia); Rufous-vented Gallinule, coined by Gould (1869) for his species name *Gallinula ruficrissa* for the obvious reason, unsurprisingly believing it to be a new species; hence also Rufous-tailed or Rufous-vented Rail, Crake or Moorhen.

Amaurornis moluccana (Wallace, 1865) [a-mow-ROR-nis mo-lʊk-KAH-nuh]: 'Moluccan dark and shadowy bird', see genus name, plus name of the Maluku (Moluccan) Islands, part of Indonesia west of New Guinea.

White-breasted Waterhen (vagrant to Christmas and Cocos (Keeling) islands)

Perhaps more associated with pools than other genus members, but not dramatically so; it is the only member of the group with a white breast.

Amaurornis phoenicurus (Pennant, 1769) [a-mow-ROR-nis foy-ni-KOOR-ʊs]: 'crimson-tailed dark and shadowy bird', see genus name, plus Greek *phoinikeos*, purple-red, crimson, and *-oura*, tail. Certainly the rump and tail are coloured, though perhaps more rusty or cinnamon than purple-red or crimson.

Baillon's Crake (breeding resident)

Louis Baillon (1778–1855) was a French professional naturalist who lived through the Revolution and was assistant naturalist at the Jardin des Plantes when Napoleon came to power. He sent specimens to Louis Vieillot, including this species from northern France, which Vieillot named *Rallus bailloni* for him in 1819. It later transpired that Peter Pallas had long since named the bird from Dauria in far eastern Russia. By then, however, Baillon's name had become permanently associated with the bird, in French and English.

Other names: Marsh or Water Crake, old names, the second one used by Gould in 1848, though Marsh Crake was still the formally recommended name in 1926 (RAOU 1926) and was in wide use until the 1978 RAOU list opted firmly for the internationally used Baillon's Crake; Little, Little Water, Little Marsh Crake, for its minute size, reflected in the species name – Australian Little Crake was used by RAOU in 1913; Little Spotted Crake, as it does resemble a small version of the next species; Pallas' Crake, rarely used in Australia, for Pallas, an eminent German 18th-century naturalist who did much Russian work (as discussed).

Porzana pusilla (Pallas, 1776) [por-ZAH-nuh pʊ-SIL-luh]: 'insignificant little crake', see genus name, and from Latin *pusillus*, very small or insignificant – paltry, in fact! Hardly fair.

Australian Crake (breeding resident)

A tug of war continues over the name of this bird. The name nearly always used in Australia is Australian Spotted Crake, a clear descriptive name for the white speckles, and long-used (e.g. Spotted Water Crake, Gould 1848). RAOU (1978) thought the name 'unnecessarily long' but Christidis and Boles (2008) did not. Now IOC seems to disagree again. It is certainly endemic, but there the usefulness of the current name seems to end.

Other names: Australian Spotted Crake, Spotted Crake or Water Crake.

Porzana fluminea Gould, 1843 [por-ZAH-nuh floo-MI-ne-uh]: 'river crake', see genus name, plus Latin *flumineus*, riverine, from *flumen*, river. Is it fonder of rivers than its genus-mates? Probably not.

Ruddy-breasted Crake (vagrant)

Another crake with lovely rufous underparts.

Porzana fusca (Linnaeus, 1766) [por-ZAH-nuh FʊS-kuh]: 'dusky crake', see genus name, plus Latin *fuscus*, dark or dusky. Not a name that makes the most of the bird's most striking feature – see common name.

Spotless Crake (breeding resident)

A distinctively plain sooty crake (in contrast with the Australian Spotted Crake), but the name derives directly from Gould's name *P. immaculata* (he didn't realise it was the same as Gmelin's Tongan specimen).

Other names: Leaden Crake for the colour; Spotless Water Crake; Little Swamphen, Native-hen or Waterhene; Swamp Rail (!); Motor-car Bird, for its revving call.

Porzana tabuensis (Gmelin JF, 1789) [por-ZAH-nuh ta-boo-EN-sis]: 'Tongan crake', see genus name, and from Tongatapu, the main island of the Kingdom of Tonga plus the suffix *ensis*, indicating the place of origin of the type specimen.

White-browed Crake (breeding resident)

For its white eyebrows (above and below the eyes!).

Other names: White-eyebrowed Water Crake, the formal name introduced by Gould in 1848, from his own name *Porzana leucophrys*.

Porzana cinerea (Vieillot, 1819) [por-ZAH-nuh si-NE-re-uh]: 'ashy-grey crake', see genus name, and from Latin *cinereus*, like ashes.

Chestnut Rail (breeding resident)

An obvious name for a hulking great chestnut-coloured bird, though the name only became adopted by RAOU (1926).

Other names: Chestnut-breasted or Chestnut-bellied Rail, from the species name, and the name used until 1926.

Eulabeornis castaneoventris Gould, 1844 [yoo-la-be-OR-nis kas-ta-ne-o-VEN-tris]: 'cautious or discreet chestnut-bellied bird', see genus name, and from Latin *castanea*, chestnut or chestnut tree, hence chestnut-coloured, and *venter/ventris*, belly. But how apt can 'discreet' be when you're hulking and great, unless perhaps this bird needs to be even more cautious than its fellow rails because of its size?

Watercock (vagrant to Christmas and possibly Cocos (Keeling) islands)

Perhaps in reference to its spectacular, arguably roosterish, red comb-like frontal plate.

Gallicrex cinerea (Gmelin JF, 1789) [GAL-li-kreks si-NE-re-uh]: 'ashy-grey cock-crake', see genus name, and from Latin *cinereus*, like ashes.

Australasian Swamphen (breeding resident)

Perhaps surprisingly – given that the species-complex is found over much of the world – the name swamphen seems to have arisen spontaneously in Australia. Gould (1848) used the term Azure-breasted or Black-backed Porphyrio (he regarded eastern and western forms as different species), but recorded that the name Swamp-Hen was used by the 'Colonists of Western Australia'; that seems to be its first recorded usage. It is tempting to see the 'Purple' as a distinguisher from Dusky Moorhen in Australia, but in fact it is the name used throughout the bird's Mediterranean, southern African and Asian range (though the different subspecies are of varying degrees of purplitude).

Other names: Purple, Blue-breasted, Eastern, Western Swamphen (the latter reflecting the different south-western subspecies); Purple, Black-backed, Azure-breasted Gallinule (see White Gallinule); Bald Coot or Baldcoot, Blue Bald Coot, a term that goes back to 1829 (*Australian National Dictionary*) and was still used into the 1950s, presumably referring to the big frontal shield but curious in that 'bald' applied to coot traditionally meant white – see Eurasian Coot, and White-headed Pigeon; Black-backed, Purple or Macquarie Water-hen, but we have no idea if the Macquarie reference was to governor, river or other; Redbill, for obvious reasons, and a reflection of its tameness and familiarity. (Other names used in New Zealand, including Pukeko and Tarler Bird, are mentioned here because they sometimes turn up as synonyms in Australian texts.)

Porphyrio melanotus (Linnaeus, 1758) [por-FI-ri-o me-la-NOH-tʊs]: 'black-backed water-hen', see genus name, and from Greek *melas*, black, and *nōton*, the back.

White Swamphen (Lord Howe Island, extinct)

The White was inevitable; this was a stunning all-white flightless bird.

Other names: White Gallinule.

Porphyrio albus (Shaw, 1790) [por-FI-ri-o AL-bʊs]: 'white water-hen', see genus name, plus Latin *albus*, dull white.

Common Moorhen (vagrant)

See Dusky Moorhen for 'moorhen'; this species is common throughout much of the world.

Gallinula chloropus (Linnaeus, 1758) [gal-LIN-yoo-luh KLOH-ro-pʊs]: 'little green-footed hen', see genus name, and Greek *khlōros*, green and *pous*, foot.

Dusky Moorhen (breeding resident)

Moor had the secondary meaning of a marsh, now obsolete but still in use into the late 19th century, especially in Britain. The Australian bird took its name from the closely related worldwide Common

Moorhen. 'Dusky' is perhaps more by way of contrast with the Australasian Swamphen in Australia than with other moorhens.

Other names: Moorhen, Black or Sombre Moorhen; Black or Sombre Gallinule, the latter used by Gould (1848) from his species name.

Gallinula tenebrosa Gould, 1846 [gal-LIN-yoo-luh te-ne-BROH-suh]: 'dark or gloomy little hen', see genus name, plus Latin *tenebrosus*, from *tenebrae*, darkness or shadows or places to lurk, so this could apply to the bird's behaviour as well as its appearance.

Black-tailed Nativehen (breeding resident)

This is unquestionably a hen-like genus of birds, with erect bantam-type tail; the black trailing edge of the tail of this species contrasts strongly with the brown back. The form Native-hen is more usual in Australia.

Other names: Black-tailed Tribonyx, used by Gould (1848) from the genus name; Native or Barcoo Bantam, the latter for the Barcoo River of south-west Queensland; Moor-hen ('of the Colonists', Gould 1848), Native Hen, Swamphen, Water Hen, Black-tailed Water-hen, all reflections of the familiarity of the bird in inland Australia; Gallinule.

Tribonyx ventralis (Gould, 1837) [tri-BO-niks ven-TRAH-lis]: 'worn-claw with a belly', see genus name, and from Latin *ventralis*, of the belly, presumably referring to the highlighting of the black belly by the white streaks of the flank feathers, though Gould (1836b) did not specify.

Tasmanian Nativehen (breeding resident)

Endemic to Tasmania. The form Native-hen is more usual in Australia.

Other names: Mortier's Tribonyx, a classic Gouldism, straight from the species name; Native Hen, Water Hen; Narkie, probably for the hard nasal aggression calls and somewhat squabbly nature, though it could also be from an Indigenous language; Turbo Chook, an affectionate Tasmanian folk name, for its impressive burst of speed – 'chook' is an old Australianism (apparently from Old English 'chuck' or 'chucky', referring to cluck) for domestic fowls.

Tribonyx mortierii Du Bus de Gisignies, 1840 [tri-BO-niks mor-ti-ER-i-ee]: 'Mortier's worn claw', see genus name, and for Bartholomé (or Barthélémy) Dumortier (or Du Mortier) (1797–1878), an eminent Belgian botanist and invertebrate zoologist who was involved in the struggle for Belgian independence from the Dutch and who sat in the first independent Belgian Parliament. He was later a driving force behind the formation of, and inaugural (and tyrannical) director of, the National Botanic Gardens in Brussels. It is not clear that he had any great interest in birds at all, let alone the Tasmanian Nativehen, but he was influential enough that Bernard Aimé Léonard du Bus de Gisignies (another Belgian ornithologist, politician and director of major Belgian museums) would want to honour him. (How the first specimen got to Belgium to be described is another story!)

Eurasian Coot (breeding resident)

Coot is onomatopoeic, for the harsh staccato calls; it dates from at least the beginning of the 14th century. (However, it was also being used for Common Guillemot *Uria aalge* well into the 18th century.) It is found across most of the Old World (and northern Africa).

Other names: Coot, Australian Coot for the local subspecies.

Fulica atra Linnaeus, 1758 [FƱ-li-kuh AH-truh]: 'black coot', see genus name, and from Latin *ater*, dull black.

GRUIDAE (GRUIFORMES): cranes

Gruidae Vigors, 1825 [GROO-i-dee]: the Crane family, from the genus name *Grus* of Brisson, 1760. The Latin word *grus* (Greek *geranos*), meaning crane, was mentioned by several Roman writers, including Cicero, Lucretius, Horace and Pliny, not least because the bird was regarded as a delicacy. Pliny (77–79 AD), in his usual style, entertained other information too: 'During the night, also, [the cranes] place sentinels on guard, each of which holds a little stone in its claw: if the bird should happen to fall asleep, the claw becomes relaxed, and the stone falls to the ground, and so convicts it of neglect', alerting the other birds at the same time. Wonderful stuff, natural history!

Crane is from Old English *cran* and is onomatopoeic for the wonderful brassy call (e.g. of the Eurasian Crane *Grus grus*).

The genus

Antigone Reichenbach, 1853 [an-TI-go-ne]: 'crane', from the specific name chosen by Linnaeus (1758) *Ardea antigone* (see species name).

The species

Sarus Crane (breeding resident)

From the Hindi *saras* for the bird. (There is an unsupported suggestion that the name is associated with the Sar Us River of Outer Mongolia, but there is no reason to suppose that the bird ever occurred anywhere near there.)

Antigone antigone (Linnaeus, 1758) [an-TI-go-ne an-TI-go-ne]: 'Crane crane', see genus name, and from the Greek proper name 'Antigone'. This is not Antigone (daughter of Oedipus and Jocasta and sister of Polynices) who was put to death by Creon for trying to give her brother proper burial, but rather the daughter of Laomedon, king of Troy. Her hair was turned into snakes by the goddess Hera, with whom this Antigone had compared herself. The other gods had different ideas and changed her into a stork instead, which gets us no nearer to a crane. But it seems (not surprisingly) that Linnaeus got a bit mixed up with all these ancient Greeks turning into birds, as there was another girl that Hera changed into a crane, so she was possibly meant to be the one (Jobling 2010).

Brolga (breeding resident)

Of Aboriginal origin, *buralga* from the Kamilaroi language of north-west Queensland according to the *Australian National Dictionary*. On the face of it, it's a relatively recent arrival in Australian English, replacing Australian Crane or Native Companion in general use as late as the early 20th century; the *Australian National Dictionary* reports the first usage as 1896. However, we've found a description in the *Maitland Mercury and Hunter River General Advertiser* of 8 February 1869 of an Aboriginal dance, wherein they moved 'as only emus and brolgas can dance', so the word was well enough known. (The following year a racehorse named Brolga made the news regularly.) However, Hall (1899) was still using Native Companion but by 1911 Leach used Brolga; its formalisation came with the recommendation by RAOU (1926).

Other names: Australian Crane. Native Companion, reported by Gould as being a name 'of the Colonists' but not why. 'Companion of the natives'? Surely not. A native bird as companion? Just possibly, given that Gould reported tame birds in Parramatta and Camden.

Antigone rubicunda (Perry, 1810) [an-TI-go-ne roo-bi-KƱN-duh]: 'ruddy crane', well, ruddy crane's head at least – the rest is grey. See genus name, plus Latin *rubicundus*, ruddy or red.

PODICIPEDIDAE (PODICIPEDIFORMES): grebes

Podicipedidae Bonaparte, 1831 [po-di-si-PE-di-dee]: the Bum-foot family, to quote from HANZAB (1990–2006, vol. 1a). From Latin *podex/podicis*, meaning anus or vent, and *pes/pedis*, foot. Not really quite that – just that their feet are set quite remarkably far back on their bodies, on the principle of the outboard motor, to create maximum power and efficiency for swimming and diving.

Grebes are an ancient group and similar species are found on all continents except the Antarctic. The French word *grèbe*, of unknown origin, was introduced into the language by the 18th century British naturalist and author Thomas Pennant; it eventually replaced older words including diver and dabchick.

The genera

Tachybaptus Reichenbach, 1853 [ta-ki-BAP-tƱs]: 'fast-dipper' from Greek *tachys*, fast and *bapto*, to dip in water (think *tachy* as in tachycardia and *bapt* as in baptism). It's a good description of the disappearing trick these birds invariably perform when you try to point them out to a companion.

Poliocephalus Selby, 1840 [po-li-o-SE-fa-lƱs]: 'grey-head' from Greek *polios*, grey and *kephalē*, head.

Podiceps Latham, 1787 [PO-di-seps]: 'bum-foot' (see family) from *podiceps*, a contraction of 'Podicipes', the name offered as an alternative to the old name '*Colymbus*' by Willughby and Ray (1676), Ray (1713) and Linnaeus (1758), and used by Catesby and Edwards (1731), who spelled it 'Prodicipes'. Linnaeus (1758) used the abbreviated word 'Podiceps', not for the Great Crested Grebe as now but as an alternative name for *Colymbus auritu* (*P. minor*) and as a species name *C. podiceps*, (*P.*

minor rostro vario), names he applied to the Horned Grebe, now *P. auritis*. (So nothing to do with *-ceps*, meaning headed, as in biceps, or indeed *ruficeps*, red-headed, *atriceps*, black-headed: a good thing, perhaps?)

The species

Tricoloured Grebe (vagrant)

See family introduction. From the species name.

Tachybaptus tricolor (Gray GR, 1861) [ta-ki-BAP-tʊs TRI-ko-lor]: 'three-coloured fast-dipper', see genus name, and from Latin *tricolor*. The three colours described by Gray are 'a deep aeneous black', 'deep rufous', and 'aeneous black mottled with rufous-white' (Gray 1860). In case you are puzzled by 'aeneous' it is from Latin *aeneus* meaning of bronze or copper – so the black has a bronze lustre.

Australasian Grebe (breeding resident)

See family introduction. Despite having been described as a separate species in 1826, for much of the subsequent time it has been regarded as a race of the widespread Little Grebe. The current name is a relative neologism to reflect that.

Other names: Australian Little Grebe; Diver; Dabchick, which is an old English name for the Little Grebe; Black-throated Grebe (used by Gould 1848 and still in use in the 20th century e.g. RAOU 1913), Diver or Dabchick, presumably in contrast to the Little Grebe, which has a rufous throat; Red-necked Grebe, for the rufous ear-stripe of the breeding plumage.

Tachybaptus novaehollandiae (Stephens, 1826) [ta-ki-BAP-tʊs no-veh-hol-LAN-di-eh]: 'New Holland fast-dipper', see genus name, and from Modern Latin *Nova Hollandia*, New Holland, the old name for Australia.

Hoary-headed Grebe (breeding resident)

See family introduction. Hoary literally means frosted but has come to mean greyed with age, in unkind reference to the snappy brushed-back courtship coiffure of the species. Used by Gould, with reference to the species name.

Other names: Diver or Dabchick (see Australasian Grebe); Hoary-headed Dabchick; Tom Pudding, an old English name for the Little Grebe, presumably an affectionate but uncomplimentary reference to its dumpy appearance, though it may be relevant that it was also applied to towed barges. It is not clear why this latter term is not applied to the Australasian Grebe.

Poliocephalus poliocephalus (Jardine & Selby, 1827) [po-li-o-SE-fa-lʊs po-li-o-SE-fa-lʊs]: 'grey-headed grey-head', an apt if repetitive description.

Great Crested Grebe (breeding resident)

See family introduction. A species found through much of the Old World; it is one of the world's largest species and the black crest is unique, making the name self-evident.

Other names: Diver (again!), reported by Gould to be used by the 'Colonists', and Loon, the latter being an old English name for the unrelated Northern Hemisphere divers, now in general use only in North America. Great Grebe, the name of choice of RAOU (1913), though now reserved for the South American species *P. major*. Other names are old English ones brought to Australia to refer to a familiar species. Carr Goose, where carr is an English east coast term for a marsh; Gaunt, a Lincolnshire term, originally referring to a goose. Tippet Grebe, used by Gould (1848), presumably reflecting its use in England. It is of uncertain origin although there are a couple of obvious explanations: one is the verb tippet, meaning tiptoe, which could well be a reference to the dramatic rushing display, where pairs patter rapidly across the water surface. A more likely connection is the usage of 'tippet' for a fur scarf or ruff, which could refer to the bird's crest or to the fact that, even before the iniquitous feather trade of the late 19th and early 20th centuries, Great Crested Grebes were hunted for their breast skin. The dense underfeathers of this made them valued for shoulder capes and muffs.

Podiceps cristatus (Linnaeus, 1758) [PO-di-seps kri-STAH-tʊs]: 'crested bum-foot' – see genus name plus *cristatus*, tufted or crested.

Great Crested Grebe ***Podiceps cristatus***

PHOENICOPTERIDAE (PHOENICOPTERIFORMES): flamingoes

Phoenicopteridae Bonaparte, 1831 [foy-ni-ko-TE-ri-dee]: the Crimson-wing family, see genus name *Phoenicopterus*.

The genus

Phoenicopterus Linnaeus, 1758 [foy-ni-ko-TE-rʊs]: 'crimson-wing' from Greek *phoiniko-pteros* used (at least by Aristophanes) to mean a waterbird, perhaps the flamingo, and derived from *phoinix*, crimson, dark red or purple, and *pteros*, winged.

The species

Greater Flamingo (vagrant to Cocos (Keeling) Islands)

'Flamingo' apparently comes from Latin *flamma* (a flame) via at least Provençal and Portuguese, for the colour. (The appealing suggestion that it referred to pink-skinned Flemings, denizens of Flanders, seems sadly to belong firmly in the field of folk etymology, though the French word is almost the same for both person and bird: *Flamand* and *flamant*.) 'Greater' by comparison with the Lesser Flamingo, the only other Old World species.

Phoenicopterus ruber Pallas, 1811 [foy-ni-ko-TE-rʊs ROO-behr]: 'red crimson-wing', see genus name, and from Latin *ruber*, red. We can take it that this bird is definitely red (well, at least pink).

TURNICIDAE (CHARADRIIFORMES): buttonquails

Turnicidae Gray GR, 1840 [toor-NI-si-dee]: the Incomplete Quail family, see genus name *Turnix*.

The sense of 'button' as used here appears to be the implication of smallness and roundness. It is a relatively recent term, in fact very recent in Australia, with the more formal hemipod or hemipode being favoured before that, from an early genus name *Hemipodius*. It appears in the 1890s; Newton's *Dictionary of Birds* (1896) is unequivocal that it was Anglo-Indian. 'Bustard-quail' also appears then, from the same source, and this initially gained more favour in Australia. Leach (1911) used it as a group name, though he continued to use simply 'quail' for the individual species, despite having explained why it is wrong to do so. This unfortunate and confusing usage of 'quail' was perpetuated by Cayley (1931) and as recently as the first edition of Slater's *Field Guide to Australian Birds* (1970) and McDonald (1973). Tellingly, CSIRO (1969) didn't even list 'button-quail' as having been used in Australia until that time, except as an alternative name for Little Button-quail, and there without an adjective. RAOU (1978) reported that the usage was widespread elsewhere in the world and recommended it; by 1980 Pizzey had adopted button-quail for his first Australian bird field guide. IOC removes the hyphen.

Both hemipod(e) and bustard-quail are references to the foot structure, in particular the lack of a hind-toe, which distinguishes buttonquails from true quails. They share this characteristic with bustards; 'hemipod' means 'half-foot'.

All but one Australian species are named for colours, which primarily refer to the more brightly coloured (and larger) females. Most of these species are quite similar to each other, so the names are generally ambiguous.

The genus

Turnix Bonnaterre, 1791 [TOOR-niks]: 'not the full quail', from a shortened form of the genus name *Coturnix* (from Latin *coturnix*, a quail), the buttonquails being a hind-toe short of being a quail (see family introduction).

The species

Red-backed Buttonquail (breeding resident)

See family introduction. For the female's brick-red shoulders.

Other names: Red-backed Button-quail, the form usually used in Australia; Black-spotted Quail or Turnix, for the spotted flanks, actually a good distinguishing feature; Orange-breasted Quail, another feature; Black-backed Quail, for the male in particular, which has dark upperparts – Black-backed Hemipode (from the then genus name *Hemipodus*) was the name used by Gould.

Turnix maculosus (Temminck, 1815) [TOOR-niks ma-koo-LOH-sʊs]: 'spotted shortened quail', see genus name, and from Latin *macula*, spot. They're all pretty spotty, really, but see common name for the flanks.

Black-breasted Buttonquail (breeding resident)

See family introduction. For the unique feature (which extends to the head and nape), from the species name.

Other names: Black-breasted Button-quail, the form usually used in Australia; Black-breasted Hemipode, per Gould (1848) from the then genus name; Black-breasted Turnix; Black-fronted Quail.

Turnix melanogaster (Gould, 1837) [TOOR-niks me-la-no-GAS-tehr]: 'black-bellied shortened quail', see genus name, and from Greek *melas*, black, and *gastēr*, stomach; see also common name.

Chestnut-backed Buttonquail (breeding resident)

See family introduction. For the rich rufous upperparts.

Other names: Chestnut-backed Button-quail, the form usually used in Australia; until relatively recently this species and the next were commonly regarded as the same, so common names were interchangeable – even when inappropriate. Chestnut-backed Hemipode, per Gould (1848) from the then genus name; Thick-billed Quail, reported by Gould to be used by 'the Colonists'.

Turnix castanotus (Gould, 1840) [TOOR-niks kas-ta-NOH-tʊs]: 'chestnut-backed shortened quail', see genus name, and from a combination of Greek *kastanōn*, chestnut, and *nōton*, the back.

Buff-breasted Buttonquail (breeding resident)

See family introduction. For the plain buffy breast.

Other names: Buff-breasted Button-quail, the form usually used in Australia. See Chestnut-backed Button-quail. Buff-backed Quail, not descriptive of either taxon! Olive's Quail, from the species name.

Turnix olivii Robinson, 1900 [TOOR-niks o-LI-vi-ee]: 'Olive's shortened quail', see genus name, and for Edmund Olive who settled in Cooktown during the 1875 gold rush and became an expert on local fauna and a prolific collector, including of the first specimen of this rare species in 1899.

Painted Buttonquail (breeding resident)

See family introduction. For the colourful chestnut, black and white patterning. Gould (1848) reported that Painted Quail was used by the 'Colonists of Van Diemen's Land and Swan River', so our name presumably derived from that.

Other names: Painted Button-quail, the form usually used in Australia; Varied Hemipode, per Gould (1848) from the then genus name and species name; Butterfly Quail, Scrub, Speckled Quail; New Holland Partridge, a real flight of fancy and one apparently coined by the eminent English ornithologist John Latham (1801) who happened to be on watch when the first flood of bird specimens was coming out of the new colony; Varied Quail, another offering from Latham (1823), Varied Turnix; Sparkling Hemipode, by Gould (1848) from his species name *Hemipodus scintillans*, for birds from the Abrolhos.

Turnix varius (Latham, 1801) [TOOR-niks VAH-ri-ʊs]: 'variegated shortened quail', see genus name, plus Latin *varius*, for the chestnut, rufous, white, grey and black colours.

Red-chested Buttonquail (breeding resident)

See family introduction. For the rufous-buff breast.

Other names: Red-chested Button-quail, the form usually used in Australia; Chestnut-breasted, Red-breasted Quail or Turnix; Red-chested Hemipode, per Gould (1848) from the then genus name; Yellow Quail, origin not at all clear.

Turnix pyrrhothorax (Gould, 1841) [TOOR-niks pi-ro-THOH-raks]: 'shortened quail with a red breast-plate', see genus name, and from Greek *purrhos*, flame-coloured or red, and *thōrax*, a breastplate or chest covering.

Little Buttonquail (breeding resident)

See family introduction. A small bird, but effectively the same size as Red-chested and Red-backed. Little Quail was, according to Gould (1848), the name 'of the Colonists' and this doubtless influenced the current name; it could well be that the comparison was with Stubble and Brown Quail, both much larger birds.

Other names: Little Button-quail, the form usually used in Australia; Butterfly Quail, Button Quail, Dotterel Quail; Swift-flying Quail or Turnix or Hemipode, from the species name, both scientific and common names being coined by John Gould (with genus *Hemipodus*).

Turnix velox (Gould, 1841) [TOOR-niks VE-loks]: 'speedy shortened quail', see genus name, and Latin *velox*, swift (like velocity, of course, or even velocipede). Is it faster than its fellows? It certainly can scuttle if alarmed and explodes rocket-like from the ground.

(And a small thought: '... when the new namers came across the ... button-quail they called that a 'Turnix' because it was smaller than a Coturnix, the Little Button-quail unfortunately not qualifying for separate generic status where it might have been known as a 'Nix'' (Dabb 2005)).

BURHINIDAE (CHARADRIIFORMES): stone-curlews

Burhinidae Mathews, 1912 [boo-RI-ni-dee]: the Bull-nose family, see genus name *Burhinus*. 'Stone-curlew' arose in Britain, being known from the 17th century for the closely related Eurasian Stone-curlew *Burhinus oedicnemus*. The first part refers to its dryland habitat, compared with the shorelines favoured by 'true' Curlews (see Family Scolopacidae); the curlew link is in the wonderful wailing call. The form Stone Curlew is also widely used in Australia.

The ugly alternative group name – 'thick-knee' – arose from the Eurasian bird's species name (from Greek *oidēma*, swelling, and *knēmē*, lower leg), and is still favoured in parts of the world for this widespread family. The RAOU tried to formalise it in 1978, but was resisted.

The genera

Burhinus Illiger, 1811 [boo-REE-nʊs]: 'bull-nose', from Greek *bous*, an ox, and *rhis*, nose.

Esacus Lesson, 1831 [EE-sa-kʊs]: 'aquatic bird'. Another unidentified bird from Greek mythology – this time named for Aisakos (Aesacus), son of Priam (of Trojan war fame), who took a fancy to the lovely Hesperia. She was not so keen, and when he was chasing her through the woods she was bitten by a snake and died. In his grief and guilt over her death, Aesacus threw himself off a cliff but was saved by Thetis, a sea nymph, and turned into a bird. In Ovid's (c. 8 AD) version of this in the *Metamorphoses*, he insisted that Aesacus was transformed into a diving bird and repeatedly tried to seek death by diving into the depths of the sea. This is not an entirely accurate description of this genus' usual behaviour.

The species

Bush Stone-curlew (breeding resident)

In Australia the two species of stone-curlew segregate themselves neatly by habitat; this species is primarily an inland bird of drier habitats.

Other names: Curlew (a widespread folk name) or Land or Scrub Curlew; Stone or Southern Stone Plover (the name used by Gould in 1848, after Latham's designation of it as a true plover); High-legged, Brown, Grizled [sic] or Bridled Plover, all used by Latham (1801, 1823) for its very long legs and dark face stripe; Willaroo or Weeloo, delightfully onomatopoeic names and likely to be influenced by Indigenous words; Bush Thick-knee, the name used by HANZAB (1990– 2006, vol. 2) as recently as 1993.

Burhinus grallarius (Latham, 1801) [boo-REE-nʊs gral-li-NAH-ri-ʊs]: 'bull-nose on stilts,' see genus name, and from Latin *grallarius* from *grallae*, stilts. If anything, rather shorter legged than many other shorebirds. Its gait varies from a very slow stalk (perhaps where the stilts come in) to twinkling along at high speed, especially at dusk.

Beach Stone-curlew (breeding resident)

Exclusively a coastal species; the name was formalised by RAOU (1926).

Other names: Great-billed Plover (Latham 1801) or Large-billed Plover (Gould 1848), both from the species name; Reef Thick-knee; Long-billed Stone Plover, Large-billed Shore Plover, Australian Long-billed Plover, Long-billed Stone-Curlew, recommended by RAOU (1913); Beach Curlew (!).

Esacus magnirostris (Vieillot, 1818) [EH-sa-kʊs mag-ni-ROS-tris]: 'large-billed aquatic bird', see genus name, and from Latin *magnus*, large, and *rostrum*, bill – this bird certainly has an impressive bill, much larger than that of its inland cousin.

CHIONIDAE (CHARADRIIFORMES): sheathbills

Chionidae Lesson, 1828 [kee-O-ni-dee]: the Snowbird family, see genus name *Chionis*.

Sheathbills comprise just two species of Antarctic and sub-Antarctic birds, characterised by a tough sheath which partially covers the upper bills.

The genus

Chionis Forster JR, 1788 [kee-OH-nis]: 'snow bird', from Greek *khiōn*, snow, for its habitat.

The species

Black-faced Sheathbill (Heard Island)

The only sheathbill with a black face.

Other names: Lesser Sheathbill, from the scientific name, but misleading – although this bird is slightly less massive than Pale-faced Sheathbill (*C. alba*) the dimensions of the two species are almost identical; Kerguelen Sheathbill, for the site of the largest breeding population; Paddy, reputedly for its thieving habits – perhaps a name that could be left behind!

Chionis minor Hartlaub, 1841 [kee-OH-nis MEE-nor]: 'lesser snowbird', see genus name, and from Latin *minor*, comparative of *parvus*, small, so meaning smaller or lesser.

HAEMATOPODIDAE (CHARADRIIFORMES): oystercatchers

Haematopodidae Bonaparte, 1838 [hee-ma-to-PO-di-dee]: the Blood-red Foot family, see genus name *Haematopus*.

The name oystercatcher (originally oyster catcher) arose in North America in the early 18th century (where it wasn't particularly applicable) and transferred to Britain very soon afterwards (where it was even less applicable), supplanting old names including Sea Pie. Australian oystercatchers are very closely related to the Eurasian Oystercatcher *H. ostralegus*.

The genus

Haematopus Linnaeus, 1758 [hee-MA-to-pʊs]: 'blood-red foot', from Greek *haima*, blood, and *pous/pod*, foot. 'Blood-red' seems to be going a bit far for the pinkish-red of the legs of all these species.

The species

South Island Oystercatcher (vagrant)

Native to the South Island of New Zealand.

Haematopus finschi Martens GH, 1897 [hee-MA-to-pʊs FIN-shee]: 'Finsch's blood-red foot', see genus name, and for Friedrich Finsch (1839–1917), a Prussian ornithologist who worked at the Dutch National Museum in Leyden then became curator of the Bremen Museum. He led a major expedition to Siberia, then eventually resigned to travel through the Pacific, including Australia and New Zealand, where it is likely that he collected the type specimen that Gustav Martens later described. (Finsch was later sent back to the New Guinea area as Imperial Commissioner and was instrumental in north-eastern New Guinea becoming a German protectorate, 'Kaiser Wilhelm's Land'.)

Pied Oystercatcher (breeding resident)

Pied appears commonly in black-and-white bird names, referring ultimately to the European Magpie. Until recently Australian preceded it, but although there are other pied oystercatchers there are no other Pied Oystercatchers.

Other names: White-breasted Oystercatcher, used by Gould (1848); Oystercatcher or Black-and-white Oystercatcher; Seapie, see family introduction; Redbill or Red-bill, for perhaps its most striking feature.

Haematopus longirostris Vieillot, 1817 [hee-MA-to-pʊs lon-gi-ROS-tris]: 'long-billed blood-red foot', see genus name, and from Latin *longus*, long, and *rostrum*, bill.

Sooty Oystercatcher (breeding resident)

The only black Australian oystercatcher, and the name used by Gould (1848) from his species name; it was formalised by RAOU (1926).

Other names: Black Oystercatcher ('Colonists of New South Wales, Van Diemen's Land and Port Essington', Gould 1848), still used by RAOU (1913); Redbill or Black Redbill ('Colonists of Western Australia', Gould 1848).

Haematopus fuliginosus Gould, 1845 [hee-MA-to-pʊs fʊ-li-gi-NOH-sʊs]: 'sooty blood-red foot', see genus name, and from Latin *fuliginosus*, sooty, from *fuligo*, soot. Sooty indeed, and making a very handsome contrast with the bill and legs.

RECURVIROSTRIDAE (CHARADRIIFORMES): stilts and avocets

Recurvirostridae Bonaparte, 1831 [re-coor-vi-ROS-tri-dee]: the Curved-back Bill family, see genus name *Recurvirostra*.

'Stilt' is an obvious reference to the fact that these birds have the longest legs, for their body size, of any waders. Note that the genus *Stiltia*, based on the same word, is reserved for phalaropes.

'Avocet', for the Pied Avocet *Recurvirostra avosetta* of Europe, made an unlikely appearance in England in the late 17th century, having been consciously adopted from the Italian *avosetta* by influential England ornithologists to replace a range of traditional English names. The Italian etymology is unknown. Linnaeus sealed it in 1758 when he applied it as a species name.

The genera

Himantopus Brisson, 1760 [hi-MAN-to-pʊs]: 'spindle-shank'. The Greek prefix *himanto-* refers to leather thongs, from this, with Greek *pous*, foot, came a word to describe spindly legs – slender and pliant as a strip of leather. From there it was applied (though only attested once) to a spindly legged waterbird. Pliny (77–79 AD) also included it (Latin *himantopus*).

Cladorhynchus Gray GR, 1840 [kla-do-RIN-kʊs]: 'twig-bill', from Greek *klados*, a branch or shoot of a tree, and *rhunkhos*, bill.

Recurvirostra Linnaeus, 1758 [re-coor-vi-ROS-truh]: 'curved-back bill', from Latin *recurvus*, bent backwards, and *rostrum*, bill.

The species

Pied Stilt (breeding resident)

Until recently regarded – albeit uneasily – as part of a worldwide species (*Himantopus himantopus*), and indeed some still hold that view. Now that species has been divided into four, with Pied Stilt endemic to Australia.

Other names: Black-winged Stilt, recently used, though it doesn't help distinguish it from the next species; White-headed Stilt, because most other species have some black on the face – this was in Gould's mind when he named it *H. leucocephalus*; Long-legged Plover, Longshanks, Stilt-bird, all for the very bizarre-looking legs; Dog-bird, for the very yappy calls at nesting colonies.

Himantopus leucocephalus (Linnaeus, 1758) [hi-MAN-to-pʊs loo-ko-SEF-a-lʊs]: 'white-headed spindle-shank', see genus name, and from Greek *leukos*, white, and *kephalē*, head.

Banded Stilt (breeding resident)

For the distinctive chestnut breastband; the name was already used by Gould in 1848, based on the then name, *C. pectoralis* (a non-specific reference to the breast).

Other names: Rottnest Snipe, for Rottnest Island off Perth, where it is common – Snipe is just a confusion with another long-billed wader; Bishop Snipe, a mystery in that while it is widely cited we can find no text that actually uses it (not a valid test of a folk name, of course).

Cladorhynchus leucocephalus (Vieillot, 1816) [kla-do-RIN-kʊs loo-ko-SE-fa-lʊs]: 'white-headed twig-bill', see genus name, and from Greek *leukos*, white, and *kephalē*, head. All true, but what about that gorgeous chestnut band? Vieillot (1816) didn't mention it; his description stated that the bird is 'completely white, with the exception of the wings, which are black'. We assume he was looking at a bird in non-breeding plumage, without even the faded mottled band they often have, or a juvenile bird, in which the breastband had not yet developed.

Red-necked Avocet (breeding resident)

The only avocet with rusty head and neck, but the name was based on a former species name, *R. rubricollis*.

Other names: Avocet; Cobbler, Cobbler's Awl, for the awl-shaped bill; Scooper, for the feeding habit; Trumpeter, Yelper, for the trumpety call (all these names are old English ones); Painted Lady, obscure and little used, though Leach (1911) recorded it.

Recurvirostra novaehollandiae Vieillot, 1816 [re-coor-vi-ROS-truh no-veh-hol-LAN-di-eh]: 'New Holland backward-curved bill', see genus name, and from Modern Latin *Nova Hollandia*, New Holland, the old name for Australia.

CHARADRIIDAE (CHARADRIIFORMES): plovers and dotterels

Charadriidae Leach, 1820 [ka-ra-DREE-i-dee]: the Plover (or is it?) family, see genus name Charadrius.

The origin of 'plover' is disputed. Dictionaries such as the Oxford and the Macquarie attest that it has its origin in Latin *pluvia*, for rain, via *pluviarius* and Old French *plouvier* (c. 1150), the birds being known as 'rain birds' in French and early English. The form *pluuer* is known from the 14th century, with 'plover' following soon afterwards. Lockwood (1984) disputed this, dismissing it as folk etymology and maintaining that it is of onomatopoeic origin; part of his reason seemed to be that no link with rain is known, but we would note that species in other families – including in Australia – are known as rainbird or stormbird with no evident reason.

Again, there are conflicting views on the origin of 'dotterel'. Dictionaries maintain that is based on the old 'dote', a simpleton (dotage, dotty), plus the originally French suffix '-rel' or '-erel', which is a diminutive implying some contempt. The basis of this is an observation (or belief) that the bird, the Eurasian Dotterel *Charadrius morinellus*, was easily approached and captured. Perhaps this referred to nesting birds; in Britain nesting is now largely restricted to high Scottish mountains but it was probably formerly much more widespread. There is also an old verb dotter, a form of totter, and we wonder if there was an association with the truly pathetic staggering distraction displays near the nest. Lockwood insisted that this is all coincidental and that the name was again onomatopoeic, from 'dot' and the '-erel' suffix. 'Dottrel' also appears from time to time, and was the recommended form by RAOU (1913).

The term dotterel has traditionally been used for smaller members of the family, but the current trend is to use plover widely and restrict dotterel to a few (arbitrarily) selected species.

'Lapwing' is known from at least the 11th century in England, in the form *hleapewince*, from two Old English verbs meaning to leap, and to totter or waver (according to the *Oxford Dictionary*), and referred to the Northern Lapwing, *Vanellus vanellus*. The Oxford asserted that this was from the flight, but we wonder if it might not instead have been a reference to the long crest, waving in the breeze? Lockwood (1984) also believed this; he maintained the word is some 300 years older at least and that the components are a noun (for the bird) and a word which also became 'winch', and meant to raise and lower. Either way it became corrupted over time as the original meanings were lost, and made more 'familiar', hence lapwing. In recent times the name has been used for the two Australian species of *Vanellus* (both of which are crestless, as are virtually all of the world's 24 species); previously they were just 'plover', as indeed they still are to many people.

Many of the Australian species in this family are non-breeding migrants from the Northern Hemisphere, and were known and named before European occupation. A spin-off is that some of the characteristics for which the birds were named are not seen in Australia, when they are in eclipse plumage.

The genera

Vanellus Brisson, 1760 [va-NEL-lʊs]: 'lapwing', from mediaeval Latin *vanellus* (used by Aldrovandus 1600 and Willughby and Ray 1676; cf. French *vanneau* used by Belon 1555), a diminutive of Latin *vannus*, a winnowing-fan, said to refer to the slow and rather floppy flight action of the northern original, *Vanellus vanellus*.

Erythrogonys Gould, 1838 [e-rith-ro-GO-nis]: 'red-knee', from Greek *eruthros*, red, and *gonu*, knee. In his description, Gould actually stated 'thighs and tarsi purplish-red; knee purplish-grey'.

Peltohyas Sharpe, 1896 [pel-to-HY-as]: 'shield plover', from Greek *peltē*, a small, light shield, and the old genus *Hyas* (Gloger, 1827), the Egyptian Plover. The shield is the mark described by Sharpe as 'round the hind neck a black collar descending onto the chest and skirting the foreneck in the shape of a triangle'. Gloger (1827) was working from Buffon's *pluvianus*, though he dismissed it as just a French word with a Latin ending. Seeking a Greek word, he wrote that it simply required a translation into Greek, and chose *hyas*, expressing rain he said, as did *pluvianus*. Hyas of Greek mythology (possibly a later invention) was an archer whose sisters (or perhaps daughters) wept so much at his untimely death that they were changed into rain-nymphs providing moisture to the earth (the Hyades). Zeus placed them among the stars, where they remain in the constellation Taurus and herald rainy seasons.

Pluvialis Brisson, 1760 [ploo-vi-AH-lis]: 'rain-bringer', from Latin *pluvialis*, rain-bearing or belonging to rain. See family discussion.

Charadrius Linnaeus, 1758 [ka-RA-dri-ʊs]: 'we'll call it a plover', from Greek *kharadrios*, probably a stone-curlew, but often translated as 'plover'. This mysterious nocturnal bird may have lived in mountain gullies (*kharadrē* being the word for such a gully), building its nest in ravines and on cliffs, according to Aristotle (c. 330 BC), who considered it pretty ordinary in terms of plumage and voice. It is sometimes described as a yellowish bird, which may be why looking it in the eye was held to be a cure for jaundice (Arnott 2007) (presumably on the basis of like curing like). It was also 'known' to be a glutton (Liddell and Scott 1940). So we have a greedy, yellow, gully-dwelling bird with medicinal properties – stone-curlew or plover? Anybody want to take a guess?

Thinornis Gray GR, 1844 [thi-NOR-nis]: 'beach bird', from Greek *this*, beach or sandbank, *ornis*, bird.

Elseyornis Mathews, 1914 [el-see-OR-nis]: 'Elsey's bird', from the older genus name *Elseya*. This was named for Joseph Ravenscroft Elsey (1834–58), English surgeon and naturalist on a north Australian exploring expedition in 1855–56. He obtained at least three new species of birds: the Purple-crowned Fairy-wren, the Buff-sided Robin and the Golden-shouldered Parrot, all given their scientific names by Gould, who expressed appreciation for the detailed information supplied by Elsey. Elsey sounds like a disarming young man – he wrote to his parents, 'You cannot imagine what delight my work as a naturalist affords me. Not a day passes but some wonder or novelty shows itself' (Chisholm 1972).

The species

Grey-headed Lapwing (vagrant)

The lovely grey head and neck are unique within its Asian range, but not unique in *Vanellus*.

Vanellus cinereus (Blyth, 1842) [va-NEL-lʊs si-NE-re-ʊs]: 'ashy-grey lapwing', see genus name, plus Latin *cinereus*, like ashes.

Banded Lapwing (breeding resident)

See family introduction for lapwing. Banded for the broad black breastband, the name introduced by RAOU (1926).

Other names: Black-breasted Pewit, Gould's name (1848) – pewit (or more commonly peewit) was and is a familiar term in Britain for the Lapwing, based on the call; Banded, Black-breasted Plover, the latter used by RAOU (1913); Brown Plover, a 'yes, but …' name; Plain Plover for its inland habitat; Flock Plover for its gregarious habits.

Vanellus tricolor (Vieillot, 1818) [va-NEL-lʊs TRI-ko-lor]: 'three-coloured lapwing', see genus name. As for the specific name, the three colours of the Latin word *tricolor* were said by Vieillot (1816–19) to be black, white and grey. Adding red for the wattles above the bill might make it easier to remember, as might mentioning that when spread in flight the wings are clearly three-coloured, unlike the Masked Lapwing's, which are two-coloured.

Masked Lapwing (breeding resident)

For the huge yellow face wattle. Note that previously the southern populations, which have much less conspicuous wattles, were regarded as a separate species, Spur-winged Plover.

Other names: Wattled Pewit and Masked Pewit used by Gould (1848) for the southern and northern races, respectively (see previous species for 'pewit/peewit'); Wattled Sandpiper (Latham 1801) and Wattled Lapwing (Jardine and Selby 1826–35, vol. 2); Masked or Wattled Plover; Spur-wing, Spur-winged, Australian Spur-winged Plover, for the sharp bony spur on each leading wing-edge in southern birds only. The 'Australian' is in acknowledgment of the African Spur-winged Lapwing (*V. spinosus*); Alarm-bird or Alarmbird, for its frenetically loud and active behaviour when nesting – Gould (1848) reported that it was a name 'of the Colonists'.

Vanellus miles (Boddaert, 1783) [va-NEL-lʊs MEE-lehz]: 'soldier lapwing', see genus name, plus Latin *miles*, soldier, because the bird is armed with its spurs – though the fact that it stands tall and is known to sometimes attack may also evoke a certain image.

Red-kneed Dotterel (breeding resident)

Undeniably pinkish-red 'knees', as well as the rest of the upper legs (though the knees are really ankles, anatomically speaking), above serviceable brown lower legs.

Other names: Banded Red-knee, Gould's name (1848) straight from his species name.

Erythrogonys cinctus Gould, 1838 [e-rith-ro-GO-nis SINK-tʊs]: 'belted red-knee', see genus name, and from Latin *cinctus*, a girding (for a toga) or belt, referring to the deep breastband.

Inland Dotterel *Peltohyas australis* and Red-kneed Dotterel *Erythrogonys cinctus*

Inland Dotterel (breeding resident)

A bird of the dry inland, though in the south and west of its range it follows the deserts to the sea.

Other names: Australian Dotterel; Desert Plover, Inland Plover.

Peltohyas australis (Gould, 1841) [pel-to-HY-as ost-RAH-lis]: 'southern shield plover', see genus name, plus Latin *australis*, southern. The 'shield' is described by Gould as the 'crescent-shaped mark of blackish brown' below the throat.

Pacific Golden Plover (non-breeding migrant)

The golden plovers have beautifully gold-spangled upperparts in breeding plumage; this species breeds in Siberia and Alaska and migrates south through the Pacific.

Other names: This species and the next were regarded as one until the early 1980s (as Eastern Golden Plover, *P. dominicus*), so common names in use until that time apply to both. Australian Golden Plover, used by Gould (1848); Lesser or Least Golden Plover (it is the smallest of the three Golden Plover species); Asiatic Golden Plover.

Pluvialis fulva (Gmelin JF, 1789) [ploo-vi-AH-lis FƱL-vuh]: 'tawny rain-bringer', see genus name, and from Latin *fulvus*, tawny, deep or reddish-yellow. See description under common name.

American Golden Plover (vagrant)

Normally restricted to the Americas, breeding in the far north and wintering in Patagonia.

Other names: See Pacific Golden Plover.

Pluvialis dominicus (Müller PLS, 1776) [ploo-vi-AH-lis do-MI-ni-kƱs]: 'Dominican rain-bringer', see genus name, plus Modern Latin *dominicus* for Santo Domingo (Müller 1776). The Müller in question is Philipp Ludwig Statius Müller (1725–76), German zoologist (not Salomon Müller, 1804–64, German naturalist).

Grey Plover (non-breeding migrant)

Closely related to the Golden Plovers, but lacks the golden spangles.

Other names: Black-bellied Plover, only in breeding birds; Grey Sandpiper; Maycock, probably a reference to its time of return to breeding grounds; Swiss Sandpiper, from Linnaeus' other name *Tringa helvetica*, from a Swiss specimen which he evidently thought was of a different species.

Pluvialis squatarola (Linnaeus, 1758) [ploo-vi-AH-lis sk(w)a-ta-ROH-luh]: 'plover-like rain-bringer', see genus name, plus from *sgatarola*, apparently a Venetian word meaning a kind of plover – Willughby in his *Ornithologia* (1676) explained helpfully that this bird is common in Venice.

Common Ringed Plover (vagrant)

For the nifty black breeding bib, which has shrunk and faded to brown by the time it reaches Australia.

Other names: Ringed Dotterel; Ringed Plover.

Charadrius hiaticula Linnaeus, 1758 [ka-RA-dri-Ʊs hi-a-TI-kƱ-luh]: 'cleft-dwelling plover', see genus name, and from Latin *hiatus*, a cleft, and *-cula* a dwelling (cf. *-cola*, dweller).

Semipalmated Plover (vagrant)

See family introduction, and descriptive of the feet, from the species name.

Charadrius semipalmatus Bonaparte, 1825 [ka-RA-dri-Ʊs se-mi-pal-MAH-tƱs]: 'half-webbed plover', see genus name, plus Latin *semi*, half *palma*, palm, and *palmatus*, webbed.

Little Ringed Plover (vagrant, or perhaps a rare migrant)

A very similar bird to the Common Ringed Plover, but noticeably smaller.

Charadrius dubius Scopoli, 1786 [ka-RA-dri-Ʊs DOO-bi-Ʊs]: 'doubtful plover', see genus name, plus Latin *dubius*, usually indicating some doubt about the taxonomy.

Kentish Plover (vagrant)

A very parochial name; this was the stronghold of the English breeding population from where the species was recognised and described. Sadly, the recognition brought elimination from there.

Other names: Snowy Plover, for the much whiter American subspecies.

Charadrius alexandrinus Linnaeus, 1758 [ka-RA-dri-ʊs a-lek-san-DREE-nʊs]: 'Alexandrian plover', see genus name, plus Latin *Alexandrinus*, of Alexandria in Egypt. Gives the lie to the parochial common name, but Egypt is one of the Kentish plover's many and widespread breeding grounds, which cover almost every continent!

Red-capped Plover (breeding resident)

Distinctive and unique among common Australian plovers.

Other names: Red-capped Dotterel; Red-necked Plover; Sand-lark, for its familiarity and active behaviour (used by the 'Colonists of Swan River', Gould 1848); Kentish Plover, for its similarity to the previous closely related species.

Charadrius ruficapillus Temminck, 1822 [ka-RA-dri-ʊs roo-fi-ca-PIL-lʊs]: 'red-haired plover', see genus name, plus Latin *rufus*, red or reddish, and *capillus*, hair on the head (remember capillary, a hair-like blood vessel).

Double-banded Plover (non-breeding trans-Tasman migrant)

For beautiful broad dark chocolate and chestnut breastbands when breeding in New Zealand; both are much faded and smaller in Australia.

Other names: Double-banded Dotterel; Banded Dotterel; Chestnut-breasted Dotterel.

Charadrius bicinctus Jardine & Selby, 1826 [ka-RA-dri-ʊs bi-SINK-tʊs]: 'plover with two belts', see genus name, plus Latin *bi*, two, and *cinctus*, a girding (for a toga) or belt, see common name description.

Lesser Sand Plover (non-breeding migrant)

Sand Plover sounds a bit of a tautology but the reference is to the desert breeding grounds in central Asia; Lesser is by comparison with the following very similar and closely related species.

Other names: Mongolian Sand-dotterel or Sand-plover, for the species name and its Mongolian breeding grounds; Mongolian Dotterel or Plover; Allied Dotterel (or Dottrel), by Gould for his belief that it was 'nearly allied to the *Hiaticula wilsonii* [Wilson's Plover, *Charadrius wilsonia*] of North America'.

Charadrius mongolus Pallas, 1776 [ka-RA-dri-ʊs mon-GO-lʊs]: 'Mongolian plover', see genus and common names.

Greater Sand Plover (non-breeding migrant)

See previous species.

Other names: Large, Great or Large-billed Sand-dotterel or Dotterel; Large Sand Plover.

Charadrius leschenaultii Lesson, 1826 [ka-RA-dri-ʊs le-shuh-NOH-ti-ee]: 'Leschenault's plover', see genus name, and for Jean Baptiste Louis Claude Théodore Leschenault de la Tour (1773– 1826), French botanist on the 1801–03 voyage of the *Géographe* and *Naturaliste* with Baudin. René Lesson was an enormously influential naturalist, but was 20 years Leschenault's junior and doubtless looked up to him. There was a more direct connection – Leschenault had supplied Lesson's father-in-law, lawyer and ornithologist Charles Henri Frédéric Dumont de Sainte-Croix, with a series of bird type specimens from Java, which Sainte-Croix named, so it is highly likely that Leschenault and Lesson knew each other socially. We know that Leschenault brought the specimen from Pondicherry in southern India, and it is quite possible that he gave it directly to Lesson. It is probably no coincidence that Lesson named this plover in the year that Leschenault died.

Caspian Plover (vagrant)

Breeds around the Caspian Sea.

Charadrius asiaticus Pallas, 1773 [ka-RA-dri-ʊs a-si-A-ti-kʊs]: 'Asian plover', see genus name, and Latin for Asiatic. This species name very commonly refers to India, but in this case goes a little further afield (see common name).

Oriental Plover (non-breeding migrant)

Breeds in central east Asia.

Other names: Oriental, Eastern or Asiatic Dotterel; Eastern Sandplover; Brown Plover used, not very helpfully, by Gould (1848).

Charadrius veredus Gould, 1848 [ka-RA-dri-ʊs ve-REH-dʊs]: 'not so much a plover, more a swift horse', see genus name, and from Latin *veredus*, a swift horse, referring to the bird's fast running speed – well, maybe it's not quite **that** fast.

Hooded Plover (breeding resident)

For the full black hood through which red-rimmed eyes peep out.

Other names: Hooded Dotterel.

Thinornis cucullatus (Vieillot, 1818) [thi-NOR-nis kʊ-koo-LA-tʊs]: 'hooded beach bird', see genus name, and from Latin *cucullatus*, hooded, from *cucullus*, a hood or cowl.

Black-fronted Dotterel (breeding resident)

Apt, remembering that a bird's front (or frons) is its forehead.

Other names: Guttersnipe, a colourful name (not in common usage!) which did not arise in Australia, being found in English dialect for the Common Snipe *Gallinago gallinago*. The mystery of how it became associated with the entirely dissimilar and unrelated Australian dotterel is beyond our power to explain.

Elseyornis melanops (Vieillot, 1818) [el-see-OR-nis MEL-a-nohps]: 'Elsey's black-faced bird', see genus name, and from Greek *melas*, black, and ōps, the face (or eye).

ROSTRATULIDAE (CHARADRIIFORMES): painted-snipes

Rostratulidae Mathews, 1913 [ros-tra-TOO-li-dee]: the Little Curved Bill family, see genus name *Rostratula*.

Although the term 'painted snipe' had been around since the early 19th century (Williamson 1809), Gould didn't use it (he called the Australian species the Australian Rhynchaea). The first unequivocal published use of it for this genus seems to have been in Englishman Alfred Newton's *Dictionary of Birds* (1896), though his account suggested the name was relatively familiar by then (he referred to 'so-called Painted Snipes'). The 'painted' is doubtless for the relative colourfulness of the female in particular, compared with true snipe. Oddly, given the IOC's stated antipathy to hyphenated names (see Introduction), they have opted to hyphenate this one, which in Australia at least has generally been rendered as two words.

The genus

Rostratula Vieillot, 1816 [ros-TRA-tʊ-luh]: 'little curved bill' from Latin *rostrum*, a bill (often curved or hooked), and *rostratus*, billed or beaked (also applied to ships – those with a projection from the prow, handy for ramming the enemy). Birds in this family have long curved bills (with a 'swollen' tip), by contrast with the Australian *Gallinago* snipes, whose bills are straight and extremely long. The diminutive *-ulus* is something of a puzzle (the only clue Vieillot (1816–19) gave is that the bill is 'longer than the head' and 'thin') but it does remind us that these birds' bills are shorter in proportion to body-length than the positively unwieldy-looking ones of the true snipes.

The species

Australian Painted-snipe (breeding resident)

The Australian population has recently been recognised as separate from the African and Asian Greater Painted Snipe *R. benghalensis*.

Other names: Australian Painted Snipe, the more usual form in Australia; Australian Rhynchaea, the old name used by Gould, for the original genus name; Painted Snipe, from when it was regarded as part of a very widespread species; Greater Painted Snipe, for no discernible reason, though the Australian birds are reported to be longer winged.

Rostratula australis (Gould, 1838) [ros-TRA-tʊ-luh ost-RAH-lis]: 'southern little-curved-bill', see genus name, and from Latin *australis*, southern.

JACANIDAE (CHARADRIIFORMES): jacanas

Jacanidae Chenu & Des Murs, 1854 [ja-KA-ni-dee]: the Jacana family.

'Jacana' comes from a Portuguese word *jaçana*, in turn from a Tupi-Guaraní word from Brazil, variously reported as *iacana*, *jacana* or *nahana*. This was the Wattled Jacana *Jacana jacana*, the only South

American member of the group. The genus *Jacana* was named in 1760 by French zoologist Mathurin Brisson, not for the Wattled Jacana but the related Northern Jacana *J. spinosa* from Central America. The apparent anomaly is explained by the fact that 'jacana' had been known in English (for the southern species) for at least a decade before that.

The genera

Irediparra Mathews, 1911 [ie-er-di-PA-ruh]: 'Iredale's Parra' (or 'Iredale's bird of ill omen') for Tom Iredale, an English ornithologist who came to Australia as Gregory Mathews' amanuensis in 1923 and became a most significant ornithologist, conchologist and zoological polymath in his own right, working at the Australian Museum until his death in 1972. *Parra* was the genus name by which it was known into the 20th century (RAOU 1913); from Latin *parra*, a bird variously believed to have been a Barn Owl, a Green Woodpecker or possibly a Lapwing. It was apparently considered a bird to be watched for signs of what fate might have in store.

Hydrophasianus Wagler, 1832 [hi-dro-fa-zi-AH-nʊs]: 'water-pheasant', from Greek *hudōr/hudro*, water, and *phasianos*, pheasant.

The species

Comb-crested Jacana (breeding resident)

For the distinctive red comb-like forehead crest.

Other names: Jacana; Lotus Bird, for the lily pond habitat; Lily-trotter, Skipper, Jesus-bird, all for the typical – though remarkable – jacana habit of walking on lily pads (not on water!) with the extraordinarily long toes; Comb-crested Parra, for the old genus name; Water Pheasant (?); Crested Snipe, because the people who named the birds weren't ornithologists!; Gallinaceous Parra, used by Gould (1848), directly from the then scientific name *Parra gallinacea*.

Irediparra gallinacea (Temminck, 1828) [ie-er-di-PA-ruh gal-li-NAH-se-uh]: 'Iredale's hen-like bird of ill omen', see genus name, and from Latin *gallina*, a domestic hen.

Pheasant-tailed Jacana (vagrant)

A bit of hyperbole, but the breeding adult **is** the only jacana with a long tail.

Hydrophasianus chirurgus (Scopoli, 1786) [hi-dro-fa-zi-AH-nʊs ki-ROOR-gʊs]: 'surgeon water-pheasant', see genus name, and from Latin *chirurgus*, surgeon (cf. Greek *kheirurgia*, handicraft or surgery, from *kheir*, hand), for the sharp carpal spurs (wing-spikes), which supposedly look like a scalpel.

PEDIONOMIDAE (CHARADRIIFORMES): Plains-wanderer

Pedionomidae Bonaparte, 1856 [pe-di-o-NO-mi-dee] the Plains-dwelling family, see genus name Pedionomus.

The genus

Pedionomus Gould, 1840 [pe-di-O-no-mʊs]: 'plains-dweller', from Greek *pedionomos*, dwelling in the plains (or as HANZAB 1990–2006, vol. 2 more poetically put it, 'haunting the fields'), from *pedion*, a plain, and *nomos*, habitation (also a pasture, so could almost do double duty).

The species

Plains-wanderer (breeding resident)

A name used by Gould (1848), who commented on its strong legs and referred to it as a 'little Wanderer'. He may well have coined the word, because Charles Sturt (1849) wrote that Gould 'first discovered [it] on the plains of Adelaide'. (Sturt dismissed it as a 'stupid little bird' for allowing his dog to catch it.)

Other names: Plain Wanderer or Collared Plain Wanderer, the term used by Gould, for the female's broad checked collar; Turkey Quail, perhaps from Gould's comparison with a 'diminutive Bustard'; Plain Wanderer Quail (!).

Pedionomus torquatus Gould, 1840 [pe-di-O-no-mʊs tor-KWAH-tʊs]: 'plains-dweller with a necklace', see genus name, plus Latin *torquatus* from *torquis*, a twisted neck-chain or collar. All true, and interesting that in this species the female is the more ornamented bird and therefore the one for which the name was chosen.

Plains-wanderer *Pedionomus torquatus* and Stubble Quail *Coturnix pectoralis*

SCOLOPACIDAE (CHARADRIIFORMES): 'waders'– snipes, sandpipers, curlews, etc.

Scolopacidae Rafinesque, 1815 [sko-lo-PA-si-dee]: 'the Snipe/Woodcock family', from Latin *scolopax*, snipe or woodcock (cf. Greek *askolōpas*, woodcock).

The comments relating to plovers and dotterels, family Charadriidae, apply even more strongly here. Every species of 'wader', used strictly in this sense, which has been recorded in Australia is a non-breeding summer migrant and, other than a few tinges in birds newly arrived or dressing up for departure, all are in their drab non-breeding plumage. All cats may be grey in the dark, but all scolopacids are indubitably grey in Australia! It is interesting how many of the names are, or are alleged to be, of onomatopoeic origin; most of us wouldn't primarily identify waders thus, but perhaps the names reflect a time when birds were commoner and the world was quieter.

Most, but not all, of the birds were already known and described, because birds breeding in the far Northern Hemisphere tend, very sensibly, to fly due south on migration, so in general most overwintering waders in Australia are more likely to have flown from Siberia rather than from Europe – some of these represent different subspecies or even full species from the European ones the early Australian ornithologists were used to.

'Snipe' appears in English (as 'snype') from the 14th century, deriving, according to Lockwood (1984) from an older word implying long and thin – like a snipe's bill, in fact. The rather deliciously Potteresque 'snite' is also recorded. The *Oxford Dictionary* felt unable to reach a conclusion.

'Godwit' is also of uncertain origin; Lockwood (1984) asserted that it is onomatopoeic in origin but the *Oxford Dictionary*, more circumspectly, simply stated 'origin obscure'. It seems to have appeared in English in the 16th century.

'Curlew' seems to have arrived in England from French in the 14th century, as *corlue* or *curlu*. The evocation of the calls seems evident.

'Sandpiper' is as straightforward as it sounds – someone who pipes on the sand – though it didn't enter the literature until the 17th century.

'Tattler' is one who tattles or chatters, yet another name for a perceived call, but it does not appear in print until the 19th century. It seems to have arisen in North America, appearing first as 'tatler' in the second volume of John Richardson, William Kirby and William Swainson's *Fauna Boreali-Americana* in 1831. It didn't gain any traction in Australia until RAOU (1926).

'Knot' is regarded by the *Oxford Dictionary* (with its commendable but sometimes frustrating caution) as of 'obscure origin', but Lockwood (1984) again insisted that it is onomatopoeic; given the distinctive grunting notes, this is a persuasive proposal. The word appears on a 15th-century Oxford University menu.

'Stint' is yet again regarded as obscure in origin by the Oxford, but Lockwood averred fairly convincingly that it is of the same root as the verb, implying a small amount, logical given the tiny nature of the birds relative to other shorebirds. It is known, in the form 'stynte', from the 15th century.

'Dowitcher' seems to have entered English in North America in the early decades of the 19th century, from the Iroquois language, presumably that people's name for the bird.

'Phalarope' is directly from the genus name; the birds were so rare in Britain that no popular name had evolved.

The genera

Bartramia Lesson, 1831 [BAR-tra-mi-uh]: 'Bartram's bird', after William Bartram (1739–1823), who was a US all-round naturalist, professional botanist and ornithologist, dubbed the 'grandfather of American ornithology'. The type specimen was shot more or less outside his back door and was originally named *Tringa longicauda* in 1812 (while Bartram was still alive), but after his death the French ornithologist René Lesson erected this mono-specific genus.

Numenius Brisson, 1760 [noo-MEH-ni-ʊs]: 'new-moon birds', from Greek *noumēnios* (*neomēnios*), from *neos*, new, and *mēnē*, moon, the crescent moon indicating a curved bill. Hesychius is the only definite user of the word and Arnott (2007) suggested that it could have referred to either the Eurasian Curlew *Numenius curvata* or the Pied Avocet *Recurvirostra avosetta*, presumably depending which way you're looking at the moon!

Limosa Brisson, 1760 [li-MOH-suh]: 'muddy birds', from Latin *limosus*, muddy or slimy, for their preferred marshy or tidal habitat.

Arenaria Brisson, 1760 [a-re-NAH-ri-uh]: 'sand-bird', from Latin *arenarius*, to do with sand (*harena* or *arena*), for the bird's habitat by seashores.

Calidris Merrem, 1804 [ka-LEE-dris]: 'speckled waterbird', from Greek *skalidris*. Aristotle (c. 330 BC) described this bird as having ashen-grey, speckled plumage (not all that helpful). It is translated as 'probably redshank' or sometimes 'sandpiper', but Arnott (2007) pointed out that 11 local birds fitted that description! *Calidris* was originally the specific name of the Red Knot *Tringa calidris*, now *Calidris canutus*.

Limnodromus zu Wied-Neuwied, 1833 [lim-no-DROH-mʊs]: 'marsh-racer', from Greek *limnē*, a marshy lake for its habitat, and *dromos*, a race or race-course.

Gallinago Brisson, 1760 [gal-li-NAH-go]: 'hen-like bird', from Modern Latin *gallinago*, snipe or woodcock, from Latin *gallina*, hen, and *-agus*, resembling.

Xenus Kaup, 1829 [ZEH-nʊs]: 'stranger', from Greek *xenos*, a stranger, foreigner or guest. Like a migrant bird, for example.

Phalaropus Brisson, 1760 [fa-LA-ro-pʊs]: 'coot-footed', from Greek *phalaris*, coot, and *pous*, foot. Coots' feet, like those of phalaropes, are lobed rather than webbed.

Actitis Illiger, 1811 [ak-TEE-tis]: 'coast-dweller', from Greek *aktitēs* (from *aktē*, headland, high rugged sea-coast).

Tringa Linnaeus, 1758 [TRIN-guh]: 'waterbird', from Greek *trungas*, a wader described by Aristotle as the size of a thrush, translated as 'white-rump' and 'wagtail'. This is thought to have indicated a sandpiper, though later in the same paragraph Aristotle also mentioned *trochilos*, translated as sandpiper. The word was also used by Aldrovandus (1600) as Modern Latin *tringa*, initially for the Green Sandpiper *Tringa ochropus*.

The species

Upland Sandpiper (vagrant, one very old record, 1848)

See family introduction. From the 'inland' sense of 'upland', for its breeding grounds in central North America.

Other names: Bartram's Sandpiper, from the genus name, as used by Gould (1869) and still in use by RAOU (1913).

Bartramia longicauda (Bechstein, 1812) [BAR-tra-mi-uh lon-gi-KOW-duh]: 'Bartram's long-tailed bird', see genus name, and from Latin *longus*, long, and *cauda*, tail.

Whimbrel (non-breeding migrant)

Believed to have derived from a word associated with whimper, for the somewhat querulously twittering call, plus the French suffix '-erel' (as also in dotterel), implying diminutive – this may have been a comparison with the much bigger curlew. Appears in the 16th century in the form 'whympernel'.

Other names: Eurasian, Siberian (for the breeding grounds) Whimbrel; Australian Whimbrel, the name used by Gould (1848), who described how it differed from the European species – to be fair, he was looking at a different subspecies, which breeds in Siberia; Little Curlew (but see next species); Jack Curlew, from a now rarely used sense of jack meaning something smaller than normal; Jack Titterel, from the previous plus an onomatopoeic name; Mayfowl, probably from the time the birds returned to Britain, on passage to their northern breeding grounds; Shipmate – did migrating birds rest on ships?

Numenius phaeopus (Linnaeus, 1758) [noo-MEH-ni-ʊs FEH-o-pʊs]: 'dusky-footed new-moon bird', see genus name, and from Greek *phaios*, dusky, dun or grey, and *pous*, foot – Linnaeus (1758) described the bird as having bluish feet. Also from Mediaeval Latin *phaeopus*, a whimbrel.

Little Curlew (non-breeding migrant)

See family introduction. By far the smallest curlew (*Numenius*).

Other names: Little Whimbrel, by the same logic; Pygmy Curlew; Siberian Baby (!) Curlew, presumably by comparison with the Far Eastern Curlew (*N. madagascariensis*), the only other curlew to breed in Siberia – see species profile.

Numenius minutus Gould, 1841 [noo-MEH-ni-ʊs mi-NOO-tʊs]: 'minute new-moon bird', see genus name, and from Latin *minutus* from *minuo*, make smaller. See common name description.

Far Eastern Curlew (non-breeding migrant)

See family introduction. Breeds in Siberia and northern China, mostly to the east of the next species.

Far Eastern Curlew *Numenius madagascariensis*

Other names: Eastern Curlew, as it was until a recent move to reduce apparent confusion with the eastern race of Eurasian Curles (*N. arquata*); Australian Curlew, the name used by Gould (1848), who was seeing a species new to him, not realising that Linnaeus had long ago named it; Sea Curlew, to distinguish in Australia from stone-curlews.

Numenius madagascariensis (Linnaeus, 1766) [noo-MEH-ni-ʊs ma-da-gas-ka-ri-EN-sis]: 'not really a Madagascar new-moon bird', see genus name, and from Modern Latin for Madagascar plus the suffix *-ensis*, usually indicating the place of origin of the type specimen. Unfortunate, as it seems the type locality was in fact Makassar, Sulawesi. Linnaeus (1767) was following Brisson (1760), who stated that the bird lived in 'Madagascaria'.

Eurasian Curlew (vagrant)

See family introduction. Breeds right across central Eurasia.

Numenius arquata (Linnaeus, 1758) [noo-MEH-ni-ʊs ar-KWAH-tuh]: 'bowed new-moon bird', see genus name, and from Mediaeval Latin *arquata*, a curlew, from Latin *arcuatus*, bow or rainbow. In his description of the bird, Linnaeus wrote 'rostro arcuato' (i.e. with a bowed bill, rather than the other possibility of a rainbow one). It has been suggested that the word refers to the colour of the bird, given that this may evoke *morbus arcuatus*, the rainbow-coloured disease (i.e. jaundice), but this seems unlikely to be the case.

Bar-tailed Godwit (non-breeding migrant)

See family introduction. The barred tail conveniently distinguishes it from Black-tailed Godwit, the only other godwit regularly seen in Australia.

Other names: Barred-tailed Godwit; Barred-rumped Godwit, as used by Gould (1848) and RAOU (1913); Pacific Ocean Godwit, despite the fact that it has the same (non-Pacific) range as Black-tailed Godwit.

Limosa lapponica (Linnaeus, 1758) [li-MOH-suh lap-PO-ni-kuh]: 'muddy Lapland bird', see genus name, and from Modern Latin *Lapponia*, meaning Lapland, the type locality. The breeding grounds are above the Arctic Circle.

Bar-tailed Godwit *Limosa lapponica*

Black-tailed Godwit (non-breeding migrant)

See family introduction. The black tail neatly distinguishes it from Bar-tailed Godwit, the only other godwit regularly seen in Australia; the name, however, is based on Gould's name *L. meluranoides*, as he thought it different from the European birds.

Other names: Large Godwit, marginally larger than Bar-tailed Godwit.

Limosa limosa (Linnaeus, 1758) [li-MOH-suh li-MOH-suh]: 'muddy, muddy bird', not so much a criticism as a statement of fact. See genus name.

Hudsonian Godwit (vagrant)

See family introduction. Breeds in northern North America, including Hudson Bay; normally heads due south to South America.

Limosa haemastica (Linnaeus, 1758) [li-MOH-suh hee-MAS-ti-kuh]: 'bloody muddy bird', another statement of fact rather than an insult – see genus name, and from Greek *haima*, blood, and *haimatikos*, bloody, for the colour (it seems there was an error by Linnaeus in his addition of an 's' but once having appeared in print, it stands thus). Not a colour we can usually enjoy in Australia, because it describes only the breeding plumage of the male (see family introduction).

Ruddy Turnstone (non-breeding migrant)

'Turnstone' goes back to the 17th century in England, a nice descriptor of its foraging habit. Its distinctive rufous back, even in non-breeding plumage, is in contrast to the only other species, which is black.

Other names: Turnstone; Beachbird (well, yes!); Sea-dotterel, similar comment; Calico-bird, presumably from the US usage of calico, meaning multi-coloured.

Arenaria interpres (Linnaeus, 1758) [a-re-NAH-ri-uh in-TEHR-pres]: 'translator sand-bird', see genus name, and from Latin *interpres*, a messenger, translator or intermediary. One explanation of how it came by this name is a rather complicated tale told by Jobling (2010): apparently Linnaeus was in Gotland and mistakenly thought that *Tolk*, the Gotland word for a Common Redshank, was not only the name for the Ruddy Turnstone but was also Swedish *tolk*, and therefore meant translator. An easy mistake to make, but a bit of a howler to endure for 250 years. HANZAB (1990–2006, vol. 3), however, made a different suggestion: that *interpres* may 'refer to going in between the tide lines' (the intermediary meaning, perhaps).

Great Knot (non-breeding migrant)

See family introduction. Larger than the only other knot (see next species).

Other names: Greater, Slender-billed (from the scientific name) Knot; Stripe-crowned Knot, true, but so is the Red Knot, though in breeding attire this species does have more dramatic black-on-white stripes; Japanese Knot, a mystery as it only appears in that country on passage; Eastern Knot, recommended by RAOU (1913) and equally mysterious; Large or Great Sandpiper, it is relatively massive (the latter was the name used by Gould 1848). ('Slender-bellied Knot', cited by CSIRO 1969, is surely a misprint for Slender-billed.)

Calidris tenuirostris (Horsfield, 1821) [ka-LEE-dris te-noo-i-ROS-tris]: 'speckled slender-billed waterbird', see genus name, and from Latin *tenuis*, slender or thin, and *rostrum*, bill. Horsfield (1821) pointed out: 'The beak is more slender than in the European species of this genus' (i.e. the redshanks, then *Totanus*), to which the Great Knot was then also considered to belong.

Red Knot (non-breeding migrant)

See family introduction. Beautiful rich russet in breeding plumage.

Other names: Knot; Common, Lesser, Grey-crowned Knot, the latter unclear; Knot-snipe, again the vague usage of snipe for any shore-wader; East Siberian Sandpiper, some populations do breed there, but so do many other sandpipers; Iceland Sandpiper, very much analogous with the curious 'Japanese' for the previous species, because it occurs there only on passage.

Calidris canutus (Linnaeus, 1758) [ka-LEE-dris ka-NOO-tʊs]: 'speckled Canute waterbird', see genus name, and from the name of the famous 11th-century King Canute (Cnut) of Denmark, England and various other places. John Ray (1713), whom Linnaeus (1758) cited in his description, called the bird '*Canuti avis*, ie Knot'. This link between king and bird dates back at least a further century to William Camden (1607), who in *Britannia* made a connection between the two words: 'Knotts, ie Canute birds, which I believe are thought to fly from Denmark' (to Lincolnshire). It seems that Linnaeus, basing his assumptions on the earlier writers, erroneously believed there was some genuine connection between king and bird, and chose his species name accordingly. We think it unlikely he chose it, as is sometimes suggested, from the story of King Canute sitting on the beach defying the tide – the rationale is that knots are supposed to ignore the tide when they are feeding. As for the idea of the name coming about because Canute really liked Red Knots, and liked them well fattened, perhaps the less said the better.

Ruff and Reeve (non-breeding migrant)

Ruff is a name that appears in 17th-century England, for the male's extraordinary great coloured feathery ruff during breeding. A remarkable thing about this species is that there are quite different names for the sexes, and the plain-coloured female is a Reeve; this name appears at the same time as Ruff. An older form, Ree, dates back another 100 years. This could well be the same word as Scottish ree, meaning crazy or alcoholically agitated, in reference to the Ruff's frenetic behaviour on the breeding lek. In this case it was the original name, pre-dating Ruff, and may have been relegated to describing the female as the meaning was lost (in England) and Ruff became favoured for the male.

Calidris pugnax (Linnaeus, 1758) [ka-LEE-dris PƱG-naks]: 'aggressive speckled waterbird', see genus name and from Latin *pugnax*, fond of fighting. The behaviour on the lek can involve bloodshed and has 'all the aggression and posturing theatre of a heavyweight title bout' (Cocker and Mabey 2005).

Broad-billed Sandpiper (non-breeding migrant)

See family introduction. For the distinctively widened bill.

Calidris falcinellus (Pontoppidan, 1763) [ka-LEE-dris fal-si-NEL-lʊs]: 'little sickle-shape speckled waterbird', see genus name, and *falcinellus*, diminutive from Latin *falx/falcus*, sickle, for the bill which is straight to start with but, as well as widening, takes a sudden downward curve at the end.

Sharp-tailed Sandpiper (non-breeding migrant)

See family introduction. From the species name, but that just begs the question. Even in the hand it would be hard to justify this name – we are baffled by it.

Other names: The range of alternative names reflect its familiarity. Australian Tringa, used by Gould (1848) who thought it was an indigenous species; Siberian or Asiatic Pectoral Sandpiper, see profile – although both species share Siberian breeding grounds, Sharp-tails don't breed in North America; Siberian Sandpiper (like most of the rest!); Little Greenshank, requires a lot of imagination; Marsh Tringa or Marsh Sandpiper, like the 'other' Marsh Sandpiper this species favours inland swamps (as well as the coast); Sharp-tailed Stint; Brown-eared Sandpiper, hmm; New Holland Knot, by Jardine and Selby (1826–35, vol. 2) for the skin of a bird in partial breeding plumage they received from Australia and called *Tringa australis*; Sharpie, an affectionate name in wide use among birders.

Calidris acuminata (Horsfield, 1821) [ka-LEE-dris a-koo-mi-NAH-tuh]: 'pointy speckled waterbird', see genus name, and from Latin *acumen*, a point to prick or sting with (see common name). Horsfield's description is not illuminating – he just said that the tail feathers are pointed, which puts us back where we started.

Stilt Sandpiper (vagrant)

See family introduction. From the scientific name, for its long thin legs.

Calidris himantopus (Bonaparte, 1826) [ka-LEE-dris hi-MAN-to-pʊs]: 'speckled stilt waterbird', see genus name, and from the generic name of the stilts *Himantopus*, Latin for stilt (see family Recurvirostridae). It does have long thin legs, and it seems to stand very tall.

Curlew Sandpiper (non-breeding migrant)

See family introduction. For the down-curved bill.

Other names: Curlew-stint, Pygmy Curlew, both variations on the theme of the primary name (Gould 1848 reported that the latter was used by 'British Ornithologists' but didn't indicate if it was also used in Australia); Redcrop, see Dunlin, from which this name distinguishes this bird, which has brick-red undersides when breeding.

Calidris ferruginea (Pontoppidan, 1763) [ka-LEE-dris fer-rʊ-GI-ne-uh]: 'rusty speckled waterbird', see genus name, and from Latin *ferrugineus*, of the colour of rusty iron (*ferrugo*, rust), though we are unlikely to see that lovely colouring while the bird is in Australia.

Temminck's Stint (vagrant)

See family introduction. For Coenraad Temminck (1778–1858), a Dutch ornithologist who directed the Dutch National Natural History Museum in Leiden for the last 38 years of his life; an honorific by Leisler, a German naturalist who was his friend.

Calidris temminckii (Leisler, 1812) [ka-LEE-dris TEM-min-ki-ee]: 'Temminck's speckled waterbird', see genus and common names.

Long-toed Stint (non-breeding migrant)

See family introduction. It is possible to see the unusually long middle toe with luck and a good telescope, but it mostly has the air of a name taken from a skin.

Other names: Long-toed Sandpiper; Middendorf's Stint, for Alexander von Middendorf, a 19th-century Russian biologist of German-Estonian heritage who had wide interests and travelled much in the hard Russian north, including far eastern Siberia where he collected this species.

Calidris subminuta (Middendorf, 1853) [ka-LEE-dris sŭb-mi-NOO-tuh]: 'a bit like a tiny speckled waterbird', see genus name, and from Latin *minutus*, tiny, plus *sub-*, somewhat (i.e. somewhat like *C. minutus*, see Little Stint). Indeed, the differences between the two, explained by von Middendorf (1853) in detail, are also minute!

Red-necked Stint (non-breeding migrant)

See family introduction. In breeding plumage it has a distinctive rufous neck and head, upper breast and back, features not much seen in Australia.

Other names: Little Sandpiper, used by Gould (1848); Land-snipe or Land Snipe, again the use of snipe to indicate any shore-wader, but the implication of 'land' is unclear – this, plus Least Sandpiper, are names recorded by Gould from the 'Colonists of Western Australia'; Rufous-necked Stint; Red-necked Sandpiper; Little Dunlin, also for no clear reason; Eastern Little Stint, for its breeding grounds, which are wholly in Siberia – though Little Stint breeds as far east, it also does so well to the west of that.

Calidris ruficollis (Pallas, 1776) [ka-LEE-dris roo-fi-KOL-lis]: 'speckled red-necked waterbird', see genus name, and from Latin *rufus*, red, and *collum*, the neck (see common name).

Sanderling (non-breeding migrant)

One of the meanings of the suffix -ling is simply 'associated with'; we would expect then that the form 'sandling' might have preceded the current word, and indeed that is known from the 17th century. Nonetheless, the first appearance, in the very early 17th century, is of the current form. (Another implication of -ling is a diminutive, but it doesn't seem necessary to invoke it in this instance.)

Calidris alba (Pallas, 1764) [ka-LEE-dris AL-buh]: 'speckled white waterbird', see genus name, and from Latin *albus*, dull white. That whitish bird on the shore we mention under Common Sandpiper – it really could be this!

Dunlin (vagrant)

A variation on Dunling. The suffix -ling has a couple of implications (see Sanderling for another). In this case the probable usage was to imply a diminutive – think of gosling or duckling. So, a 'little dun (coloured) one'.

Other names: Blackcrop, using crop as a British dialect word for stomach, referring to breeding plumage.

Calidris alpina (Linnaeus, 1758) [ka-LEE-dris al-PEE-nuh]: 'speckled waterbird of the mountains', see genus name, and from Latin *alpinus*, of the Alps (meaning mountains in general). Linnaeus (1758) said its habitat is in 'Lapponia' (Lapland), and there are certainly plenty of mountains there. The bird breeds in the Arctic, then migrates to warmer climes in Asia for the winter.

Baird's Sandpiper (vagrant)

See family introduction, and from the species name.

Calidris bairdii (Coues, 1861) [ka-LEE-dris BEHR-di-ee]: 'Baird's speckled waterbird', see genus name, and for Spencer Baird, a highly influential 19th-century US ornithologist who was a protégé of Audubon, a relentlessly hard-working secretary of the Smithsonian Institution and a mentor to many younger biologists.

Little Stint (non-breeding migrant)

See family introduction. Given the apparent original meaning of 'stint', this name is a strong statement! In dimensions (but perhaps not in mass), it is just the smallest of the tiny stints.

Other names: Sparrow-shorebird, a delightful nod to the size and a name that may have arisen only in Australia, because it seems not to be known in England or North America. (Having said that, we can't find an Australian reference outside HANZAB 1990–2006, vol. 3 either.)

Calidris minuta (Leisler, 1812) [ka-LEE-dris mi-NOO-tʊs]: 'tiny speckled waterbird', see genus name, and from Latin *minutus*, tiny, from *minuo*, make smaller. (See common name. *Calidris minutilla*, the Least Sandpiper of the Americas, though not a stint, is even smaller.)

White-rumped Sandpiper (vagrant)

See family introduction. Very few American sandpipers (and this species breeds in the North American Arctic and winters in Patagonia) have white rumps; in particular this distinguishes it from the very similar Baird's Sandpiper.

Other names: Bonaparte's Sandpiper – in 1844 Hermann Schlegel named the bird *Tringa bonapartii*, presumably in ignorance of its previous description. This was for Prince Charles Lucien Bonaparte, exiled nephew of Napoleon and a very eminent ornithologist who did a lot of work in the US and sponsored Audubon's career. His name was widely used in the 19th century in North and South America.

Calidris fuscicollis (Vieillot, 1819) [ka-LEE-dris fʊs-ki-KOL-lis]: 'dusky-necked speckled waterbird', see genus name, and from Latin, *fuscus*, dusky, and *collum*, neck. The name Vieillot (1816–19) used in French is '*le Tringa à cou brun*' – perhaps by comparison with his next entry, which is '*le Tringa à cou roux*', see Red-necked Stint.

Buff-breasted Sandpiper (vagrant)

See family introduction. The lovely buffy breast is very distinctive, and remarkably for this family it doesn't change plumage seasonally.

Other names: Yellowcrop, see Dunlin and Curlew Sandpiper.

Calidris subruficollis (Vieillot, 1819) [ka-LEE-dris sʊb-roo-fi-KOL-lis]: 'not quite as rufous-necked speckled waterbird', see genus name, and from Latin *sub-*, slightly, *rufus*, red, and *collum*, neck. Vieillot (1816–19) and his source (Azara 1805) described the bulk of the head and neck as being '*blanc roussâtre*' (russet-white) and '*blanca acanelada*' (cinnamon-white), respectively.

Pectoral Sandpiper (non-breeding migrant)

See family introduction. The name is apparently a somewhat clumsy reference to the diagnostically sharp demarcation between streaked buffy breast and white belly, which helps distinguish it from the very similar Sharp-tailed Sandpiper.

Other names: American Pectoral Sandpiper, not obvious because while it does breed in far North America and winters in southern South America, it also breeds in Siberia (whence it comes to Australia). The point seems to be that the Sharp-tailed Sandpiper has been referred to as Siberian or Asiatic Pectoral Sandpiper, so perhaps this name was invented to make the distinction.

Calidris melanotos (Vieillot, 1819) [ka-LEE-dris me-la-NOH-tos]: 'black-backed speckled waterbird', see genus name, and from Greek *melas*, black, and *nōton*, the back. Vieillot (1816–19) took the name from the *chorlito lomo negro* (black-backed plover) of Azara (1805); although not a feature that springs readily to the eye in the field, the feathers of the bird's back are largely black, albeit with lighter fringes. The spelling of *melanotos* (Greek rather than the usual Latinised version with *-us*) is Vieillot's own.

Asian Dowitcher (non-breeding migrant)

See family introduction. The other two dowitcher species are essentially American.

Other names: Asiatic Dowitcher; Snipe-billed and Asiatic Snipe-like Godwit, for the indisputably snipe-like appearance, due to the long straight bill.

Limnodromus semipalmatus (Blyth, 1848) [lim-no-DROH-mʊs se-mi-pal-MAH-tʊs]: 'half-webbed marsh-racer', see genus name, plus Latin *semi*, half, *palma*, palm, and *palmatus*, webbed. True, though not a feature that is much help with identification, since the bird does a lot of standing about in ankle-deep water and mud!

Short-billed Dowitcher (vagrant)

See family introduction. This is by contrast with the other American dowitcher, Long-billed Dowitcher (though there's not much in it).

Limnodromus griseus (Gmelin JF, 1789) [lim-no-DROH-mʊs GREE-se-ʊs]: 'grey marsh-racer', see genus name, plus Mediaeval Latin *grisus/griseus*. Hmm, grey again – we did tell you.

Latham's Snipe (non-breeding migrant)

The connection of the great turn-of-the-19th-century ornithologist John Latham with this bird is most unclear – the species was named in 1831 (not 1931 as suggested by HANZAB 1990–2006, vol. 3) by the Englishman John Gray. The clue seems to be in Gould (1848), who called the bird *Scolopax australis* after Latham (1801). He also listed Gray's *Scolopax hardwickii*, so clearly Latham's name was later deemed

Latham's Snipe *Gallinago hardwickii* and Australian Painted-snipe *Rostratula australis*

invalid. Given the nebulous nature of the connection and the fact that Gould did not use it (he called it New Holland Snipe), it is curious that the name has persevered. Leach (1911) used Australian Snipe, with '(Latham, Japanese)' as alternatives.

Other names: Snipe, Common Snipe, Latham Snipe, New Holland Snipe, Jack Snipe, the last being the name of a similar English bird (the word implying 'small'); Bleater, presumably for the voice, though not especially evocative; Longbill, clear enough (and harking back to one purported original meaning of 'snipe', see family introduction).

Gallinago hardwickii (Gray JE, 1831) [gal-li-NAH-go hard-WICK-i-ee]: 'Hardwicke's snipe/woodcock', see genus name, and for Charles Browne Hardwicke (1788–1851), an English naval lieutenant who settled in Tasmania. He was a horse-racing enthusiast who founded the Tasmanian Turf Club and became a horse breeder: so far, so good. However, he was also a farmer who declared what is now regarded as some of Tasmania's best farming land to be so poor that he hardly dared speak of it lest his remarks discourage settlement in the area, and a chief constable convicted of selling illicit alcohol. Perhaps not the most reliable of witnesses, and what he knew about woodcocks, or indeed snipe, is anyone's guess; however, given that the type specimen originated from Tasmania, our guess is that he shot it.

Pin-tailed Snipe (vagrant, or perhaps a rare migrant)

See family introduction. A classic example of a name derived solely from a (presumably dead) bird in the hand, it refers to the diagnostic short narrow stiff outer tail feathers, which have evolved to make a characteristic whirring noise in display flight.

Other names: Asiatic Snipe, most unhelpful!

Gallinago stenura (Bonaparte, 1830) [gal-li-NAH-go ste-NOO-ruh]: 'narrow-tailed snipe/woodcock', see genus name, and from Greek *stenos*, narrow, and *oura*, tail. Bonaparte (1830) described the tail as being wedge-shaped, with 'aborted lateral awls' (the 'pin' feathers already described).

Swinhoe's Snipe (non-breeding migrant)

See family introduction. Robert Swinhoe was a mid 19th-century British diplomat and collecting naturalist who spent 19 very profitable years in China and provided a substantial number of the specimens for Gould's *Birds of Asia*. He collected and named this species.

Gallinago megala Swinhoe, 1861 [gal-li-NAH-go ME-ga-luh]: 'big snipe/woodcock', see genus name, and from Greek *megas*, great. Swinhoe considered that this bird was markedly larger than *G. stenura*.

Terek Sandpiper (non-breeding migrant)

See family introduction. The Terek River flows from Georgia through Russia into the Caspian Sea; the type specimen was collected in 1774 from near the Terek mouth and *Terekia* was an early genus name.

Other names: Terek Godwit, used by Gould (1848), presumably a reference to the up-tipped bill; Terek Snipe (Latham 1823).

Xenus cinereus (Güldenstädt, 1775) [ZEH-nʊs si-NE-re-ʊs]: 'ashy-grey stranger', see genus name, and from Latin *cinereus*, like ashes. Grey, like many another.

Wilson's Phalarope (vagrant)

See family introduction. Alexander Wilson was a Scot who was effectively exiled (for allegedly writing scurrilous verse about factory owners during a labour conflict) and went to the US in 1794, where he became a self-taught giant of American ornithology. The bird was originally named *Phalaropus wilsonii* in tribute by the Englishman Edward Sabine, who was unaware that Vieillot had already done so; the name stuck as an English vernacular name. Curiously, Sabine was apparently unaware that Wilson had collected and painted the bird; this was discovered only after his death.

Phalaropus tricolor (Vieillot, 1819) [fa-LA-ro-pʊs TRI-ko-lor] 'three-coloured coot-foot', see genus name, and the three colours are … Vieillot (1816–19) gave a detailed description of all the brown, white and black bits.

Red-necked Phalarope (non-breeding migrant)

See family introduction. For the brick-red neck-sides of the breeding female.

Other names: Northern Phalarope, for its breeding grounds across the Arctic (compared with those of Wilson's Phalarope in central North America).

Phalaropus lobatus (Linnaeus, 1758) [fa-LA-ro-pʊs lo-BAH-tʊs]: 'lobed coot-foot', see genus name – does that mean double lobes, one for the feet like a coot and one for mediaeval Latin *lobatus*, lobed (from Greek *lobos*)? Just making sure we notice.

Red Phalarope (vagrant)

See family introduction. For the brick-red neck and underparts of the breeding female.

Other names: Grey Phalarope; hmm, in non-breeding plumage all phalaropes are grey, though this species is 'clearer' and less scalloped on the back than the others.

Phalaropus fulicarius (Linnaeus, 1758) [fa-LA-ro-pʊs fʊ-li-KAH-ri-ʊs]: 'coot-like coot-foot', double coots this time, a Greek one (see genus name) and a Latin one, from *fulica*, coot, and *-arius*, relating to.

Common Sandpiper (non-breeding migrant)

See family introduction. It is not clear that this bird was ever especially common, even in Britain, but perhaps as a bird of inland waters it was more familiar to more people than the coastal species.

Other names: Summer Snipe, 'snipe' a commonly used catch-all for waders and 'summer' for the time it appeared (like most other waders in Australia, including snipe!); Fairy Sandpiper, used by Gould (1848), who believed it to be a different species from the European one and called it *A. empusa* (though *empusae* are spectres, monsters and incubae, daughters of Hecate – perhaps not quite the fairies Gould had in mind?); Green Sandpiper, reported by Gould to be used by the 'Colonists of Port Essington' (Darwin's prototype, on the far north coast) – see also next species.

Actitis hypoleucos (Linnaeus, 1758) [ak-TEE-tis hi-po-LOO-kos]: 'whitish coast-dweller', see genus name, and from Greek *hupo-*, somewhat or slightly below, and *leukos*, white. A name that is not only bland but somewhat misleading – if you're looking at a whitish bird on the shore, it won't be one of these.

Green Sandpiper (vagrant)

See family introduction. Best explained by a vivid imagination – in the right light it might appear greenish-brown ...

Tringa ochropus Linnaeus, 1758 [TRIN-guh OH-kro-pʊs]: 'sallow-footed waterbird', see genus name, and from Greek ōkhros, sallow or pale yellow, and *pous*, foot. The legs are more often described as greenish-grey or even dark green – perhaps they tend to reflect the colour of the surroundings.

Wandering Tattler (non-breeding migrant)

See family introduction. Indeed it wanders (from the northern Pacific to the south-west Pacific) just as the Grey-tailed does, the only difference being that the latter wanders even further south. 'True but unhelpful' category.

Other names: American Grey-rumped Sandpiper, used by Lucas and Le Soeuf (1911), in direct contra-distinction to the next species; American Ashen Tringine Sandpiper – according to HANZAB 1990–2006, vol. 3! – but we can't claim to have heard anyone drop it into the conversation in a bird hide.

Tringa incana (Gmelin JF, 1789) [TRIN-guh in-KAH-nuh]: 'hoary light-grey waterbird', see genus name, and from Latin *canus*, meaning white (or hoary), so *incanus* is 'not white', hence grey (also hoary!), as is the bird.

Grey-tailed Tattler (non-breeding migrant)

See family introduction. As useful as the previous name: both tattlers have grey tails.

Other names: Grey-rumped Sandpiper, the name used by Gould (1848), long before 'tattler' arrived from North America (he called it *Totanus griseopygius*, from which he translated the name) – RAOU was still recommending it in 1913.

Tringa brevipes (Vieillot, 1816) [TRIN-guh BRE-vi-pehz]: 'short-footed waterbird', see genus name, and from Latin *brevis*, short, and *pes*, the foot (also leg, in Late Latin). It may be short-legged in relation to *Tringa* species whose feet protrude beyond their tail-tip when flying, but it is not the only one whose legs are shorter than the tail.

Lesser Yellowlegs (vagrant)

Yellowlegs for the distinctly yellow legs; Lesser to distinguish it from the similar but larger Greater Yellowlegs, which has not yet been reported in Australia.

Tringa flavipes (Gmelin JF, 1789) [TRIN-guh FLAH-vi-pehz]: 'yellow-legged waterbird', see genus name, and from Latin *flavus*, golden or reddish-yellow (from *flagro*, burn), and *pes/pedis*, foot (also leg, in Late Latin). The legs are indeed deep yellow, and remain so throughout the year.

Common Redshank (non-breeding migrant)

The Redshanks are two species of sandpiper with long dark red legs. This one is far commoner in Britain than the Spotted Redshank (see profile); it breeds there, while the Spotted is just a passage migrant.

Tringa totanus (Linnaeus, 1758) [TRIN-guh TOH-ta-nʊs]: 'redshank waterbird', see genus name, and from one of the Italian words for redshank, *tótano*.

Marsh Sandpiper (non-breeding migrant)

See family introduction. Marsh from the scientific name, which reflects its preference for freshwater rather than coastal habitats.

Other names: Little Greenshank – apt, as it does resemble a small fine Greenshank.

Tringa stagnatilis (Bechstein, 1803) [TRIN-guh stag-NAH-ti-lis]: 'pond waterbird', see genus name, plus Latin *stagnatilis*, belonging to ponds or pools (see common name).

Wood Sandpiper (non-breeding migrant)

See family introduction. For its habitat preference of inland ponds with dead fallen trees.

Tringa glareola Linnaeus, 1758 [TRIN-guh gla-RE-o-luh]: 'little gravelly waterbird', see genus name, and from Latin *glarea*, gravel, with diminutive *-olus*. Linnaeus' (1758) note on this did not mention habitat but gave a description of the bird that includes 'spotted white body', so 'gravel' has sometimes been taken to indicate the spots. However, Brisson (1760) cited Kaspar Schwenckfeld as using the word as a genus name (*Glareola secunda*) in *Theriotropheum Silesiae* of 1603; the common names attached to this are *Sand-vogel* and *Koppriegerle* (i.e. some kind of wader which lives on gravel or sand), so habitat may be the real key. (Because Schwenckfeld published long before the magic date of 1 January 1758 – see the Introduction section on 'How scientific names of birds are derived' – it was not a valid name and was thus later available for the pratincole genus.)

Common Greenshank (non-breeding migrant)

The two Greenshanks have greenish legs, this one's more convincingly so than the very restricted Nordmann's Greenshank. This species is also far more common and widespread.

Other names: Greenshank; Greater Greenshank, it is larger than Nordmann's; Large Tringine Sandpiper (Tringine from the genus); Australian Greenshank, used by Gould (1848) even though he knew it was part of the worldwide species.

Tringa nebularia (Gunnerus, 1767) [TRIN-guh ne-bʊ-LA-ri-uh]: 'misty waterbird', see genus name, and from Latin *nebula*, mist or cloud. Two explanations are given: it either refers to the colour of the bird (grey again!) or indicates its marsh habitat. In the second case, it is connected with the Old Norwegian name, *Skoddeføll*, mist-foal, used by Johan Gunnerus (1718–73), Norwegian botanist, bishop and collector, in his description (Leem and Gunnerus 1787). He also mentioned other names for the bird, the Ploughman and the Breakerman. He suggested that all those names referred to the bird's arrival in spring, when the land is ploughed or broken by horses. We can find no evidence for a suggestion that the foal part of the name refers to the bird's 'whinnying' call – and in any case we feel that quite a deal of imagination would be required.

Spotted Redshank (vagrant)

See Common Redshank; this species has striking white-spangled black breeding plumage.

Tringa erythropus (Pallas, 1764) [TRIN-guh e-RITH-ro-pʊs]: 'red-footed waterbird', see genus name, plus Greek *eruthropous* (from *eruthros*, red, and *pous*, foot). Now these really are red legs!

Nordmann's Greenshank (vagrant)

See Common Greenshank. Alexander von Nordmann was a 19th-century Finnish zoologist and botanist who spent much of his working life in Russia and described (and perhaps collected, though it is not clear that he went that far east) the type specimen from Okhotsk in the far east.

Tringa guttifer (Nordmann, 1835) [TRIN-guh Gʊ-ti-fehr]: 'droplet-bearing waterbird', see genus name, and from Latin *gutta*, a droplet or spot, and *-ferus*,-bearing. This refers to the bird in breeding plumage, when its breast and head are heavily spotted and the upper parts are dark and mottled with white edges to the feathers.

GLAREOLIDAE (CHARADRIIFORMES): pratincoles

Glareolidae Brehm, 1831 [gla-re-O-li-dee]: Little Gravelly family, see genus name *Glareola*. 'Pratincole' is from a coined Latin word meaning 'meadow-, or plain-dweller'. Adopted into English in the 18th century, when used by the very influential Thomas Pennant, Welsh ornithologist (among his other interests), in the monumental *British Zoology*.

The genera

Stiltia Gray GR, 1855 [STIL-ti-uh]: 'bird with long legs', from Middle English *stilt*, indicating the length of the bird's legs. Note that this genus does not apply to the birds we call stilts, though its name is based on the same word.

Glareola Brisson, 1760 [gla-RE-o-luh]: 'little gravelly bird', from Latin *glarea*, gravel, with diminutive suffix *-olus*. It is unclear what this refers to, because Brisson (1760) didn't say. He seems to have got the name from Kaspar Schwenckfeld's *Theriotropheum Silesiae* of 1603, rather than from Linnaeus' species name for the Wood Sandpiper *Tringa glareola*, published 2 years previously. The Schwenckfeld reference he gives is clearly to a sandpiper of some kind (*Sand-vogel* and *Koppriegerle* are given as common names). HANZAB suggested that *Glareola* is a reference to 'the supposed habitat of the Collared Pratincole *G. pratincola*' (1990–2006, vol. 3). If *pratincola* means living in meadows, perhaps they were gravelly meadows.

The species

Australian Pratincole (breeding resident)

See family introduction. The only Australia-breeding species of the family.

Other names: Pratincole; Swallow-plover, a delightful coining that attempts to combine its terrestrial habits with its aerial skills; Road-runner, for a common behaviour, probably influenced by the North American cuckoo of that name; Australian Courser, coursers being the other branch of the family, not occurring in Australia; Nankeen Plover, for the colour (see Nankeen Kestrel and Night Heron); Arnhem Land Grouse, for no obvious reason beyond it being a ground-based bird that occurs there.

Stiltia isabella (Vieillot, 1816) [STIL-ti-uh i-sa-BEL-luh]: 'pale yellow stilt', see genus name, and French *isabelle*, pale yellow. This colour name is said ('apocryphal', is the word that springs to mind) to be from Queen Isabella I of Castile and was adopted because, when she and her husband besieged Granada in 1491, she swore not to change her underwear (*chemise*, at least) till the town was taken. The siege began in the European spring and lasted until the end of the year … the less said the better, really.

Oriental Pratincole (non-breeding migrant)

See family introduction. For its breeding grounds, which encompass virtually all of southern Asia.

Other names: Swallow-plover, see Australian Pratincole; Eastern Collared Pratincole, to distinguish it from the very similar Collared Pratincole (*G. pratincola*) of Africa; Grasshopper Bird, for a major food source; Swarmer, for its vast flocks; Little Stormbird, presumably for its arrival in the Top End in the wet (though it's unclear which other stormbird it's smaller than).

Glareola maldivarum Forster JR, 1795 [gla-RE-o-luh mal-di-VAH-rʊm]: 'little gravelly bird from the Maldives', see genus name, and for the type specimen, which was taken at sea near the Maldive Islands.

LARIDAE (CHARADRIIFORMES): gulls, terns and noddies

Laridae Rafinesque, 1815 [LAH-ri-dee]: the Ravenous Seabird family, see genus name *Larus*.

As we might expect of a family of seabirds, this family is found throughout the world, so the basic names associated with it long pre-dated the discovery of Australian species; many species also range well beyond Australia, so were known before their local recognition.

'Noddy' is an old English word meaning a simpleton. It is presumed that insult was directed at the bird for being trusting of marauding sailors in its breeding colonies. We have form for such things – see also Booby.

Oriental Pratincole *Glareola maldivarum* and Australian Pratincole *Stiltia isabella*

'Tern' probably comes from the old Norse word *perno* – there are words very similar to 'tern', for the same birds, in all modern Scandinavian languages. We know of the old English *stern* – the Erfurt Glossary (Latin/Anglo-Saxon; early 9th century) gave '*gavia, avis qui dicitur stern saxonice*' (*gavia*, a bird that is called *stern* by the Saxons) (Lindsay 1921). It seems curious, however, that 'tern' should not have appeared in English until the 17th century. There is also a suggestion that it is onomatopoeic, with a source word similar to the contemporary English dialect 'starn', but these explanations are not mutually exclusive.

'Gull' comes by an etymologically interesting route. It is Gaelic in origin, coming into English in the 15th century from Cornish or Welsh, which had similar words, 'gullen' and 'gwylan', respectively, for gull; the origin is a word meaning one who weeps. Why it should have replaced established English words is unclear – Gaelic was certainly not a conquering language in England, as Norman French had been. Whatever the reason, it was interpreted as plural (like children or oxen), so by back-formation the artificial singular 'gull' was coined.

The genera

Anous Stephens, 1826 [AH-noos]: 'stupid bird', from Greek *anous*, foolish (having no 'nous', in fact: see family introduction).

Gygis Wagler, 1832 [GEE-gis]: 'mystery bird', another Ancient Greek bird (*gugēs*) described by Dionysius as eating 'amphibious birds' at night and having a call that is like its name, from which information Arnott (2007) concluded it was probably a bittern. There is a previous suggestion that the word is connected with a Lithuanian word for stork (Thompson 1895). We assume that Wagler was unaware of either of these possibilities when he applied it to a tern.

Xema Leach, 1819 [ZEH-muh]: 'meaningless', the name, that is, not the bird. Leach gave no clues to his thinking (Ross 1819). Suggestions (Richmond 1902) have been: *khēmē*, yawning (once formally proposed as an emendation of *Xema*, but dismissed or ignored) and *xenē*, the feminine form of *xenos*, a foreigner, but in that case why the 'm'? We wonder about *zema*, one of whose meanings is lewdness, which would perhaps fit with ancient ideas of some birds being randier than others. Unfortunately, the consensus is that this is a totally meaningless made-up name! It's what you do, apparently, when you've run out of real names to use (and when you're tired of making anagrams, perhaps: Leach was also the author of the anagram name *Dacelo*. See also Sabine's Gull for more of this sorry tale.)

Chroicocephalus Eyton, 1836 [kroy-koh-SEF-a-lʊs]: 'stained-head', from Greek, *khrōizō*, to touch or to tinge or stain, and *kephalē*, head. Eyton (1836) said that such gulls have 'head only, or head and upper part of the neck, dark coloured in the summer state of plumage'.

Leucophaeus Bruch, 1853 [loo-ko-FEH-ʊs]: 'white and grey [bird]', from Greek *leukos*, white, and *phaios*, dusky, dun, grey. Sounds like a gull, really.

Larus Linnaeus, 1758 [LAH-rʊs]: 'ravenous gull', from Greek *laros*, Latin *larus*, a ravenous seabird, perhaps a gull.

Gelochelidon Brehm, 1830 [ge-lo-KE-li-dohn]: 'laughing swallow', from Greek *gelaō*, laugh, and *khelidōn*, a swallow. The Gull-billed Tern is the only species in this genus, and only with a modicum of imagination can we interpret its call as 'laughing'. Swallow is an obvious comparison, for the rapid flight and forked tail.

Hydroprogne Kaup, 1829 [hi-dro-PROG-ne]:'water swallow', from Greek *hudōr/hudro*, water, and Latin *progne or procne*, a swallow (see also *Gelochelidon*). Procne was a daughter of Pandion, king of Athens (see genus *Pandion* under Accipitridae), sister of Philomela and wife of Tereus. In a ghastly tale of rape, sequestration, child murder and involuntary cannibalism (in which some of the details are disputed), Procne's tongue was cut out by Tereus in case she discovered his rape of her sister and spoke of it. At the end of the story, Tereus was changed into a hoopoe and the sisters into a nightingale (Philomela) and a swallow (Procne) with blood on her breast feathers (Graves 1996).

Thalasseus Boie F, 1822 [tha-LAS-se-ʊs]: 'fisherman', from Greek *thalasseus*.

Sternula Boie F, 1822 [STEHR-nʊ-luh]: 'little tern', from genus name *Sterna*, with Latin diminutive ending *-ulus*.

Onychoprion Wagler, 1832 [o-ni-ko-PREE-ohn]: 'saw with claws', from Greek *onux*, talon or nail, and *priōn*, saw, referring to common name Prion (see Procellariidae).

Sterna Linnaeus, 1758 [STEHR-nuh]: 'tern', Modern Latin *sterna* but from Old English *stern* – see family introduction.

Chlidonias Rafinesque, 1822 [kli-DOH-ni-as]: 'swallow-like [bird]', from Greek *khelidōn/khelidonios*, a swallow/swallow-like, but apparently with a spelling mistake? (See also *Gelochelidon.*)

The species

Brown Noddy (breeding resident)

See family introduction; to distinguish from Black and Grey Noddies (see Grey Ternlet).

Other names: Noddy, Noddy Tern (used by Latham 1823 and Gould 1848); Greater Noddy, because it is the largest noddy (in particular, larger than the similar Black Noddy, with which it often occurs);

Common Noddy, certainly the most widespread noddy species, and thus the one sailors most often encountered.

Anous stolidus (Linnaeus, 1758) [AH-noos STO-li-dʊs]: 'really stupid bird', see genus name, and from Latin *stolidus*, dull, stupid (see also family name).

Lesser Noddy (breeding resident)

See family introduction. Should logically be Least Noddy, but in fact it only overlaps in range with Brown Noddy – the name seems to have been coined by Gould (1848).

Anous tenuirostris (Temminck, 1823) [AH-noos te-noo-i-ROS-tris]: 'stupid slender-billed bird', see genus name, and from Latin *tenuis*, slender, thin, and *rostrum*, bill. Has a long fine bill, especially by contrast with the Common Noddy.

Black Noddy (breeding resident)

See family introduction. Not really black, but darker brown than Brown Noddy.

Other names: White-capped Noddy, a widespread name and appropriate, though not much more so than for the Brown Noddy – it is taken directly from Gould's name *A. leucocapillus* (he called it White-capped Tern); Lesser Noddy (i.e. than Brown Noddy, but see previous species); Titerack, a Norfolk Island term, supposedly for the call.

Anous minutus Boie F, 1844 [AH-noos mi-NOO-tʊs]: 'smaller stupid bird', see genus name, and from Latin *minutus* from *minuo*, make smaller.

Blue Noddy (mainland vagrant; breeds on Lord Howe and Norfolk islands)

See family introduction. The grey does have a bluish tinge.

Other names: Grey or Blue-grey Noddy (see Grey Noddy, with which this species has often been combined); Little Grey Tern, Grey or Blue Ternlet, it is a tiny (though not the smallest) grey-bodied tern; Blue-grey Fairy-tern or ternlet; Blue Billy, a Lord Howe Island name; Patro, a name used on Norfolk Island, though we can't find the origin.

Anous ceruleus (Bennett FD, 1840) [AH-noos se-ROO-le-us]: 'blue stupid bird', see genus name, plus from Latin *caeruleus* used for a variety of different shades of blue including greenish-blue and azure, and perhaps including bluish-tinged grey!

Grey Noddy (Lord Howe and Norfolk Islands)

See family introduction. Certainly grey above, but ...

Other names. This species was formerly regarded as a pale morph of Blue Noddy, so most of its names have been applied to this one too.

Anous albivitta (Bonaparte 1856) [AH-noos al-bi-VIT-tuh]: 'white-banded stupid bird', from genus name and Latin *vitta*, a ribbon or band.

White Tern (mainland vagrant; breeds on Lord Howe, Norfolk and Cocos (Keeling) islands)

See family introduction. Couldn't be whiter!

Other names: White Noddy, seemingly strange given that 'noddy' is generally regarded as referring to the large dark *Anous* terns, but *Anous*, *Procelsterna* and *Gygis* form a (somewhat disputed) subgroup; Angel or Fairy Tern, Little, Common, Atlantic Fairy-tern, perhaps best explained by Graham Pizzey, who in Pizzey and Knight (2003) described it as 'ethereal' ('Fairy Tern' is particularly used on Norfolk Island, but see also Fairy Tern profile); Norfolk Island Tern; Love or Lover Tern, has been used throughout the Pacific for some decades at least and is explained by Wakelin (1968) as 'because they cuddle close and appear to whisper in each other's ears' – which they certainly do!; White Bird.

Gygis alba (Sparmann, 1786) [GEE-gis AL-buh]: 'white mystery-bird', see genus name, and from Latin *albus*, dull white.

Sabine's Gull (vagrant)

See family introduction. From the species name.

Xema sabini (Sabine, 1819) [ZEH-muh sa-BEE-nee]: 'Sabine's meaningless name', see genus name, and for Sir Edward Sabine (1788–1883), soldier, astronomer and President of the British Royal Society;

the bird was collected by him on one of the numerous Arctic expeditions in search of the North-West Passage and named for him by his brother Joseph as *Larus sabini* (Sabine 1819) – it was mere chance that Joseph thus immortalised his own name too (!). Sir John Ross, the leader of that expedition, had a good old grumble about Edward and certainly wouldn't have supported any honours for him. Sabine, in spite of much having been made of his (and his absent ornithologist brother Joseph's) knowledge of natural history, declined to write a report of the expedition's findings or to provide any illustrations of the specimens, claiming, rather belatedly, a lack of expertise. He called in William Leach of the British Museum, who hadn't been on the expedition, to help. The report was eventually written, largely by the expedition's surgeon and assistant surgeon and, based on the information that Sabine's notes ('very rough', he admitted) and memory could provide, was revised and corrected by Leach, who thus had the opportunity to name the genus *Xema* (Ross 1819). Even that was a source of frustration, and not just to Ross – see *Xema*. Sabine's memory and confidence improved sufficiently for him to later publish his own account of the expedition's bird discoveries (Mearns and Mearns 1992).

Silver Gull (breeding resident)

See family introduction. Hmm, maybe the grey upper-parts could look a bit silvery in the right light. Gould (1848) reported that it appeared on a 'List of Birds in the Tasmanian Journal', so it was a spontaneously arising name.

Other names: Crimson-billed Gull (Latham 1823); Seagull, as the archetypal Australian gull and, courtesy of the cricket, probably the most televised bird in Australia; Kitty Gull, never widely used, perhaps by association with kittiwake, perhaps an affectionate name for a familiar bird, rather like Willie Wagtail; Mackerel or Red-billed Gull, both used in New Zealand, the former a name used for various gull species elsewhere in the world; Red-legged Gull, accurate enough though red is a popular leg colour among smaller gulls; Sea-pigeon, doubtless for its ubiquity and scavenging habits, often with some of the previous adjectives; Little Gull, used by the 'Colonists of Western Australia' (Gould 1848). Jameson's Gull, for *Larus jamesoni*, a name used (including by Gould 1848) in the 19th century for Professor Robert Jameson (1774–1854), who taught natural history to Charles Darwin at Edinburgh University; it was bestowed by his colleague James Wilson (they co-authored a *Narrative of Discovery and Adventure in Africa*) in *Illustrations of Zoology* in 1831, but by then the bird had already been described.

Chroicocephalus novaehollandiae (Stephens, 1826) [kroy-koh-SE-fa-lʊs no-veh-hol-LAN-di-eh]: 'New Holland stained-head', see genus name (even though this bird can only be so described in its juvenile state) and from Modern Latin *Nova Hollandia*, New Holland, the old name for Australia.

Black-headed Gull (vagrant)

See family introduction. The only black-headed gull commonly seen in Britain.

Chroicocephalus ridibundus (Linnaeus, 1766) [kroy-koh-SE-fa-lʊs ri-di-BʊN-dʊs]: 'laughing stained-head', see genus name (and this bird definitely is!), and from Latin *rideo/ridibundus*, laugh/laughing.

(So will the real laughing gull please stand up? Is it this chap or the Laughing Gull, *Leucophaeus atricilla*? Well, a funny thing happens with gull's names, as pointed out by Stephen Poley (2012) on his website uk.rec.birdwatching: Mediterranean Gull is *Ichthyaetus melanocephalus*, which means black-headed gull [fish-eagle really, but let's not spoil his joke]. Black-headed Gull is *Chroicocephalus ridibundus*, which means laughing gull. Laughing Gull is *Leucophaeus atricilla*, which means black-tailed gull. Black-tailed Gull is *Larus crassirostris*, which means large- (or thick-) billed gull. Large-billed Gull is *Larus pacificus*, which means … well at this point, as Poley said, 'a disappointing touch of sanity intervenes, because Pacific Gull is another name for *Larus pacificus*'.)

Laughing Gull (vagrant)

See family introduction. More a sort of sarcastic cackle, really.

Leucophaeus atricilla (Linnaeus, 1758) [loo-ko-FEH-ʊs a-tri-SIL-luh]: 'black-tailed gull', see genus name, and from Latin *ater*, dull black, and Modern Latin *cilla*, a tail. Black-tailed? Well, no, it's actually white-tailed (except in its first year when its tail is dark). Linnaeus' description was of a white gull, with head and tips of wings black, so it seems he might have got confused (perhaps when transcribing notes) with *atricapilla* (black-haired, i.e. black-capped), which would have fitted. Jobling (2010) suggested an alternative, that Linnaeus could have been misled by Catesby and Edwards' (1731) plate of the bird, which 'could give the impression of a black-tailed bird'. This is a kindness on Jobling's part as the picture is actually pretty clear and is accompanied by text which points out that while the ends of the wings are dusky black, 'all the rest of the Body is white, as is the Tail'.

Franklin's Gull (vagrant)

See family introduction. From an old species name, *L. franklinii*, which HANZAB (1990–2006, vol. 3) asserted is for Major James Franklin, a British military man and ornithologist. His activities seem to have been limited to India, though that doesn't rule him out. Other sources (Mearns and Mearns 1988, 1992) favour Sir John Franklin, British naval officer, Governor of Tasmania and eternal seeker for the North-West Passage, who eventually vanished in the Arctic – the evidence for him seems incontrovertible. Stroud (2000) is insistent that the type specimen was part of a collection made by Dr John Richardson on two of Sir John's Arctic voyages, and was described and named *L. franklinii* by William Swainson in 1830. This date would mean that it took precedence over Wagler's name, but though the cover of Richardson and Swainson's *Fauna Boreali-American* clearly shows the date as 1831, it appears that the volume did not appear in print until February 1832 (Mearns and Mearns 1992). The only concern is that the gull is not found in the Arctic – it breeds in inland North America and migrates to South America in the northern winter – but Mearns and Mearns (1988) reported that the specimen was taken on the Saskatchewan River. It must be recalled that much of Franklin and Richardson's exploratory work was conducted from the land, going overland from Hudson Bay; Richardson overwintered on the Saskatchewan in 1820.

Leucophaeus pipixcan (Wagler, 1831) [loo-koh-FEH-ʊs pi-PISH-can]: 'pipixcan gull', see genus name, and from Nahuatl (Mexican indigenous language) *pipixcan*, believed to be their name for this particular gull. Bernardino de Sahagún (1499–1590), a priest and ethnographer evangelising the Mexicans, recorded the bird (and the word) in his massive collection of the 'things of New Spain', also known as the Florentine Codex (de Sahagún 1829), as did Francisco Hernandez (1651) in his thesaurus of medical discoveries from 'New Spain'.

Pacific Gull (breeding resident)

See family introduction. A misleading name for what it is essentially a Southern Ocean bird, with Indian and Pacific Ocean extensions up the west and east coasts; probably a misunderstanding by John Latham, who named it.

Other names: Larger Gull (i.e. relative to the Silver Gull) reported by Gould (1848) to be used by 'the Colonists'. Jack Gull, an odd one in that 'jack' in a bird name generally implied something small (e.g. Jack Snipe); in this case it was more likely to have been a familiar name, like Willie Wagtail (although Jack Gull is used in its diminutive sense for the quite dissimilar Northern Hemisphere Black-legged Kittiwake *Rissa tridactyla*). Mollymawk and Nelly, by confusion with small albatross or giant-petrel (see Albatross family introduction, and Southern Giant-Petrel); Large-billed Gull.

Larus pacificus Latham, 1802 [LAH-rʊs pa-SI-fi-kʊs]: 'Pacific gull', see genus name, and Latin *pacificus*, peaceful, for the Pacific Ocean (but see common name for how misleading this is).

Black-tailed Gull (vagrant)

It is. See also family introduction.

Other names: Japanese Gull, that being the centre of its natural range. Temminck's Gull, for Coenraad Temminck (1778–1858), a Dutch ornithologist who directed the Dutch National Natural History Museum in Leiden for the last 38 years of his life; he has no discernible connection with this bird and it seems never to have borne his name in the scientific sense. It is possible that he supplied the specimen from which the Frenchman Louis Vieillot named the species in 1818 – Coenraad had inherited a vast bird collection from his father, who worked for the Dutch East India Company.

Larus crassirostris Vieillot, 1818 [LAH-rʊs kras-si-ROS-tris]: 'thick-billed gull', see genus name, and from Latin *crassus*, thick, and *rostrum*, bill.

Mew Gull (vagrant)

See family introduction. A tautologous name, because mew is one of the old English names that 'gull' supplanted (see family introduction); it was a first-choice name until the 17th century, and presumably is based on the call.

Other names: Common Gull – though not the commonest British gull, it is a very familiar one, in part due to its willingness to forage far inland; this doubtless explains the retention of 'mew' in its name.

Larus canus Linnaeus, 1758 [LAH-rʊs KAH-nʊs]: 'grey gull', see genus name, and from Latin *canus*, white or sometimes grey (usually in terms of grey hair). The bird is grey above and white below (pretty much a gull, really), and as Linnaeus (1758) described its colours as *canus* and *albus*, we can see he was using *canus* as meaning 'grey'.

Kelp Gull (breeding resident)

See family introduction. For habitat certainly, but it is not a kelp specialist.

Other names: Dominican Gull, from species name; Southern Black-backed Gull, for its similarity to the Lesser Black-backed Gull (and others in the group); Antarctic Gull, because it is found on many sub-Antarctic islands.

Larus dominicanus Lichtenstein MHK, 1823 [LAH-rʊs do-mi-ni-KAH-nʊs]: 'Black Friar gull', the specific name apparently chosen for the similarity of the bird's plumage to the white habit and black cloak (*cappa*) worn by Dominican monks, who were missionaries in Latin America where the type specimen was obtained.

Slaty-backed Gull (vagrant)

See family introduction. From the species name.

Larus schistisagus (Stejneger, 1884) [LAH-rʊs shi-sti-SA-gʊs]: 'gull with a slaty cloak', from the rock schist (similar to slate, and from Greek *schistos*, able to be split or divided, Late Latin *schistus*), plus Greek *sagos* (Latin *sagum*), a coarse woollen cloak. Stejneger describes the colour as 'pure bluish slate-grey'.

Lesser Black-backed Gull (vagrant)

See family introduction. By direct comparison with the similar but larger Great Black-backed Gull *L. marinus*, with which it co-exists in Europe. The other common British Gull, the Herring Gull *L. argentatus*, is grey-backed.

Larus fuscus Linnaeus, 1758 [LAH-rʊs FʊS-kʊs]: 'dusky gull', see genus name, plus Latin *fuscus*, dark or dusky.

Gull-billed Tern (breeding resident)

See family introduction. For the relatively short deep (gull-like) bill.

Other names: Long-legged Tern, for another characteristic, obvious when the birds are standing with other tern species; hence also Great-footed Tern, by Gould (1869) for his species name *G. macrotarsa*, thinking it a new species.

Gelochelidon nilotica (Gmelin JF, 1789) [ge-lo-KE-li-dohn nee-LO-ti-kuh]: 'laughing Nile swallow', see genus name, and for the river Nile. Although the species is very widespread, Egypt was where the type specimen was collected.

Caspian Tern (breeding resident)

See family introduction. From the species name.

Other names: Powerful Tern, by Gould (1848) from his name *Sylochelidon strenuus* – he thought the Australian birds were different from, and even larger than, the Caspian Terns he knew; Taranui, the name used in New Zealand and cited in Australia, being the Māori name for the bird.

Hydroprogne caspia (Pallas, 1770) [hi-dro-PROG-ne KAS-pi-uh]: 'Caspian water-swallow', see genus name, and for the Caspian Sea, type locality of this very widespread species.

Greater Crested Tern (breeding resident)

See family introduction. One of six species of crested terns in the genus *Thalasseus*, and marginally the 'greatest' of them.

Other names: Crested Tern, widely used in parts of Australia where there are no other familiar crested terns – the name dates from Latham (1823); Swift Tern, mostly used in Africa; Yellow-billed Tern, a surprisingly uncommon feature in terns, as the perceptive 'Colonists' recognised (Gould 1848), though the name is formally applied to a South American species *Sternula superiliaris*; Torres Strait Tern, for the bird described as *Thalasseus pelecanoides* by Lieutenant Phillip Parker King (and accepted as a good species by Gould 1848); Bass Strait Tern, for Gould's *T. poliocercus* (not now recognised).

Thalasseus bergii (Lichtenstein MHK, 1823) [tha-LA-se-ʊs BEHR-gi-ee]: 'Bergius' fisherman', see genus name, and for Karl (or Carl) Heinrich Bergius (1790–1818), a German collector who worked for the Berlin Museum in southern Africa for the last 3 years of his short life, where he collected the type specimen.

Lesser Crested Tern (breeding resident)

See family introduction. Distinctly crested, but smaller than the previous (similar) species.

Other names: Indian Tern, for the origin of the type species; Allied Tern, an old name drawn directly from an early invalid (and unexplained) species name *Sterna affinis*; Torres' Tern, for Gould's species name *T. torresii* for birds from Port Essington, believing them to represent a new species.

Thalasseus bengalensis (Lesson, 1831) [tha-LA-se-ʊs ben-ga-LEN-sis]: 'Bengal fisherman', see genus name, and sort of for the type location (the coasts of India). 'Bengal' was at times used as a catch-all name for any species coming from the Indian subcontinent, but it actually covered much of Northern India and Bangladesh (see 'Other names').

Little Tern (breeding resident)

See family introduction. Indeed a tiny tern, and the smallest in our region.

Other names: Sea-swallow, a widely used name for terns (cf. French *hirondelle de mer*); White-shafted Ternlet, not obvious given that black outer primaries are a characteristic, but it was the recommended name of RAOU (1913); Black-lored Tern, for a character that distinguishes it from Fairy Tern in breeding plumage; Least Tern, though there is a Northern Hemisphere species of the same name (*S. antillarum*), which disputes the claim.

Sternula albifrons (Pallas, 1764) [STEHR-nʊ-luh AL-bi-frons]: 'little white-fronted tern', see genus name, and from Latin *albus*, dull white, and *frons*, forehead. It's not so much the white forehead that is distinctive, as the pointy black cap that adjoins it in breeding plumage.

Saunders's Tern (vagrant)

See family introduction. From the species name, for Howard Saunders (1835–1907), much-travelled British merchant banker and publishing ornithologist, an acknowledged expert on gulls and terns.

Sternula saundersi (Hume, 1877) [STEHR-nʊ-luh SAWN-der-zee]: 'Saunders's tern', see genus and common names.

Fairy Tern (breeding resident)

See family introduction. Perhaps for the image of the hovering bird, clear white from below.

Other names: Little Tern, used by the 'Colonists of Western Australia' according to Gould (1848), but see Little Tern entry; Nereis Tern, from the species name; Sea Swallow or Little Sea Swallow, see Little Tern; Ternlet or White-faced Ternlet, for the white lores (see Little Tern).

Sternula nereis Gould, 1843 [STEHR-nʊ-luh ne-REH-is]: 'little sea-nymph', see genus name, and from the Greek Nereides, nymphs of the sea (the Mediterranean) and daughters of Nereus.

Bridled Tern (breeding resident)

See family introduction. For the black eye-line that extends to the nape.

Other names: Brown-winged or Smaller Sooty Tern, in contradistinction to the very similar but black-winged and slightly larger Sooty Tern; Dog Tern, for its yapping call; Panayan Tern, for the central Philippines island of Panaya where it was first collected in the late 18th century – this was the name by which Gould (1848) knew it.

Onychoprion anaethetus (Scopoli, 1786) [o-ni-ko-PREE-ohn an-eh-THEH-tʊs]: 'senseless clawed saw', see genus name, and from Greek *anaisthētos*, without sense or feeling – another allegedly silly seabird. Trying to find out where the 's' of the Greek word went has defeated us; perhaps it was just a misprint. Indeed, Newton, in his edition of Scopoli's writings, said that 'ordinary misprints ... are numerous' in it, and ascribed them 'to [Scopoli's] failing sight or the carelessness of his printer'. This is a kindness: the incorrect or missing references that Newton then listed are also numerous (Newton 1882).

Sooty Tern (breeding resident)

See family introduction. From the species name.

Other names: Wide-awake or Wide-awake Tern, for the breeding colonies, which never sleep (and don't stay awake quietly!); Egg-bird, a West Indian name, because of the significance of the egg harvest at colonies (colonies of most tern species have suffered thus, but this species was particularly targeted in the Caribbean for the eggs' supposed aphrodisiac qualities); Whale Bird, for its reported association with feeding whales, especially Humpbacks.

Onychoprion fuscatus (Linnaeus, 1758) [o-ni-ko-PREE-ohn fʊs-KAH-tuh]: 'dusky clawed saw', see genus name, and from Latin *fuscus*, dusky, *fuscatus*, darkened, for the black upper parts.

Roseate Tern (breeding resident)

See family introduction. For the beautiful rosy flush to the undersides when breeding.

Other names: Graceful Tern, true but we don't like the implication that other terns aren't – it comes from Gould's name *S. gracilis*, because he thought they were different from the Northern Hemisphere birds; Dougall's Tern, for Dr Peter McDougall, Scottish doctor and naturalist who recognised that the birds he saw on the Clyde in 1812 were different from those he was used to seeing, shot them and sent them to George Montagu who described them. It is not clear why Montagu didn't like the 'Mc' part of his name (though one of us suspects a Sassenach bias).

Sterna dougallii Montagu, 1813 [STEHR-nuh DOO-gal-li-ee]: 'McDougall's tern', see genus and common names.

Black-naped Tern (breeding resident)

See family introduction. For the distinctive breeding plumage of black nape extending forwards in a narrow band to the eye.

Other names: Apparently a good enough name not to need alternatives!

Sterna sumatrana Raffles, 1822 [STEHR-nuh soo-ma-TRAH-nuh]: 'Sumatran tern', see genus name, and for the type locality.

White-fronted Tern (non-breeding summer migrant)

See family introduction. Distinguishing this species (when breeding) from the similar Roseate Tern, bearing in mind that a bird's 'front' (or frons) is its forehead.

Other names: Black-billed Tern, again by contrast with the red-billed breeding Roseate Tern, and from Gould's name *S. melanorhyncha*; Southern Tern, for its limited range around south-eastern Australia and New Zealand.

Sterna striata Gmelin JF, 1789 [STEHR-nuh stree-AH-tuh]: 'striped tern', see genus name, and from Latin *stria*, a furrow. Named from Latham's common name Striated Tern, which described and depicted a juvenile bird with its barred plumage (Latham 1785).

Common Tern (non-breeding summer migrant)

See family introduction. The most familiar (but not the commonest) British tern, being widespread around southern England.

Other names: Asiatic Common Tern, not clear why, because it breeds right across the Northern Hemisphere mid-latitudes; Sea-swallow.

Sterna hirundo Linnaeus, 1758 [STEHR-nuh hi-RŬN-do]: 'swallow tern', see genus name, and from Latin *hirundo*, swallow. As in the genus names that use the term swallow, it is the rapid flight, long wings and forked tail that invite the comparison.

Arctic Tern (non-breeding summer migrant)

See family introduction. Breeds right across far northern latitudes.

Sterna paradisaea Pontoppidan, 1763 [STEHR-nuh pa-ra-di-SEH-uh]: 'paradise tern', see genus name, and from Greek *paradeisos*, enclosed park or garden, though a secondary meaning was 'stupid fellow'. The bird was named (without explanation) by a pietist bishop, Erik Pontoppidan, who among other things wrote a book entitled *The Truth about God-fearing*, so may have had an interest in paradise. What might there be of paradise in this little bird? Perhaps the suggestion in HANZAB (1990–2006, vol. 3) will appeal – that it relates to 'the ethereal plumage and buoyant flight'.

Antarctic Tern (vagrant)

See family introduction. Breeds as far south as the Antarctic mainland.

Other names: Wreathed Tern, for the Macquarie Island subspecies, though the reference is unclear; Kingbird, another Antarctic term, which the *Antarctic Dictionary* (Hince 2000) traced back to 1831, though the origin is again unknown; Sub-Antarctic Tern.

Sterna vittata Gmelin JF, 1789 [STEHR-nuh vi-TAH-tuh]: 'banded tern', see genus name, and from Latin *vittatus*, a ribbon or band. Gmelin (1789) described a 'white line' between the black head and the grey neck and body.

Whiskered Tern (breeding resident)

See family introduction. For the white cheeks of the breeding birds, highlighted between the black cap and grey chin; it first appears in an authoritative publication in RAOU (1926), though had already been used (e.g. in the *Forbes Advocate* in 1918 in a report of an expedition to Lake Cowal).

Other names: Marsh Tern, widely used, for its inland habitat – Gould (1848) in using the name noted that it is a 'denizen of inland waters, rather than those of the sea-coast', and it was still first-choice name of RAOU (1913) but only a back-up name in the second list (RAOU 1926); Black-fronted Tern, true in breeding plumage, but this is the norm rather than the exception among terns.

Chlidonias hybrida (Pallas, 1811) [kli-DOH-ni-as HEE-bri-duh]: 'hybrid swallow-like bird', see genus name, and Latin *hybrida*, a half-breed – a noun in apposition, so not needing to agree with the masculine genus name. This refers to apparent similarities between this bird and the Black Tern *Chlidonias niger* and Common Tern *Sterna hirundo*.

White-winged Tern (non-breeding summer migrant)

See family introduction. True, but not as helpful as the widely used first alternative (see 'Other names').

Other names: White-winged Black Tern, widely used and a precise description to contrast with the two closely related species. White-tailed Tern, true but not very diagnostic; White-winged Marsh Tern, see Whiskered Tern; White-winged Sea-swallow, see Little Tern; Black Tern, but see next species.

Chlidonias leucopterus (Temminck, 1815) [kli-DOH-ni-as loo-ko-TE-rʊs]: 'white-winged swallow-like bird', see genus name, and from Greek *leukopteros*, from *leukos*, white, and *pteron*, wing. See common name.

Black Tern (vagrant)

See family introduction. Very like the previous species, but with grey rather than white wings.

Chlidonias niger (Linnaeus, 1758) [kli-DOH-ni-as NEE-gehr]: 'black swallow-like bird', see genus name, and Latin *niger*, gleaming black.

STERCORARIIDAE (CHARADRIIFORMES): skuas and jaegers

Stercorariidae Gray GR, 1870 [stehr-ko-ra-REE-i-dee]: the Dung-bird family, see genus name *Stercorarius*.

'Skua' is of Faroese origin, reported and interpreted in the early 17th century by a Norwegian, then Latinised! The route was 'skuvur' to 'sku' to skua, which was at one time used as a specific name. Applied until recently to the whole family, it now tends to be reserved for the largest species, formerly of genus *Catharacta*, with 'jaeger' used for the three smaller species in genus *Stercorarius*. Even more recently, all species have been lumped into one genus (*Stercorarius*) so the distinction is somewhat pointless, but we are stuck with it for now at least.

'Jaeger' is German (Jäger) for a hunter; its use in this context is recorded in Britain in the early 19th century, though why the German was adopted is unclear. The name seemed to have gained popularity in North America. For instance, Newton didn't use it in England in his *Dictionary of Birds* (1896) in a long entry on skuas, and it still hasn't fully taken on there, whereas Coues (1882) already included it (spelled Jäger) in his *Checklist of North American Birds*. The RAOU recommended the adoption of the North Americanism here in 1978.

The genus

Stercorarius Brisson, 1760 [stehr-ko-RAH-ri-ʊs]: 'dung-bird', from Latin *stercus*, dung, *stercorarius*, of dung. So-called because skuas are often kleptoparasites, stealing the catches of other seabirds by making them disgorge their food. This food was for some reason thought to be (or equated with?) excrement.

The species

Skua taxonomy is notoriously complex and unclear, and until recently many authorities regarded many or all of the big southern '*Catharacta*' skuas as being subspecies of the Great Skua *C. skua*, now regarded as being a bird of north Atlantic coasts. Accordingly, some of the names used were not necessarily linked to either of the 'Australian' skuas.

South Polar Skua (non-breeding offshore winter migrant)

See family introduction. Breeds on the Antarctic mainland.

Other names: Antarctic Skua; McCormick's Skua, from the scientific name.

Stercorarius maccormicki Saunders, 1893 [stehr-ko-RAH-ri-ʊs mak-KOR-mi-kee]: 'McCormick's dung-bird', see genus name, and for Robert McCormick, British naval surgeon and natural historian who sailed on the *Beagle*, but left in a huff when he found that Darwin had the job of making the zoological collections, instead of him (as was traditional for the surgeon). By today's standards McCormick got an apparently unseemly pleasure out of shooting things, and perhaps felt that Darwin was robbing him of his perks. He did seem to spend a lot of potentially useful natural history time in interesting places complaining about the conditions; any climate more benign than that of Britain or the polar regions was not to his liking. He accompanied the Ross expedition to the Antarctic from 1839 to 1843 and collected the type specimen on Possession Island, but it was only named 50 years later in 1893 (3 years after his death at the very active age of 90). In case you're wondering, many of the 'Mc' names are Latinised to 'mac' in their species-name use.

Brown Skua (breeds Heard Island; non-breeding Australian winter migrant)

See family introduction. More uniformly dark brown than other large skuas, though all are brown.

Other names: Skua Gull, used by Gould (1848); Dark or Dark Southern Skua, for the blackish dark morph; Great or Antarctic Skua; Sea Hawk, Robber Gull, because all skuas pirate other seabirds' catches; Sea-hen, Skua-hen, Port Egmont Hen, for an early British settlement on the Falkland Islands, where they would have encountered this (or another!) species. The familiar 'hen' was probably due to its brazen scavenging around settlements and fishing vessels.

Stercorarius antarcticus (Lesson, 1831) [stehr-ko-RAH-ri-ʊs an-TARK-ti-kʊs]: 'southern dung-bird', see genus name, and Latin *antarcticus*, southern, as opposed to *arcticus* (Greek *arktikos*), pertaining to the constellation of the Bear (Greek *arktos*, Latin *ursa*) Ursa Major, hence northern. See common name.

Pomarine Jaeger (non-breeding offshore winter migrant)

See family introduction. From the species name.

Other names: Pomarine Skua, see family introduction and species name; Twist-tailed Skua, for the characteristically twisted spathulate central tail feathers – perhaps a more useful name, in fact.

Stercorarius pomarinus (Temminck, 1815) [stehr-ko-RAH-ri-ʊs po-ma-REE-nʊs: 'dung-bird with covered nose', see genus name, and from Greek *pōma*, a lid or cover, and *rhis/rhin-*, nose. This refers to a soft sheath covering the nostrils, which is not confined to this species.

Parasitic Jaeger (non-breeding winter migrant)

See family introduction. True – like all skua and jaeger species – for the lifestyle of thieving other seabirds' catches (and from the species name).

Other names: Arctic Skua or Jaeger, see family introduction, and it breeds across the Arctic – like all jaegers; Parasitic Skua or Gull, Robber Gull, Sea Pirate – see common name. Richardson's Skua, from an old species name (*Lestris richardsoni* by William Swainson in 1831, long after Linnaeus had already named it), for Sir John Richardson, Scottish Arctic explorer and friend of Sir John Franklin for whom he later searched; Richardson supplied the specimen that Swainson named. Even though the species name was dropped, RAOU (1913) recommended this as the common name.

Stercorarius parasiticus (Linnaeus, 1758) [stehr-ko-RAH-ri-ʊs pa-ra-SI-ti-kʊs]: 'parasitic dung-bird', see genus name, plus Latin for parasitic (see genus and common names). A pretty unpleasant combination, really!

Long-tailed Jaeger (non-breeding offshore winter migrant)

See family introduction. From the species name.

Other names: Long-tailed Skua, see family introduction.

Stercorarius longicaudus Vieillot, 1819 [stehr-ko-RAH-ri-ʊs lon-gi-KOW-dʊs]: 'long-tailed dung-bird', see genus name, and from Latin *longus*, long, and *cauda*, tail, for the distinctive long central tail feathers.

PHAETHONTIDAE (PHAETHONTIFORMES): tropicbirds

Phaethontidae Brandt, 1840 [feh-TON-ti-dee]: the Shiner family, see genus name *Phaethon*.

'Tropicbird' appears in literature by the late 17th century, based on early observations – not entirely accurate, as we now know – that the birds were to be seen only in the tropics.

'Bosunbird' (or Bos'n-bird) is a long-established sailors' name, though there is dispute as to whether it refers to the courtship calls – like a bosun's pipe – or to the tail streamers, supposedly resembling the bosun's marlin-spike.

The genus

Phaethon Linnaeus, 1758 [FEH-tohn]: 'shiner' (or 'cocky teenager') from Greek *phaethōn* meaning shining, also the name of a son of the sun god Helios/Phoebus. This boy pestered his father until, against his better judgment, Helios allowed his son to drive the chariot of the sun across the heavens for a day. Phaethōn was too weak to control the horses, got into a panic and set the Earth on fire, with many dire consequences, including his own death. A sad but familiar tale. 'Phaeton' was also used (from the mid 18th century) as the name of a light horse-drawn carriage. However, according to Newton (1896), Linnaeus gave the genus this name 'in allusion to its attempt to follow the path of the sun' – a reference to the nomadic habits of the bird. Unfortunately, Linnaeus (1758) did not confirm this or offer anything else. Maybe it was named just for the long tail trailing behind it.

The species

Red-billed Tropicbird (vagrant)

See family introduction. It does have a red bill, but so does Red-tailed; this one, however, was probably the first one encountered by Europeans, off the west coast of Africa and in the Red Sea.

Phaethon aethereus Linnaeus, 1758 [FEH-tohn eh-THE-re-us]: 'ethereal shiner', see genus name, and from Latin *aetherius*, Greek *aitherios*, heavenly, or pertaining to the upper air.

Red-tailed Tropicbird (breeding resident)

See family introduction. This is the only tropicbird with red central tail plumes, and Latham was using the name in 1801.

Other names: New Holland Tropicbird (Latham 1821); Red-tailed or Silver Bosun (Bos'n) Bird; Straw-tail, though little evidence of written use.

Phaethon rubricauda Boddaert, 1783 [FEH-tohn roo-bri-KOW-duh]: 'red-tailed shiner', see genus name, and from Latin *ruber*, red and *cauda*, tail.

White-tailed Tropicbird (resident Australian waters, breeding Christmas Island)

See family introduction. Unambiguous in Australia, but not so elsewhere in its range, where the Red-billed Tropicbird shares the characteristic.

Other names: White-tailed Bosun (Bos'n) Bird; Yellow-billed Tropicbird, a distinguishing character; Longtail or Marlin-spike; Golden Tropicbird or Bosunbird, for the golden morph found around Christmas Island.

Phaethon lepturus Daudin, 1802 [FEH-tohn lep-TOO-rʊs]: 'slender-tailed shiner', see genus name, and from Greek *leptos*, slender or delicate, and *-ouros*, tailed.

SPHENISCIDAE (SPHENISCIFORMES): penguins

Spheniscidae Bonaparte, 1831 [sfeh-NIS-ki-dee]: the Wedge family, see genus name *Spheniscus*.

Despite assertions that the name penguin derives from the Latin *pinguis*, fat, the general consensus is that the origin is unknown. It was originally applied to the superficially similar and now extinct Great Auk off Newfoundland in the 16th century. When penguins were discovered in the Magellan Strait the name was applied to them too.

The genera

Aptenodytes Miller JF, 1778 [a-teh-no-DEE-tehz]: 'flightless diver', from Greek *aptēnos*, wingless or unfledged, and *dutēs*, a diver. Named for its inability to fly (not literally without wings) – 'useless for flight', as Miller puts it, in Shaw (1796).

Pygoscelis Wagler, 1832 [pi-go-SKE-lis]: 'bum-leg', from Greek *pugē*, rump, and *skelos*, the leg (apparently indicating that the legs of a penguin are set far back on its body – the outboard motor principle also remarked on in the grebes, though Wagler (1832a) did not mention this feature). He said that one of the most striking characteristics of this bird is the considerable length of the tail, which serves as a rudder; indeed in German he called it tail-penguin. This was some of the last work Wagler did – he shot himself accidentally while collecting in August 1832, dying at the age of only 32.

Eudyptes Vieillot, 1816 [yoo-DIP-tehz]: 'good diver', from Greek *eu*, good, and *duptēs*, a diver.

Eudyptula Bonaparte, 1856 [yoo-DIP-tʊ-la]: 'good little diver', diminutive of *Eudyptes*.

Spheniscus Brisson, 1760 [sfeh-NIS-kʊs]: 'wedge-shaped [bird]', from Greek *sphēniskos*, a wedge-shaped plug, apparently referring to the shape of a swimming penguin (though Jobling 2010 suggested it refers to the shape of the flippers). Brisson (1760) explained his French name for the bird *Manchot*, (a person with one arm, or sometimes meaning with two hands missing), saying he called it that because of the shortness of its wings, but all he said about '*Spheniscus*' was that this is the name he gives to the birds.

The species

King Penguin (Heard and Macquarie islands and vagrant to Australian mainland)

See family introduction. Doubtless the sheer size of the bird – around 13 kg and nearly 1 m tall – was enough to induce superlatives in the form of the royalism.

Aptenodytes patagonicus Miller JF, 1778 [a-teh-no-DEE-tehz pa-ta-GON-i-kʊs]: 'Patagonian flightless diver', see genus name, with place-name Patagonia (the southern tip of South America), where the type specimen was found (see Gentoo Penguin).

Emperor Penguin (vagrant to Heard Island)

See family introduction. It seems that emperors outrank kings in this particular hierarchy: this species can be up to 1.3 m tall and weigh close to 40 kg.

Aptenodytes forsteri Gray GR, 1844 [a-teh-no-DEE-tehz FOR-ste-ree]: 'Forster's flightless diver', see genus name, and after Johann Reinhold Forster, naturalist aboard the *Resolution* under Captain Cook from 1772 to 1775. This species is one of several penguins he reported for the first time; it's unclear why John Gray in 1844 chose to so honour Forster, because no-one else seemed to have a good word to say for his personality (Gray wouldn't have met him). To be fair, Forster's natural history credentials are generally seen as very sound.

Gentoo Penguin (Heard and Macquarie islands and vagrant to Australian mainland)

See family introduction. The *Oxford Dictionary* defined a Gentoo as 'a pagan inhabitant of Hindostan, opposed to Mahommedan; a Hindu; in South India, one speaking Telugu'. If this is indeed the origin of the name, as is generally asserted, the connection seems likely to remain one of the great mysteries of the universe.

Pygoscelis papua (Forster JR, 1781) [pi-go-SKE-lis PAH-pʊ-uh]: 'Papuan bum-leg', see genus name. *Papua* is a misnomer, due to the mistaken or possibly fraudulent claim of French naturalist Pierre Sonnerat (1776) to have discovered it in New Guinea, where neither he nor the penguin had ever been. He also 'discovered' two other penguins while he wasn't there, one of which he called the New Guinea penguin (known to the modern world as *Aptenodytes patagonicus*, the King Penguin).

Adélie Penguin (vagrant to Heard and Macquarie islands, and occasionally to the mainland)

See family introduction. From the species name.

Pygoscelis adeliae (Hombron & Jacquinot, 1841) [pi-go-SKE-lis a-DEH-li-eh]: 'Adélie bum-leg', see genus name, and for Adélie Land. In 1840 Jules Sébastien César Dumont D'Urville claimed for France a sliver of Antarctic territory south-west of Macquarie Island. He named it *Terre Adélie* (Adélie Land) for his wife; the penguin was named for the place, not directly for her.

Chinstrap Penguin (vagrant to Australian mainland and to Heard Island)

See family introduction. For the thin black 'hat strap' that runs around the throat from ear to ear.

Pygoscelis antarcticus (Forster JR, 1781) [pi-go-SKE-lis ant-ARK-ti-kʊs]: 'Antarctic bum-leg', see genus name, plus Latin *antarcticus*, southern, as opposed to *arcticus* (Greek *arktikos*), pertaining to the constellation of the Bear (Greek *arktos*, Latin *ursa*) Ursa Major, hence northern.

Fiordland Penguin (regular visitor)

See family introduction. Breeds only in Fiordland, in New Zealand's South Island.

Other names: Thick-billed Penguin, widely used and descriptive, based on the species name; Victoria Penguin, perhaps by confusion with the Southern Rockhopper Penguin (or vice versa, see profile) but that doesn't help us; Crested Penguin (again – see Southern Rockhopper Penguin).

Eudyptes pachyrhynchus Gray GR, 1845 [yoo-DIP-tehz pa-ki-RIN-kʊs]: 'good diver with a thick bill', see genus name, and from Greek *pakhus*, thick, large, massive, and *rhunkhos*, bill, though this is not diagnostic even within the genus. Also has a yellow crest, neater and shorter than others of the genus.

Snares Penguin (vagrant)

See family introduction. For the Snares Islands, 200 km south of New Zealand, where the type specimen was taken.

Eudyptes robustus Oliver, 1953 [yoo-DIP-tehz ro-BUS-tʊs]: 'robust good diver', see genus name, and from Latin *robustus*, originally 'made of oak' (*robur*) but here meaning strong or robust.

Erect-crested Penguin (Macquarie Island, vagrant to Australian mainland)

See family introduction. Has enormous yellow backwardly erect bushy 'eyebrows'; those of other members of the genus are either much flimsier or downswept.

Other names: Sclater's Penguin, from the scientific name; Big-crested Penguin.

Eudyptes sclateri Buller, 1888 [yoo-DIP-tehz SKLEH-te-ree]: 'Sclater's good diver', see genus name, and it is often unclear which Sclater – Philip Lutley or his son William Lutley – are being honoured with *sclateri* (several genera include the name). Both were very eminent British ornithologists, who between them edited *Ibis*, journal of the British Ornithologists' Union, for the first 72 years of its life (barring a short 12-year interregnum). In this case, however, Buller makes clear that it was for Philip, who drew Buller's attention to two of the birds newly arrived at Regent's Park Zoo (and who had an interest in New Zealand zoology).

Southern Rockhopper Penguin (Heard and Macquarie islands and vagrant to Australian mainland)

See family introduction. Unlike others of their genus, Rockhoppers come ashore to their rocky wave-pounded breeding grounds 'hopping' with both feet together (not the usual definition of hopping!). In recent times the species has been split into two, with this one breeding on islands to the south of the next species.

Other names: Crested, Drooping-crested, Tufted Penguin, all descriptive – of this and every other member of the genus; Jackass Penguin, for the donkey-like call, which again it shares with many other penguin species; Victoria Penguin, a total mystery – we might suppose it referred to Victoria Land, part of the Antarctic mainland, but this penguin does not visit there.

Eudyptes chrysocome (Forster JR, 1781) [yoo-DIP-tehz kri-so-KOH-meh]: 'good diver with golden hair', see genus name, and from Greek *khrusos/khruso-*, gold/golden, and *komē*, hair. *Khruso-komos* was used by Herodotus (c. 435 BC) to describe birds 'with golden plumage'; it also evokes the sun god Apollo, being an epithet regularly applied to him. Here it refers to the bird's yellow crest, with its multi-directional plumes.

Northern Rockhopper Penguin (vagrant)

See family introduction. See comments under previous species.

Other names. All names of previous species have probably been applied to birds of this species also.

Eudyptes moseleyi Mathews and Iredale, 1921: see genus name and for Professor Henry Nottidge Moseley, 19th century British zoologist and member of the 4-year HMS *Challenger* expedition of the 1870s, travelling and studying the world's oceans.

Macaroni Penguin (Heard and Macquarie islands and vagrant to Australian mainland)

See family introduction. This rather foppishly dressed penguin reminded someone of the Macaronis of late 18th-century London: elegantly dressed dandies with continental European pretensions. The Macaroni Club was named, without irony, for exotic cuisine!

Eudyptes chrysolophus (von Brandt, 1837) [yoo-DIP-tehz kri-so-LO-fʊs]: 'good diver with a golden crest', see genus name, and from Greek *khrusos/khruso-*, gold/golden, and *lophos*, a crest or tuft, for the splendid growth of yellow-orange plumes – the longest of the lot in this genus.

Little Penguin (breeding resident)

See family introduction. This is the world's smallest penguin, and the common name (as used by Gould 1848) mirrors the species name.

Other names: Fairy Penguin, used by Gould for his species *Spheniscus undina* of Van Diemen's Land, now known to be the same species; Little Blue Penguin, for the bluish plumage.

Eudyptula minor (Forster JR, 1781) [yoo-DIP-tʊ-la MEE-nor]: 'smaller good little diver' – yes, it really is very small. See genus name, plus Latin *minor* (comparative of *parvus*, small).

Magellanic Penguin (vagrant)

See family introduction. A South American species whose breeding range centres on the Strait of Magellan, at the southern tip of the continent.

Spheniscus magellanicus (Forster JR, 1781) [sfeh-NIS-kʊs ma-ge-LA-ni-kʊs]: 'little Magellanic wedge', see genus name, plus the Latinised place-name.

OCEANITIDAE (PROCELLARIIFORMES): southern storm petrels

Oceanitidae Forbes, 1882 [o-se-a-NI-ti-dee]: the Sea-Nymph family, see genus name *Oceanites*.

These elegant little birds have long been regarded by sailors as harbingers of storms and held in some dread. Indeed the word 'storm-petrel' has entered the language in a metaphorical sense, though it did not appear in writing until the late 18th century, in the form 'stormy petrel'. All storm petrels were originally included in the one family Hydrobatidae, but recently the mostly Southern Hemisphere genera – already recognised as a separate subfamily – were separated into the family Oceanitidae. All were until recently called 'storm-petrels' (now two words).

The genera

Oceanites Keyserling & Blasius, 1840 [o-se-a-NEE-tehz]: 'Ocean-Nymph'; the Nymphs were low-ranked goddesses associated with different habitats or areas of nature. The Sea-Nymphs were made up of Oceanides, nymphs of the ocean and daughters of Oceanus, and Nereides, nymphs of the sea (the Mediterranean) and daughters of Nereus (cf. Greek ōkeanos, sea outside the Mediterranean).

Garrodia Forbes, 1881 [ga-ROD-i-uh]: 'Garrod's [bird]' after Alfred Garrod (1846–1879), pioneering English bird anatomist who died of tuberculosis aged just 33. The genus was named in 1881 by his friend English zoologist William Forbes, who himself died on the Niger River the following year. In 1879 he had succeeded Garrod as prosector (he prepared specimens for dissection) for the Zoological Society of London.

Pelagodroma Reichenbach, 1853 [pe-la-go-DROH-muh]: 'sea-racer' from Greek *pelagos*, the sea, and *dromas*, racing.

Fregetta Bonaparte, 1855 [fre-GET-tuh]: 'small frigate-bird', a diminutive of Fregata (see family Fregatidae).

Nesofregetta Mathews, 1912 [neh-so-fre-GET-tuh]: 'small island frigate-bird', see *Fregetta*, and from Greek *nēsos*, island. The island referred to is apparently Samoa.

The species

Wilson's Storm Petrel (migrant to Australian waters)

See family introduction. For Alexander Wilson, a US ornithologist, artist and engraver of the late 18th and early 19th centuries and author of the seven-volume *American Ornithology*. It was Wilson who originally named the species *Procellaria pelagica* in 1808, but this name was invalid (it was taken). The Frenchman Charles Lucien Bonaparte, nephew of Napoleon, corrected the error and named the bird *Thalassidroma wilsoni* for him in 1823; it subsequently came to light that Heinrich Kuhl's poorly illustrated publication of the name *P. oceanicus* from 3 years previously referred to the same species (as Bonaparte himself acknowledged) so that name took precedence. Wilson's name stuck, however, as a common name. (It is sometimes asserted that Wilson named the small warbler genus *Wilsonia* for himself – very naughty – but Bonaparte was responsible for that one too.)

Other names: Wilson's Storm-Petrel, the more usual form in Australia; Yellow-webbed Storm-petrel, descriptive, but not of a character that we can normally observe on a storm petrel – it was used by Gregory Mathews (1913). Mother Carey's Chicken, an old name, but one that only appears in writing at around the same time as stormy petrel. It is agreed that it refers to the biblical Mary, perhaps with the implication of warding off evil, but the derivation is debated. It is widely asserted that it derived from '*mater cara*' ('dear mother'), but the more obvious origin, from Mother Mary (with the Carey substituted to avoid blasphemy, in the manner of 'cripes') is also proposed.

Oceanites oceanicus (Kuhl, 1820) [o-se-a-NEE-tehz o-se-A-ni-kus]: 'ocean ocean-nymph', see genus name, plus Mediaeval Latin *oceanicus*, oceanic. Undoubtedly a seabird.

Grey-backed Storm Petrel (regular visitor to Australian waters)

See family introduction. Descriptive, but not uniquely so.

Garrodia nereis (Gould, 1841) [ga-ROD-i-uh ne-REH-is]: 'Garrod's sea-nymph', see genus names *Garrodia* and *Oceanites*.

White-faced Storm Petrel (breeding migrant to southern Australian islands)

See family introduction. Uniquely descriptive, and in use by Gould's time (1848).

Other names: White-faced Storm-Petrel, the more usual form in Australia; Frigate Petrel, used by Latham (1823) but not particularly obvious – it is not noted for following ships but it used to be in the genus *Fregetta*; White-breasted Storm-Petrel, another descriptive name; Mother Carey's Chicken, see Wilson's Storm Petrel.

Pelagodroma marina (Latham, 1790) [pe-la-go-DROH-muh ma-REE-nuh]: 'Sea sea-racer', see genus name, and from Latin *marinus*, of the sea. Another definite seabird, and apparently a fast one.

White-bellied Storm Petrel (Lord Howe Island, vagrant to Australian mainland waters)

See family introduction. See previous species; both Grey-backed and White-faced Storm Petrels also have white bellies. On the other hand, some individuals of this species do not. Gould named it *Thalassidroma leucogaster* (hence his common name, which we still use), again unaware of the pre-existing name, from a South American specimen.

Other names: White-bellied Storm-Petrel, the more usual form in Australia; Vieillot's Storm-Petrel, for Louis Vieillot, the 18th- and 19th-century prolific and pioneering French ornithologist who described it.

Fregetta grallaria (Vieillot, 1817) [fre-GET-tuh gral-LAH-ri-uh]: 'small frigate-bird on stilts', see genus name, plus Latin *grallae*, stilts. Vieillot (1816–19) simply said the bird has 'long spindly feet [legs]', an observation that he doesn't make about the other storm petrels. HANZAB (1990–2006, vol. 1a) had it as referring to the bird's 'habit of dangling the feet and treading the water' (like the other storm petrels).

Black-bellied Storm Petrel (passage migrant through Australian waters)

See family introduction. Certainly not unique because four of the species have entirely black undersides, but Gould was alluding to the black central belly with white surrounds when he named it *Thalassidroma melanogaster* from a bird from the southern Indian Ocean ('when viewed from the ship [it] is at once distinguished from all the other Petrels by the broad black mark which passes down the centre of the abdomen': Gould 1848). Curiously, he had forgotten that he'd already named it, (see species name), from the Atlantic Ocean.

Other names: Black-bellied Storm-Petrel, the more usual form in Australia; Gould's Storm-Petrel (see also Gould's Petrel).

Fregetta tropica (Gould, 1844) [fre-GET-tuh TRO-pi-kuh]: 'small tropical frigate-bird', see genus name, and from Latin *tropicus*.

New Zealand Storm Petrel (vagrant)

See family introduction. Described by Gregory Mathews from 19th-century museum specimens from New Zealand but not seen alive until very recently (2003); a population is now known off northern New Zealand.

Fregetta maoriana Mathews, 1932 [fre-GET-tuh ma-o-ri-AH-nuh]: 'small Māori frigate-bird', see genus name plus Modern Latin *maorianus*, for the Māori peoples of *Aotearoa* (New Zealand).

Polynesian Storm Petrel (vagrant to Norfolk Island)

See family introduction. Found throughout much of the southern Pacific.

Nesofregetta fuliginosa (Gmelin JF, 1789) [neh-so-fre-GET-tuh fʊ-li-gi-NOH-sʊs]: 'small sooty island frigate-bird', see genus name, and from Latin *fuliginosus*, sooty, from *fuligo*, soot. Although the bird has white underparts and throat (and used to be called *albigularis* for this) the rest of it is certainly very sooty.

DIOMEDEIDAE (PROCELLARIIFORMES): albatrosses

Diomedeidae Gray GR, 1840 [dee-o-me-DEH-i-dee]: the Diomedes bird family, see genus name *Diomedes*.

We might expect the word albatross to be based on *alba* (white), but it seems that this is quite incidental. The word chain seems to have begun in old Portuguese with *alcatruz*, the bucket of a water-lifting device, which was applied to pelicans in accordance with the belief that they carried water to their young in their bills. It seemed that one seabird was regarded as much like another and by the 17th century it had appeared in English as alcatras and applied to frigate-birds – which of course are black! The intermediate form algatross subsequently appeared but albatross is not known until the late 18th century, by then applied in its modern form. It may be that the 'white' notion influenced the change from 'alg' to 'alb', but this is not known. It is of interest that the meticulous Gould (1848) consistently spelt it 'albatros'.

'Mollymawk' – subsequently varied to mollyhawk – derives from old Dutch words meaning 'foolish gull' and was applied originally to the Northern Fulmar. Later it was used more widely, coming particularly to refer to the smaller albatrosses.

There have been major and ongoing upheavals in albatross taxonomy in recent times; using various DNA analyses, some authors recently began raising several subspecies to full species status, in part to emphasise the conservation importance and fragile status of many albatross taxa. After a somewhat tumultuous period, with increasing evidence that approach is gaining general acceptance.

The genera

Phoebastria Reichenbach, 1853 [foy-BAS-tri-uh]: 'prophetess' from Greek *phoibas*, a prophetess. As for what she might be prophesying: Elliott Coues (1882) suggested that the word *Phoebastria* and its fellow genus name *Phoebetria* are 'with great propriety and correct sentiment applied to albatrosses, the import of whose weird presaging will be felt by one who reads Coleridge's 'Antient Mariner' [sic] or himself goes down the deep in ships'.

Diomedea Linnaeus, 1758 [di-o-me-DEH-uh]: 'Diomedes [bird]' – Diomedes was the friend and companion of Odysseus in Homer's *Iliad*. His story is full of adventures, fighting, slaughter and good deeds. In the course of these escapades, he managed to alienate Aphrodite by wounding her with a spear – not a clever move, because she repaid him by causing his wife to be unfaithful to him. Much later, Diomedes' friends also offended her by taunting her and cockily suggesting that she'd have trouble doing them any harm, so she turned them into large white birds (unspecified) just to teach them a lesson. In another version of the story, they were transformed, after Diomedes' death, into 'gentle and virtuous birds' (Graves 1996). So what were these birds? Well, they turn out, according to Arnott's (2007) convincing explanation, to be shearwaters (see genus *Ardenna*), so it's a bit of a pity the name was applied by Linnaeus to albatrosses.

Phoebetria Reichenbach, 1853 [foy-BET-ri-uh]: 'prophet' from Greek *phoibētēs*, a prophet (see *Phoebastria*).

Thalassarche Reichenbach, 1853 [tha-la-SAR-keh]: 'ruler of the sea' from Greek *thalassa*, the sea, and *arkhos*, ruler. An appropriate name for birds of such majesty.

The species

Laysan Albatross (vagrant to Norfolk Island)

See family introduction. A major breeding colony is on Laysan Island, north-west of Hawaii.

Phoebastria immutabilis (Rothschild, 1893) [foy-BAS-tri-uh im-moo-TAH-bi-lis]: 'unchangeable prophetess', see genus name, and Latin *immutabilis*, unchangeable. Rothschild hinted at the meaning of the name by saying the bird is 'distinguished by attaining the coloration of the adult bird in the first plumage' (instead of a series of intermediate stages in several moults).

Wandering Albatross (regular visitor to Australian offshore waters)

See family introduction. Early basic experiments with marking and releasing birds confirmed the ability of this species to travel thousands of kilometres in days. Note that this species has recently been split into four – see the next three species, which were formerly subspecies of this one.

Other names: White-winged or Snowy Albatross, because only this species (as previously broadly defined) and the Royal Albatross have white upperwings (at least in old birds); this, however, is not useful in distinguishing from the Royals. Cape Sheep, a delightful traditional term perhaps for the appearance of the birds when nesting ashore (though they do not breed on either Cape). Given the traditional sailors' fear of harming albatrosses, the possibility that it refers to their eating characteristics seems remote, but in fact the taboo seems not to have been universal. Marsh (1948), in his account of an expedition to the sub-Antarctic, claimed that it was because 'early sailors used to catch albatrosses in the waters south of the Cape in order to make feather rugs and other articles out of their skins'. Man-of-War Bird, perhaps for its ship-following predilections.

Diomedea exulans Linnaeus, 1758 [di-o-me-DEH-uh eks-OO-lans]: 'banished friend of Diomedes', see genus name, plus Latin *exsulans*, exiled, presumably for the bird's imagined lonely wandering existence, though Linnaeus does not mention this.

Antipodean Albatross (regular visitor to Australian offshore waters)

See family introduction and notes under Wandering Albatross. From the species name, for the southern Pacific Antipodes Islands, the type locality; see also species name.

Other names. New Zealand Albatross; all Wandering Albatross alternatives potentially apply here too.

Diomedea antipodensis Robertson and Warham 1992 [di-o-me-DEH-uh an-ti-po-DEN-sis]: 'southern friend of Diomedes', see genus name and common name. From Greek *antipodes* literally meaning opposite feet, and now referring broadly to a point on the globe viewed from the opposite point, plus the suffix *-ensis*, usually indicating the place of origin of the type specimen.

Amsterdam Albatross (visits southern Australian waters)

See family introduction and notes under Wandering Albatross. From the species name, for the southern Indian Ocean Amsterdam Island, the type locality and sole breeding colony.

Other names. All Wandering Albatross alternatives potentially apply here.

Diomedea amsterdamensis Roux, Jouventin, Mougin, Stahl & Weimerskirch, 1983 [di-o-me-DEH-uh am-ster-da-MEN-sis]: 'Amsterdam friend of Diomedes', see genus and common name.

Tristan Albatross (visits southern Australian waters)

See family introduction and notes under Wandering Albatross. From Tristan da Cunha in the southern Atlantic, where it formerly bred.

Other names. All Wandering Albatross alternatives potentially apply here.

Diomedea dabbenena Mathews, 1929 [di-o-me-DEH-uh dab-ben-NE-nuh]: 'Dabbene's friend of Diomedes' see genus name, and for Roberto Dabbene (1863–1938), a much-published Italian-Argentinian ornithologist who developed the ornithological collection of the Museo Nacional in Buenos Aires.

Southern Royal Albatross (visitor to Australian waters)

See family introduction. The 'royal' is presumably in deference to its size, matched only by the Wandering. Recently the Royal Albatross was split in two; the breeding grounds of this species (on Campbell and Auckland Islands) are further south than those of the next.

Diomedea epomophora Lesson, 1825 [di-o-me-DEH-uh e-po-MO-fo-ruh]: 'friend of Diomedes with shoulder straps', see genus name, plus Greek *epōmis*, point of the shoulder (cf. ōmos, shoulder and upper arm) or a shoulder-strap, and *phoros*, bearing. Lesson (1825) wrote of the broad white marks on the elbow of each wing in his description of the '*Albatros à epaulettes*'.

Northern Royal Albatross (uncommon visitor to Australian waters)

See family introduction and notes on previous species. This species breeds on Chatham Island and the New Zealand mainland, well to the north of the previous one.

Diomedea sanfordi Murphy, 1917 [di-o-me-DEH-uh SAN-for-dee]: For Doctor Leonard Sanford (1868–1950), a US amateur ornithologist associated with the American Museum of Natural History. Murphy wrote: 'I take pleasure in naming this handsome bird in honor of Dr. Leonard C. Sanford, whose ornithological enthusiasm has led to the assembling of a unique collection of South American water birds.'

Sooty Albatross (uncommon visitor to Australian waters)

See family introduction. With the next species, the only all-dark albatrosses in Australian waters.

Other names: Dark-mantled Sooty Albatross, in contrast with the next species.

Phoebetria fusca (Hilsenberg, 1822) [foy-BET-ri-uh FŬS-kuh]: 'dusky prophet', see genus name, and Latin *fuscus*, dark or dusky.

Light-mantled Albatross (uncommon visitor to Australian waters)

See family introduction and previous species; this albatross has a distinctive pale grey neck and back.

Other names: Light-mantled Sooty Albatross, Grey-mantled Albatross.

Phoebetria palpebrata (Forster JR, 1785) [foy-BET-ri-uh pal-pe-BRAH-tuh]: 'prophet with eyelids', see genus name, and from Latin *palpebra*, eyelid. The name is based on the fact that the bird's eyes have a white or grey partial ring around them, though this is not a unique feature: the Sooty Albatross has it too.

Black-browed Albatross (breeds Heard and Macquarie Islands, common non-breeding migrant to Australian mainland waters)

See family introduction. From the species name. This species was recently split in two, giving rise to the next one as well.

Other names: Black-eyebrowed Albatross, used by Gould (1848); Mollymawk, Black-browed Mollymawk, see family introduction.

Thalassarche melanophris (Temminck, 1828) [tha-la-SAR-keh me-LA-no-fris]: 'black-browed ruler of the sea', see genus name, and from Greek *melas*, black, and *ophrus*, eyebrow. This is a distinctly frowny albatross, though probably no more so than the Yellow-nosed and Shy groups.

Campbell Albatross (present in Australian mainland waters)

See family introduction and previous species, of which this species was formerly regarded as a race. Breeds on Campbell Island, south of New Zealand.

Other names. Names assigned to previous species are also relevant here.

Thalassarche impavida (Mathews, 1912) [tha-la-SAR-keh im-PAV-i-duh]; 'fearless ruler of the sea', see genus name, and from Latin *impavidus*, fearless, intrepid. Mathews did not specify why he chose this name. However, judging by his report on this bird, we can guess it might have been by contrast with the similar but allegedly very shy *T. cauta*, or based on descriptions from others of the bird following ships, being unmoved by the presence of people in the breeding grounds, and being fierce in retaliation when actually booted off its position, grabbing its attacker and taking a 'piece out of trousers, hose and skin' (Dougall, in Mathews 1912b).

Shy Albatross (breeding migrant)

See family introduction. The origin is not obvious – the bird is quite willing to approach boats – but it is based on the scientific name, which of course explains it not at all. This species was recently split in three, raising races to species status, giving rise to the next two species.

Other names: Cautious Albatross, used by Gould (1848) from his species name; Shy Mollymawk, see family introduction.

Thalassarche cauta (Gould, 1841) [tha-la-SAR-keh KOW-tuh]: 'cautious ruler of the sea', see genus name, and from Latin *cautus*, cautious or heedful, but see common name. Gould (1840) did not explain his choice of name.

Chatham Albatross (vagrant)

See family introduction and under Shy Albatross. From the Chatham Islands, east of New Zealand, where it breeds.

Thalassarche eremita (Murphy, 1930) [tha-la-SAR-keh e-re-MEE-tuh]: 'hermit ruler of the sea', see genus name and from Latin *eremita*, hermit. Murphy described the bird as 'highly sedentary', 'apparently confined to a single small breeding area' (Pyramid Rock in the Chathams, an extremely remote location). Sounds ideal for a hermit!

Salvin's Albatross (uncommon visitor)

See family introduction and Shy Albatross. From the species name, for Osbert Salvin (1835–98), English mathematician-turned-naturalist and prolific shooter of birds, specialising in central America. He recognised that some of Rothschild's specimens – of this species – differed from Shy Albatross.

Thalassarche salvini (Rothschild, 1893) [tha-la-SAR-keh sal-VEE-nee]: 'Salvin's ruler of the sea', see genus and common names.

Grey-headed Albatross (uncommon visitor)

See family introduction. Descriptive, but not uniquely, especially compared to the Shy Albatross group.

Other names: Culminated Albatross, used by Gould (1848) for his species name *Diomedea culminata*, not realising that it was the same as Forster's bird; Grey-headed Mollymawk, see family introduction; Flat-billed Albatross, possibly true, certainly unhelpful; Grey-mantled Albatross; Yellow-billed Albatross, for the same character as the 'other' yellow-billed albatrosses. Gould's Albatross, for John Gould, collector, publishing ornithologist and natural history entrepreneur extraordinaire, though the connection is unclear – the bird was described well before his time, but it probably relates to his later description of the bird, as already noted.

Thalassarche chrysostoma (Forster JR, 1785) [tha-la-SAR-keh kri-so-STO-muh]: 'golden-mouthed ruler of the sea', see genus name, and from Greek *khrusos/khruso-*, gold/golden, and *stoma*, mouth, for the yellow line along the top and bottom edges of its bill.

Atlantic Yellow-nosed Albatross (vagrant)

See family introduction. The next species was recently split from this one, which breeds and forages in the south Atlantic. The yellow stripe down the top of the black bill is very distinctive.

Other names: Yellow-nosed Albatross, before the split; Yellow-billed Albatross, used by Gould (1848); Western Yellow-nosed Albatross; Yellow-nosed Mollymawk, see family introduction.

Thalassarche chlororhynchos (Gmelin JF, 1789) [tha-la-SAR-keh kloh-ro-RIN-kos]: 'green-billed ruler of the sea', see genus name, and from Greek *khlōros*, greenish-yellow, and *rhunkhos*, bill.

Indian Yellow-nosed Albatross (regular visitor to Australian waters)

See family introduction and previous species. Breeds in the south Indian Ocean but ranges as far east as New Zealand.

Other names. See previous species.

Thalassarche carteri (Rothschild, 1903) [tha-la-SAR-keh CAR-tuh-ree]: see genus name, and for Thomas Carter, a one-eyed Yorkshire adventurer who came to Western Australia from England in 1887 and bought a station near Cape Range, virtually commuting between Yorkshire and Australia (including a trip to England to get married) and doing much collecting of bird specimens, some of which, including this one, went to Rothschild.

Buller's Albatross (non-breeding migrant)

See family introduction. From species name.

Other names: Buller's Mollymawk, see family introduction; Buller's Shearwater.

Thalassarche bulleri (Rothschild, 1893) [tha-la-SAR-keh BŬL-ler-ee]: 'Buller's sea-ruler', see genus name, and for Sir Walter Buller, 19th-century New Zealand barrister who represented Māori interests in complex 1890s land claims, and publishing ornithologist who gained some notoriety for his rapacious collection of rare species. Beolens and Watkins (2003) also describe him as a cabinet minister, but other authorities (e.g. National Library of New Zealand 1966; Encyclopaedia of New Zealand 1966) do not support that assertion. The fact that Lord Rothschild named the bird in honour of Buller is ironic, because shortly afterwards Rothschild was outraged when he determined that a specimen of the now extinct Laughing Owl (*Sceloglaux albifacies*), originating with Buller and which he was intending to buy, was doctored (Worthy 1997). Among other pretty scornful criticisms, Rothschild (1907) wrote 'The tail, however, is 'skillfully' (as Buller calls it, though I should use a less complimentary adverb), stuck in, and does not belong to a *Sceloglaux*, but to an Australian *Ninox*, and also some feathers on the neck are foreign.'

HYDROBATIDAE (PROCELLARIIFORMES): northern storm petrels

Hydrobatidae Mathews, 1913 [hi-dro-BA-ti-dee]: the Water-walker family, from the genus name *Hydrobates* Boie, 1822 [hi-dro-BAH-tehz] 'water-walker', from Greek *hudōr/hudro*, water, and *batēs*, one that treads (cf. bainō, walk or step), for the way these birds appear to step or dance on the sea surface.

See Family comments under Oceanitidae. The members of Hydrobatidae are mostly Northern Hemisphere species. Note that all were until recently called 'storm-petrels' (now two words).

The genus

Oceanodroma Reichenbach, 1853 [o-se-a-no-DRO-muh DRO-muh] 'Ocean racer' from Greek *ōkeanos*, ocean, and *dromas*, racing.

The species

Band-rumped Storm Petrel (vagrant)

See family introduction. For the complete white band above the black tail, a characteristic shared by only one other member of the genus.

Oceanodroma castro (Harcourt, 1851) [o-se-a-no-DRO-muh CAS-tro]: 'castro ocean racer', see genus name, and after the name used by the inhabitants of Madeira for this bird. Harcourt explained it thus: 'There is another petrel called by the natives 'Roque de Castro,' and pronounced 'Roque de Crasto,' which is likewise an inhabitant of the Dezerta Islands ... I have called it *Thalassidroma castro*'. Jobling (2018) offers the following: 'The late Alec Zino (in litt.) told me 'roque de castro' means 'rock of the castle' in Old Portuguese, but found it difficult to believe that illiterate fishermen would have invented such a meaningless name. He pointed out, however, that the fishermen of the village of Machico on Madeira (one of whose ancestors doubtless supplied Harcourt with his specimen) still habitually metathetise letters in common speech, especially 'r' (e.g. Prediz for Perdiz partridge; Trocaz for Torcaz or Torquaz pigeon; Crasto for Castro).' So far so good – makes a lot of sense, though it gives no clue to the meaning. Next, however, Alec Zino suggested that 'the name is probably an onomatopoeia for one of the birds' incessant brooding calls, 'rrrrrrr oquedecastro''. Listen to some recordings online and see what you think.

Swinhoe's Storm Petrel (vagrant)

See family introduction. Robert Swinhoe was a mid 19th-century British diplomat and collecting naturalist who spent 19 very profitable years in China and provided a substantial number of the specimens for Gould's *Birds of Asia*. It was Swinhoe who collected and named this species.

Oceanodroma monorhis (Swinhoe, 1867) [o-se-a-no-DRO-muh MON-o-ris]: 'single-nostrilled ocean racer', see genus name, plus Greek *monos*, single, and *rhis*, nostril. The tubular nostrils of these birds are joined and this nasal passage is what enables them to smell – storm petrels are now known to have an extremely powerful sense of smell. Perhaps a misleading name though, in that the characteristic is not confined to this species.

Leach's Storm Petrel (uncommon migrant to Australian waters)

See family introduction. For William Leach, an early 19th-century multidisciplinary zoologist at the British Museum, who purchased a specimen in a job-lot when adventurer-collector William Bullock sold up. (One version of the story says that they were bought from his estate, but Bullock lived for perhaps 30 years after the sale.) This bird was initially described by Temminck in 1820 with the specific name *leachii*, but it transpired that the species had already been named by Vieillot, for a northern French specimen belonging to Louis Baillon (see Baillon's Crake).

Other names: Fork-tailed Storm-petrel, descriptive, though also true of the next species.

Oceanodroma leucorhoa (Vieillot, 1818) [o-se-a-no-DRO-muh loo-ko-ROH-uh]: 'white-rumped ocean racer', see genus name, plus Greek *leukos*, white, and *orrhos*, rump. This species is far from being the only white-rumped storm petrel, though it does have a distinctive dark dividing line through the white.

Tristram's Storm Petrel (vagrant)

See family introduction. From the species name.

Oceanodroma tristrami Salvin, 1896 [o-se-a-no-DRO-muh TRIS-tra-mee]: 'Tristram's ocean racer', see genus name, and for the Rev. Henry Tristram (died 1906), who was a collector of bird specimens, a Darwinian naturalist, traveller and writer.

Matsudaira's Storm Petrel (rare visitor to Australian waters)

See family introduction. From the species name.

Other names: Sooty Storm-petrel, descriptive for the only all-dark storm petrel in Australian waters (except for dark phase White-bellied Storm Petrel).

Oceanodroma matsudairae (Kuroda, Nagamichi, 1922) [o-se-a-no-DRO-muh ma-soo-DIE-reh]: 'Matsudaira's ocean racer', see genus name, and for Yorikatsu Matsudaira, a Japanese ornithologist from the first half of the 20th century who collected the original specimen.

PROCELLARIIDAE (PROCELLARIIFORMES): petrels, prions, shearwaters

Procellariidae Leach, 1820 [pro-sel-la-REE-i-dee]: the Storm-bird family, see genus name *Procellaria*.

The word 'petrel' has an interesting and somewhat vexed history. For a start, it is unusual in that it seems to have arisen in English and been adopted by other European languages. It is first recorded in 1602; the great English sailor cum pirate cum biologist cum polymath William Dampier used the word in 1703 and claimed that it was derived from Peter – the one who was supposed to have walked on water (Dampier 1703). Dampier's description indicates that he was referring to storm petrels, which patter across the ocean surface. Lockwood (1984) was adamant, however, that Dampier was retrospectively constructing the meaning. Lockwood contended that the original form was pitteral – a diminutive formed (analagously to dotterel) from 'pitter patter'; this word, however, does not appear in writing until 1676. Lockwood insisted that it must have around for much longer and that it was probably being used by illiterate sailors long before it was recorded. We are never likely to know for sure.

'Prion' is from Greek for a saw, referring to the comb-like lamellae on the upper bill for filtering plankton; although it was widely used elsewhere in the 19th century, dove-petrel was the preferred group name in Australia until the RAOU formalised prion in the 20th century (RAOU 1926).

'Shearwater' comes from the characteristic superbly controlled flight of the birds, often with a wingtip cutting the water. It was first known to be used by the pioneering English biologist John Ray in the 17th century.

'Diving-Petrel', because although many petrels plunge-dive for food, the diving-petrels are specialists at hunting underwater, swimming with their wings. Disturbed on the water, they will often dive rather than fly. (Many authoritative accounts place the diving-petrels in their own family.)

The genera

Macronectes Richmond, 1905 [mak-ro-NEK-tehz]: 'big swimmer' from Greek *makros*, large, and *nēkhō*, to swim, *nēktos*, swimming. Why its particular swimming skills were singled out is unclear.

Fulmarus Stephens, 1826 [fʊl-MAH-rʊs]: 'nasty gull', from the common name, for its regurgitating habits (see Southern Fulmar).

Thalassoica Reichenbach, 1853 [tha-la-SOY-kuh]: 'sea-dweller' from Greek *thalassa*, sea, and *oikeo*, to inhabit.

Daption Stephens, 1826 [DAP-ti-on]: 'painted diving bird'. Generally considered to be an anagram of *pintado*, one of the common names of the Cape Petrel *D. capense*. Stephens himself gave no clue (Stephens and Shaw 1826), though he referred to 'the slight notice of *Daption capensis* in the *Règne Animal*'; this makes it sound as if he got the name from Cuvier (1817), but Cuvier did not use the word. Coues (1882) preferred the Latin form *Daptium* and had it as deriving from 'Gr. *daption* or *duption*, a diminutive of *duptēs* or *dutēs* a diver', a derivation which he justified at some length. We may never know.

Pagodroma Bonaparte, 1856 [pa-go-DROH-muh]: 'ice racer', from Greek *pagos*, rock, ice or frost, and *dromas*, racing.

Halobaena Bonaparte, 1856 [ha-lo-BEH-nuh]: 'sea-walker', from Greek *hals* or *halos*, the salt sea, and *baino*, to walk or go.

Pachyptila Illiger, 1811 [pa-ki-TIL-uh]: 'bird with thick down', from Greek *pakhus*, thick, and *ptilon*, feather, down. This refers to the very dense downy plumage of the prions.

Aphrodroma Olson, 2000 [aff-ro-DROH-muh]: 'foam racer', from Greek *aphros*, foam or froth, and *dromas*, racing, 'from the habitat of the species in stormy seas' as Olson wrote.

Pterodroma Bonaparte, 1856 [te-ro-DROH-muh]: 'winged racer', from Greek *pteron*, wing, and *dromas*, racing.

Pseudobulweria Mathews, 1936 [syoo-do-bʊl-WE-ri-uh]: 'false Bulwer's [bird]', from Greek *pseudo*, false, *Bulwer's* as in Bulwer's Petrel. An imposter – Bulwer's seems prone to imitation (see also Jouanin's Petrel).

Procellaria Linnaeus, 1758 [pro-sel-LAH-ri-uh]: 'Storm-bird', from Latin *procella*, a violent wind or storm. These birds are at home in violent storms, and even fly more powerfully in stronger winds.

Calonectris Mathews & Iredale, 1915 [ka-lo-NEK-tris]: 'beautiful swimmer', from Greek *kalos*, beautiful, and *nēkhō*, to swim, *nēktos*, swimming.

Ardenna Reichenbach, 1853 [ar-DEN-nuh]: 'Ardenna'. From 'Artenna': Aldrovandus (1603) explained that the Augustine friars, inhabitants of the Tremiti islands (in the Adriatic Sea), known from antiquity as the islands or boulders of Diomedes (see family Diomedidae) use the word *Artenna* for what he calls

Diomedean birds – not that his discussion of what constitutes one of these is very conclusive. Arnott (2007) enlightened us: details in the classical references to this bird point to Cory's Shearwater, *Calonectris diomedea*, which still nests in that area. Reichenbach (1852) cited Aldrovandus but does not explain his reasoning for the change from 't' to 'd'. Penhallurick and Wink (2004) retained that change in their renaming of several shearwater species.

Puffinus Brisson, 1760 [pʊf-FEE-nʊs]: 'the puffin-or-is-it?' Well, no, it's a shearwater. A small mystery, with many an inventive explanation. The simplest we can come up with is this: the Middle English word *puffin* and its variants (*paphin*, *pophin*, *pupin*, etc.) had been around for centuries, meaning a chubby young nestling, much favoured for eating. It seems that came to be used in particular for two birds: the Puffin (genera *Fratercula* and *Lunda*) (John Caius 1570) and the Manx Shearwater (Willughby and Ray 1676; Ray 1713). Caius (1570) was very definitely talking about the Puffin – bill, red and pale ochre and with four grooves, reddish feet, nesting in rabbit burrows not far from the sea. Willughby and Ray were very definitely talking about the Shearwater (long pointed black bill with a hook, dark feet, etc.), calling it *Puffinus anglorum* and Puffin of the Isle of Man (i.e. Manx Shearwater). Brisson (1760) was talking about the Shearwater and his specimen came from St Kilda, which was one of the centres of the centuries-long trade in Manx Shearwater meat and eggs, which were much valued (Cocker and Mabey 2005). Both these birds were collected and eaten, and both are burrow-nesting species (Cocker and Mabey 2005). Both were common on the Calf of the Isle of Man (Ralphe 1905). Whether the use of *Puffinus* for the shearwater was, as some assert, a mistake of Ray's in the first place or simply an overlap of meaning – as seems to us more likely, especially given the evidence that the Puffin was often called the Sea-parrot rather than Puffin (Ralphe 1905) – we shall probably never know.

Pelecanoides Lacépède, 1799 [pe-le-ka-NOY-dehz]: 'something like a pelican', from Greek *pelekan*, a pelican, and *-oeidēs*, resembling (referring to the Diving-Petrel's throat pouch).

Bulweria Bonaparte, 1843 [bʊl-WE-ri-uh] 'Bulwer's [bird]', for the Rev. James Bulwer, a 19th-century English naturalist who 'collected' this species at Madeira off Morocco, where he was wont to winter, apparently for his health. His interests in general were in shells rather than birds, however. Beolens and Watkins (2003) described him as a Scot, but more detailed accounts leave little doubt that he was born in Norfolk, lived later in Ireland and eventually returned to Norfolk (Mearns and Mearns 1988, 1992).

The species

Southern Giant Petrel (Macquarie Island, regular migrant to Australian coastal waters)

See family introduction. The two Giant Petrels (or Giant-Petrels, as they are generally known) are indeed the giants of the family, of near-albatrossian proportions; the name has been used since at least the early 19th century. From an Australian perspective the 'southern' seems perverse, since it is found further north on both east and west coasts than the Northern Giant Petrel! However, the name refers to its total range, which includes the Antarctic land mass.

Other names: Southern Giant-Petrel; Giant Petrel (the two species were separated in 1966); Giant Fulmar (for similarity to a related genus of much smaller petrels, the northern species of which would have been familiar to English sailors); Stinker (a defensive habit of both adults and chicks is to regurgitate thoroughly – and much of the diet comprises carcasses); Nelly (unknown, but presumably a human name given to a familiar bird that regularly follows ships); Vulture of the Seas (for its largely carrion diet on land); Mother Carey's Goose (see 'Mother Carey's Chicken' under Hydrobatidae – presumably the 'goose' was to emphasise the size difference).

Macronectes giganteus (Gmelin JF, 1789) [mak-ro-NEK-tehz gi-GAN-te-ʊs]: 'giant big swimmer', see genus name, with Latin *giganteus*, describing one of a fabulous race of monsters.

Northern Giant Petrel (Macquarie Island, regular migrant to Australian coastal waters)

See family introduction and previous species; this species does not range as far south as Antarctica.

Other names: Northern Giant-Petrel, see previous species; Hall's Giant-Petrel, for Robert Hall who in 1899 published a key to Australian birds with habitats and distribution. He was a foundation member of the RAOU in 1901 and later was curator of the Tasmania Museum and Botanic Gardens. In 1912 and 1923 he was president of the RAOU. Since the species was recognised as distinct from the Southern Giant-Petrel only in 1966, all these names have also been applied to this species, which was formerly subspecies *halli*.

Macronectes halli Mathews, 1912 [mak-ro-NEK-tehz HAW-lee]: 'Hall's big swimmer', see genus name and common name.

Southern Fulmar (irregular visitor to Australian coastal waters)

From Gaelic *fulmair* (Hebridean Norn dialect), deriving in turn from Old Norse *full* meaning foul, and *mar*, a gull; like many seabirds, it defensively regurgitates food (only when provoked!). The Southern Fulmar is a bird of Antarctic and associated waters; the Northern Fulmar has an equivalent Arctic and sub-Arctic range. Note that the name is commonly applied to other petrel species by non-birding sailors.

Other names: Antarctic Fulmar; Silver-grey Petrel (used by Gould 1848 and RAOU 1913, but see next species) or Fulmar; Slender-billed Fulmar (cf. the Northern Fulmar).

Fulmarus glacialoides (Smith A, 1840) [fʊl-MAH-rʊs gla-si-a-LOY-dehz]: 'nasty gull like the Northern Petrel', see genus name, and from *glacialis*, the species name of the Northern Petrel (Latin meaning, icy) and Latin *-oides*, resembling that bird.

Antarctic Petrel (vagrant)

See family introduction. The species rarely leaves the limits of the pack ice; among the petrels only the Snow Petrel is so tightly tied to the Antarctic.

Other names: Fulmar (see previous species); Silver-grey Petrel, confusingly used by RAOU (1926), see previous species.

Thalassoica antarctica (Gmelin JF, 1789) [tha-la-SOY-kuh ant-ARK-ti-kuh]: 'southern sea-dweller', see genus name, plus Latin *antarctica*, southern (see common name).

Cape Petrel (migrant to Australian coastal waters)

See family introduction. A distinctive and common species across southern oceans, which would have been associated by sailors with journeys around Cape Horn or the Cape of Good Hope, though Linnaeus made it clear he was referring to the African cape (see species name). The name was well established by the early 19th century – it was the name used by Gould (1848).

Other names: Pintado Petrel, from the Spanish for painted; Cape Pigeon, used by 'voyagers' (Gould 1848); Cape Fulmar; Pied, Spotted or Black-and-white Petrel.

Daption capense (Linnaeus, 1758) [DAP-ti-on ka-PEN-se]: 'painted diving Cape bird', see genus name, with Cape plus *-ense* indicating a place of origin, in this case the Cape of Good Hope (Linnaeus 1758).

Snow Petrel (vagrant)

See family introduction. A snow-white species that rarely leaves the Antarctic pack ice.

Pagodroma nivea (Forster G, 1777) [pa-go-DROH-muh ni-VEH-uh]: 'snowy ice racer', see genus name, plus Latin *niveus*, white or snowy. A beautiful and apt name.

Blue Petrel (Macquarie Island and migrant to Australian offshore waters)

See family introduction. Distinctly blue-grey above; the name was used by Latham (1801).

Halobaena caerulea (Gmelin JF, 1789) [ha-lo-BEH-nuh seh-ROO-le-uh]: 'blue sea-walker', see genus name, plus Latin *caeruleus*, used for a variety of different shades of blue including greenish-blue and azure, and connected with both sky and sea.

Broad-billed Prion (uncommon visitor to Australian coastal waters)

See family introduction. The plankton-filtering bill adaptation of the group, alluded to in the family introduction, is carried to an extreme in this species. Gould (1848) used this name.

Other names: Forster's Prion, for the name's author, as used by Jardine and Selby (1826–35, vol. 1); Broad-billed Petrel, used by Latham (1801); Dove Petrel (also Blue or Broad-billed Dove Petrel), for the relatively small and delicate appearance and blue-grey colouring; Whale-bird, because as plankton feeders they doubtless associate with whales; Ice-bird as they are found far to the south, though this hardly makes them unique! Note that because the prions are notoriously hard to distinguish from each other, all these names have been applied to the other species.

Pachyptila vittata (Forster G, 1777) [pa-ki-TIL-uh vit-TAH-tuh]: 'banded bird with thick down', see genus name, plus Latin *vittatus*, a ribbon or band. Forster described it in his diary as 'the blue petrel, so called from its having a bluish-grey colour, and a band of blackish feathers across the whole wing'. A week later he saw the blue petrel again and named it *Procellaria vittata* (Thomas and Berghof 2000).

Salvin's Prion (visitor to Australian coastal waters)

See family introduction. From the species name.

Other names: Medium-billed Prion, for contrast with Broad-billed Prion, but not very useful! See also names for Broad-billed Prion.

Pachyptila salvini (Mathews, 1912) [pa-ki-TIL-uh SAL-vi-nee]: 'Salvin's bird with thick down', see genus and for Osbert Salvin, a prominent 19th-century English naturalist, collector and writer, who concentrated especially on Latin America.

Antarctic Prion (uncommon visitor to Australian coastal waters)

See family introduction. The only member of the group regularly found breeding in continental Antarctica.

Other names: See names under Broad-billed Prion; the name Ice-bird probably applied initially to this species and was then transferred to the others. Also Snowbird; Banks' Prion, Petrel or Dove Petrel, from an earlier specific name honouring the famous botanist Sir Joseph Banks.

Pachyptila desolata (Gmelin JF, 1789) [pa-ki-TIL-uh de-so-LAH-tʊs]: 'Desolation Islands bird with thick down', see genus name, plus Latin *desolata*, abandoned, empty of people, and here referring to the name given by Captain Cook to the islands we now know as Kerguelen. There is another Desolation Island, off Tierra del Fuego and also close to Antarctica.

Slender-billed Prion (uncommon visitor to Australian coastal waters)

See family introduction. By contrast with the Broad-billed Prion in particular, and while not very useful in the field, the name does reflect that this species is a less-specialised plankton filterer than most other prions.

Other names: See names for Broad-billed Prion. Also Thin-billed or Narrow-billed Prion.

Pachyptila belcheri (Mathews, 1912) [pa-ki-TIL-uh BEL-cher-ee]: 'Belcher's bird with thick down', see genus name, and after Sir Charles Belcher (1876–1970), a judge and zoologist born in Victoria, who was a founder member of the RAOU.

Fairy Prion (breeding resident)

See family introduction. The smallest of the prions.

Other names: See names for Broad-billed Prion. Dove-like Prion, used by Gould (1848); Fairy Dove Petrel; Blue Petrel, a confusing name, given the 'other' Blue Petrel and the fact that the Fairy Prion's plumage is essentially the same colour as that of other prions.

Pachyptila turtur (Kuhl, 1820) [pa-ki-TIL-uh TOOR-toor]: 'Dove with thick down', see genus name, plus Latin *turtur*, a turtle-dove, presumably so named for its dainty appearance, though Kuhl (1820a) did not say.

Fulmar Prion (Heard Island, vagrant to Australian coastal waters)

See family introduction. It is not obvious to us what makes this prion any more fulmar-like than any of the others! (See Southern Fulmar.)

Other names: See names for Broad-billed Prion.

Pachyptila crassirostris (Mathews, 1912) [pa-ki-TIL-uh kras-si-ROS-tris]: 'thick-billed bird with thick down', see genus name, plus Latin *crassus*, thick, and *rostrum*, bill. Although this bird is seen as very like the Fairy Prion, it has a thicker, deeper bill.

Kerguelen Petrel (regular visitor to Australian coastal waters)

See family introduction. The species breeds on Kerguelen Island (some 500 km north-west of Heard Island) and on several other sub-Antarctic islands. Kerguelen was named for Captain Yves-Joseph de Kerguelen-Trémarec, who discovered the island in 1771.

Aphrodroma brevirostris (Lesson, 1831) [aff-ro-DROH-muh bre-vi-ROS-tris]: 'short-billed foam racer', see genus name, and from Latin *brevis*, short, and *rostrum*, bill.

Great-winged Petrel (breeding migrant)

See family introduction. A large powerful petrel, but its wings are not especially 'greater' than those of some other species; based on the species name.

Other names: Great-winged or Long-winged Fulmar (see Southern Fulmar); Grey-faced Petrel or Fulmar (true, but not unique).

Pterodroma macroptera (Smith A, 1840) [te-ro-DROH-muh mak-ro-TE-ruh]: 'large-winged winged racer', see genus name, and from Greek *makros*, large, and *pteron*, wing. See common name for wing size.

White-headed Petrel (non-breeding migrant)

See family introduction. Strictly the head is pale grey, but among dark-winged Australian petrels only the Kermadec Petrel could lay claim to even a similar description; it was already in use in Gould's time (1848).

Other names: White-headed Fulmar (see Southern Fulmar).

Pterodroma lessonii (Garnot, 1826) [te-ro-DROH-muh LES-so-ni-ee]: 'Lesson's winged racer', see genus name, and after René Lesson (1794–1849), French naval surgeon and highly influential ornithologist and botanist who travelled widely on French expeditions of discovery.

Atlantic Petrel (vagrant)

Normally restricted to the South Atlantic.

Pterodroma incerta (Schlegel, 1863) [te-ro-DROH-muh in-SER-tuh]: 'uncertain winged racer', see genus name, plus Latin *incertus*, doubtful or uncertain. The doubt was about where the bird really fitted in taxonomically.

Providence Petrel (breeds Lord Howe Island, regular visitor to Australian coastal waters)

See family introduction. The history of this name refers to events not the least providential to the bird. They bred on Norfolk Island and when the convict colony was set up there the birds (unwillingly) provided food for convicts and soldiers for three months in 1790 when the supply ships failed. A record of the killings reveals that some 170 000 birds died in that time, an onslaught (followed by another by feral pigs) from which the population never recovered. It was gone for good by 1800. At least the colony expressed its gratitude in this name.

Other names: Bird of Providence; Solander's Petrel or Fulmar, from the specific name; Brown-headed Petrel or Fulmar (descriptive, though not very usefully so, but still recommended by RAOU 1913). Big Hill Muttonbird, which we assume refers to Big Hill Saddle on Mt Gower, Lord Howe Island, where the species breeds; we suggest that Bill Hill Muttonbird, which also appears, is a corruption of this. (For discussion of muttonbird, see Wedge-tailed Shearwater.)

Pterodroma solandri (Gould, 1844) [te-ro-DROH-muh so-LAN-dree]: 'Solander's winged racer', see genus name, and for Daniel Solander, Swedish naturalist, a star student of Linnaeus and member of Joseph Banks' staff on the *Endeavour* voyage of 1770. There appears to be no particular connection of Solander with the bird, but Gould honoured him with the name in 1844, which was 62 years after the Swede's early death.

Magenta Petrel (vagrant)

See family introduction. From the species name, for the steamship *Magenta*, which, under the command of the Italian Captain Vittorio Arminjon sailed around the world in 1865 collecting specimens, including this one. Enrico Giglioli, one of the authors of the species, was a naturalist on the ship.

Pterodroma magentae (Giglioli & Salvadori, 1868) [te-ro-DROH-muh muh-DJEN-teh]: Magenta winged racer', see genus and common names.

Juan Fernandez Petrel (apparently rare visitor to Australian coastal waters)

See family introduction. Breeds only on Isla Alejandro Selkirk (formerly Isla Mas Afuera), the second-largest of the Juan Fernandez islands, west of Chile.

Other names: Pacific Petrel; Black-capped Petrel, descriptive, emphasised by the broad white collar. Often regarded as conspecific with White-necked Petrel.

Pterodroma externa (Salvin, 1875) [te-ro-DROH-muh eks-TEHR-nuh]: 'outside winged racer', see genus name, and from Latin *externus*, for Mas Afuera (see common name) meaning 'more outside' (i.e. 'further out') in Spanish.

Vanuatu Petrel (vagrant)

See family introduction. Described in 2001 from specimens collected from Vanuatu 74 years previously; suspected breeding in Vanuatu confirmed only in 2009. Very difficult to distinguish from White-necked Petrel at sea.

Other names: Until 2001 this was regarded as a population of White-necked Petrel; Falla's Petrel for Sir Robert Falla, a New Zealander who went to the Antarctic with Mawson and later was president of the RAOU. It was he who 'found' the museum specimen that became the type.

Pterodroma occulta Imber & Tennyson, 2001 [te-ro-DROH-muh ok-KƱL-tuh]: 'hidden winged racer', see genus name, and Latin *occultus*, hidden. Was it hidden among all the White-necked Petrels? Or just in the specimen collection from which it was unearthed in 2001? Imber and Tennyson (2001) put it this way: 'Latin *occultus/occulta*, meaning 'without being observed or known', in reference to its having been collected yet remaining unidentified for almost 50 years till exposed by Falla (1976); and 'from a hidden or mysterious source', in reference to the continuing obscurity of its breeding place or places'.

Kermadec Petrel (breeds Lord Howe and Norfolk islands, uncommon visitor to Australian coastal waters)

See family introduction. Breeds mostly on the Kermadec islands north-east of New Zealand.

Pterodroma neglecta (Schlegel, 1863) [te-ro-DROH-muh ne-GLEK-tuh]: 'overlooked winged racer', see genus name, and from Latin *neglectus*, ignored. According to HANZAB (1990–2006, vol. 1a), the species name results from the fact that the bird was initially mislabelled in 1854 as a different species (*Aestrelata diabolica*) by one of the Verreaux brothers (19th-century French naturalists), only to be recognised as a separate species 9 years later by Hermann Schlegel, German ornithologist and author (1804–84).

Herald Petrel (breeding migrant)

See family introduction. From the species name; the type specimen was taken in 1859 in the Coral Sea in the course of a surveying expedition by *HMS Herald* under the command of Captain Henry Denham.

Other names: Long considered conspecific with Trinidade, Trinidade Island or Trinidad Petrel *P. arminjoniana*, for Ilha da Trinidade, some 1200 km east of Brazil; Round Island Petrel, for Round Island near Mauritius.

Pterodroma heraldica (Salvin, 1888) [te-ro-DROH-muh he-RAL-di-kuh]: 'herald winged racer', see genus and common names.

Barau's Petrel (vagrant)

See family introduction. From species name.

Pterodroma baraui (Jouanin, 1964) [te-ro-DROH-muh ba-ROH-ee]: 'Barau's winged racer', see genus name, and for Armand Barau, land-owner and birder of Réunion (east of Madagascar), who co-wrote the first bird study of the island in 1982. He identified this petrel as a separate species in 1963 and his friend Christian Jouanin named it – see Jouanin's Petrel for information on him.

Mottled Petrel (uncommon visitor to Australian coastal waters)

See family introduction. The alternative name Scaled Petrel might be more descriptive, though the faint scaling of the head and belly in particular is not very conspicuous.

Other names: Scaled Petrel; Rainbird, reason not known; Peale's Petrel, for Titian Ramsay Peale, a US naturalist who collected specimens south of South America in 1848, though it had already been described by then.

Pterodroma inexpectata (Forster JR, 1844) [te-ro-DROH-muh in-eks-pek-TAH-tuh]: 'unexpected winged racer', see genus name, and from Latin *inexpectatus*, because the first sight of a new species gave such unexpected joy to its namer, Johann Reinhold Forster, naturalist aboard Captain Cook's *Resolution* from 1772 to 1775 (Forster 1844).

White-necked Petrel (visitor to Australian coastal waters)

See family introduction. Descriptive of a unique feature among dark-headed petrels (except for Juan Fernandez Petrel, with which it is often regarded as conspecific).

Other names: Black-capped Petrel (see Juan Fernandez Petrel); Sunday Petrel for Sunday Island (also called Raoul Island) in the Kermadec group north-east of New Zealand.

Pterodroma cervicalis (Salvin, 1891) [te-ro-DROH-muh sehr-vi-KAH-lis]: 'winged racer with a neck', see genus name, and from Latin, *cervicalis*, from *cervix*, neck, the outstanding feature already described.

Black-winged Petrel (Norfolk and Lord Howe islands; visitor to Australian coastal waters)

See family introduction. Wings darkened by a broad longitudinal bar top and bottom, distinctive but not unique, though the name is based on the species name.

Pterodroma nigripennis (Rothschild, 1893) [te-ro-DROH-muh nig-ri-PEN-nis]: 'black-feathered winged racer', see genus name, plus Latin *niger*, gleaming black, and *penna*, a feather. See common name.

Gould's Petrel (breeding migrant)

See family introduction. For John Gould, a very successful and creative natural history entrepreneur, who collected vast numbers of specimens for his taxidermy business and as models for high-quality and successful illustrated natural history texts. His financial motivations did not prevent him from being a very good zoologist, who described many new bird species including this one. The name was formally adopted by RAOU (1978).

Other names: White-winged Petrel or Fulmar, from species name (see also Southern Fulmar); Black-capped Petrel.

Pterodroma leucoptera (Gould, 1844) [te-ro-DROH-muh loo-ko-TE-ruh]: 'white-winged winged racer', see genus name, plus Greek *leukopteros*, from *leukos*, white, and *pteron*, wing. For the white underwings, which are not unique among small petrels, though they do contrast particularly strongly with the black wing bar and cap.

Collared Petrel (vagrant to Norfolk Island)

See family introduction. Unlike the closely related Gould's Petrel (of which it was long regarded as a subspecies), the white throat and undersides are separated by a black band.

Pterodroma brevipes (Peale, 1849) [te-ro-DROH-muh BREH-vi-pehz]: 'short-footed winged racer', see genus name, plus Latin *brevis*, short, and *pes*, the foot (or leg also, in Late Latin).

Cook's Petrel (uncommon visitor to Australian offshore waters)

See family introduction and from the species name.

Other names: Blue-footed Petrel, descriptive but only likely to be of use in the hand.

Pterodroma cookii (Gray GR, 1843) [te-ro-DROH-muh KŬK-i-ee]: 'Cook's winged racer', see genus name, and for Captain James Cook, famed among many other things as the 'discoverer' of eastern Australia in the *Endeavour* in 1770. He had no particular connection with the petrel – this is an honorific name and he was long dead by the time of its naming.

Stejneger's Petrel (vagrant)

See family introduction. Leonhard Stejneger (1851–1943), born in Norway, worked at the US Smithsonian Institution where he was best known as a herpetologist. Fascinated by the 18th German naturalist explorer Georg Steller, he followed some of his routes to the North Pacific, where he collected and later described this species.

Pterodroma longirostris (Stejneger, 1893) [te-ro-DROH-muh lon-gi-ROS-tris], see genus name, and from Latin *longus*, long, and *rostrum*, bill.

Tahiti Petrel (regular visitor to Australian coastal waters)

See family introduction. A tropical Pacific petrel that breeds on Tahiti (and elsewhere).

Pseudobulweria rostrata (Peale, 1848) [s-yoo-do-bŭl-WEH-ri-uh ro-STRAH-tuh]: 'beaked false Bulwer's', see genus name, and from Latin *rostrum*, beak. Peale described the bird's bill as 'deeply furrowed, very strong and black', which may account for his choice of specific name. However, he also recounted the following: 'The native who carried it, finding its bite to be too severe to be often repeated, submitted the poor creature to a most cruel remedy, often practised by the island bird-catchers, which was to thrust the sharp point of the upper mandible through the lower, thus yoking it with its own jaw; in this state it was kept until the party reached the coast' (Peale 1848).

Grey Petrel (vagrant)

See family introduction. It is greyish, but also brownish and darkish. (Mind you, we wouldn't want to have to come up with useful and unambiguous names for petrels either!)

Other names: Great Grey Petrel, used by Gould (1848), who regarded his specimen as a separate species from Grey Petrel and called it by Kuhl's 1820 name *P. hasitata*; Pediunker, widely used, derives from its name on Tristan da Cunha, part-way between Cape Town and Buenos Aires (HANZAB 1990–2006, vol. 1a); Black-tailed Petrel or Shearwater (notably unhelpful); Brown Petrel, somewhat perplexing but not to RAOU (1913), which recommended it; Nighthawk, presumably a sailor's name from the observation that the bird tends to be nocturnal when breeding – it is also prone to following ships.

Procellaria cinerea Gmelin JF, 1789 [pro-sel-LAH-ri-uh si-NE-re-uh]: 'ashy-grey storm-bird', see genus name, and from Latin *cinereus*, like ashes. See common name for bird's colour.

White-chinned Petrel (uncommon visitor to Australian coastal waters)

See family introduction. A white chin is unique among Australian all-dark petrels, but unfortunately the feature is very small and, worse, it isn't always present.

Other names: Shoemaker, a delightful sailor's name referring to its loud calling from its nesting burrow: Robert Cushman Murphy in *Oceanic Birds of South America* (1936) recounted a whaler reporting 'he sits in his shop and sings'! Cape Hen, a reflection of its commonness and distribution throughout southern oceans and its habit of following ships; Spectacled Petrel, referring to a race with white eye rings, currently proposed as a separate species, *P. conspicillata* – this is an Atlantic Ocean taxon that is unlikely to appear in Australia.

Procellaria aequinoctialis Linnaeus, 1758 [pro-sel-LAH-ri-uh eh-kwi-nok-ti-AH-lis]: 'storm-bird of the equinox', see genus name, plus *aequus*, equal, and *nox*, *noctis*, night, meaning those days of the year when night and day are equal in length. Perhaps an even stormier stormy petrel, given the legendary power of equinoctial storms.

Black Petrel (uncommon visitor to Australian coastal waters)

See family introduction. A dark petrel, but perhaps no more so than other, commoner, Australian petrels.

Other names: Parkinson's Petrel, from the specific name; Black Fulmar (see Southern Fulmar).

Procellaria parkinsoni Gray GR, 1862 [pro-sel-LAH-ri-uh PAR-kin-son-ee]: 'Parkinson's storm-bird', see genus name, and for Sydney Parkinson, an artist who was employed by Sir Joseph Banks on the *Endeavour* expedition but died during the voyage.

Westland Petrel (vagrant)

See family introduction. Its sole breeding grounds are on the Westland coast of New Zealand's South Island.

Other names: Westland Black Petrel, for its extreme similarity and relationship to the previous species.

Procellaria westlandica Falla, 1946 [pro-sel-LAH-ri-uh west-LAN-di-kuh]: 'Westland storm-bird', see genus and common names.

Streaked Shearwater (non-breeding migrant to Australian coastal waters)

See family introduction. The streaked hindcrown and nape are distinctive.

Other names: White-fronted or White-faced Shearwater or Petrel, both descriptive, though not uniquely.

Calonectris leucomelas (Temminck, 1835) [ka-lo-NEK-tris loo-ko-MEL-as]: 'beautiful light and dark swimmer', see genus name plus Greek *leukos*, white, and *melas*, black. Whitish-black is maybe too extreme a description – brownish-white would be closer.

Wedge-tailed Shearwater (breeding migrant)

See family introduction. The long tapering tail is indeed wedge-shaped (unlike that of the eagle of the same name!); Gould used the name *sphenurus*, which means just that – he was unaware of Gmelin's name for a type specimen from the Kermadec Islands.

Other names: Muttonbird or Wedge-tailed Muttonbird for the flesh of the chicks, which are still harvested for food. The name was apparently first applied to the Providence Petrel on Norfolk Island – other reports suggest that with regard to flavour, the name may have been wishful thinking! Note that all shearwaters are commonly referred to as muttonbirds. Little Muttonbird, presumably in contrast to the Flesh-footed Shearwater, or Big Muttonbird, which can be twice the weight.

Ardenna pacifica (Gmelin JF, 1789) [ar-DEN-nuh pa-SI-fi-kuh]: 'Pacific Ardenna', see genus name, and from Latin *pacifica*, peaceful, for the ocean of that name.

Buller's Shearwater (visitor to Australian offshore waters)

See family introduction. From species name.

Other names: Grey-backed or Ashy-backed Shearwater, both appropriate names – it is a greyer bird than the Grey Petrel, for instance! New Zealand Shearwater, because it breeds exclusively on the Poor Knights Islands, off the east coast of far northern New Zealand.

Ardenna bulleri (Salvin, 1888) [ar-DEN-nuh BŬL-le-ree]: 'Buller's Ardenna', see genus name, and for Sir Walter Buller, controversial late 19th-century New Zealand ornithologist who collected the type specimen of this species; it seems that in this instance the bird was not shot but found on a beach. (For more on Buller see Buller's Albatross.)

Sooty Shearwater (breeding migrant)

See family introduction. Certainly sooty-brown, but then so are most other shearwaters; adopted by RAOU (1926).

Other names: New Zealand Muttonbird, because its main breeding – and harvesting – grounds are there (see Wedge-tailed Shearwater); King Muttonbird presumably reflects its importance to this trade; Sombre Shearwater or Petrel, the name of choice of RAOU (1913).

Ardenna grisea (Gmelin JF, 1789) [ar-DEN-nuh GREE-se-uh]: 'grey Ardenna', see genus name, and from Mediaeval Latin *grisus/griseus*, grey. See common name on colour.

Short-tailed Shearwater (breeding migrant)

See family introduction. An unhelpful name – the tail is not noticeably shorter in the field than that of other shearwaters. However, it was based on the name applied by Gould (1848) for his species name *Puffinus brevicaudus*, not realising that Temminck had already named the bird from a specimen from Japan.

Other names: Tasmanian Muttonbird, because this is the bird harvested industrially in Bass Strait (see Wedge-tailed Shearwater); Short-tailed Petrel; Slender-billed Petrel or Shearwater, which is accurate but not a useful identifier either; Solitary Shearwater, used briefly (RAOU 1913) with the short-lived species name *Puffinus intermedius* from New South Wales; Sealbird, which is unclear though the birds were very likely using the same islands as seals when the sealers arrived.

Ardenna tenuirostris (Temminck, 1835) [ar-DEN-nuh te-noo-i-ROS-tris]: 'slender-billed Ardenna', see genus name, and from Latin *tenuis*, slender or thin, and *rostrum*, bill, see common name.

Pink-footed Shearwater (vagrant)

See family introduction. Descriptive, with pinker feet than the next species.

Ardenna creatopus (Coues, 1864) [ar-DEN-nuh kre-A-to-pŭs]: 'Greek fleshy-footed Ardenna', see genus name, and from Greek *kreas*, flesh or meat, and *pous*, the foot.

Flesh-footed Shearwater (breeding migrant)

See family introduction. The pink feet are distinctive in the field, though they are identical to those of the Wedge-tailed Shearwater; from the species name.

Other names: Flesh-footed Petrel; Fleshy-footed or Pale-footed Shearwater or Petrel (Gould 1848 used Fleshy-footed Petrel); Big Muttonbird (see Wedge-tailed Shearwater).

Ardenna carneipes (Gould, 1844) [ar-DEN-nuh kar-NEH-i-pehz]: 'Latin fleshy-footed Ardenna', see genus name, and from Latin *carneus*, flesh-coloured, and *pes*, the foot. However, HANZAB (1990–2006, vol. 1a) suggested that the name refers to the feet 'but ambiguously as regards colour or nature' so it may come instead from *carnosus*, fleshy or pulpy. The two words are related, both coming from *caro/carnis*, flesh.

Great Shearwater (vagrant)

See family introduction. Although not particularly 'longer' than some other shearwaters, it is more massive.

Ardenna gravis (O'Reilly, 1818) [ar-DEN-nuh GRAH-vis]: 'heavy Ardenna', see genus name, and from Latin *gravis*, heavy or weighty.

Manx Shearwater (vagrant)

See family introduction. John Ray, notable 17th-century English biologist, described specimens from the Isle of Man.

Puffinus puffinus (Brünnich, 1764) [pʊf-FEE-nʊs pʊf-FEE-nʊs]: 'shearwater-shearwater', see genus name. With genus and species name being the same, do we get double the confusion?

Newell's Shearwater (vagrant)

See family introduction. From the species name.

Other names: Hawaiian Shearwater, because it breeds exclusively on the Hawaiian islands. Long regarded as conspecific with Manx Shearwater. On the other hand, it is also often regarded as a subspecies of *P. auricularis*, Townsend's Shearwater. Charles Townsend was a US fish biologist and explorer who discovered the bird in 1889 and described it.

Puffinus newelli Henshaw, 1900 [pʊf-FEE-nʊs NYOO-wel-lee]: 'Newell's shearwater', see genus name, and for Matthias Newell, at the end of the 19th century a missionary to Hawaii, where the bird breeds; Henshaw explains that he provided the type specimen.

Fluttering Shearwater (non-breeding migrant to Australian coastal waters)

See family introduction. Descriptive of the distinctive flight, though this is common to other members of *Puffinus* too. Used in Australia since RAOU (1926).

Other names: Forster's Petrel or Shearwater, for Johann Forster who replaced Joseph Banks as naturalist on Cook's second Pacific expedition (after Banks pulled out in a huff at the last minute). Forster described this shearwater but it is not clear who admired him enough to name it, as no-one seems to have had a good word to say for him – at least at a personal level, since his professional abilities seem unchallengeable. The RAOU apparently did, because it recommended the name in RAOU (1913). Brown-backed Shearwater, presumably because worn dorsal plumage fades to brown; Brown-beaked Shearwater, unclear – perhaps a mistranscription of Brown-backed?

Puffinus gavia (Forster JR, 1844) [pʊf-FEE-nʊs GAH-vi-uh]: 'seabird shearwater', see genus name, and from Latin *gavia*, an unidentified seabird mentioned by Pliny, translated as sea-mew, mew being an old English word for a gull. *Gavia* is also the current genus name for the loons or divers, originally used by Forster in 1788.

Hutton's Shearwater (non-breeding migrant to Australian coastal waters)

See family introduction. From the species name.

Other names: Long regarded as conspecific with Fluttering Shearwater.

Puffinus huttoni Mathews, 1912 [pʊf-FEE-nʊs HUT-ton-ee]: 'Hutton's shearwater', see genus name, and for Frederick Hutton, 19th-century English soldier and zoologist who moved to New Zealand and published a catalogue of New Zealand birds.

Audubon's Shearwater (vagrant)

See family introduction. For the great John Audubon, 19th-century US ornithologist and collector, artist and author. At one stage this shearwater bore his name as a specific name.

Puffinus lherminieri Lesson, 1839 [pʊf-FEE-nʊs lehr-meen-YEHR-ee]: 'L'Herminier's shearwater', see genus name. For Félix Louis L'Herminier (1779–1833), a French chemist and naturalist who worked as a pharmacist in Guadaloupe, became an expert on Caribbean natural history and later worked as director of the natural history museum in Charleston, South Carolina. The bird was described by his countryman René Lesson in 1839 from a specimen from Bermuda, so the specimen may have been collected by L'Herminier (or his son Ferdinand, also a biologist, who stayed in Guadaloupe after his father returned to France in 1829); Lesson (1839a) didn't explain and it may well be that his intent was to honour both father and son.

Tropical Shearwater (vagrant)

See family introduction. Atypically for shearwaters, found in the tropical waters of the Pacific and Indian Oceans.

Puffinus bailloni Bonaparte, 1857 [pʊf-FEE-nʊs by-OH-ni]: 'Baillon's Shearwater', see genus name and for Louis Baillon, French naturalist and collector of bird specimens, who died 2 years before Bonaparte named this species. See Baillon's Crake for a little more on him.

Heinroth's Shearwater (vagrant)

See family introduction. From the species name, for Oskar Heinroth who went on a voluntary collecting expedition to the south-west Pacific for the Berlin Zoo in 1900–01 and collected this species. The bird was a newly fledged juvenile, and Heinroth wasn't prepared to commit himself to describing it as a new species. Fortunately Anton Reichenow later was. See Long-tailed Finch for a little more on Heinroth.

Puffinus heinrothi Reichenow, 1919 [pʊf-FEE-nʊs HYNE-roh-ti]: 'Heinroth's shearwater', see genus and common names.

Little Shearwater (breeding migrant)

See family introduction. Notably smaller than other Australian shearwaters; in use in Australia at least since RAOU (1926).

Other names: Allied Shearwater or Petrel, from the specific name, still used by RAOU (1913); Dusky Shearwater, true but unhelpful; Gould's Shearwater, for John Gould (see Gould's Petrel), who described the species.

Puffinus assimilis Gould, 1838 [pʊf-FEE-nʊs as-SIM-i-lis]: 'similar shearwater', see genus name, and from Latin *adsimilis* or *assimilis*. Gould (1838b) considered it similar to *Puffinus obscurus* ('which it so much resembles, and to which it is so nearly allied'; in Gould (1838a) he was more specific: 'Very nearly allied to *Puffinus obscurus*; but considerably smaller than that species'). *P. obscurus* was the then name for what is now known as Audubon's Shearwater *P. lherminieri*.

South Georgia Diving Petrel (Heard Island; vagrant to Australian mainland)

Breeds on sub-Antarctic South Georgia Island off South America. But so does the Common Diving-Petrel, and furthermore the South Georgian Diving-Petrel also breeds on several other islands.

Other names: South Georgia Diving-Petrel, the usual form in Australia; Georgian Diving-Petrel.

Pelecanoides georgicus Murphy & Harper, 1916 [pe-le-ka-NOY-dehz ge-OR-gi-cʊs] 'Georgian something like a pelican', see genus name, plus, as in the common name, for the South Georgian Islands, themselves named for George III.

Common Diving Petrel (breeding resident)

This species is common in the sense of being much more widespread than the other three of its genus.

Other names: Common Diving-Petrel, the usual form in Australia; Diving Petrel (used by Latham (1801) and Gould (1848)).

Pelecanoides urinatrix (Gmelin JF, 1789) [pe-le-ka-NOY-dehz oo-ree-NAH-triks]: 'something like a pelican that dives', see genus name, and from Latin *urinator* – no, it means a diver (like *urinari* and *urinare*, to plunge under water, to dive).

Bulwer's Petrel (vagrant)

See family introduction. See genus name; Bulwer scored the trifecta, in also bagging both common and species names!

Bulweria bulwerii (Jardine & Selby, 1828) [bʊl-WE-ri-uh bʊl-WE-ri-ee]: 'Bulwer's Bulweria', see genus name.

Jouanin's Petrel (vagrant to north-western Australian waters – from the Arabian Sea!)

See family introduction. For Christian Jouanin, a giant of French ornithology for most of the 20th century and a committed conservationist. It was he who described this bird in 1955, having recognised it as a distinct species from the more widespread Bulwer's Petrel, by studying museum skins.

Bulweria fallax Jouanin, 1955 [bʊl-WE-ri-uh FAL-laks]: 'false Bulwer's' from Latin *fallax*, false or deceitful – masquerading as *Bulweria bulwerii*.

CICONIIDAE (CICONIIFORMES): storks

Ciconiidae Sundevall, 1836 [si-ko-NEE-i-dee]: the Stork family, from Latin *ciconia*, stork.

'Stork' is an Old English word (from Old German) meaning a stick, applied as a nickname to the nesting storks (White Stork in Britain), which commonly roost on one leg. Lockwood (1984) helpfully noted a secondary meaning in Old German (presumably low Old German!) of penis – which, he claimed, is why storks are said to bring babies.

The genus

Ephippiorhynchus Bonaparte, 1855 [ef-ip-pi-o-RIN-kʊs]: 'saddle-bill', from Greek *ephippios*, for 'putting on a horse' (a saddle-cloth or saddle), and *rhunkhos*, bill. The 'saddle bill' is much more obvious in the only other species in this genus, *E. senegalensis* (the Saddle-billed Stork), whose large red and black large bill has a bright yellow frontal shield – the so-called saddle.

The species

Black-necked Stork (breeding resident)

See family introduction. The only black-necked stork outside Africa (apart from the nearly all-black Black Stork). It seems likely that the Australian and New Guinea populations will eventually be treated as a separate species from the Asian populations.

Other names: New Holland Jabiru, used by Latham (1801), Jabiru being from the Brazilian Tupi-Guaraní language for a large South American stork, though it is not clear why he used it, given that the latter bird has an all-white body and is thus quite different in appearance; hence Jabiru, widely used to the present day; Australian Mycteria, a real Gouldism, used by him in 1848 from the then genus name; Policeman-bird, a lovely folk name for its tall, still posture while hunting, and apparently stern visage (especially of the yellow-eyed female).

Black-necked Stork *Ephippiorhynchus asiaticus*

Ephippiorhynchus asiaticus (Latham, 1790) [ef-ip-pi-o-RIN-kʊs a-see-A-ti-kʊs]: 'Asian saddle-bill', see genus name, plus Latin *asiaticus*, Asian (see common name).

FREGATIDAE (SULIFORMES): frigatebirds

Fregatidae Degland & Gerbe, 1867 [fre-GA-ti-dee]: the Frigate family, see genus name *Fregata*. 'Frigatebird' has the same derivation as *Fregata*.

The genus

Fregata Lacépède, 1799 [fre-GAH-tuh]: 'forked-tail [bird]'. It seems to be commonly accepted that the name refers to a vessel, and there is only some doubt about which aspect of the ship metaphor applies. The *Oxford Dictionary* quoted from Albin's 1738 *The History of Birds* 'The Indians call it so, because of the Swiftness of its Flight' (and we won't be deflected by the concept of the 'Indians' giving the bird the English name of a European vessel!). On the other hand, a frigate also came to imply a warship, and we are tempted by the analogy with the birds' overtly piratical habits.

However, we must mention the hypothesis put forward in the *Italian Dictionary of Etymology* of Ottorino Pianigiani (Polaris 1993) that the name of the bird *fregata* originates by metathesis from Italian *coda forcata*, forked tail, and comes ultimately from Spanish *rabo forcado*, also meaning forked tail. This was the name given the bird by Columbus and mentioned in his account of his first journey to the Americas (1492–93) (Dunn and Kelley 1991), and it is generally understood, in that context, to refer to the frigatebird, though Linnaeus (1758) pointed to Aldrovandus' (1603) use of the term '*Avis Rabos forcados*', claiming that it means the Tropicbird. Unfortunately for Linnaeus, Aldrovandus' charming illustrations of the bird do not confirm this meaning (but could definitely be taken for the frigatebird).

The species

Christmas Island Frigatebird (Christmas Island)

See family introduction. Breeds only on Christmas Island.

Other names: Christmas Frigatebird; Andrews' Frigatebird, from the specific name.

Fregata andrewsi Mathews, 1914 [fre-GAH-tuh AN-droo-zee]: 'Andrews' forked-tail bird', see genus name, and for Cater Andrews, a British Museum geologist who was commissioned by the Christmas Island Phosphate Company to collect there in the first years of the 20th century. He collected Abbott's Booby, but not this species it seems.

Great Frigatebird (breeding resident)

See family introduction. Given the specific name, we might reasonably think this is a somewhat strange one. It makes sense in Australia (though pedantically we might prefer Greater) but curiously this is its accepted name internationally. See species name for the denouement.

Other names: Man-o'-War Bird, analagous to Frigatebird, see family introduction; Sea Hawk, likewise.

Fregata minor (Gmelin JF, 1789) [fre-GAH-tuh MEE-nor]: 'smaller forked-tail bird', see genus name. As for the specific name *minor*, there is a tale attached! Pizzey and Knight (2007) noted that, when it was originally described in 1789, this bird was named *Pelecanus minor*, the Lesser Pelican. It was indeed clearly 'lesser' than even the smallest Pelican. And, as Pizzey pointed out, 'Under rules governing scientific names, the specific name *minor* must stand, though a smaller frigatebird, Lesser, was later described'.

Lesser Frigatebird (breeding resident)

See family introduction. Straightforward in Australia. Internationally this is its preferred name too, though it is actually the least of the five species. Never mind.

Other names: Least Frigatebird; Small Frigatebird, used by Gould in 1848; see also names of previous species.

Fregata ariel (Gray GR, 1845) [fre-GAH-tuh AH-ri-el]: 'Ariel forked-tail bird', see genus name. Apart from its biblical uses (for the city of Jerusalem and the name of an angel), the best-known use of the name Ariel was by Shakespeare in 1611 for his 'aryie spirit' in *The Tempest*. Gray (1849), under his genus name *Atagen*, gave a wonderful description of the bird's powers of flight, but as far as the species name goes, he just said: 'A. Ariel (Gould's MSS)' so we must assume Gould chose the name but failed to publish it.

PHALACROCORACIDAE (SULIFORMES): cormorants

Phalacrocoracidae Reichenbach, 1850 [fa-la-kro-kor-A-si-dee]: the Bald Raven family, see genus name *Phalacrocorax*.

'Cormorant' comes to us via French from the Latin *corvus marinus*, a 'sea raven' for the sooty-coloured Great Cormorant, the only European bird called cormorant in English.

'Shag' is the name for the other European cormorant, the European Shag *P. aristotelis*. It is used in the sense of matted cloth, for the somewhat scruffy crest. The term is used fairly haphazardly elsewhere – in our part of the world it is commoner among fishing people and New Zealanders than elsewhere. There have been some recent attempts to formalise the word for the genus *Leucocarbo*, but even the recognition of this genus is far from universal, so good luck with that.

The genera

Microcarbo Bonaparte, 1856 [mik-ro-KAR-bo]: 'little charcoal cormorant', from Greek *mikros*, small, and from the species name of the Great Cormorant (Latin *carbo*, charcoal).

Phalacrocorax Brisson, 1760 [fa-la-kro-KOR-aks]: 'bald raven', from Greek *phalakros*, bald, and *korax*, raven. Brisson did not explain why he chose the 'bald' part of this and it remains a puzzle, because the bird doesn't even have a white head (remember that 'bald' originally meant white, especially in reference to animals' heads).

Leucocarbo Bonaparte, 1856 [loo-ko-KAR-bo]: 'white and charcoal [bird]' from Greek *leukos*, white, and Latin *carbo*, charcoal (and see Great Cormorant).

The species

Little Pied Cormorant (breeding resident)

See family introduction. Pied literally means black and white, like a (European) magpie – the more general sense of mixed colours came later. This is (just) the world's smallest pied cormorant.

Other names: Little Cormorant or Shag (the latter used by 'Colonists of Swan River', Gould 1848); Little Black-and-white Cormorant or Shag; Frilled Cormorant or Shag, for the little bristly crest; Pied Cormorant, perplexingly used by Gould (1848) – see Pied Cormorant.

Microcarbo melanoleucos (Vieillot, 1817) [mik-ro-KAR-bo me-la-no-LOO-kos]: 'little black and white charcoal cormorant', see genus name, and from Greek *melas*, black, and *leukos*, white.

Black-faced Cormorant (breeding resident)

See family introduction. The only Australian black and white cormorant with a black hood in front of the eyes; the name was adopted by RAOU (1978) 'by consensus' (undefined).

Other names: Black-faced Shag; White-breasted Cormorant, true but not very useful, though it was taken directly from Gould's name *P. leucogaster* – he didn't realise that Vieillot had already named it from Tasmania. Despite the application of the present species name, RAOU (1926) was still using White-breasted.

Phalacrocorax fuscescens (Vieillot, 1817) [fa-la-kro-KOR-aks fʊs-SES-sens]: 'darkish bald raven', see genus name, and from Latin *fuscus*, dark or dusky, and *-escens*, growing or becoming '-ish'. 'Black and white' and 'variegated' being already spoken for, we are left with 'darkish' (i.e. not all dark – there's quite a lot of white in there too!).

Little Black Cormorant (breeding resident)

See family introduction. Usefully descriptive in Australia.

Other names: Little Black Shag; Groove-billed Cormorant, used by Gould (1848), direct from the species name.

Phalacrocorax sulcirostris (von Brandt JF, 1837) [fa-la-kro-KOR-aks sʊl-see-ROS-tris]: 'bald raven with a furrowed bill', see genus name, and from Latin *sulcus*, furrow, and *rostris*, bill, indicating the grooved bill of this species.

Australian Pied Cormorant (breeding resident)

A comprehensively true but unhelpful name.

Other names: Pied Shag, used by Latham (1801); Pied Cormorant, already being used by Gould in 1848; Large Pied Cormorant, arguably more useful in the Australian context; Black and White Shag, used by the 'Colonists of Western Australia' according to Gould (1848); Yellow-face Cormorant or Shag, distinguishes from all other pied cormorants (not only Australian).

Phalacrocorax varius (Gmelin JF, 1789) [fa-la-kro-KOR-aks VAH-ri-ʊs]: 'bald variegated raven', see genus name, and from Latin *varius*, variegated, indicating in this case the black and white plumage, but not stepping on the toes of the Little Pied, *Microcarbo melanoleucos*.

Great Cormorant (breeding resident)

See family introduction. A species (or, as seems increasingly likely, a group of species) common throughout much of the world; in most of this range it is clearly the largest cormorant species.

Other names: Australian Cormorant, per Gould (1848); Black Cormorant or Shag (!); Large or Big Black Cormorant or Shag.

Phalacrocorax carbo (Linnaeus, 1758) [fa-la-kro-KOR-aks KAR-bo]: 'bald charcoal raven', see genus name, and from Latin *carbo*, charcoal. *Pelecanus carbo* was the name given by Linnaeus (1758) to this bird, which he called a pelican with a black body.

Imperial Shag (Macquarie and Heard islands; but see Kerguelen Shag)

See family introduction. The name probably originates with the crest and the crown-like orange fleshy caruncles between the eye and the bill, both in breeding plumage only.

Other names: King Shag, gives weight to the suggestion in the previous paragraph, rather than a dedication to any particular emperor or king; Blue-eyed Shag, descriptive, but only of a group of species.

Leucocarbo atriceps King PP, 1828 [loo-ko-KAR-bo AT-ri-ceps]: 'black-headed white and charcoal [bird]', see genus name, and from Latin *ater*, dull black, and *-ceps*, -headed (*caput*, head): yet another way of saying 'black-and-white'.

Kerguelen Shag ('ship-assisted vagrant' to Western Australia and Heard Island)

See family introduction. Endemic to the Southern Indian Ocean Kerguelen Islands.

Other names: Has been regarded – and by some authorities still is – as a subspecies of Imperial Shag.

Leucocarbo verrucosus (Cabanis, 1875) [loo-ko-KAR-bo ver-rʊ-KOH-sʊs]: 'warty white and charcoal [bird]', see genus name, with Latin *verrucosus*, from *verruca*, a wart, for the orange-yellow tubercles on the bird's face – at last a name to differentiate this one.

ANHINGIDAE (SULIFORMES): darters

Anhingidae Reichenbach, 1849 [an-HIN-gi-dee]: the Devil-bird family, see genus name *Anhinga*.

The genus

Anhinga Brisson, 1760 [an-HIN-guh]: 'devil-bird'. Brisson takes the name from Marcgrave (1648), who gave a detailed description of the bird *Anhinga anhinga*, as well as a slightly convincing illustration. Marcgrave called it *Anhinga Brasiliensibus Tupinambis*, meaning 'Anhinga from the Brazilian Tupinamba' (the latter meaning first or oldest people), some of the indigenous people of what came to be known as Brazil after its colonisation by the Portuguese. The Tupi word for demon or devil was something like *ayinga* or *anhanga*, transcribed as *anhinga* by the Portuguese, whose priests apparently made use of it when spreading their ideas about hell (Nash 1926). The bird is frequently compared with a snake for its habit of swimming with only the sinuous neck above water, for example, in another indigenous language, Guaraní, which calls it *mbiguá-mboí*, the snake cormorant (Mouchard 2011).

The species

Australasian Darter (breeding resident)

Said to be for the hunting habit, in which a neck adaptation allows the head and pointed bill to be hurled forward to stab larger prey or seize smaller. This is the most likely explanation, though the process takes place underwater. The term first appeared in English in 1825, in Richard Gore's translation from German of Johann Blumenbach's monumental work, rendered as *Manual of the Elements of Natural History*. He referred to it as both Anhinga (see genus) and Darter, and explained that it has a 'very long neck, which the animal can roll up spirally, and by that means dart at the fish it wishes to seize'. (That isn't the adaptation referred to earlier, incidentally.) Australasian distinguishes it from the Oriental Darter *A. rufa*, with which it has been lumped in very recent times.

PLOTUS NOVÆ HOLLANDIÆ
New Holland Darter

Australasian Darter ***Anhinga novaehollandiae***

Other names: New Holland Darter, used by Gould (1848); Diver; Snake-bird (see genus name); Needle-beaked Shag, see cormorants; Anhinga, see genus name – this has been applied here at times when the world darter populations have been regarded as conspecific.

Anhinga novaehollandiae (Gould, 1847) [an-IN-guh no-veh-hol-LAN-di-eh]: 'New Holland devil-bird', see genus name, and from Modern Latin *Nova Hollandia*, New Holland, the old name for Australia.

SULIDAE (SULIFORMES): boobies and gannets

Sulidae Reichenbach, 1849 [SOO-li-dee]: the Sharp-sighted, Rapacious Gannet family (to use all the possibilities), see genus name *Sula*.

'Gannet' is from the same Old English root as gander, though that word implied 'goose' and was used independently of gender. It became applied to the Northern Gannet as well, that usage being retained when the old goose one faded, but gannet did not become the universal term until well into the 19th century. Gannet tends to be used for the three cool-climate *Morus* species and Booby for the rest, which are essentially tropical.

'Booby' was used in the sense of a 'foolish fellow', presumably for the poor beasts' trusting habits that allowed them to be easily slaughtered by sailors. The word itself appears in English from the very beginning of the 17th century, apparently from the Spanish *bobo*, a fool.

The genera

Morus Vieillot, 1816 [MOH-rʊs]: 'fool', from Greek *mōros*, dull or stupid (= English moron, of course), see family introduction.

Papasula Olson & Warheit, 1988 [pa-pa-SOO-luh]: 'father-gannet'. *Papasula* was split from *Sula* in 1988 to accommodate Abbott's Booby, and Olson and Warheit (1988), who proposed and justified the change, explained the name as follows: '*Etymology*. Greek *papas*, father, plus *sula*, the type genus of Sulidae. The name refers both to the patronym of the type species (abbot, from Hebrew, *abba*, father) and to the fact that this genus represents an ancient lineage in the family.' The type species is Abbott's Booby.

Sula Brisson, 1760 [SOO-luh]: 'gannet'. There are various theories about the origins of the word *Sula*. It is said to be from Old Norse (it also appears in modern Northern European languages as a word for the Northern Gannet *Morus bassanus*, including Norwegian *Havsule*, Finnish *Suula*, Danish *Sule*, Swedish *Havssula* and Icelandic *Sùla*). It may also be connected with Gaelic *suilaire*, sharp-sighted, or to the Greek word *sulē*, plural *sulai* or *sula*, the right of seizure – and both of these definitions could be said to fit with the gannet family's behaviour.

The species

Cape Gannet (vagrant)

See family introduction. From southern African coasts, centred on the Cape of Good Hope.

Other names: South Africa, African Gannet.

Morus capensis (Ridgway, 1893) [MOH-rʊs ka-PEN-sis]: 'fool from the Cape', see genus name, and Latin *Cape* (see common name) plus the suffix *-ensis*, usually indicating the place of origin of the type specimen.

Australasian Gannet (breeding resident)

See family introduction. Restricted to southern Australia and New Zealand.

Other names: Australian Gannet; Booby (see 'booby'); Diver, for conspicuous feeding behaviour; Solan Goose, being an old name for the Northern Gannet *M. bassanus*, Solan coming from Old Norse via Scottish Gaelic, and undoubtedly connected with *Sula* (see genus name).

Morus serrator (Gray GR, 1843) [MOH-rʊs ser-RAH-tor]: 'fool with a saw', see genus name, and from Latin *serra*, saw, *serrator*, sawyer, for the saw-edged bill common to the Sulidae.

Abbott's Booby (Christmas Island)

See family introduction. For William Abbott, a US collector specialising in Asia, who collected the type specimen.

Papasula abbotti (Ridgway, 1893) [pa-pa-SOO-luh AB-bot-tee]: 'Abbott's father-gannet', see genus and common names.

Masked Booby (breeding resident)

See family introduction. The bare blue-black facial skin against white feathers is conspicuous and unique.

Other names: Masked Gannet; Blue-faced Booby; White Booby – certainly the whitest booby, black flight feathers and tail notwithstanding; Whistling Booby, for the male's thin descending call constantly heard at breeding colonies, which is different from that of other family members.

Sula dactylatra Lesson, 1831 [SOO-luh dak-ti-LAH-truh]: 'black-fingered gannet', see genus name, and from Greek *dactulos*, finger, and Latin *ater*, dull black – a hybrid word – for the spreading tips of the wings.

Red-footed Booby (Cocos (Keeling) and Christmas islands; visits Australian mainland waters)

See family introduction. An evidently apposite name.

Other names: Red-footed or Red-legged Gannet.

Sula sula (Linnaeus, 1766) [SOO-luh SOO-luh]: 'gannet-gannet' – one of those repetitive and quintessential ones. See genus name.

Brown Booby (breeding resident)

See family introduction. The only all-brown booby, other than the white belly and lower breast.

Other names: Black or Brown Gannet; White-bellied Booby or Gannet, as per scientific name; Common Booby, a fair comment in Australia at least.

Sula leucogaster (Boddaert, 1783) [SOO-luh loo-ko-GAS-tehr]: 'white-bellied gannet', see genus name, plus Greek *leukos*, white, and *gastēr*, stomach.

THRESKIORNITHIDAE (PELECANIFORMES): ibis and spoonbills

Threskiornithidae Poche, 1904 [thres-kee-or-NI-thi-dee]: the Sacred Bird family, see genus name *Threskiornis*.

'Ibis' must be one of the few words in common English use to come to us from the Egyptian, albeit via Greek and Latin. The bird was of great significance to the ancient Egyptians, as witnessed by the name Sacred Ibis for the African species; they arrived with the annual life-bringing Nile floods, though the Egyptians interpreted the cause and effect differently, crediting the ibis with bringing the floods.

'Spoonbill' was introduced by John Ray, the sadly underappreciated 'father of British natural history', as Spoon-bill when he translated Francis Willughby's *Ornithologia* from Latin in 1678, replacing various folk names including shoveller.

The genera

Threskiornis Gray GR, 1842 [thres-kee-OR-nis]: 'sacred bird', from Greek *thrēskos*, religious, and *ornis*, bird (see 'Ibis') .

Plegadis Kaup, 1829 [pleh-GAH-dis]: 'sickle-shape', from Greek *plēgas*, sickle or pruning-knife.

Platalea Linnaeus, 1758 [pla-TA-le-uh]: 'spoonbill' from Latin *platalea*, spoonbill, mentioned by Cicero (45 BC) in *De Natura Deorum*.

The species

Australian White Ibis (breeding resident)

See family introduction. Can't argue with that. The only one of the three very closely related 'black-headed white' ibis found in Australia.

Other names: Sacred Ibis, from when it was thought to be a race of the African Sacred Ibis (*T. aethiopica*); White Ibis, used by Gould (1848), who noted that it was also the name used by the 'Colonists of New South Wales'; Black-necked Ibis, reported by Gould to be the name of the 'Colonists of Port Essington' (the predecessor to Darwin, to the north of there); Sickle-bill. In recent times (primarily this millennium) this rapidly urban-adapting species has achieved a spectacular media, and especially social media, presence, including coming second in a popular vote for Australian Bird of the Year in 2017 (run by Birdlife Australia and *The Guardian*). Two widely used names associated with this presence are Bin Chicken and Tip Turkey; others include Sandwich Snatcher, Picnic Pirate, Dump Chook, Trash Vulture, Bankstown Bin Diver, Garbage Flier, Winged Waste Lizard, Refuse Raptor, Dreg-Beaked Scrap Fowl, Detritus Drake and Bottom-Feeder Biddy. Many of these names were consciously coined and may not have much actual usage, but with oral tradition it's hard to be sure as we've commented many times elsewhere and they are being promulgated in popular news articles.

Threskiornis molucca (Cuvier, 1829) [thres-kee-OR-nis mo-LUK-kuh]: 'sacred Moluccan bird', see genus name, and from the Maluku (Moluccan) Islands, part of Indonesia.

Australian White Ibis *Threskiornis molucca*

Straw-necked Ibis (breeding resident)

See family introduction. Directly from the species name, used by Gould (1848).

Other names: New Holland Ibis (Latham 1823); Letter-bird, a source of considerable frustration since, although everyone cites it, nobody suggests why – we think it might go back to a quote from John Gould (1848) who described how in flight they 'arrange themselves in a form of figure or letter similar to that so frequently observed in flights of geese or ducks'; Farmer's Friend, a nice nod to the vast numbers of insects, especially ground crickets or grasshoppers, that a flock can consume. Dryweather Bird, probably because, like other Australian waterbirds, they disperse widely as waters dry up but do so in large and conspicuous numbers; this one too may be down to Gould, who made special mention of vast numbers on the Liverpool Plains following the drought of 1830, or perhaps it (and Letter-bird) simply arose spontaneously among country people making similar observations.

Threskiornis spinicollis (Jameson, 1835) [thres-kee-OR-nis spee-ni-KOL-lis]: 'sacred spiny-neck bird', see genus name, and from Latin *spina*, a spine or thorn, and *collum*, the neck, for the straight yellow breast plumes.

Glossy Ibis (breeding resident)

See family introduction. The universal name for the species, which is scattered throughout six continents. Its wing feathers are beautifully iridescent.

Other names: Black Curlew, an old English name for the same species – to people unused to ibis, this was a good general description.

Plegadis falcinellus (Linnaeus, 1766) [pleh-GAH-dis fal-si-NEL-lʊs]: 'little sickle-shaped sickle-shape', see genus name, and *falcinellus*, diminutive from Latin *falx/falcus*, sickle, for the curved bill.

Royal Spoonbill (breeding resident)

See family introduction. Presumably this was intended as a compliment to the bird's very natty breeding garb; whether this or the specific name came first is unclear – Gould named it formally in 1838.

Other names: Black-billed Spoonbill, clear and unambiguous – and probably fairer to the apparently plebeian Yellow-billed Spoonbill. It was still recommended by RAOU (1913), but its successor firmly put Royal back on top in 1926.

Platalea regia Gould, 1838 [pla-TA-le-uh REH-gi-uh]: 'royal spoonbill', see genus name, plus Latin *regius*, royal.

Yellow-billed Spoonbill (breeding resident)

See family introduction. The only such in the world.

Other names: Yellow-legged Spoonbill, an equally apt name and the one used by Gould, taken directly from his specific name. It is unclear when the current name was adopted but it was in use by the end of the 19th century, when it was the only name noted by Morris (1898).

Platalea flavipes Gould, 1838 [pla-TA-le-uh FLAH-vi-pehz]: 'yellow-footed spoonbill', see genus name, and from Latin *flavus*, golden or reddish-yellow (from *flagro*, burn), and *pes/pedis*, foot, but in Late Latin it could also indicate the leg.

ARDEIDAE (PELECANIFORMES): herons, egrets, bitterns

Ardeidae Leach, 1820 [ar-DEH-i-dee]: the Heron family, see genus name *Ardea*.

'Heron' finds its origin way back in European languages. For instance, when the French *hayroun* entered the English language at the beginning of the 14th century, the Old English *hragra* that it replaced had the same linguistic roots. These apparently were based onomatopoeically in the harsh croaking of the bird (Grey Heron in that case, but none of them sing).

'Egret' is of a similar complex origin. The word is from Old Provençal *aigron*, heron, which also appears in mediaeval French and may still persist as a dialect word. The diminutive *aigreta* (Old Provençal) appears as mediaeval French égreste (14th century) and is well established as *aigrette* by the 16th century. Belon (1555) gave a full description of the bird under that name, saying that the bird is so named because of the raucousness of its voice, 'much more powerful than the heron's'. (*Aigrette* has a secondary meaning – a bundle of feathers, especially egrets' plumes, used for decoration: we found reference to an aigrette that used 150 feathers.) 'Egret' first appeared in English in the early 15th century.

In vernacular, egret is generally used simply to denote herons that happen to be white and it has no taxonomic relevance: for example, White-faced Heron is in the genus *Egretta* and Great and Intermediate Egrets are now *Ardea* ('herons').

'Night Herons' (generally presented as Night-Heron) are indeed mostly nocturnal, or at least crepuscular.

'Bittern' is also ultimately onomatopoeic, via a complex route. It is present in 12th-century French as *butor*, derived via a presumed Vulgar Latin link *buti-taurus* from Latin (see genus *Botaurus*). From there it enters 14th-century English as *botors* or *bitore*, appearing in Chaucer's Wife of Bath's tale (when Midas' wife whispers his secret to the rushes 'as a bitore bombleth in the myre').

The genera

Botaurus Stephens, 1819 [boh-TOR-ʊs]: 'bittern-bull' from the Mediaeval Latin word *botaurus*, a combination of Latin *butio*, bittern, and *taurus*, bull. Pliny (77–79 AD), often cited in this connection, wrote of a bird '*taurus*', found in the region round Arles; he said it imitates the sound of oxen and that it is a small bird – perhaps not a bittern, though Pliny may have been referring to the Little Bittern *Ixobrychus minutus* and 'small' may have been a relative term.

Ixobrychus Billberg, 1828 [ik-so-BREE-kʊs]: 'reed-bellower', from Greek *ixos* or *ixias*, mistletoe or plant, and *brukhomai*, to bellow or roar. Billberg did not explain, so we can only conjecture that he named this one based on a folk belief that the bittern produced its sound by bellowing through a reed. Unfortunately, he used Greek *ixos*, mistletoe, rather than *ixias*, a plant whose identity is somewhat vague but that may be reed-like (perhaps from the iris or lily family). Billberg also gives the Latin *arundo* (also spelled *harundo*) meaning reed, which sheds light on his intention here.

Dupetor Heine & Reichenow, 1890 [doo-PE-tor]: 'clatterer', from Greek *doupētōr*, clattering.

Gorsachius Bonaparte, 1855 [gor-SUH-ki-ʊs]: 'Japanese night-heron'. *Gorsachius* is apparently from *goisagi*, Temminck's (1838) species name for the Japanese Night Heron. Temminck said: 'Its Japanese name is *Awogoisagi* or just *Goisagi*, which we will keep for this beautiful species.' [In Japanese, we understand it would be pronounced more like go-i-suh-gi, with 'suh' very short.]

Nycticorax Forster T, 1817 [nik-ti-KOR-aks]: 'nightraven' from Greek *nuktikorax*, nightraven, from *nux*, *nukt*, night, and *korax*, raven. There is a lot of confusion about which bird it refers to: Aristotle (c. 330 BC) said it is a name given by some to the eared owl, and the King James Bible had 'owl' for what appears as Latin *nycticorax* in the Vulgate (Psalm 102, verse 6). Gesner (1555) had a long discussion about all the different birds the word has been used for, including a variety of owls and even the hoopoe, but concluded by describing something we can recognise as a night heron, and his illustration confirms that. He described the voice of this bird as inharmonious and like someone vomiting.

Butorides Blyth, 1852 [boo-TOR-i-dehz]: 'bittern-family [bird]', from the genus name *Butor* (none in Australia) and from Greek *-ides*, belonging to a family.

Ardeola Boie, 1822 [ar-de-OH-lʊh]: 'little heron', diminutive of Ardea (see genus name *Ardea*).

Bubulcus Bonaparte, 1855 [bʊ-BʊL-cʊs], from Latin *bubulcus*, an ox driver or herdsman.

Ardea Linnaeus, 1758 [ar-DEH-uh]: 'heron', from Latin *ardea*, a heron. Pliny (77–79 AD) mentioned it in his *Natural History* (usually using the diminutive *ardeola/ardiola*), and provided as usual entertaining and imaginative descriptions of its behaviour, including the information that the males of the 'ash-coloured' kind of heron experience great anguish when they mate, even bleeding from the eyes during the process. Ardea was also the name of the chief town of the Rutili, a tribe led by Turnus. Ovid (c. 8 AD), in the *Metamorphoses*, told that when this town fell to the Trojans under Aeneas and was burned, 'From the midst of the ruin and confusion a bird of a kind never seen before flew up into the air, shaking off the cinders from its beating wings. It retained the leanness and the pallor, the mournful cries, and all the other characteristics of a captured town. Even the name of the city survived in that of the bird, and Ardea, changed into a heron, beats itself with its own wings, and bemoans its fate.'

Egretta Forster T, 1817 [e-GRET-tuh]: 'little heron'. Simply a Latinisation of egret – see family introduction.

The species

Australasian Bittern

See family introduction. The only bittern restricted to Australia and New Zealand.

Other names: Australian Bittern, applied by Gould (1848); Brown Bittern, formerly widely used, presumably in contradistinction to Black Bittern; Boomer, Bullbird, Bunyip, all evocative folk names for the roaring nocturnal call, which doubtless reinforced bunyip stories among European settlers.

Botaurus poiciloptilus (Wagler, 1827) [boh-TOR-ʊs poy-si-lo-TI-lʊs]: 'mottled-feathered bittern', see genus name, with Greek *poikilos* many-coloured or spotted or mottled, and *ptilon* feather, down. The bird is hardly spotty or many-coloured, so we'll settle for mottled.

Black-backed Bittern (breeding resident)

See family introduction; the adjective is accurate, but not unique even within the genus.

Other names: Australian Little Bittern, from when it was first split from the widespread Old World Little Bittern *I. minutus* – it is not clear what prompted the change to the current name; Spotted Heron, used by Latham (1801) from his own species name *maculata* (apparently for a young bird); Leech Bittern, presumably referring to a dietary preference, though the literature doesn't seem to support the observation. Minute Bittern, used by Gould (1848) – a bit of hyperbole and also a bit odd: although it might seem to have been drawn straight from the specific name *minutus*, because it was before the Australian bird was separated from the Old World Little Bittern, Gould actually called it *Ardea pusilla*, meaning 'very small', while also referring to the European species *Ardetta minuta*. Both species and common name were still recognised by RAOU (1913).

Ixobrychus dubius Mathews, 1912 [ik-so-BREE-kʊs DOO-bi-ʊs]: 'doubtful reed-bellower', see genus name, with Latin *dubius*, indicating Mathews' doubt about the taxonomy.

Yellow Bittern (vagrant)

See family introduction. Another piece of hyperbole, though it is certainly buffy-yellowish.

Other names: Little Yellow Bittern, cited by RAOU (1913, 1926).

Ixobrychus sinensis (Gmelin JF, 1789) [ik-so-BREE-kʊs si-NEN-sis]: 'Chinese reed-bellower', see genus name, plus Modern Latin *sinensis*, from China.

Von Schrenk's Bittern (vagrant to Christmas Island)

See family introduction. For Leopold von Schrenck (note the disparity in spelling), director of the St Petersburg Imperial Academy of Sciences in the late 19th century, who worked in north-eastern Russia for some time, in often very difficult conditions. In his account of eastern Siberian birds he noted this species but mistook it for an immature Cinnamon Bittern. When Robert Swinhoe (British diplomat-collector) described the species, he generously acknowledged Schrenck's primacy.

Other names: Little Bittern; Shrenk's or Schrenck's Bittern, the latter being the correct spelling.

Ixobrychus eurhythmus (Swinhoe, 1873) [ik-so-BREE-kʊs yoo-RITH-mʊs]: 'well-proportioned reed-bellower', see genus name, and from Greek *eurhuthmos*, rhythmical or well proportioned. We don't really know what well proportioned might mean to a bittern, but judging by a couple of the appearances of this one on You-tube, it's a very handsome bird!

Cinnamon Bittern (vagrant)

See family introduction. Straight from the scientific name, an apt name for this richly coppery-cinnamon coloured bird.

Ixobrychus cinnamomeus (Gmelin JF, 1789) [ik-so-BREE-kʊs si-na-mo-MEH-ʊs]: 'cinnamon reed-bellower', see genus name, and from Latin *cinnamomum*.

Black Bittern (breeding resident)

See family introduction. A descriptively appropriate name, both in Australia and throughout its wide South and South-East Asian range, where it is the only black bittern.

Other names: Yellow-necked Heron, used by Latham (1801) from his species name; Yellow-necked Bittern, as used by Gould; Mangrove Bittern, for one of its habitats – this is the only Australian bittern to use mangroves.

Dupetor flavicollis (Latham, 1790) [doo-PE-tor flah-vi-KOL-lis]: 'yellow-necked clatterer', see genus name, and from Latin *flavus*, golden or reddish-yellow (from *flagro*, burn) and *collum*, neck, for the bold yellow streak down the side of the neck.

Japanese Night Heron (vagrant)

See family introduction. The species breeds in Japan (and over-winters in the Philippines).

Gorsachius goisagi (Temminck, 1836) [gor-SUH-ki-ʊs go-i-suh-gi]: 'Japanese night heron', see genus name.

Malayan Night Heron (vagrant to Christmas Island)

See family introduction. Actually found throughout South and South-East Asia.

Gorsachius melanolophus (Raffles, 1822) [gor-SUH-ki-ʊs me-la-no-LOH-fʊs]: 'black-crested Japanese night-heron', see genus name, plus Greek *melas*, black, and *lophos*, tuft or crest, which is indeed a feature of this bird.

Black-crowned Night Heron (vagrant to Cocos (Keeling) Islands)

See family introduction. A name used throughout its near-worldwide range, though it does nothing to distinguish the species from many others.

Nycticorax nycticorax (Linnaeus, 1758) [nik-ti-KOR-aks nik-ti-KOR-aks]: 'nightraven-nightraven', see genus name.

Nankeen Night Heron (breeding resident)

See family introduction. 'Nankeen' derives from Nankin or Nanking, major town of Kiangsu province, China. The town gave its name to a widely used cheap yellowish cotton cloth manufactured there, which in turn came to be used for the colour (see also Nankeen Kestrel). It was in use early – in 1836 James Backhouse referred to it as the Nankin-bird and Gould (1848), in using the current name, reported that Nankeen-bird was a name 'of the Colonists'.

Other names: Nankeen Night-Heron, the usual form in Australia; Rufous Night Heron, for the colour, also widely used; Nankeen Crane, see White-faced Heron, though there is little that is crane-like about the stocky night herons; Quaker, also reported by Gould (1848) to be a name 'of the Colonists', but without explanation – pity about that; New Holland Night Heron and Caledonian Night Heron, both used by Latham (1823), the latter being a young bird that he mistook for a separate species.

Nycticorax caledonicus (Gmelin JF, 1789) [nik-ti-KOR-aks ka-le-DO-ni-kʊs]: 'Scottish nightraven' – well, no, not really! See genus name, plus Latin *Caledonia*, the province of the Ancient Britons/the highlands of Scotland, but in this case standing for New Caledonia, the source of the type specimen.

Striated Heron (breeding resident)

See family introduction. An unhelpful name that in Australia only applies to immature birds; it is drawn from the specific name, which in turn refers to a South American subspecies. This five-continent species in fact has 30 widely varying subspecies!

Other names: Mangrove Heron, widely used, for the habitat; Green or Green-backed Heron, again relevant to some subspecies, but not in Australia; Johnny Mangrove or Mangrove Jack, an affectionate tribute to its familiarity to fishing people; Mangrove Bittern, and indeed it is very bittern-like in appearance; Red Mangrove Heron, perhaps for the rufous subspecies in north-western Australia, but this could also refer to a dominant mangrove species (*Rhizophora stylosa*); Thick-billed Green Bittern, applied by Gould (1848) from his species name *Ardetta macrorhyncha* for a bird from Gosford; Thick-billed Mangrove-Bittern, for the same species, recognised by RAOU (1913); Little Grey Bittern, adopted from the name 'of the Colonists', again by Gould for his Port Essington species *Ardetta stagnatilis*, which we no longer recognise; Little Mangrove-Bittern (RAOU 1913) for the same species, which they still accepted.

Butorides striata (Linnaeus, 1758) [boo-TOR-i-dehz stree-AH-tuh]: 'streaked bittern-family [bird]', see genus name, and Latin *striatus*, streaked or furrowed.

Chinese Pond Heron (vagrant)

See family introduction. Four of the six species of *Ardeola* are known as Pond Herons, for their preference for small waterbodies. This species is found throughout China, but also in South-East Asia and western Indonesia.

Ardeola bacchus (Bonaparte, 1855) [ar-de-OH-lʊh BAK-kʊs]: 'little wine-coloured heron', see genus name, plus the Roman name of Bacchus (Greek Bakkhos), god of wine. This refers to the striking reddish-purple colour of the male bird's head and neck in breeding plumage, what Bonaparte calls 'dark wine colour'.

Javan Pond Heron (vagrant)

See family introduction and previous species. The type specimen was taken on Java but it is found throughout much of Indonesia and into South-East Asia.

Ardeola speciosa (Horsfield, 1821) [ar-de-OH-lʊh spe-see-OH-suh]: 'splendid little heron', see genus name, and Latin *speciosus*, handsome, splendid, showy. Certainly an attractive bird in its orange, white and charcoal breeding plumage.

Eastern Cattle Egret (breeding resident)

See family introduction. Until recently, part of a near cosmopolitan species named for its foraging behaviour of accompanying large herbivores, but recently split in two. This one is found in East Asia and Australasia.

Other names: Cattle Egret; Buff-backed Heron, pertinent, but only in breeding plumage.

Bubulcus coromandus Boddaert 1783 [bʊ-BʊL-cʊs co-ro-MAN-dʊs]: 'Coromandel ox driver', see genus name, and from the place-name Coromandel. Boddaert notes that the bird was called by Brisson 'Crabier de Coromandel'. The Coromandel referred to is in south-eastern India and is named after the ancient kingdom of Cholamandalam.

Grey Heron (vagrant)

See family introduction. The common heron of Europe (and much of the rest of the Old World) and undeniably mostly grey.

Ardea cinerea Linnaeus, 1758 [ar-DEH-uh si-NE-re-uh]: 'ashy-grey heron', see genus name, and from Latin *cinereus*, like ashes – a double dose of ashes here!

White-necked Heron (breeding resident)

See family introduction. Straightforward descriptive name (though does not distinguish it from the Pied Heron). It seems to be home-grown, in that Gould (1848) reported that it was already the name 'of the Colonists', though he eschewed it.

Other names: Pacific Heron, for no obvious reason other than it was apparently taken from the scientific name (it is not regularly found outside Australia, where it occurs on the north, south and west coasts, as well as the Pacific coast) – Gould (1848) adopted the name, though Latham (1801) had already used it; White-necked Crane, see White-faced Heron. Bull-ra-gang Heron from the scientific name *bullaragang* proposed by Wagler in 1827, honouring an Indigenous name – what a pity he was too late!

Ardea pacifica Latham, 1801 [ar-DEH-uh pa-SI-fi-kuh]: 'Pacific heron', see genus name, and from Latin *pacificus*, from the Pacific region (see common name).

Great-billed Heron (breeding resident)

See family introduction. Not especially pertinent relative to its size, but Gould (1848) used it and it stuck.

Other names: Alligator or Croc Bird, for the roaring call; Giant Heron, certainly the largest Australian heron; Sumatran Heron, for the species name and type locality, but not much used in Australia.

Ardea sumatrana Raffles, 1822 [ar-DEH-uh soo-ma-TRAH-nuh]: 'Sumatran heron', see genus and common names.

Purple Heron (vagrant)

See family introduction. Pertinent; from the species name.

Ardea purpurea Linnaeus, 1766 [ar-DEH-uh pʊr-pʊr-EH-uh]: 'purple heron', see genus name, and from Greek *porphuroeis*, purple or red.

Great Egret (breeding resident)

See family introduction. The largest of the white herons.

Other names: Eastern Great Egret, for those who prefer to split this cosmopolitan species (or species group) into two or more species; White Egret, not in the least helpful, despite it being the then species name, but the recommended name of RAOU (1926); White Crane, see White-faced Heron; Australian Egret, applied by Gould (1848) and recommended by RAOU (1913).

Ardea alba Gray JE, 1831 [ar-DEH-uh AL-buh]: 'white heron', see genus name, plus from Latin *albus*, dull white.

Intermediate Egret (breeding resident)

See family introduction. A tempting name that is in wide international use and is intended to create a size name continuum from Great to Little Egret, but in fact this species is scarcely larger than the Little Egret.

Other names: Plumed Egret, formerly widely used in Australia, for the superb breeding plumes which are more dramatic than those of the other Australian species; Yellow-billed Egret, a useful distinction from Great and Little Egrets.

Ardea intermedia Wagler, 1829 [ar-DEH-uh in-ter-MEH-di-uh]: 'intermediate heron' (we did say heron and egret were interchangeable!), see genus name, plus Latin *intermedius*, intermediate.

Pied Heron (breeding resident)

See family introduction. Pied literally means black and white, like a (European) magpie; Gould coined the name, from his species name.

Other names: Pied or White-headed Egret, the latter applicable only to immature birds; Aru Egret, for *Notophoyx aruensis*, now part of this species, from the Aru Islands.

Egretta picata (Gould, 1845) [e-GRET-tuh pi-KAH-tuh]: 'little pied heron', see genus name, and *picatus* from *pica*, a jay or magpie (i.e. black and white).

White-faced Heron (breeding resident)

See family introduction. A neatly descriptive name, a unique characteristic even beyond Australia. It was formalised by RAOU (1926).

Other names: Blue Crane, a traditional confusion with a superficially similar but unrelated family which is unlikely to mislead anyone in Australia, where Brolga is the most familiar crane; White-fronted Heron, accurate but the accepted name is more so – Gould followed Latham (1801) in using this name and RAOU (1913) followed him; Matuka, a Māori name used in New Zealand.

Egretta novaehollandiae (Latham, 1790) [e-GRET-tuh no-veh-hol-LAN-di-eh]: 'little New Holland heron', see genus name, and from Modern Latin *Nova Hollandia*, New Holland, the old name for Australia.

Little Egret (breeding resident)

See family introduction. A worldwide name for a species found nearly everywhere except the Americas. Not totally helpful even in Australia, where Cattle Egrets are smaller and Intermediate and Pacific Reef Herons (or Egrets) barely larger.

Other names: Lesser Egret; Spotless Egret, sounds unnecessary in the absence of any spotted egrets, but Gould applied it from his species name *Herodias immaculata*; White Crane, reported by Gould to be the name 'of the Colonists' (of Port Essington, that being the only place from which he knew of it).

Egretta garzetta (Linnaeus, 1766) [e-GRET-tuh gar-ZET-tuh]: 'little heron little heron', see genus name, and *garzetta*, a diminutive of *garza* (heron), a Spanish (and earlier Italian) word, whose derivation seems to be a bit of a puzzle. It is variously reported as possibly derived from a potential earlier Spanish form *gardea* (cf. *ardea*) or from Arabic (and certainly it does not seem to be a Latin word), or as onomatopoeic from the bird's call. *Garzetta* is the Italian name for this bird, and *Garzette* has been used as a common name for it in French.

Western Reef Heron (vagrant)

See family introduction. Reef is for the habitat, though they also patrol beaches. Western in contrast to the following species, this one lives around Africa, the Middle East and India.

Egretta gularis (Bosc, 1792) [eh-GRET-tuh gʊ-LA-ris]: 'throated egret', see genus name, and from Latin *gula*, the throat. The bird, according to Bosc, is dark to blackish, with three inches of white on the throat.

Pacific Reef Heron (breeding resident)

See family introduction. 'Pacific' is in distinction from the previous species.

Other names: Eastern Reef or Reef Heron or Egret; White Reef Heron (for the white morph); Blue (Reef) Heron (for the dark morph) – Gould had uncharacteristic trouble with this species, as he didn't recognise the morphs and described them as separate species, in addition to his Sombre Egret; Sacred Heron, for the supposed relationship of this bird to some Polynesian peoples, taken straight from the species name; Sombre Egret, used by Gould for his name *Herodias pannosus*, for a dark morph bird from Port Stephens – he was well aware of the species, however, and it is unclear why he decided that the specimen warranted separate species status, other than that it had more white on the chin; Blue Crane, reported by Gould (1848) to be used by the 'Colonists of Port Essington' (see also White-faced Heron).

Egretta sacra (Gmelin JF, 1789) [eh-GRET-tuh SAK-ruh]: 'little sacred egret' and Latin *sacer*, holy or sacred (also used poetically to mean the opposite, surely not relevant here). See also Other names.

PELECANIDAE (PELECANIFORMES): pelicans

Pelecanidae Rafinesque, 1815 [pe-le-KA-ni-dee]: the Pelican family, see genus name *Pelecanus*. 'Pelican' is of the same origin as *Pelecanus*.

The genus

Pelecanus Linnaeus, 1758 [pe-le-KAH-nʊs]: 'pelican', from Greek *pelekan* (also *pelekanos*, *pelekinos*), meaning pelican. There are other similar words in Greek, such as *pelekan* (a part of the verb *pelekaō*, to hew wood) and *pelekus* an axe, which has led some to suggest that this is a word describing a bird that cut wood with its bill. We are not sure how this applies to a pelican ... but then there IS the word '*pelekas*', apparently meaning a woodpecker. It seems the Ancient Greeks were not above having a little joke based on the similarity (Arnott 2007).

The species

Australian Pelican (breeding resident)

See family introduction. Limited to Australia (and more recently New Guinea).

Other names: Spectacled Pelican, from the species name, for the ring of bare skin around the eyes.

Pelecanus conspicillatus Temminck, 1824 [pe-le-KAH-nʊs con-spi-si-LAH-tʊs]: 'spectacled pelican' see genus name, and from Latin *conspicio*, look [at] carefully, and *conspicillum*, a place to look from. Temminck (1838) described the 'two large, more or less circular bare patches round the eyes, not to be found in any known pelican', calling the bird the *Pélican à lunettes*.

PANDIONIDAE (ACCIPITRIFORMES): ospreys

Pandionidae Sclater & Salvin, 1873 [pan-dee-O-ni-dee]: the Osprey family, from the genus name *Pandion*.

The genus

Pandion Savigny, 1809 [pan-DEE-on]: 'sea-eagle's father', from Greek proper name Pandion. There are most likely two Pandions in Greek mythology (though some say they were the same person). They were both kings of Athens and between them had three children who were turned into birds. One of these children was Nisus, who was transformed into a 'tawny-winged sea-eagle' in Ovid's (c. 8 AD) version of the story involving Nisus' daughter Scylla and Minos, King of Crete – another tale of lust and violence. Scylla herself was also turned into a bird – a common problem, apparently – well, this was the *Metamorphoses*, after all. This doesn't explain why the Osprey got *Pandion* instead of *Nisus*, but we should note that the Eurasian Sparrowhawk is *Accipiter nisus*, so Nisus didn't miss out altogether.

The species

Eastern Osprey (breeding resident)

The word osprey comes to us via a long and improbable journey. It seems to have begun in Latin as *ossifraga*, a bone-breaker – this is not a habit of ospreys, and instead strongly suggests the magnificent Lammergeier or Bearded Vulture, *Gypaetus barbatus* (which uses gravity and rock outcrops for the purpose). It is unclear how the confusion occurred, but it became Old French *ospres*, attached to our bird, and thence into English. Perhaps surprisingly it is the 'eastern' part of the name that is controversial. Until recently it was accepted that the Osprey represented just one worldwide species; recently two species have been recognised (some would say three): one in Australia and into Indonesia, the other everywhere else. This break-up of the species is not universally supported.

Other names: Fish Hawk, used by the 'Colonists of Swan River', and Little Fish Hawk ('Colonists of New South Wales'), perhaps to distinguish from White-bellied Sea-eagle, though this isn't made clear – both are cited by Gould (1848); White-headed Osprey, a reference to the fact that the Eastern Osprey lacks the black nape of the Northern Hemisphere birds – this was also the name used by Gould (1848) from his own name *P. leucocephalus*, believing it to be different from the European birds. He has been vindicated, though his name was pre-empted by Vieillot.

Pandion cristatus (Vieillot, 1816) [pan-DEE-on kri-STAH-tʊs]: 'crested sea-eagle's father', see genus name, and Latin *cristatus*, crested, which the Osprey is, albeit slightly.

Western Osprey (vagrant)

See family introduction and previous species.

Pandion haliaetus (Linnaeus, 1758) [pan-DEE-on ha-li-EH-tʊs]: 'sea-eagle father of a sea-eagle', see genus name, and Greek *haliaetos* sea-eagle, (from *hali-*, sea, and *aetos*, eagle).

ACCIPITRIDAE (ACCIPITRIFORMES): eagles, hawks and allies

Accipitridae Vigors, 1824 [ak-si-PIT-ri-dee]: the Capturing family, see genus name *Accipiter*.

'Kite' comes from an old English word *cyta*, apparently onomatopoeic for the shrill call of the Red Kite (*Milvus milvus*), until recently very rare in Britain but formerly abundant even in cities. (The good news is that, due to conservation measures, Red Kite numbers are beginning to recover.) The word is now applied rather randomly and indiscriminately to a range of other genera, some of which have little similarity to *Milvus* kites. In particular it is applied to all four very similar and closely related members of the genus *Elanus*, which are small mostly white hovering rodent specialists, including two in Australia – totally different in almost every particular from the big scavenging *Milvus* kites. (In fact, there are suggestions that they merit their own family.)

'Buzzard' (properly applied to the Old World genus *Buteo*, found across much of the world but not Australia) comes to English via Old French *busard*, apparently from Latin. (The poor bird was regarded as so useless at its ordained purpose of being a hunting pet of the nobility that a secondary meaning of the word arose, meaning a stupid person.)

'Eagle' comes via Old French *egle*, probably from Latin *aquila*, and appears in English in that form from the 14th century. Eagle began to appear some 200 years later. Applied most purely to the worldwide genus *Aquila* but also to many other large diurnal raptors, often with feathery 'trousers'.

'Hawk', from old English *hafoc*, just meant any raptorial bird. It is still used today with great imprecision.

'Goshawk' was originally goose hawk, from Old English *goshafoc* (meaning the same thing), implying a bird that was used to hunt geese. However, this was unlikely to have been an *Accipiter* (the genus to which the name is now applied, including in Australia), and is thought more likely to have referred to a Peregrine Falcon. The transfer of name is again something of a mystery (as for Osprey), the problem being that we have no way of knowing which bird is referred to at any stage of the process.

'Sparrowhawk' is generally used for smaller members of the genus *Accipiter*; in England there is one large and one small species. Old English was *spearhafoc*, again meaning literally sparrowhawk. The bird certainly hunts sparrows, but the name could also have been intended to carry the implication of smallness (cf. the goshawk).

'Harrier', one who harries (raids or ravages), be it human or bird. Almost exclusively used for the genus *Circus*.

The genera

Elanus Savigny, 1809 [e-LAH-nʊs]: 'kite', from Greek *elanos*, one of the words for kite.

Pernis Cuvier, 1816 [PEHR-nis]: 'hawk', from Greek *pternis*, a hawk.

Lophoictinia Kaup, 1847 [lo-foyk-TI-ni-uh]: 'crested kite', from Greek *lophos*, crested, and *ictinos*, kite. According to HANZAB (1990–2006, vol. 2), the bird has 'a short occipital crest', though this might not be observed in the field.

Hamirostra Brown, 1846 [ha-mi-ROS-truh]: 'hook-bill', from Latin *hamus*, a hook, and *rostrum*, bill.

Aviceda Swainson, 1836 [a-vi-SEH-duh]: 'bird-killer', from Latin *avis*, bird, and *caedere*, to cut or kill.

Hieraaetus Kaup, 1844 [hi-e-ra-EH-tʊs]: 'hawk-eagle', from Greek *hierax*, hawk or falcon, and *aetos*, eagle.

Aquila Brisson, 1760 [A-qui-luh]: 'eagle', from Latin *aquila*.

Erythrotriorchis Sharpe, 1875 [e-rith-ro-tri-OR-kis]: 'red hawk' from Greek *eruthros*, red (as in 'erythrocytes', red blood cells) and *triorkhes*, a kind of hawk, possibly a buzzard. *Tri-orkhes* also means having three testicles – metaphorically, being very lecherous. Why was this applied to the Greeks' hawk or buzzard? It seems the ancient belief that some birds of prey had a high level of sexual activity (as compared with most humans) has been borne out in modern studies. For example, Peregrine Falcons have been reported as copulating up to 15 times a day, sometimes even after eggs have been laid (Olsen 1995).

Accipiter Brisson, 1760 [ak-SI-pi-ter]: 'captor', from Latin *accipere*, which derives from *capere*, to seize or capture.

Circus Lacépède, 1799 [SEER-kʊs]: 'circling-hawk', from Greek *kirkos*, an unspecified hawk named for its wheeling flight (*kirkos* also means a circle or ring, just like a circus!).

Milvus Lacépède, 1799 [MIL-vʊs]: 'nasty kite', from Latin *miluus/milvus*, a kite. The word *milvus* was also used by the Romans to mean a rapacious or exploitative person. This finds an echo in the use of 'kite' itself; for example, in Shakespeare's *King Lear*, Lear exclaims to his horrible daughter Goneril, 'Detested kite, thou liest'.

Haliastur Selby, 1840 [ha-li-AS-toor]: 'sea-hawk', from Greek *hals*, *hali-*, sea-, with Latin *astur*, a species of hawk.

Haliaeetus Savigny, 1809 [ha-li-EH-tʊs]: 'sea-eagle', from Greek *haliaetos* (from *hali-*, sea, and *aetos*, eagle).

The species

Black-shouldered Kite (breeding resident)

See 'kite'. An unhelpful name, in that all four *Elanus* species share the characteristic of a large black shoulder on a white body.

Other names: Australian Black-shouldered Kite, to distinguish it from the Black-shouldered Kite (*Elanus caeruleus*) – now generally called Black-winged – of southern Africa and Asia.

Elanus axillaris (Latham, 1801) [e-LAH-nʊs ak-sil-LAH-ris]: 'kite with armpits', see genus name, and Latin *axilla*, armpit. This may seem a strange name, because the axilla area of this bird is simply white like much of the rest of it. Although Latham's naming of the bird has been accepted as fitting his description of the black carpal patches on its underwings, Mathews (1915) wrote very persuasively that the description fitted better with the Letter-winged Kite. It is true that in his English version, Latham wrote 'a large long patch of black also occupies the whole of the inner part of the wing when closed', sounding much like a Letter-winged Kite, but this has not convinced those who determine these things, and the Black-shouldered Kite remains *Elanus axillaris*.

Letter-winged Kite (breeding resident)

See 'kite'. Refers to the roughly M-shaped bar across the underwings and was introduced by Gould (1848) from his own species name.

Elanus scriptus Gould, 1842 [e-LAH-nʊs SKRIP-tʊs]: 'kite with writing', see genus name, and Latin *scribere/scriptus*, write/written, for the underwing markings – this bird, unlike its cousin *axillaris*, has very distinctive 'armpits'.

Crested Honey Buzzard (vagrant)

See 'buzzard'. The honey buzzards (or honey-buzzards), three species of *Pernis*, specialise in wasp and bee larvae – doubtless their interest in the nests was misinterpreted as honey-seeking.

Other names: Crested Honey-buzzard; Oriental or Eastern Honey-buzzard or Honey Buzzard; lives in far eastern Asia, moving south in winter.

Pernis ptilorynchus (Temminck, 1821) [PEHR-nis ti-lo-RIN-kʊs]: 'feather-billed hawk', see genus name, and from Greek *ptilon*, feather or down, and *rhunkhos*, bill. Note that in the Latinised species name there is, unusually, no 'h' after the 'r'.

Square-tailed Kite (breeding resident)

See 'kite'. An accurate description of its flight silhouette from below, and used by Gould as an interpretation of his own species name.

Other names: Long-winged Kite, equally descriptive; Kite, the name 'of the Colonists' according to Gould (1848), indicating that it was formerly much more common than it is today.

Lophoictinia isura (Gould, 1838) [lo-foyk-TI-ni-uh i-SOO-ruh]: 'equal-tailed crested kite', see genus name, and from Greek, *isos*, equal (think isosceles triangles), and *oura*, tail. Equal refers to the square tail (all feathers of equal length).

Black-breasted Buzzard (breeding resident)

See 'buzzard'. No relation to the 'true' buzzards, but shares the general characteristic of being a soaring generalist hunter – along with many other species! Gould (1848) believed it to be a true *Buteo*, and named it thus. The black breast of the adult is unique among Australian accipitrids (with the exception of old Wedge-tailed Eagles).

Other names: Black-breasted Kite, as cheerfully inaccurate as Buzzard; Broad-tailed Mountain Eagle, used by early Western Australian settlers, apparently in the same sense as 'Mountain Duck' for Australian Shelduck, and 'Mountain Devil' for the lizard *Moloch horridus*, purportedly implying 'across the mountains'.

Hamirostra melanosternon (Gould, 1841) [ha-mi-ROS-truh me-la-no-STER-non]: 'black-chested hook-bill', see genus name, and from Greek *melas*, black, and *sternon*, chest (see common name). The hooked bill is certainly a feature of this bird, but perhaps not more markedly than in several other large birds of prey.

Black-breasted Buzzard *Hamirostra melanosternon*

Pacific Baza (breeding resident)

Baza is taken from an older genus name (still used by Cayley in 1931); it is a Latinised form of an Arabic word (via Hindi), referring to a goshawk. RAOU (1978) introduced the term to Australia, on the basis that it is 'the simplest group name in wide use for the genus'. HANZAB (1996– 2006, vol. 2) noted that we adopted the name baza here because that is 'the standard group name in se. Asia, where the genus is best-developed'; we note simply that in addition to the Australian species there are only two South-East Asian species, plus one African and one Madagascan. Pacific presumably refers to the fact that, in addition to north and east Australia, the species is found from Sulawesi east through New Guinea to the Solomons.

Other names: Crested Hawk, obvious and in wide usage in Australia from Gould's time until recently. Pacific or Gurney's Cuckoo-falcon, this being an alternative group name, perhaps for the barred undersides? John Gurney was a 19th-century English banker, parliamentarian and amateur ornithologist; his connection with this species is unclear, because it appears never to have borne his name in the scientific version and he did not collect the type species, though he did write on the genus *Baza* for *Ibis* (British Ornithologists' Union) in 1880. Pacific or Crested Lizard-hawk, but, although lizards and frogs form part of its diet, it is essentially an insectivore.

Aviceda subcristata (Gould, 1838) [a-vi-SEH-duh sŭb-kris-TAH-tuh]: 'slightly crested bird-killer', see genus name, and from Latin *sub*, below (meaning not fully), and *cristatus*, crested – just a small crest, then (which it is, if you compare it with, say, that of a Sulphur-Crested Cockatoo). See common name for feeding preferences – the genus name is not entirely a misnomer, because this baza does take nestlings and very small birds on occasion.

Little Eagle (breeding resident)

See 'eagle'. Much smaller than the Wedge-tailed Eagle, with which it was being compared. (It is suggested that the Little Eagle isn't really an eagle, but there are other 'eagles' that are not *Aquila*; other members of the genus *Hieraaetus* include the Booted and Wahlberg's Eagles of Europe and Africa.)

Other names: Little Australian Eagle, used by Gould (1848); Australian Little Eagle, though we are unaware of any other Little Eagle.

Hieraaetus morphnoides (Gould, 1841) [hi-e-ra-EH-tŭs morf-NOY-dehz]: 'eagle-like hawk-eagle', see genus name, and from Greek *morphnos*, eagle, and *-oeidēs* (Latin *-oides*), resembling. Well, if it's not an eagle (see common name), this is one way of pointing out that it's very much like one (though maybe a bit like a hawk too)!

Gurney's Eagle (vagrant to Torres Strait)

See 'eagle'. For John Gurney, see Pacific Baza.

Aquila gurneyi Gray GR, 1861 [A-kwi-luh GUR-ni-ee]: 'Gurney's eagle', see genus name and Pacific Baza.

Wedge-tailed Eagle (breeding resident)

See 'eagle'. Misleading, in that the tail is much more diamond- than wedge-shaped. It was Gould (1848) who noted the 'lengthened and wedge-shaped form of the tail' and applied the name. However, it has been suggested to us that from the side, the tail could be seen as wedge-like.

Other names: Eaglehawk or Eagle Hawk, a name that Gould said was in use from early times by the 'Colonists of New South Wales', perhaps implying that Australia couldn't be home to a real eagle! Mountain Eagle, used by Collins (1802) in illustrating a specimen from New South Wales, presumably referring to the Blue Mountains. Bold Vulture is a now-quaint name that was in very early usage; it is tempting to explain it as the direct translation of Latham's 1801 description of the species as *Vultur audax*, except that there is a picture in the National Library of Australia by 'Port Jackson Painter' dated as 'between 1788 and 1797'. In fact Latham used the picture to make his description (which would have been interesting, given the picture!), and simply took the name accompanying it. (He had form for this; he also described the Australian Brush-turkey as 'New Holland Vulture' based on a drawing.)

Aquila audax (Latham, 1801) [A-qui-luh OW-daks]: 'bold eagle', see genus name, with Latin *audax*, bold, audacious. Why this should be the specific name remains something of a mystery, and indeed 'bold' is said by Debus (1998) to be 'a misnomer for this normally shy and wary bird'. On the other hand, we can well imagine early settlers being somewhat intimidated by this imposing and magnificent creature, and wedgies being less wary of men with guns until they learned better.

Red Goshawk (breeding resident)

See 'goshawk' and 'hawk', but note that it is not closely related to *Accipiter*, although its hunting methods are reminiscent of them. It is a beautiful russet red colour.

Other names: Red, Rufous-bellied Buzzard; Red-legged Goshawk, a briefly recognised species from north-western Australia (described by Campbell in White (1911)), recognised by RAOU 1913 but subsumed by RAOU 1926), referring to its rufous 'trousers'; Radiated Falcon (Latham 1823) or Goshawk (Gould 1848), from the scientific name.

Erythrotriorchis radiatus (Latham, 1801) [e-rith-ro-tri-OR-kis ra-di-AH-tŭs]: 'barred red hawk', see genus name, plus Latin *radiatus*, with rays (i.e. barred). So is this bird, with its alleged three testes, more lustful than others? Cupper and Cupper (1981) carefully observed a male of the species frequently alighting on the female's back for a moment as she incubated her eggs, and interpreted this as 'most probably a sort of mock or symbolic copulation – pair bonding behaviour' rather than the 'real thing'.

Chinese Sparrowhawk (vagrant)

See 'sparrowhawk'. Breeds in China and Korea, migrating south for winter.

Wedge-tailed Eagle *Aquila audax*

Accipiter soloensis (Horsfield, 1821) [ak-SI-pi-ter so-lo-EN-sis]: 'captor from Solo River', see genus name, and from the name of the longest river in the Indonesian island of Java plus the suffix *-ensis*, usually indicating the place of origin of the type specimen. Horsfield tells us the bird is called Allap-Allap in the language of the Javanese.

Grey Goshawk (breeding resident)

See 'goshawk' and 'hawk'. One colour morph is a beautiful pale grey, the other pure white.

Other names: New Holland White Eagle (Latham 1801), Australasian White Eagle (Shaw 1819), and Lacteous Eagle (Latham 1823); Fair Falcon (Latham 1801), apparently based on his name *Falco clarus* – not that that explains much, though he described the white and grey morphs as separate species; White Goshawk or White Hawk, the latter the name of 'the Colonists' (Gould 1848); Grey-backed Goshawk; Milk-white Hawk, in 'Mr Caley's MS Notes to Birds in collection of Linn. Soc.' (Jardine and Selby 1826–35, vol. 1); New Holland Goshawk, from the species name; Variable Goshawk, from when it was regarded as a subspecies of that bird (see next species).

Accipiter novaehollandiae (Gmelin JF, 1789) [ak-SI-pi-ter no-veh-hol-LAN-di-eh]: 'New Holland captor', see genus name plus Modern Latin *Nova Hollandia*, New Holland, the old name for Australia. A rather ordinary name for such a stunning bird.

Brown Goshawk (breeding resident)

See 'goshawk' and 'hawk'. The Australian Brown Goshawk is indeed browner than its generally grey congenerics elsewhere in the world.

Other names: Australian Goshawk; Grey-headed or Chestnut-collared Goshawk, both true but not distinctive in a wider sense; Chicken-hawk, as a bird-hunting specialist it doesn't distinguish between wild and domestic prey; West Australian or Western Goshawk, for Gould's *Astur cruentus* from the south-west – it is no longer recognised, though RAOU (1913) did so, under the name Lesser Goshawk; Christmas Island Goshawk for race *natalis*, formerly assigned to *A. hiogaster*

Accipiter fasciatus (Vigors & Horsfield, 1827) [ak-SI-pi-ter fas-si-AH-tʊs]: 'banded captor', see genus name plus Latin *fascia*, band or bandage. This refers to the barring on the bird's underparts.

Japanese Sparrowhawk (vagrant)

See family introduction. Breeds in Japan (but also west into Korea, north China and Russia).

Accipiter gularis (Temminck & Schlegel, 1844) [ak-SI-pi-ter gʊ-LA-ris]: 'throated captor', see genus name, and from Latin *gula*, throat, referring to the fine black line running down the throat in both male and female birds.

Collared Sparrowhawk (breeding resident)

See 'sparrowhawk' and 'hawk'. True, but it's by no means the only collared sparrowhawk in the world; however, the name was introduced by Gould (1848) directly from his species name *A. torquatus* (not valid because it had already been named by Vieillot).

Other names: New Holland Sparrowhawk (Latham 1823); Sparrowhawk or Australian Sparrowhawk; Chicken-hawk (see Brown Goshawk); Little Hawk, of the 'Colonists of Western Australia' (Gould 1848).

Accipiter cirrocephalus (Vieillot, 1817) [ak-SI-pi-ter si-ro-SE-fa-lʊs]: 'tawny-headed captor', see genus name, and from Greek *kirros*, orange-tawny, and *kephalē*, head. Nothing to do with Latin *cirrus/cirrhus* meaning wispy tufts or curls (like the cloud form).

Swamp Harrier (breeding resident)

See 'harrier'. Swamp for its typical hunting and breeding habitat; Swamp-Harrier was recommended by RAOU (1926).

Other names: Swamphawk; Marsh Harrier, also a name used in Britain for a very similar species, *C. aeruginosus*, which at times was suspected of being the same species as the Swamp Harrier; Pacific Marsh Harrier; Allied Harrier again, to indicate its close relationship to the European Marsh Harrier – this was Gould's first-choice name (Gould 1848), but see the note under Spotted Harrier for the confusion associated with Gould's treatment of harriers; Allied Swamphawk or Swamp-hawk, used by RAOU (1913); Wheat Hawk, for the typical harrier habit of hunting low over cultivated ground. Gould's Harrier, for John Gould, from the time when the species was widely known as *C. gouldii*, named by Bonaparte in 1850, but 2 years after Peale had already done so – RAOU was using this name as late as 1913 and although we might think that it should have noted the error, the mix-up in Gould's treatment, and the fact that Peale named his species from a Fijian specimen, exonerate it!

Circus approximans Peale, 1849 [SEER-kʊs ap-PROK-si-manz]: 'quite like a circling-hawk', see genus name, and from Latin *approximare*, to draw near, approach, approximate. Again, it's seen as pretty much like the other harriers.

Spotted Harrier (breeding resident)

See 'harrier'. For the beautifully speckled wings and underside; the name was recommended by RAOU (1926).

Other names: New Holland Harrier, used by Jardine and Selby (1826–35, vol. 1); Allied Harrier, from the species name; Spotted Swamphawk or Swamp-Hawk, to distinguish it from the Swamphawk or Swamp Harrier – this name was still in formal use in the 20th century (RAOU 1913); Smoke Hawk, perhaps for the lovely grey uppersides or perhaps because it hunts fleeing prey in front of grassfires.

Jardine's Harrier, for Sir William Jardine, the 19th-century Scottish ornithologist who co-described the species and whose 40-volume *The Naturalist's Library* brought natural history to the masses. Adding to the association, Gould called it *Circus jardinii*, presumably not realising that Jardine had already named it, though there was major confusion, as Gould also incorrectly attributed *C. approximans* to Jardine and Selby, and used Allied Harrier and *C. assimilis* for a painting that clearly shows Swamp Harriers!

Circus assimilis Jardine & Selby, 1828 [SEER-kʊs as-SI-mi-lis]: 'similar circling-hawk', see genus name. It was deemed 'similar' to *Circus* species that had been named earlier (from Latin *adsimilis* or *assimilis*). The flight of the hunting Spotted Harrier is described by Cupper and Cupper (1981): '[The birds] work in a series of arcs, taking a few wing-beats to windward of the fence, then soaring low along it and drifting to leeward, they repeat the few beats to windward again. This movement is so precise one is drawn to counting the wing-beats to see if they vary in number. They seldom have.'

Black Kite (breeding resident)

See 'kite'. A name causing some puzzlement in Australia (it's really brown), but it was originally used in Britain to distinguish it from the Red Kite (*M. milvus*).

Other names: Fork-tailed Kite, for its most obvious attribute – there are those who still prefer this name; Allied Kite, used by Gould from his own name *M. affinis*, to indicate that he thought it to be different from, though related to, the European species – used by RAOU (1913); Kite-hawk; Kimberley Kite or Hawk, though it is abundant throughout inland Australia.

Milvus migrans (Boddaert, 1783) [MIL-vʊs MEE-grans]: 'nasty migrating kite', see genus name, and from Latin *migrare*, to migrate – though the bird is opportunistic rather than migratory in Australia, because it turns up over much of the country especially where there's a rubbish tip or an abattoir. In Europe it arrives in the summer, having migrated in massive flocks from sub-Saharan Africa, though it is now only a rare visitor to Britain.

Whistling Kite (breeding resident)

See 'kite'. The pulsating whistle of this species is a characteristic sound of inland waterways.

Other names: Whistling Hawk, used by the 'Colonists of New South Wales' (Gould 1848); Whistling Swamp Eagle, per the 'Colonists of Western Australia' (Gould 1848); Whistling Eagle or Eagle-hawk; Carrion Hawk, for its scavenging habits.

Haliastur sphenurus (Vieillot, 1818) [ha-li-AS-toor sfe-NOO-rʊs]: 'wedge-tailed sea-hawk', see genus name, and from Greek *sphēn/sphēnos*, a wedge, and *oura*, tail. Wedge-tailed? Well, it's a more obvious wedge than the Wedge-tailed Eagle's!

Brahminy Kite (breeding resident)

See 'kite'. The bird has a continuous distribution from Australia to India. To Hindus it is the mount of Lord Vishnu and represents Garuda, the king of birds. The Brahmins are, among other things, religious scholars and teachers. There is an added contrast to the scavenging Black Kite, known as Pariah Kite in India.

Other names: White-breasted Rufous Eagle (Latham 1823); White-breasted Sea-eagle, somewhat confusingly used by Gould (1848), though he based it on his own name *H. leucosternus* – he believed it to be different from its 'Indian ally'; Red-backed or Rufous-backed Sea Eagle or Kite or White-headed Sea Eagle (RAOU 1913), White-headed Rufous Eagle, all descriptive of the strikingly unusual plumage.

Haliastur indus (Boddaert, 1783) [ha-li-AS-toor IN-dʊs]: 'Indian Sea-hawk', see genus name, plus Latin *indus*, Indian. Boddaert (1783) called it the *Aigle de Pondicherry*, and it was also known as the *Aigle des Grandes Indes*: French *aigle*, eagle, plus the 17th- and 18th-century French name for India.

White-bellied Sea Eagle (breeding resident)

See 'eagle'. Strongly, but not exclusively, associated with marine habitats. Six of the other seven *Haliaeetus* are called Sea Eagle or Fish Eagle (the exception being the North American Bald Eagle); this is the only one with white undersides and the name comes directly from the species name. Gould used it, but it's not clear that he coined it.

Other names: White-bellied Sea-Eagle, the most familiar form in Australia; White-bellied Eagle (Latham 1823); Sea Eagle; White-breasted Sea-Eagle or Sea Eagle; White-bellied Fish Hawk.

White-bellied Sea Eagle *Haliaeetus leucogaster*

Haliaeetus leucogaster (Gmelin JF, 1789) [ha-li-EH-tʊs loo-ko-GAS-tehr]: 'white-bellied sea eagle', see genus name, plus Greek *leukos*, white, and *gastēr*, stomach.

TYTONIDAE (STRIGIFORMES): barn owls

Tytonidae Mathews, 1912 [tee-TOH-ni-dee]: The Tu-whoo family, see genus name *Tyto*.

'Owl' is an old word, as we might expect, coming to us from Old English *ule*; there are similar words in other north European languages, all reflecting the call (the hooting of the Tawny Owl *Strix aluco* rather than the tearing screech of the Barn Owl *Tyto alba*). The family is based on the Common Barn Owl *Tyto alba*, found throughout much of Europe, Africa, Asia and the Americas; it was named in Britain for its fondness for outbuildings, both as roosts and hunting grounds for mice. Barn Owl is recorded in writing from the 17th century but was probably used before that.

The genus

Tyto Billberg, 1828 [tee-TOH]: 'owl', from Greek *tutō*, another name for the Little Owl, *Athene noctua*, also called *glaux* (see genus *Ninox* under Strigidae). The word *tutō* is said to have derived from the call of the owl but this is the 'traditional' Tawny Owl *Strix aluco* call only. The birds of this genus all lean to the screech rather than the traditional tu-whit tu-whoo.

The species

Greater Sooty Owl (breeding resident)

For the pale-speckled black plumage, and from the species name, used by Gould (1848). For a while until recently this and the following species were lumped again. A female of this species can weigh twice as much as one of the next.

Other names: Dusky or Black Barn Owl; Sooty Owl, from when only one species was recognised.

Tyto tenebricosa (Gould, 1845) [tee-TOH te-ne-bri-KOH-suh]: 'gloomy owl', see genus name, and from Latin *tenebricosus*, shrouded in darkness, full of gloom. This is for its plumage, though its call does also suggest a degree of suffering!

Lesser Sooty Owl (breeding resident)

See family introduction and previous species.

Other names. Silver Owl, for its beautifully contrasted white undersides.

Tyto multipunctata Mathews, 1912 [tee-TOH mʊl-ti-pʊnk-TAH-tuh]: 'many-spotted owl', see genus name, and from Latin, *multus*, many, and *punctatus*, pointed, from Latin pungo/punctus, prick or sting, hence a speck or spot.

Australian Masked Owl (breeding resident)

From an old species name *Strix personatus* of Nicholas Vigors (a few years after Stephens had already named it), presumably for the big black-rimmed facial disc, though the disc is characteristic of all barn owls; Gould (1848) knew it as *T. personatus* and accordingly called it the Masked Barn Owl. The 'Australian' has been added to distinguish it from other Masked Owls in islands to the north.

Other names: Masked Owl; Cave Owl, used originally for the race of Barn Owls living in the limestone sinkholes of the Nullarbor Plain by Frederick Whitlock (Whitlock 1922), but adopted by Neville Cayley for the race *troughtonii* of Masked Owl from the same habitat (for a full discussion of the convolutions, see Parker 1977); Tasmanian Masked or Chestnut-faced Owl for the dark-faced Tasmanian race *castanops*, regarded by Gould (1848) as a full species *Strix castanops*, and as late as 1926 the RAOU agreed with him. Monkey-faced Owl, a name purportedly used elsewhere for Barn Owl, though the few (online) Australian references seem to be from Queensland and refer equally to Eastern Grass Owl; Man-faced Owl (per HANZAB 1990–2006, vol. 4) presumably follows from that, though we can find no further usages of it; Maw-faced Owl, for which the only reference is CSIRO (1969), completely baffles us and leaves us suspecting a typo for the previous name.

Tyto novaehollandiae (Stephens, 1826) [tee-TOH no-veh-hol-LAN-di-eh]: 'New Holland owl', see genus name, and from Modern Latin *Nova Hollandia*, the old name for Australia.

Eastern Barn Owl (breeding resident)

Until recently regarded as part of a cosmopolitan species with some 30 subspecies, the Australian, South-East Asian and south-west Pacific birds are now recognised as forming one of three species.

Other names: Barn Owl; White Owl, which it certainly is from below; Screech Owl, for the harsh ripping-shrieking call; Delicate Owl, from an old species name *Strix delicatula*, coined by Gould who, percipiently, thought it a separate species from the British Barn Owl; Monkey-faced and Cave Owl (see Masked Owl); Ghost Owl, we have little doubt that the silent-flying shrieking white shape around church yards has been so interpreted more than once!

Tyto javanica (Gmelin JF, 1788) [tee-TOH ja-VA-ni-kuh]: 'Javan owl', see genus name, and for the type locality.

Eastern Grass Owl (breeding resident)

Forms a super-species with the African Grass Owl *T. capensis*, and indeed until recently (HANZAB 1990–2006, vol. 4) was regarded as the same species. Both species hunt over grasslands and nest under grass or sedge clumps. 'Eastern' is to distinguish this Asian-Australasian species from the African one.

Other names: Grass Owl or Australian Grass Owl; Daddy Long-legs Owl, a delightful name for its long legs, though beyond HANZAB (1990–2006, vol. 4) we have no evidence that it has ever been used, but as always this doesn't preclude it from having been in popular oral use; Monkey-faced Owl (see Masked Owl).

Tyto longimembris (Jerdon, 1839) [tee-TOH lon-gi-MEM-bris]: 'long-limbed owl', see genus name, and from Latin *longus/longi-*, long, and *membrum*, limb (see common name).

STRIGIDAE (STRIGIFORMES): typical or hawk owls

Strigidae Vigors, 1825 [STRI-gi-dee]: the Screech-owl family, from genus name *Strix* (Latin *strix/strigis*, Greek *strix*), not found in Australia. Ancient myths abounded concerning these birds, such as that they were like vampires, sucking the blood of young children. Pliny took an uncharacteristically conservative view of one of these tales – 'As for the stories that they tell, about *strix* ejecting milk from its teats upon the lips of infants, I look upon it as utterly fabulous' and while he expressed some doubts about exactly which bird was meant by the word, he said, 'from ancient times the name 'strix', I am aware, has been employed in maledictions'.

See Tytonidae for 'owl'. The Strigids are only 'typical' relative to the Barn Owls in that there are some 190 species of them, relative to only 16 barn owl species.

'Hawk Owl' is really only used outside Australia in our region, in relation to the big genus *Ninox*, found from South-East Asia to Australia and New Zealand; the reference is to the long tails and relatively slender wings, and lack of the barn owls' facial disc.

The genera

Otus Pennant 1769 p [OH-tʊs]: 'eared owl', from Greek ōtus, eared owl.

Ketupa Lesson, 1830 [kuh-TOO-puh]: 'fish-owl', Lesson's version of Horsfield's species name, *ketupu*.

Ninox Hodgson, 1837 [NEE-noks]: 'sparrowhawk-owl', a combination of genera *Nisus*, sparrowhawk, and *Noctua*, owl. See family Accipitridae for the story of Nisus. The Little Owl, *Athene noctua*, was the owl (*noctua* in Latin) sacred to the goddess Minerva. Athene and the owl *glaux* were the Greek equivalents.

The species

Oriental Scops Owl (vagrant)

See family introduction. This large genus of nearly 50 owls are all called 'scops', from a former (now invalid) genus name, based on a Greek word skōps for a type of owl. Found widely in East Asia.

Otus sunia (Hodgson, 1836) [OH-tʊs SʊN-yuh]: 'sunya eared owl', see genus name, and from the Nepali name for this owl: *sunya kusial*, written by Hodgson as 'Sunya cusyal'.

Buffy Fish Owl (Cocos (Keeling) Islands)

See family introduction. There are three species of Asian Fish Owls, which are primarily fish-eaters; the three species are called Buffy, Brown and Tawny – we suggest you look at a picture of them to see if the names help you.

Ketupa ketupu Horsfield, 1821 [kuh-TOO-puh kuh-TOO-poo]: 'fish-owl fish-owl', see genus name, and from Horsfield's (1821) 'BLO ketupuk Javanis'. We understand that the common name of this bird in Indonesian is *Beluk ketupa*.

Rufous Owl (breeding resident)

See family introduction. For the rich rusty underside bars, straight from Gould's species name – nobody seems to have felt the need to think of an alternative name.

Ninox rufa (Gould, 1846) [NEE-noks ROO-fuh]: 'rufous sparrowhawk-owl', see genus name, and from Latin *rufus*, red or reddish.

Powerful Owl (breeding resident)

See family introduction. Straight from the species name; it is an enormous owl, but not necessarily relatively more powerful than smaller ones.

Other names: Eagle Owl, a reference to the huge size (though the term is generally reserved for the Old World members of the genus *Bubo*); Great Scrub-owl.

Ninox strenua (Gould, 1838) [NEE-noks STREH-noo-uh]: 'vigorous sparrowhawk-owl', see genus name, and from Latin *strenuus*, vigorous, nimble, strenuous. Named for its very large size, but nimble enough to catch a Sugar Glider at least.

Powerful Owl *Ninox strenua*

Barking Owl (breeding resident)

See family introduction. Couldn't be called anything else – the double call is amazingly dog-like. The first recorded use of the name appears to be by Ludwig Leichhardt in south-central Queensland (Leichhardt 1847).

Other names: Winking Owl, straight from the species name applied by John Latham (though he called it a falcon!) and still used by Gould in 1848; Goora-a-Gang, which Latham asserted is 'the native name'; Northern and Western Winking Owl, from Gregory Mathews who recognised, as was his wont, various subspecies (Mathews 1913).

Ninox connivens (Latham, 1801) [NEE-noks kon-NEE-venz]: 'winking sparrowhawk-owl', see genus name, and from Latin *con(n)iveo*, close the eyes or blink (as in English connive, of course). But is the Barking Owl more likely to wink (or turn a blind eye) than any other owl? Well, possibly. Latham (1801) considered that it had 'a wonderful faculty of contracting and dilating the iris', which is almost as good as a wink. We note in passing that the question of the date and indeed the original document in which this name was first published is a contentious one – a detailed account may be found in Schodde *et al.* (2010) and Peterson (2011).

Southern Boobook (breeding resident)

See family introduction. From the two-note call, dropping on the second, one of the most familiar night calls throughout Australia. The first reference we have to the name is in the notebooks of soldier and scientist William Dawes *On the Aboriginal Language of Sydney*, which he compiled in 1790–91, in the form Bōkbōk. Gould (1848) cited George Caley as referring to it by a 'native name ... Buck-buck'. Interestingly, Latham had picked up on this when he named the species *Strix boobook* in 1801. At the time and until recently it was deemed invalid because Gmelin had already named it from New Zealand and it was regarded as the same bird; now it has been reinstated. The 'Southern' distinguishes Australian–New Guinea boobooks from other boobooks to the north.

Other names: Mopoke or Morepork, but see next species; Cuckoo Owl, for the call, which reinforced the view that Australia was a seriously perverse part of the world – this, with Brown Owl, was reported by Gould (1848) to be a name 'of the Colonists'; Fawn-bellied Owl, for no very good reason; Red Boobook (Owl) or Red Owl, for the north Queensland rufous race formerly known as *N. lurida* (RAOU 1926).

Ninox boobook (Latham, 1801) [NEE-noks BOO-bʊk]: 'boobook sparrowhawk-owl', see genus name and common name.

Southern Boobook *Ninox boobook*

Morepork (breeding resident, Tasmania, Norfolk and Lord Howe Islands)

See family introduction and previous species. Morepork, or Mopoke, for the call – the *Australian National Dictionary* traced it back to 1825, as More Pork (Mopoke is also used for Tawny Frogmouth, through confusion as to the origin of the calls).

Other names. Marbled, Spotted or Tasmanian Spotted Owl, for the smaller, spottier Tasmanian birds that Gould called *Athene maculata*. A couple of the names for Southern Boobook were likely used in Tasmania too.

Ninox novaeseelandiae (Gmelin JF, 1788) [NEE-noks no-veh-zeh-LAN-di-eh]: 'New Zealand sparrowhawk-owl', see genus name, and from Modern Latin for New Zealand where it also lives.

Northern Boobook (vagrant)

See family introduction. Found in far eastern Russia, Korea, Japan and Taiwan, further north than other boobooks.

Other names. Brown Hawk-Owl, before the current species was split from *N. scutulata*; many non-Australian *Ninox* are known as hawk-owls (see family introduction). They're also all brown.

Ninox japonica (Temminck & Schlegel, 1844) [NEE-noks ja-PO-ni-ca]: 'Japanese sparrowhawk-owl', see genus name, and from Modern Latin *japonica*, Japanese.

Christmas Island Hawk-Owl (Christmas Island breeding resident)

See family introduction. Endemic to Christmas Island.

Other names: Christmas Island Owl or Christmas Hawk-Owl; Christmas Island Boobook.

Ninox natalis Lister, 1889 [NEE-noks na-TAH-lis]: 'birthday sparrowhawk-owl', see genus name, and from Latin *natalis*, birthday, for the birth of Christ, hence the place-name, Christmas Island, where the type specimen was collected.

UPUPIDAE (BUCEROTIFORMES): hoopoes

Upupidae Leach, 1820 [oo-POO-pi-dee]: the Hoopoe family, see genus name *Upupa*.

The genus

Upupa Linnaeus, 1758 [oo-POO-puh]: 'hoopoe', from Latin *upupa* – these birds were well known to many of the ancients as a common passage migrant.

The species

Eurasian Hoopoe (vagrant)

See family introduction. Onomatopoeic, from the bubbling call (see also genus name). This species is found right across southern Eurasia; there is also a species in each of Africa and Madagascar.

Upupa epops (Linnaeus, 1758) [oo-POO-puh EH-pops]: 'hoopoe hoopoe', see genus name and from Greek and Latin *epops*. Aristophanes makes mention of *epops* in *The Birds*.

ALCEDINIDAE (CORACIIFORMES): kingfishers

Alcedinidae Rafinesque, 1815 [al-seh-DI-ni-dee]: the Kingfisher family, see genus name *Alcedo*.

The original form was King's Fisher, dating back (in forms such as Kyng's Fyshare) to 15th-century England. According to Lockwood (1984), this apparently came directly from the local French *roi pêcheur*, though that implies Fisher King; however, the subtle meaning change could easily have arisen in translation and subsequent usage. 'Kingfisher' did not appear until the mid 17th century. The fact that this entire family of some 90-odd species is called 'kingfisher', although only about a third of them fish to any significant extent, is entirely due to historical accident. It is pure chance that the only species found in Europe, from which the entire group took its name, is one of the minority that specialises in fishing. Most of them are branch-perching, wait-and-pounce invertebrate hunters.

The genera

Tanysiptera Vigors, 1825 [ta-ni-si-TE-ruh]: '[bird] with wings outstretched', from Greek *tanusipteros*, broad-winged (cf. *tanaos*, outstretched).

Dacelo Leach, 1815 [da-SEH-lo]: 'anagram [bird]', based on genus name *Alcedo*. Strickland *et al.* (1842), in the report of the committee looking into the rules of scientific nomenclature, to which Charles Darwin contributed, said, 'Such verbal trifling as this is in very bad taste, and is especially calculated to bring the science into contempt. It finds no precedent in the Augustan age of Latin, but can be compared only to the puerile quibblings of the middle ages … And it is peculiarly annoying to the etymologist, who after seeking in vain through the vast storehouses of human language for the parentage of such words, discovers at last that he has been pursuing an *ignis fatuus*.' Enough said!

Halcyon Swainson, 1821 [HAL-si-ohn]: 'kingfisher', from Greek *alkuōn*. The form *halcyon* probably arose from the ancient belief that the bird was connected with the sea, from Greek *hals*.

Todiramphus Lesson, 1827 [to-di-RAM-fʊs]: '[bird] with a bill like a tody', from Latin [*todus*] *todi*, small unidentified birds mentioned by Plautus, and Greek *rhamphos*, bill. *Todus* is now used for a genus of tiny West Indian birds, including the Jamaican Tody *Todus todus*.

Syma Lesson, 1827 [SEE-muh]: 'sea-nymph', from the name of a nymph in Greek mythology, on whom Neptune fathered a child. Lesson (1831) said, 'The only known species, whose generic name is that of a sea-nymph, lives on the shores, and catches on the banks the small fish it lives on.' There is no context to guide understanding of the French, so we don't know if he meant seashores or riverbanks (or both). However, the allusion to fish-eating is very clear, and this is a misconception on Lesson's part, because the birds are not known to eat fish. To be fair to him we should say that, as a European, Lesson may have found the idea of a non-fishing kingfisher somewhat difficult to swallow.

Alcedo Linnaeus, 1758 [al-SEH-do]: 'kingfisher', from Latin *alcedo*, kingfisher. The form *halcedo* also exists, perhaps by analogy with *Halcyon* (see next family) or from the ancient idea of the bird being connected with the sea, from Greek *hals*.

Ceyx Lacépède, 1799 [SEH-iks]: from Ceyx, in Greek mythology, son of the Morning Star and husband of Alcyone, herself the daughter of Aeolus, the guardian of the winds. They were so happy together that Alcyone jokingly referred to them as Zeus and Hera. Zeus, the greatest of the Olympian gods, and Hera, his sister and wife, heard this sacrilegious impertinence and were not best pleased. Overreacting as usual, they sent a wild storm that wrecked Ceyx's ship and drowned him. Read the very dramatic account in Ovid's (c. 8AD) *Metamorphoses*! Alcyone found Ceyx's body washing up on the beach and leapt into the sea, beside herself with grief. Both were transformed by unnamed but compassionate gods into birds (usually said to be kingfishers), and many myths of a different kind about kingfishers were born (see Common Kingfisher).

The species

Little Paradise Kingfisher (vagrant to Torres Strait)

'Paradise Kingfisher' (or more commonly 'Paradise-Kingfisher') is the rather grandiose name for the eight primarily New Guinean species, all gorgeous, of the genus *Tanysiptera*, but it is a name only recently adopted in Australia; for instance, CSIRO (1969) didn't even list it as an alternative name for this species. It seems to have been introduced for Papua New Guinea by Rand and Gilliard (1967) and recommended for Australia by RAOU (1978). It still didn't take immediately; Pizzey (1980) offered it only under 'Other names'. This is among the smallest of them.

Other names: Aru Paradise-Kingfisher, for the origin of the type specimen, Aru Island.

Tanysiptera hydrocharis Gray GR, 1858 [ta-ni-si-TE-ruh hi-dro-KAH-ris]: 'grace of the waters with outstretched wings', see genus name, and from Greek *hudōr/hudro*, water, and *kharis*, grace or charm.

Buff-breasted Paradise Kingfisher (breeding migrant)

See previous species. This is the only paradise kingfisher with buff undersides.

Other names: Buff-breasted Paradise-Kingfisher, the more common form – see previous species; White-tailed Kingfisher, the name universally used in Australia until recently; White-tailed Tanysiptera, from the genus name, as used by Gould (1848); Australian Paradise-Kingfisher, though it spends half its year in New Guinea; Long-tailed, Racquet-tailed, Silver-tailed Kingfisher, the tail being one of its most striking features, though it is the least racquet-tailed of the genus; Black-headed Kingfisher, cited in HANZAB (1990–2006, vol. 4), though this refers to race *nigriceps* from the Bismarck Archipelago off north-east Papua New Guinea.

Tanysiptera sylvia Gould, 1850 [ta-ni-si-TE-ruh SIL-vi-uh]: 'woodland bird with outstretched wings', see genus name, and from Latin *silva*, forest or woodland – rainforest is its usual habitat.

Buff-breasted Paradise Kingfisher *Tanysiptera sylvia*, adult and immature bird

Laughing Kookaburra (breeding resident)

'Kookaburra' (in various forms) was recorded from different Indigenous languages, all apparently – and unsurprisingly – based on the remarkable territorial call. Perhaps the earliest is that of Bennett (1834), who recorded that 'the Natives at Yas [sic] call the bird 'Gogera' or 'Gogobera''. The *Macquarie Aboriginal Words* (Thieberger and McGregor 1994) listed, among others, *Guuguubarra* in Wiradjuri, *Jawawoodoo* in Gooniyandi and *Kuukakaka* in Paakantyi. However, it took a while for it to be accepted as a first-choice name in English. Gould (1848), following Latham (1801), prosaically called it Great Brown Kingfisher, merely noting *Gogo-bera* as per 'Aborigines of New South Wales', while Bennett used the widespread Laughing Jackass, as did Backhouse (1843). 'Kookaburra'was known though; indeed the first record we can find is, perhaps surprisingly, in an account of proceedings of the NSW Parliament in the *Sydney Morning Herald* of 30 January 1871, in the context of a member's somewhat heavy-handed and obscure witticism. Into the 20th century, Hall (1907) still used Gould's name, with Laughing Jackass as the only alternative, though 4 years later Lucas and Le Soeuf (1911), in a more popular text, used Kookaburra as an alternative to Laughing Kingfisher, as did Leach (1911). The RAOU, in its second *Official Checklist of the Birds of Australia* RAOU (1926), took a deep breath and blessedly used Laughing Kookaburra as the recommended name, and Cayley (1931) followed with the first use as the primary name in a bird book. The 'laughing' nature of the call was noted early; Vigors and Horsfield (1827) cited George Caley as writing, 'The settlers call this bird the Laughing Jackass.'

Other names: This species has possibly attracted more folk names, many of them affectionate human names, than any other Australian bird. Jackass is at the heart of many and appeared very early; for example, David Collins in 1798 called it the Laughing Jack-Ass, the allusion being to a donkey's call, with the Jack used in its sense of a diminutive. Later, perhaps as that implication was lost or overlooked, Jack became interpreted as a familiar name and in some cases was then formalised to John (even to the somewhat ridiculous 'Johnass'). Jackass, Jack, Jacky, Jacko, John, Johnny, Laughing John or Johnass or Johnny or Jack or Jackass; Alarm Bird, Breakfast Bird, Clockbird, Settler's or Shepherd's Clock, all for the rousing early-morning territorial chorus; Ha Ha Pigeon or Woop Woop Pigeon ('pigeon' being used as a catch-all term for a bird); Ha Ha Duck, surely an ironic appellation?!; Laughing, Brown, Giant (Shaw 1819) or Great Brown Kingfisher – the last was regarded well into the 20th century as a formal name (RAOU 1913); Northern or Lesser Brown Kingfisher, for briefly recognised species *D. minor* from north Queensland, not now recognised at any level (see White 1917b and RAOU 1913, respectively, for examples of the use).

Dacelo novaeguineae (Hermann, 1783) [da-SEH-lo no-veh-GI-ni-eh]: 'New Guinea anagram', see genus name, and for the place-name, based on the erroneous belief that this was where the type specimen originated (it was actually from New South Wales and the bird is not found in New Guinea).

Blue-winged Kookaburra (breeding resident)

See Laughing Kookaburra; Blue-winged for the obvious reason.

Other names: Barking or Howling Jackass – it truly is a remarkably awful call! Fawn-breasted Kingfisher, coined by Gould (1848) for his species, now race, *cervina* (meaning fawn) from the Top End – still recognised by RAOU (1913); Leach or Leach's Kingfisher, as used by Gould (1848) – and few others except for RAOU (1913) – from the species name.

Dacelo leachii Vigors & Horsfield, 1827 [da-SEH-lo LEE-chi-ee]: 'Leach's anagram', see genus name, and for William Leach (1790–1836), doctor turned zoologist who worked at the British Museum, initially in the library then on a wide range of animal groups before ill health, possibly associated with depression, forced his premature retirement before an early death. It was he who perpetrated the contentious genus name. The authors chose to honour this 'exceptional ornithologist, who first described this genus and illustrated its characteristics'.

Black-capped Kingfisher (vagrant)

See family introduction. A most apt name.

Halcyon pileata (Boddaert, 1783) [HAL-si-ohn pi-le-AH-tuh]: 'capped kingfisher', see genus name, and from Latin *pileus*, the felt cap or 'cap of liberty' worn by Roman slaves who had been granted their freedom.

Forest Kingfisher (breeding resident)

See family introduction. Tends to be found in closed forest situations more than the other *Todiramphus* kingfishers, but overlaps habitat with both Sacred and Collared. The name apparently arose spontaneously in the 19th century, because it appears as first-choice name in Morris' 1898 *Dictionary of Austral English*.

Other names: Blue Kingfisher, sounds silly but it really is **very** blue compared with its closest relations; Bush Kingfisher, probably to distinguish it from the other widespread very blue kingfisher, the Azure, of streamlines – Gould (1848) wrote that this name is used by the 'residents at Port Essington'. Macleay or Macleay's Kingfisher or Macleay's Halcyon (used by Gould 1848, for the then genus name), from its species name, for Alexander Macleay (1767–1848), Scottish highlander, friend of the great botanist Robert Brown, entomologist and Fellow of the Royal Society, who came to Australia in 1825 as Colonial Secretary of New South Wales and brought with him one of the world's great insect collections. He was instrumental in founding the Australian Museum in 1827. Beolens and Watkins (2003) claimed that the name refers to his son William Sharp Macleay, also an eminent entomologist, but he was only 6 years old at the time and not yet very eminent. Fortunately, Jardine and Selby (1826–35) made it clear that it was indeed Alexander, who sent a specimen to the Linnean Society (see also Macleay's Fig-Parrot and Macleay's Honeyeater).

Todiramphus macleayii (Jardine & Selby, 1830) [to-di-RAM-fʊs mak-KLEH-i-ee]: 'Macleay's tody-bill', see genus and common names.

Collared Kingfisher (vagrant)

See family introduction. For the broad white collar; the name was only introduced in Australia by RAOU (1978), noting the 'inadequacies of some traditional names', notably the fact that it doesn't always occur in mangroves and that there is an African Mangrove Kingfisher. However, they were really talking about the next species, comprising the Australian population recently split from the widespread Collared Kingfisher.

Todiramphus chloris (Boddaert, 1783) [to-di-RAM-fʊs KLOH-ris]: 'green tody-bill', see genus name, and from Greek *khlōros*, greenish-yellow, in this case a rich olive-green, though much depends on the light.

Red-backed Kingfisher *Todiramphus pyrrhopygius* and Collared Kingfisher *T. chloris*

Torresian Kingfisher (breeding resident)

See family introduction. A recently coined name referring to the Australian tropical biome, following the split of the formerly very widespread Collared Kingfisher.

Other names. Collared or White-collared Kingfisher; Black-masked Kingfisher, for another (though not unique) character; Mangrove Kingfisher, long and widely used until 1978, for its primary habitat; Sordid Kingfisher, used by Gould for his own species name *Halcyon sordidus* (not realising that Pieter Boddaert had already named it, from a specimen from the Moluccas; the name was long suppressed until it was determined recently that Gould and Boddaert were talking of different species) – he meant dull-coloured rather than degenerate!

Todiramphus sordidus (Gould, 1841) [to-di-RAM-fʊs sor-DI-dʊs]: 'dirty tody-bill', see genus name, and from Latin *sordidus*, meaning dirty, foul, squalid. Gould describes the head, back and wing-coverts of the bird as 'brownish oil-green' which certainly sounds rather dull and dirty.

Sacred Kingfisher (breeding resident/migrant)

See family introduction. Generally ascribed to the veneration with which some Pacific Islanders held the bird – or at least, as we now understand it, a related one (Pacific Kingfisher *T. sacer*), which is widespread in the south-western Pacific. The name pre-dated the Sydney colony, as White (1790) already knew it as Sacred Kings-Fisher (see family introduction).

Other names: Green Kingfisher (cf. Forest); Tree or Wood Kingfisher, for its non-fishing habits; Sacred Halcyon (Gould 1848), for the genus name of the time.

Todiramphus sanctus (Vigors & Horsfield, 1827) [to-di-RAM-fʊs SANK-tʊs] 'sacred tody-bill', see genus name, and from Latin *sanctus*, holy or sacred, because of a similarity Vigors and Horsfield perceived to Gmelin's *Alcedo sacra*. Note that the 'sacred' appellation long pre-dated the scientific name (see common name).

Red-backed Kingfisher (breeding resident)

See family introduction. A significant identifying character, albeit somewhat bowdlerised, as was often the case when slightly less delicate body parts were mentioned in bird names in earlier times, used by Gould (1848) from a gentrified version of his species name.

Other names: Red-rumped Kingfisher, from the species name; Red-backed Halcyon, the name used by Gould (1848), from the then genus name; Golden Kingfisher, perhaps for the same character, but we are not very convinced, although clearly it was in use at one stage, as both Leach (1911) and Cayley (1931) offered it an alternative name; Northern Red-backed Kingfisher, for Mathews' name *Halcyon obscurus* for a bird from north-western Australia (Matthews 1912a), still used by RAOU (1913).

Todiramphus pyrrhopygius (Gould, 1841) [to-di-RAM-fʊs pi-ro-PI-gi-ʊs]: 'red-rumped tody-bill', see genus name, and from Greek *purrhos*, flame-coloured or red, and *pugē*, rump.

Yellow-billed Kingfisher (breeding resident)

See family introduction. A striking characteristic, shared among kingfishers only by its close New Guinea relative the Mountain Kingfisher *S. megarhyncha*; Gould applied the name for his species name *Halcyon flavirostris*, not realising that Lesson had beaten him to it with a New Guinea specimen.

Other names: Saw-billed Kingfisher, for the serrated tip of the upper mandible, which is really only visible in the hand; Lesser or Lowland Yellow-billed Kingfisher, both only relevant in New Guinea, to compare with Mountain Kingfisher.

Syma torotoro Lesson, 1827 [SEE-muh to-ro-TO-ro]: 'torotoro sea-nymph', see genus name, and from an Indigenous name for the bird. Lesson (1827) said that he saw them skimming along the little rivers running down to Dorey Bay, West Papua, and that the Papuans called the bird *torotoro* 'doubtless by analogy with its call'. Meanwhile, on the other side of the world and a couple of thousand years earlier, that same analogy was made by Aristophanes (414 BC) for the Common Kingfisher *Alcedo atthis*: when one of his characters in *The Birds* is trying to call birds including the kingfisher to join him, the sounds he makes include 'torotorotorotorotorotix'.

Common Kingfisher (vagrant to Christmas Island)

See family introduction. This is **the** common (indeed the only) kingfisher of Europe, though its enormous range extends via north Africa and South and East Asia to New Guinea and the Solomons.

Other names: River or European or Eurasian Kingfisher.

Alcedo atthis (Linnaeus, 1758) [al-SEH-do AT-tis]: 'Atthis kingfisher', see genus name. As far as the species name goes, Linnaeus (1758) gave no clue, except perhaps when he referred to the bird as living in Egypt. It is often said to be the name of a young woman admired by Greek poetess Sappho, to whom she wrote (if the remaining fragments are to be believed) some very beautiful poetry. Or there is Atthis or Attis or Atys, a Phrygian shepherd who in one version of the story went mad for love and castrated himself (Grant and Hazel 2002). However, there is another possible candidate, the Athis mentioned by Ovid (c. 8 AD) in the *Metamorphoses*, in a part of the story set in Æthiopia (a vague name, but possibly for an area directly to the south of Egypt). He is a beautiful 16-year-old Indian boy, son of a nymph of the Ganges. He is said to be in the flower of his strength and his beauty is enhanced by his Tyrian purple cloak and gold adornments, which perhaps evoke the colours of the bird. (We should add that Athis meets a sad and extremely violent end. Amid scenes of gang violence sparked by a dispute between two young hotheads over a girl, the urge to get involved is too much for him. He reaches for his bow and is smashed to death with blows to the face and head from a flaming brand wielded by one of the enraged lovers, Perseus; Graves 1996.) But, back to the kingfisher itself: it is the bird around which so many legends arose. Arnott (2007) reported, 'our ancient sources provide a medley of information, combining accurate observation with wild lunacy and unsolved mystery'. This medley includes the idea which Aristotle (c. 330 BC), among others, perpetuated, that the kingfisher builds a thorny nest (indeed, a red, gourd-shaped nest incorporating fish bones), which she launches on the sea before laying and hatching out her eggs in it. This is said to happen in the so-called Halcyon Days, the weeks before and after the winter solstice, in which the seas are calm: the myth explains that this is because Alcyone's father Aeolus stops his winds blowing during those days (see genus *Ceyx*).

Azure Kingfisher (breeding resident)

See family introduction. It really is a most splendid glossy deep blue above – see species name; used by Gould (1848), but it could well have been in general use before that.

Other names: Tridigitated Kingfisher (!), used by Shaw in 1805 from his own name *Alcedo tribrachys* – it was a fair observation of a feature unusual among kingfishers in general, but not unique among members of this sub-family; hence Three-toed and Azure Three-toed Kingfisher; River, Creek or Water Kingfisher, reflecting the fact that, unlike most Australian kingfishers, it really is tied to water, usually a vegetated stream line; Blue Kingfisher, unimaginatively and unhelpfully used by RAOU (1913); Purple Kingfisher, for Gould's *Alcyone pulchra* from the Top End coast – still accepted by RAOU (1913).

Ceyx azureus (Latham, 1801) [SEH-iks a-ZOO-re-ʊs]: 'azure Ceyx', see genus name, and from Mediaeval Latin *azurus/azureus/azurium/azorium*, azure-blue, from Persian *lajvard*, to Arabic *al-lazaward* (meaning lapis lazuli). All indicating a wonderful blue.

Little Kingfisher (breeding resident)

See family introduction. Vying with the African Dwarf-kingfisher for the title of world's tiniest kingfisher, possibly missing the title by the weight and width of a down feather.

Other names: Ramsay Kingfisher, for the Northern Territory race *ramsayi*, known for a while in the early 20th century as a separate species *Alcyone ramsayi* (RAOU 1913), for Edward Pierson Ramsay (1842–1916), a New South Wales zoologist specialising in birds and fish, who didn't have a university degree but was elected to the Philosophical Society at age 23. He corresponded with John Gould and was appointed curator of the Australian Museum in Sydney in 1874, sadly at the expense of the grievously wronged Gerard Krefft. The museum flourished under him and in 1888 he published a full list of the birds of Australia.

Ceyx pusillus Temminck, 1836 [SEH-iks pʊ-SIL-luh]: 'tiny Ceyx', see genus name, from Latin *pusillus*, very small or insignificant and, as with other species bearing this name, the bird is definitely the former and **not** the latter – it is a minute jewel of a bird, the size of a Silvereye, and more than a quarter of its length is in the bill.

CORACIIDAE (CORACIIFORMES): rollers

Coraciidae Rafinesque, 1815 [ko-ra-SEE-i-dee]: the Roller family, from *Coracias*, the genus name of the majority of roller species, though not the Australian one. The genus was named by Linnaeus (1758), without explanation of his choice, though previous names for the European Roller *Coracias garrulus* included *Corvus* and *Cornix*, both meaning crow, and Greek *korakias* is sometimes thought to be a type of crow or jackdaw (it is actually more likely a Chough *Pyrrhocorax pyrrhocorax* or an Alpine Chough *P. graculus*, according to Arnott 2007). It seems completely bizarre that Linnaeus continued the tradition of naming such an intensely colourful group of birds after black ones.

The rollers comprise a family of 12 species found across southern Europe, Africa and Asia, with one species found in Australia but extending through Indonesia to south-eastern Asia. They take their name from the dramatic tumbling aerial territorial and courtship displays.

The genera

Coracias Linnaeus, 1758 [ko-RA-see-as]: 'crow chough', see family introduction.

Eurystomus Vieillot, 1816 [yoo-RI-sto-mʊs]: 'wide mouth', from Greek *eurustomos*, wide-mouthed, because it is!

The species

European Roller (vagrant)

See family introduction. The only roller found in Europe.

Coracias garrulus Linnaeus, 1758 [ko-RA-see-as GA-rʊ-lʊs]: 'chattering crow-chough', see family introduction, and from John Ray's genus name for the bird: *Garrulus* – from the Latin, meaning chattering, talkative.

Oriental Dollarbird (breeding resident)

From the large white spot on the underside of each wing, seen as silver coins. According to Gould (1848) it was already a name used by 'the Colonists', though he preferred the more prosaic Australian Roller. Despite the species' very wide distribution into Asia, the name Dollarbird seems to have arisen in Australia. By 1913 even the staid Gregory Mathews was using it alongside 'Australian Roller'. 'Oriental' is a recent addition, presumably is deference to the species name, though as the only other 'dollarbird' is restricted to the Moluccas it doesn't seem to add a lot.

Other names: Dollarbird, as it is near-universally known in Australia; Pacific Roller (Latham 1801); Australian Roller; Broad-billed or Eastern Broad-billed Roller (there is an African Broad-billed Roller

Oriental Dollarbird *Eurystomus orientalis*

too); Red-billed Roller, as one of only two rollers with a red bill (the other, Azure Roller, *E. azureus*, being restricted to the northern Moluccas); Rainbird, for its summer arrival that coincides with the wet season (see also Eastern Koel); Starbird, probably for the 'dollar'.

Eurystomus orientalis (Linnaeus, 1766) [yoo-RI-sto-mʊs o-ri-en-TAH-lis]: 'Eastern wide-mouth', see genus name, and from Latin *orientalis*, eastern, for the type locality, Java (i.e. the 'East Indies').

MEROPIDAE (CORACIIFORMES): bee-eaters

Meropidae Rafinesque, 1815 [me-RO-pi-dee]: the Bee-eater family, see genus name *Merops*.

The Bee-eaters form a large and very conspicuous family of mostly closely related species found throughout southern Europe, Asia and Africa, so the term was in use long before English-speakers came to Australia, being recorded from the 17th century. They are specialist aerial insect-catchers but the name refers to only one part of their diet, albeit the part that most concerned humans.

The genus

Merops Linnaeus, 1758 [ME-rops]: 'bee-eater', from Greek *merops*, bee-eater. The birds were also known to the Romans, who called them *apiastra*, from Latin *apis*, a bee.

The species

Rainbow Bee-eater (breeding resident/migrant)

See family introduction. Although a stunning bird, 'our' bee-eater doesn't especially stand out as more rainbow-hued than many others in the family. We might suppose that the widespread 'Rainbowbird' arose in Australia and was appended to Bee-eater later, but that seems only partly to be the case. In 1801 Latham had called it Variegated Bee-eater, but White (1790) referred to it as Mountain Bee-eater (making it clear that it inhabited the Blue Mountains); Gould (1848) used simply Australian Bee-eater. In 1899 Hall just called it Bee-eater, as did Lucas and Le Soeuf (1911). Leach (1911) reverted to Australian Bee-eater, but for the first time introduced Rainbow Bird as an alternative. The word had seemingly appeared in print for the first time (in the Melbourne *Weekly Times*) just 3 years previously. The first RAOU checklist (1913) didn't use it, but the second list (1926) used Rainbow Bird alongside Australian Bee-eater. Cayley (1931) used Rainbow Bird as first choice and Slater was still using it in 1970, though the CSIRO *Index of Australian Bird Names* had used Rainbow Bee-eater as primary name the previous year. RAOU (1978) also supported this name, seeing it as a compromise between Rainbow Bird and Australian Bee-eater. Graham Pizzey, in his ground-breaking field guide (1980), seems to have been the first to cement the name in the Australian birding community.

Other names: Rainbow Bird, see common name; Bee-eater; Australian, Mountain, Variegated Bee-eater, see common name; Pin-tailed Bee-eater or Pintail or Spinetail, for the breeding male's extended central tail feathers; Golden-swallow, Gold-digger, Gold-miner, for its vigorous excavations of nesting hollows in sandy banks. Berrin-berrin, cited by CSIRO (1969), HANZAB (1990–2006, vol. 4) and the *Macquarie Dictionary*, though we can find no reference to its usage beyond Gould's 1848 reference to '*Bé-rin bé-rin*, Aborigines of the mountain district of Western Australia'; various sources, mostly quoting each other, suggest that this is a possible source of the name of the Western Australia wheatbelt town Kellerberrin. (Although it's never been adopted into English, we do like the Wembawemba word *pirrimpirr*, a beautiful rendition of the call: Thieberger and McGregor 1994.)

Merops ornatus Latham, 1801 [MEH-rops or-NAH-tʊs]: 'splendidly dressed bee-eater', see genus name, and Latin *ornatus*, in splendid attire. Need we say more – except to note that in HANZAB (1990–2006, vol. 4) the plumage is described as 'gaudy'.

FALCONIDAE (FALCONIFORMES): falcons

Falconidae Leach, 1820 [fal-KO-ni-dee]: the Falcon family, see genus name *Falco*.

'Hobby' appears in English in the 15th century from the Old French *hobe*, *hobé* and several other variants (cf. modern French *hobereau*), all meaning a small falcon, and possibly derived from the verb *hober*, to jump about or move; it is a falconer's term for the Eurasian Hobby *F. subbuteo*, perhaps referring to the agility of the hunting bird.

'Falcon' arrived in English with the Norman falconers, who referred to their birds (presumably including Peregrines, though the term was then a raptor catch-all) as *falcun* or *faucun* (cf. modern French *faucon*). Falcon is now used only for members of the genus *Falco*.

The genus

Falco Linnaeus, 1758 [FAL-ko]: 'falcon', from Latin *falco/falconis*, a falcon (cf. *falx/falcis*, sickle or pruning hook; also Byzantine Greek *phalkōn*, based on the Latin).

The species

Nankeen Kestrel (breeding resident)

'Kestrel' appeared in English (for Common Kestrel, *F. tinnunculus*) in the 16th century from the French *crécerelle* or similar, apparently ultimately of onomatopoeic origin. It is applied now to 13 species of mostly similar small hovering falcons across the world. 'Nankeen' derives from Nankin or Nanking, major town of Kiangsu province, China. The town gave its name to a widely used cheap yellowish cotton cloth manufactured there, which in turn came to be used for the colour (see also Nankeen Night Heron). Apparently the colour-word was in reasonably general use, as Gould was using Nankeen Kestril [sic] in 1848 and reported that the name used by 'the Colonists' was Nankeen Hawk.

Other names: Kestrel; Chickenhawk, this might have been a 'lumping' with Sparrowhawk, but while primarily an insectivore it does hunt small birds too; Mosquito Hawk, surely a reference to its diminutive size rather than perceived diet; Windhover, a nicely observed old English name for one of its main hunting techniques; Hoverer.

Falco cenchroides Vigors & Horsfield, 1827 [FAL-ko sen-KROY-dehz]: 'kestrel-like hawk falcon', see genus name, and from Latin *cenchris* (Greek *kerkhnē*), a type of hawk, most likely a kestrel; and *-oides* (Greek *-oeidēs*), resembling. Hedging bets again, but it gets us to a kestrel in the end.

Eurasian Hobby (vagrant)

See family introduction. It has a vast breeding range, from Spain to near the eastern end of Russia.

Falco subbuteo Linnaeus, 1758 [FAL-ko sŭb-BOO-te-o]: 'falcon less than a hawk', see genus name, and from Latin *sub*, under, beneath, of lesser rank, and *buteo*, a kind of hawk (now a genus of buzzards).

Australian Hobby (breeding resident)

See family introduction. 'Hobby' was adopted in Australia only recently and applied to this bird, because of its similarity to the Eurasian Hobby, but it was long known simply as Little Falcon.

Other names: Little Falcon, as the smallest falcon (with the Kestrel) in Australia; Duck-hawk, but although it certainly takes ducks, most of its prey are smaller and this could be a confusion with Peregrine Falcon; Little Duck-hawk, probably to distinguish it from Peregrine Falcon; Black-faced Hawk, true, in contrast to pale throat and collar (like Peregrine Falcon); White-fronted Falcon, clearly not so (a bird's 'front' is its forehead and this species' front is buff) – the imprecision of the name is unusual in that it comes from the usually accurate Gould, but he was basing it on his own name *F. frontatus* and his illustration shows a white frons, so perhaps it is simply down to a slightly unusual specimen. In any case Gould had been beaten to naming rights by Swainson, so the name sank from sight.

Falco longipennis Swainson, 1838 [FAL-ko lon-gi-PEN-nis]: 'long-winged falcon', see genus name, plus Latin *longus*, long, and *pennis*, wing, and indeed the wings are usually described as long, narrow and pointed.

Brown Falcon (breeding resident)

A safe enough name for a notoriously variably coloured bird; Gould called it Brown Hawk.

Other names: Brown Hawk (widespread and in part a reflection that it is not especially 'falcon-like', having in some ways filled the generalist hunter niche occupied by *Buteo* buzzards in the Northern Hemisphere and doing a lot of ground-hunting of reptiles) – Gould reported that this was the name of the bird among the 'Colonists of Van Diemen's Land'. Cackling Hawk, as it is indeed a noisy falcon; Chickenhawk (again), not a bird specialist, but an opportunist and would doubtless take the odd chook; Orange-speckled Hawk (a very early name, reported by Vigors and Horsfield 1827 as 'called by the settlers'); Striped, Striped Brown, White-breasted Hawk, all referring to aspects of the common paler, redder morphs; Western Brown Hawk, used by Gould (1848) for his now defunct species *Ieracidea occidentalis* from the western half of Australia.

Falco berigora Vigors & Horsfield, 1827 [FAL-ko be-ri-GOR-uh]: 'berigora falcon', see genus name, plus *berigora*. The original namers, Vigors and Horsfield (1827), simply said: 'The native name of this bird, which we have adopted as its specific name, is Berigora'. Gould (1848) mentioned 'Aborigines of New South Wales' against the word, and Morris (1898), in his *Dictionary of Austral English*, claimed it is made up of *beri*, claw, and *gora*, long. The word does not appear in a glossary of the languages spoken by Indigenous people of the Sydney region at the time of early white settlement (Troy 1994), though many other bird names do, and the bird was certainly to be found there. The most convincing clue seems to be from *biyaagaarr,* a Yuwaalaraay (north-western New South Wales) name for this bird, and to explain the slight difference in sound, Nash (2014) points out that *birraagaar* would be the expected form in the related Gamilaraay (Kamilaroi) language, as well as others from Central New South Wales (intervocalic 'y' commonly corresponds with 'r'; the final trilled 'rr' may have been heard as 'ra'). However, George Caley, who provided Vigors and Horsfield with the specimen and the name, hardly set foot outside the Sydney region, so Nash concludes that there may in fact have been a similar word in use around the Sydney region at that time, with Caley the first to record it. For further detail, see Nash (2014).

Grey Falcon (breeding resident)

The obvious name for this elusive, beautiful softly grey falcon.

Other names: Blue Hawk, Smoke Hawk, both attempts to capture the colour.

Falco hypoleucos Gould, 1841 [FAL-ko hi-po-LOO-kos]: 'not very white falcon' see genus name, plus Greek *hupo-*, somewhat, slightly, below, and *leukos*, white. Not a very nice way of putting it – we prefer it being softly grey!

Black Falcon (breeding resident)

Not quite black (no more so than very dark Brown Falcons), but certainly consistently very dark.

Falco subniger Gray GR, 1843 [FAL-ko sŭb-NEE-gehr]: 'not very black falcon', see genus name plus Latin *sub-*, somewhat or rather, and *niger*, gleaming black (see common name).

Peregrine Falcon (breeding resident)

A peregrine is a foreigner (and thus strange!), one travelling abroad or on a pilgrimage. The 'best' young falcons of this highly esteemed species (those best able to be trained to hunt on behalf of humans) were not taken from the nest as chicks, but captured as young birds flying in from elsewhere.

Peregrine Falcon *Falco peregrinus*

Other names: Peregrine; Black-cheeked Falcon, accurate, but no more so than if applied to Australian Hobby – however, it is based on Gould's name *F. melanogenys*, because he believed it was different from the European Peregrine; Blue Hawk, used by the 'Colonists of Western Australia' (Gould 1848) but see also Grey Falcon; Pigeon or Duck Hawk, probably applied by aggrieved racing pigeon fanciers and duck hunters!

Falco peregrinus Tunstall, 1771 [FAL-ko pe-re-GREE-nʊs]: 'strange and foreign and even exotic falcon', see family and common names (Latin *peregrinus* means foreign).

STRIGOPIDAE (PSITTACIFORMES): New Zealand parrots

Strigopidae Bonaparte, 1849 [stri-GO-pi-dee]: the Owl-face family, from the genus name *Strigops* (Gray 1845), from Greek *strix*, owl and ōps, face. Although Gray does not elaborate on his choice of name, the general resemblance to an owl's face can certainly be seen in the Kakapo (*Strigops habroptilus*), for which the genus was named.

The genus

Nestor Lesson, 1830 [NES-tor]: 'Nestor', from the name of the king of Pylos, who went with the other Greek heroes to the Trojan Wars. He is described in Homer's *Iliad* as great-souled and clear-voiced, with speech sweeter than honey. He was a great warrior and strategist, a good athlete and boxer and a highly effective cattle rustler and horse thief. He was considered to be very wise and was much listened to, even though he did go on a bit, especially as he got older. Robert Graves (1996) hypothesised that the king's name might come from Greek *neostoreus*, which he translated as 'newly speaking'. Its current usage was taken from the old name, *Psittacus nestor*, of the bird now known as *Nestor meridionalis*, the New Zealand Kaka.

The species

Norfolk Island Kaka (Norfolk Island, extinct)

Kaka is the Māori name for a closely related New Zealand parrot, *N. meridionalis.*

Other names: Philip Island Kaka, for the small island off Norfolk where the last surviving birds were reputed to linger in the 19th century; Wilson's Parrot, for one Thomas Wilson, in whose collection the somewhat dilettante but influential English ornithologist John Latham 'found' the specimen that he described (but as was often his wont he omitted to supply a scientific name, so Gould's description and name from 1844 is the valid one – by that time the bird may well have been extinct in the wild); Norfolk Island Parrot or Nestor, for the genus name; Long-billed Parrot, for one of its most remarkable attributes (see species name).

Nestor productus (Gould, 1836) [NES-tor pro-DʊK-tʊs]: 'long-drawn-out Nestor', see genus name, and from Latin *produco*, to lead forward or draw out. Gould (1836a) said, 'The bill was exceedingly produced, the upper mandible extending fully one half of its total length beyond the lower.' The bird itself was apparently slightly smaller in length than the surviving members of the genus, New Zealand Kaka and Kea, but its bill, if the illustrations of it are to be believed, was even longer than theirs, which is definitely saying something.

CACATUIDAE (PSITTACIFORMES): cockatoos

Cacatuidae Gray GR, 1840 [ka-ka-TOO-i-dee]: the Cockatoo family.

'Cockatoo' appears in English in the early 17th century (in the form 'Cock-a-two', though it was then used in a word-play, so may have been altered for the purpose). Its origin is the Malay word *kakatua* (which some assert is onomatopoeic in origin), which was incorporated into Dutch as *kakatoe*. Cacatoe was recorded in English in the 1630s, to make it more familiar than the Dutch 'k', and cockatooan (plural) appears later in the century. It seems that 'cock' was incorporated into the word as a familiar 'bird word', though later versions include crockadore. The current form is known from the 1730s.

'Black Cockatoo' (with 'Black-Cockatoo' still used near universally in Australia) is applied to a group of large – and black! – cockatoos of the genus *Calyptorhynchus*, which have distinctively coloured tail panels. Surprisingly, this name did not gain much if any currency until the RAOU rationalised black cockatoo names in its second checklist of recommended names (RAOU 1926).

'Corella' is doubtless from an Australian Indigenous language, but it is unclear which one. The *Australian National Dictionary* suggested *garala* from Wiradjuri; the Australian Oxford agreed but suggested that the root word is *garila*; the Macquarie was more circumspect, not suggesting a language group source but offering *caralla* as the origin. The *Macquarie Aboriginal Words* (Thieberger and McGregor 1994) listed four words for corella, none of them anything like that (but not including Wiradjuri). It's doubtful that this one will be resolved, though it is worth noting that Morris (1898), in his generally excellent *Dictionary of Austral English*, claimed that it is 'dim. of late Lat. *cora* = *korh*, a girl, doll, etc.' – we feel fairly confident in discounting that. He also noted that it 'is often used indiscriminately by bird-fanciers for any pretty little parrot, parrakeet, or cockatoo', but confused things by asserting that properly it refers to 'any parrot of the genus *Nymphicus*' – this name has only ever been applied to Cockatiel, though he made it otherwise clear that he was referring to at least Long-billed Corella. Corella came into general usage only in the early 20th century, but had been around much longer than that. For instance there is an account in the *Bell's Life in Sydney and Sporting Reviewer* of 23 December 1848 of a court case concerning 'a specimen of the parrot genus, ornithologically termed a corella' that had been stolen from its publican owner. Corella is offered as an alternative name for Long-billed Corella only (perhaps echoing Morris) by RAOU (1913), but was adopted by RAOU (1926).

The genera

Nymphicus Wagler, 1832 [NIM-fi-kʊs]: 'bride-like [bird]' from Greek *numphē*, which has an astonishing number of meanings (c. 15), but since Wagler (1832b), in his footnote, also gave the Latin *sponsalis* [sic], meaning relating to betrothal, 'bride' or 'young wife' is clearly the one we need here. He also mentioned the 'diadem' (i.e. crest).

Calyptorhynchus Desmarest, 1826 [ka-lip-toh-RIN-kʊs]: 'covered-bill', from Greek *kaluptō*, cover or hide, and *rhunkhos*, bill. The feathers partly cover the lower part of the bill, sometimes more, sometimes less. This character is by no means restricted to this genus: at times we see it very clearly, for example, in the Sulphur-crested Cockatoo, *Cacatua galerita*.

Probosciger Kuhl, 1820 [pro-BO-si-gehr]: 'snout-bearer' from Latin *proboscis*, a snout (or sometimes, particularly, an elephant's trunk), and *gero*, carry or bear.

Callocephalon Lesson, 1837 [kal-lo-SE-fa-lon]: 'beautiful head', from Greek *kallos*, beauty, and *kephalē*, head. The sole member of the genus certainly is so endowed, with its brilliant red head (grey in the female) and jaunty and slightly comic little crest.

Eolophus Bonaparte, 1854 [eh-oh-LO-fʊs]: 'dawn crest', from Greek ēōs, the dawn, and *lophos*, tuft or crest, for the pale-pink crest.

Lophochroa Bonaparte, 1857 [lo-fo-KROH-uh]: 'beautiful-coloured crest', from Greek *lophos*, tuft or crest, and *khroa/khrōs*, colour (cf. *khrōma*), and *khroos*, beautiful-coloured.

Cacatua Vieillot, 1817 [ka-ka-TOO-uh]: 'cockatoo', from Malay *kakatua* (see family introduction – Vieillot is one of those who believed the word to be based on the calls of white cockatoos).

The species

Cockatiel (breeding resident)

From the same derivative as 'cockatoo' (see family introduction), with a diminutive. This is generally said to be Dutch – as per 'cockatoo' – but it has also been suggested that the form rather implies Portuguese (albeit from the same root), that having been a *lingua franca* for sailors of the time (perhaps because many of the maps were in that language). The oldest reference we can find in Australia is in a report in the *Launceston Daily Telegraph* of 27 August 1885, in a report of the Southern Tasmanian Poultry Society Show (under the category *Canaries and Cage Birds*). First formalised by RAOU (1926), though in 1898 Morris had noted that it was 'an alternative name for the Cockatoo-Parrakeet'.

Other names: Crested Parrakeet (Latham 1823); Cockatoo-parrot, Cockatoo Parrakeet (favoured by Gould 1848) or Cockatoo Crested Parrot, names dominant in the 19th century, reflecting the uncertainty until recent times as to which family it belongs to; Quarrion, said by the Australian *Oxford Dictionary* to be from the Wiradjuri *guwarraying* and by the *Australian National Dictionary* to be from the Ngiyampaa word *guwarrayi* – these are related languages from western New South Wales; Weero, surely from *wiru* of the Yindjibarndi language of the Pilbara area (Thieberger and McGregor 1994) – the usage seems to be mainly Western Australian. We wonder too about a reference in an advertisement ('wanted to buy') in the *Sydney Morning Herald* of 27 January 1845 to Cook's Crested Parrot *Platycercus Novæ Hollandæ*, though can't find any other reference to either of these names, while noting Gmelin's old name (see species name).

Nymphicus hollandicus (Kerr, 1792) [NIM-fi-kʊs ho-LAN-di-kʊs]: 'Australian (not Dutch!) bride-like bird', see genus name, and an adaptation of New Holland, the old name for Australia. The more usual

Cockatiel *Nymphicus hollandicus*

form had been used by Gmelin (1788) for this bird (*Psittacus novae Hollandiae*) but the name was corrected by Kerr (1792): Gmelin had already used it a dozen pages before, for what he called the Blue-bellied Parrot (perhaps a race of *Trichoglossus moluccanus*, the Rainbow Lorikeet).

Red-tailed Black Cockatoo (breeding resident)

See family introduction. Note the next species too, with regard to the red tail.

Other names: Red-tailed Cockatoo or Black-Cockatoo (the usual form in Australia); Great-billed Black Cockatoo, applied by Gould to his species *C. macrorhynchus* from Port Essington (not now recognised, though RAOU 1913 still did); Banksian or Banks's Black Cockatoo, from the species name; Western Black Cockatoo, for Gould's species *C. naso* from Western Australia – now subspecies *naso* (from Latin, *nasus*, nose), generally known as Forest Red-tailed Black-Cockatoo.

Calyptorhynchus banksii (Latham, 1790) [ka-lip-toh-RIN-kʊs BANK-si-ee]: 'Banks' covered-bill', see genus and for Sir Joseph Banks, giant of 18th and early 19th century English biology whose experiences with Cook on the *Endeavour* in 1770 left him with a life-long interest in Australia. Although a specimen was collected on that expedition (possibly by Banks) in the current Cooktown area while the *Endeavour* was beached, Latham described the bird from a later specimen from the new Port Jackson colony.

With regard to the comments under *Calyptorhynchus* about facial feathers covering the lower part of the bill, in this species the bill is also covered at times by the feathers of the crest – the 'Elvis Presley' hairdo comes right down over the top of the bill.

Glossy Black Cockatoo (breeding resident)

See family introduction. An odd name, because the bird has less glossy plumage than most of the genus; it seems to have arisen spontaneously in the 19th century, as Gould (1865) seemed not to have been aware of it, but Morris' 1898 *Dictionary of Austral English* used Glossy Cockatoo as the only option, without explanation. It was in general formal usage by the 20th century.

Other names: Glossy Black-Cockatoo, the usual form in Australia; Leach or Leach's Cockatoo or Black Cockatoo or Black-cockatoo or Red-tailed Cockatoo, for British Museum zoologist William Leach, whose speciality was actually crustaceans (he wasn't the influential early 20th century Australian ornithologist John Leach), the name used by Gould in 1848; the name was a tribute by German ornithologist Heinrich Kuhl, who (re)named the bird in 1820, shortly before his own death – presumably he was unaware that Temminck had named the bird 13 years previously. Casuarina or Casuarine Cockatoo, appropriate names for a casuarina seed specialist; Nutcracker, in reference to how they spend their whole day, cracking casuarina cones to extract seeds; Latham's Cockatoo, from the species name; Banksia Cockatoo, confusingly applied by Latham (1801) – see previous species, and the next! Cook's Cockatoo, Latham again, in 1823, presumably for Vigors and Horsfield's invalid *C. cookii*; Solander's Cockatoo – this bird was chosen to honour many people – for the early but still invalid name *Psittacus solanderi*, commemorating Daniel Solander, the Swedish naturalist, a star student of Linnaeus and member of Joseph Banks' staff on the *Endeavour* voyage of 1770.

Calyptorhynchus lathami (Temminck, 1807) [ka-lip-toh-RIN-kʊs LEH-thuh-mee]: 'Latham's covered-bill', see genus and for John Latham, doctor and eminent English ornithologist who is the author of over 60 Australian bird names – and the number would have been much higher had he been willing to accept the newfangled system of Latin names earlier in his career!

Yellow-tailed Black Cockatoo (breeding resident)

See family introduction. The only black cockatoo with yellow tail panels; it seems not to have had much use, however, until RAOU (1926).

Yellow-tailed Black Cockatoo *Calyptorhynchus funereus*

Other names: Yellow-tailed Black-Cockatoo, the usual form in Australia; Black Cockatoo, for the commonest black cockatoo in the Sydney area, though it seems surprising that it was still recommended as the national name by RAOU (1913); Yellow-eared Black Cockatoo, for another distinctive character, applied by Gould (1848) to his species *C. xanthonotus* (and a translation of that) from Van Diemen's Land, now a subspecies, though it was still regarded as a good species into the 20th century (RAOU 1913); Wylah, a beautifully onomatopoeic name whose similarity to *wayilahr* (for the same bird) of the Bundjalung people of the northern rivers of NSW is inescapable (Thieberger and McGregor 1994); Funereal Cockatoo, from the scientific name, applied by Latham (1823) and Gould (1848); Banksian Cockatoo, Latham again (1801) – see previous two species!

Calyptorhynchus funereus (Shaw, 1794) [ka-lip-toh-RIN-kʊs foo-NE-re-ʊs]: 'funereal covered-bill', see genus name, and from Latin *funereus*, for the black plumage – not that it's any blacker (and possibly even a bit less black) than its genus-fellows.

Baudin's Black Cockatoo (breeding resident)

See family introduction. From the species name.

Other names: Until 1948 – and by many authors for long after that – only one white-tailed black cockatoo was recognised, so some names were used for both, especially White-tailed Black-Cockatoo or Cockatoo; Long-billed Black-Cockatoo, clearly the most helpful name in distinguishing from the co-existing next species; Baudin's Cockatoo or Black-Cockatoo, the generally preferred Australian form.

Calyptorhynchus baudinii Lear, 1832 [ka-lip-toh-RIN-kʊs boh-DI-ni-ee]: 'Baudin's covered-bill', see genus name, and for Nicolas Baudin, commander of one of the great French scientific expeditions to Australia, from 1800 to 1803. In terms of new material delivered to France, the expedition was a stunning success but Baudin did not survive the voyage; by the accounts of those who did – and the 60 or so expeditioners, including most of the scientists, who left the ships early in protest – Baudin was an appallingly rude and uncongenial leader. We have no reason to doubt this, but we note that history is written by the survivors. The type specimen was published as an illustration by Edward Lear in *Illustrations of the Family Psittacidae* in 1832 when he was just 20 years old. Known better now as the man who introduced limericks to the world, Lear was a brilliant wildlife artist, especially of parrots, and one of John Gould's artistic mainstays. The type specimen was collected on the Baudin expedition, circumstantial evidence pointing to Geographe Bay as the type locality (Saunders 1979).

Carnaby's Black Cockatoo (breeding resident)

See family introduction. For Keith Carnaby (1910–1994), a Western Australian entomologist and world expert on jewel beetles, who described this species in the *Western Australian Naturalist* in 1948.

Other names: Carnaby's Black-Cockatoo, the generally preferred form in Australia; White-tailed Cockatoo or Black Cockatoo or Black-cockatoo, see Baudin's Black Cockatoo; Short-billed Black-cockatoo, again it would seem the most useful name. Mallee Black-cockatoo or Cockatoo, for a preferred habitat, again relative to the previous species.

Calyptorhynchus latirostris Carnaby, 1948 [ka-lip-toh-RIN-kʊs la-ti-ROS-tris]: 'broad-billed covered-bill', see genus name, and Latin *latus*, broad, and *rostrum*, bill. The bill is short (see common name) as well as broad (similar to that of the Yellow-tailed Black cockatoo).

Palm Cockatoo (breeding resident)

See family introduction. For its association with Pandanus groves (which aren't palms); it was known from its range to the north of Australia before its discovery in Cape York Peninsula (Gould 1869, who used Great Palm Cockatoo), so was probably already named in English by then.

Other names: Grey Cockatoo (Latham 1823), really a very odd and insipid name in the circumstances – did he intend Great Cockatoo?; Cape York Cockatoo, for its Australian range; Great Palm, Great Black, Goliath Cockatoo – it really is a very big bird! Goliath Aratoo, recorded for instance (as Goliah Aratoo) in *The Illustrated London Reading Book* (Illustrated London News 1851) with regard to a bird brought back from Papua, so we might assume that it is from a Papuan language, but Gould (1869) recorded a name for it as Ara Noir, French for Black Macaw (*Ara* is also the genus of big macaws) and we suspect that we are seeing an awkward portmanteau of Ara and Cockatoo. Black Macaw also appears as an English name, presumably as a direct translation (and a reminder of the danger of a little knowledge).

Probosciger aterrimus (Gmelin JF, 1788) [pro-BO-si-gehr a-TE-ri-mʊs]: 'very black snout-bearer', see genus name, and from Latin *ater*, dull black, with superlative suffix *–rimus*. Some snout! It has a very chunky bill indeed.

Gang-gang Cockatoo (breeding resident)

See family introduction. Of Aboriginal origin, though as is so often the case the actual language is uncertain; the estimable *Australian National Dictionary* asserted that it is of Wiradjuri origin but the evidence does not seem to support that (and the *Macquarie Aboriginal Words*, (Thieberger and McGregor 1994), does not list it as such). The earliest recorded use is by Charles Sturt (1833) in the form 'gangan' from the Mittagong area; although this was probably within the eastern edge of Wiradjuri country, Sturt did not suggest that he learnt the word there and it is quite likely that it was known from a coastal language. George Bennett (1834) reported from the Tumut River area that the bird was 'known by the native name of 'Gang, Gang', but again didn't suggest that he learnt the name on site. It seems reasonable to suggest it was an onomatopoeic name. Gould reported the name as being used by the 'Colonists of New South Wales' and used it himself.

Other names: Red-crowned or Red-headed Cockatoo or Parrot, both being accurate descriptors – Red-crowned Parrot was used by Latham (1801); Helmeted Cockatoo, for the crest; Ganga, presumably a corruption of the name; Galah; Cockatoo Corella, for reasons beyond us.

Callocephalon fimbriatum (Grant, 1803) [kal-lo-SE-fa-lon fim-bri-AH-tʊm]: 'beautiful fringed head', see genus name, and from Latin *fimbriatus*, fringed, for that funny little hairdo.

Galah (breeding resident)

According to the *Australian National Dictionary*, from *gilaa* in the Yuwaalaraay language group of north-western NSW. Interestingly, that dictionary did not record it in English until 1862, though we can do a bit better than that, with an advertisement in the *Sydney Empire* of 14 December 1857 for a 'fine lot' of young 'Rose Cockatoo or Galah Bird'. Certainly Gould (1848) was still calling it Rose-breasted Cockatoo and didn't list Galah among the alternatives. Rose-breasted Cockatoo was regarded as a formal name well into the 20th century until the RAOU, in its *Official Checklist of the Birds of Australia* (RAOU 1926), opted for Galah.

Other names: Rose-coloured Cockatoo (Latham 1823); Rose, Rose-breasted or Roseate Cockatoo (see common name); Pallid Rose-breasted Cockatoo, one of Mathews' names for Northern Territory birds – his names are so numerous and generally so fleeting that we don't list them all, but this one was recognised by RAOU (1913); Pink-and-Grey Galah, a seemingly unnecessary embellishment, which we are told is commonplace in Western Australia; Goulie or Goolie, surely again from an Australian language for which we can find no reference, but note *girlinygirliny* from the Gooniyandi language of the Kimberley; Willock or Willie-willock, likewise but here we feel that the existence of *wilek-wilek* from the Wembawemba language of the Wimmera and Riverina is too close to be a coincidence (Thieberger and McGregor 1994).

Eolophus roseicapilla (Vieillot, 1817) [eh-oh-LO-fʊs ro-se-ee-ka-PIL-lʊs]: 'rosy-haired dawn-crest', see genus name, and from Latin *roseus*, rose-coloured, and *capillus*, hair on the head. A beautiful name for a beautiful bird (though, truth to tell, Vieillot thought it was probably from India).

Major Mitchell's Cockatoo (breeding resident)

See family introduction. For Scot Sir Thomas Mitchell, British army veteran of the peninsular wars in Spain, who came to Australia in 1827 as deputy Surveyor-General to John Oxley, whom he succeeded the following year when Oxley died, holding the position until his own death in 1855. He conducted many expeditions inland, especially in search of the mythical Kindur River that would lead him to the northern seas. He rhapsodised over this cockatoo and referred to it several times in his account of the journey as 'cockatoo of the interior' and Red-top Cockatoo; he lamented that the 'rich crest of the cockatoo of the desert could not be preserved in dead specimens, and [was] too fine to be omitted among the sketches I endeavoured to snatch from nature' (Mitchell 1838). It had already been described 7 years before this, but Mitchell's lyrical waxings helped bring it to the attention of the public.

Other names: Leadbeater's Cockatoo, from the species name, used by Gould (1848); Pink Cockatoo, an obvious name, reported by Gould to be used by the 'Colonists of Swan River'; Desert Cockatoo for one of its habitats; Chockalott, Chock-a-lock, Joggle-joggle, Cockalerina, all variations of one or more Indigenous names of unspecified origin; Wee Juggler, also of Aboriginal origin and later anglicised for familiarity. The source word is said by the *Australian National Dictionary* to be Wiradjuri *wijugla*. Though the *Macquarie Aboriginal Words* (Thieberger and McGregor 1994) didn't record that word, it did list *kilaa* from the Ngiyampaa language of central NSW.

Lophochroa leadbeateri (Vigors, 1831) [lo-fo-KROH-uh led-BEE-tuh-ree]: 'Leadbeater's beautiful-coloured-crest', see genus, and for Benjamin Leadbeater (1760–1837), London taxidermist and

Major Mitchell's Cockatoo *Lophochroa leadbeateri*

ornithologist who provided Irishman Nicholas Vigors of the London Zoological Society with the specimen on which Vigors based the species description. It is not clear where Leadbeater obtained it because inland exploration in Australia was still in its infancy, though it is likely that the bird formerly came closer to the coast – Gould (1848) recorded it breeding from Gawler, now on the outskirts of Adelaide.

Long-billed Corella (breeding resident)

See family introduction. This species is distinguished from Little Corella by its very long top mandible (for digging up tubers).

Other names: Slender-billed Cockatoo or Corella (directly from the species name), Western Slender-billed or Long-billed Cockatoo; Dampier Cockatoo, a puzzle, because although the enigmatic English pirate-naturalist included a corella in the first English descriptions of Australian birds, from north-western Australia, it could not have been this species – only Little Corellas live there. Presumably the name was attached by association or confusion.

Cacatua tenuirostris (Kuhl, 1820) [ka-ka-TOO-uh te-noo-i-ROS-tris]: 'slender-billed cockatoo', see genus name, and from Latin *tenuis*, slender or thin, and *rostrum*, bill. See common name.

Western Corella (breeding resident)

See family introduction. The only corella endemic to Western Australia (though all three species are found there).

Other names: This species was until relatively recently included in the previous one (and at times in Little Corella), so doubtless names which were applied to the others were also attached to this one. Bare-eyed Corella (like the other two); Western Long-billed Corella or Cockatoo.

Cacatua pastinator (Gould, 1841) [ka-ka-TOO-uh pas-ti-NAH-tor]: 'vineyard worker cockatoo', see genus name, and from Latin *pastinator*, one who digs and trenches the ground, especially of vineyards. The birds do dig up roots, insects, and so on. Not sure the wine-growers would necessarily appreciate the help, though!

Little Corella (breeding resident)

See family introduction. Slightly smaller than the other corellas. It seems not to have been used until well into the 20th century, apparently being formalised by RAOU (1926).

Other names: Bare-eyed Cockatoo or Corella (again), this time from an 1871 name by the British Museum's William Sclater, *C. gymnopsis*, for a northern Australian specimen, still recognised as of species status by RAOU (1913); Blood-stained Cockatoo, from the species name, the name used by Gould (1848) and still in use by RAOU (1913); Short-billed Corella or Cockatoo, to differentiate from the other two species; Dampier's Cockatoo, see Long-billed Corella; Blue-eyed Cockatoo, for the bare blue skin around the eyes (but a feature of all three species).

Cacatua sanguinea Gould, 1843 [ka-ka-TOO-uh san-GWI-ne-uh]: 'bloody cockatoo', see genus name, and from Latin *sanguineus*, bloody or blood-coloured, for the pink colouring about the head. A little confusing, because the Long-billed Corella, named well before this one, would seem to justify the name better, having far more red on it.

Sulphur-crested Cockatoo (breeding resident)

See family introduction. For the iconic yellow crest; the name become established only in the second half of the 20th century (Slater 1970) and was formalised by RAOU (1978).

Other names: Crested Cockatoo, a name that goes back to the earliest days of the colony (White 1790) and was used by Gould in 1848; Sulphur Crest, White Cockatoo, both indicative of its familiarity.

Cacatua galerita (Latham, 1790) [ka-ka-TOO-uh ga-le-REE-tʊh]: 'hooded cockatoo', see genus name, and from Latin *galeritus*, wearing a hood, from *galerum*, a pointed cap like a helmet (also *galerus*). This word was also used for birds with crests, which this bird certainly has! Pliny (77–79 AD) explained that the Lark (probably the Crested Lark *Galerida cristata*) was once called *galerita*, for the appearance of its tuft, until this was replaced by the Gallic name *alauda*.

PSITTACULIDAE (PSITTACIFORMES): Old World parrots – Australasian and Indian Ocean

Psittaculidae Vigors, 1825 [si-tuh-KU-li-dee]: the Little Parrot family, from Latin *psittacus* (Greek *psittakos*), parrot, with Latin diminutive ending *–ulus, a, um.*

These birds are referred to by Ctesias, a Greek physician in the court of the Persian king at the beginning of the 5th century BC. In his account of Persian beliefs about India, he described the parrot (which he called *bittacos*), as speaking 'Indian like a native' and, if taught, able to speak Greek as well (Ctesias c. 400 BCa). Parrots (mainly the Ring-necked Parakeet *Psittacula krameri*) were well known to the ancient world: they were used in the great processions of Ptolemy Philadelphus of Egypt in the 3rd century BC, carried in cages or tied onto branches, and caged birds were favourites with both Greeks and Romans. Aristotle (c. 330 BC) said that parrots are 'said to have a man's tongue' and that 'after drinking wine, the parrot becomes more saucy than ever'. Ovid (c. 16 BC) wrote a poem to his mistress' dead parrot – was this the inspiration for the Monty Python sketch? – in which he compared its plumage to emeralds. Pliny (77–79 AD) mentioned, 'India sends us this bird, which it calls *siptace*', though he didn't say which of the many Indian languages was his source. Perhaps coincidentally, there was apparently a town called Psittake (Sitake, Citake) on the Tigris River. The abbreviation 'psitta' was the bit that made it into Modern

Latin and appears in various genus names. In particular the type genus of African and New World parrots (in which Australian parrots were until recently included) is *Psittacus*, once a great grab-bag for new parrot species, but now including only two species of African Grey Parrots.

'Parrot' appears in English from the early 16th century, its origin being generally – if not entirely comfortably – accepted as deriving from the French man's name Pierrot, a diminutive of Pierre (i.e. 'little Peter'). There are problems, though; there is no record in French of Pierrot referring to a parrot, and no record of Pierrot being used as an English name in 16th-century England. Nor does it seem to be related to any known word in any other language. Another reputable source, not apparently available in English, maintained that Perroquet (diminutive of Pierre) was the given name of a particular parrot in 1395; until then, we are told, the usual word for parrot was *papegai* – into English as popinjay (Quémada 1988). We are largely persuaded by this.

'Lorikeet' is a derived diminutive of lory (cf. parrakeet/parrot), which is from the Malay *luri* and was used to denote any parrot. 'Lorikeet' pre-dates English-Australian usage (being applied to birds from the Malay Peninsula and New Guinea), but not by much.

'Rosella' comes via a route that sounds like an etymological tall story, but that appears to be true. The parrots now known as Eastern Rosellas, common in the woodlands at Rose Hill to the west of Sydney where Parramatta now stands, were known as Rose Hill Parrots, then Rose Hillers, eventually (as the origin was lost) eliding to Roselle and Rosella! Roselle is recorded from 1829 (*Australian National Dictionary*), but James Backhouse in 1843 referred to the Eastern Rosella as 'Rosella, Rosehill, or Nonpareil Parrot' (Backhouse 1843). Support is offered by another newspaper quote supplied by the *Australian National Dictionary* from 1860: 'The Rosehill parrot, erroneously called rosella …'. Gould (1848) didn't use the term, suggesting that it was seen as a commoners' name, preferring the more formal Rose-hill Parrakeet. Interestingly, he recorded that the Western Rosella was also known as 'Rose-hill' by the colonists (of Western Australia), presumably without an understanding of the origin. The name was restricted to Eastern Rosella (simply known as Rosella), the other rosellas being just 'parrots', until the RAOU tidied things up in 1926.

The genera

Polytelis Wagler, 1832 [po-LI-te-lis]: 'valuable [bird]', from Greek *polutelēs*, costly or extravagant (so nothing to do with 'polly').

Alisterus Mathews, 1911 [A-li-stuh-rʊs]: 'Alister's', for Alister, son (then about 4 years old) of Gregory Mathews, Australian ornithologist.

Aprosmictus Gould, 1842 [ap-ros-MIK-tʊs]: 'misery-guts', from Greek *aprosmiktos*, uncommunicative or solitary. Gould suggested that members of this genus are 'dull and sullen', unlike the Platycerci, which he says are easier to tame.

Eclectus Wagler, 1832 [ek-LEK-tʊs]: 'chosen one', from Greek *eklektos* (and Latin *electos*, as Wagler 1832b pointed out), select or chosen.

Geoffroyus Bonaparte, 1850 [zho-FRWAH-ʊs]: 'Geoffroy's bird', for Étienne Geoffroy Saint-Hilaire (1772–1844), student of theology, law and medicine, professor at age 21 at the National Museum of Natural History in Paris, world expert on mammals (he named over 100 new species and was the first to recognise monotremes as a distinct group), and one of Napoleon's 'savants'.

Psephotus Gould, 1845 [se-FOH-tʊs]: 'mosaic [bird]', from Greek *psēphōtos*, inlaid with mosaic or precious stones, from *psephos*, a pebble or mosaic stone (think of 'psephologist', an expert on elections and on voting, which was done by the Ancient Greeks using pebbles). Opinions vary on what this actually means – the pattern of the cheek-feathers (though this is not mentioned by Gould) or just the lovely glowing colours, which he described in great detail, referring to *Psephotus haematogaster*, later renamed *Northiella haematogaster*, the Blue Bonnet (Gould 1848).

Northiella Mathews, 1912 [nor-thi-EL-luh]: 'North's little [bird]', a diminutive of the name of Alfred John North (1855–1917), Australian jeweller and ornithologist, specialist in nests and eggs and Assistant in Ornithology at the Australian Museum from 1891 to 1917.

Psephotellus Mathews, 1913 [se-fo-TEL-lʊs]: 'little mosaic [bird]', see *Psephotus*, with Latin diminutive ending *-ellus, a, um*.

Purpureicephalus Bonaparte, 1854 [pʊr-pʊr-e-i-SE-fa-lʊs]: 'purple or red head', from Greek *porphuroeis*, purple (or red), and *kephalē*, head. The word *porphuroeis* does mean purple, but it also means varying shades of red. It was originally the species name of the same bird *Psittacus purpureocephalus*.

Platycercus Vigors, 1825 [pla-ti-SEHR-kʊs]: 'broad-tail', from Greek *platukerkos*, broad-tailed, from *platus*, flat or broad, and *kerkos*, tail. Vigors (1825) said that he followed Kuhl (1820b) in distinguishing these birds from others in the parrot family who have 'that member narrowed and cuneted' (wedge-shaped, tapering).

Barnardius Bonaparte, 1854 [bar-NAR-di-ʊs]: 'Barnard's [bird]', for Edward Barnard (1786– 1861), a senior British civil servant with responsibility for the British Crown Colonies and with a strong interest in natural history; it is an honorific name, rather than indicating that Barnard had an association with the bird. It was originally the species name when the bird was regarded as a rosella.

Lathamus Lesson, 1830 [LEH-tha-mʊs]: 'Latham's [bird]', for John Latham (1740–1837), English naturalist and ornithologist, who arrived on the scene just in time to catch the wave of new species flooding out of Australia, naming (somewhat erratically and inconsistently) as he went.

Cyanoramphus Bonaparte, 1854 [si-a-no-RAM-fʊs]: 'blue bill', from Greek *rhamphos*, bill, and *kuanos/kuaneos*, dark blue – the colour of blue steel, a swallow's wing, a porpoise's skin or the deep sea. Also the colour of the enamel on Achilles' breastplate, of lapis lazuli, even of cornflowers. Only a pale version is seen on the bills of these parakeets, but perhaps the one it was named for (Black-fronted Parakeet, *C. zealandicus*, of Tahiti but now extinct) had a bill that was richer in colour, though the very few old museum specimens do not assist with this.

Pezoporus Illiger, 1811 [pe-zo-POR-ʊs]: 'walking [bird]', from Greek *pezoporos*, going by land.

Neopsephotus Mathews, 1912 [ne-o-se-FOH-tʊs]: 'new mosaic-bird', see genus name *Psephotus*, with Greek *neos*, new. Originally under *Euphema*, then *Neophema*, then *Psephotus*. Mathews had already foreshadowed a split in the genus *Psephotus*, suggesting that, if this was necessary, *Neopsephotus* would be the name he proposed.

Neophema Salvadori, 1891 [ne-o-FEH-muh]: 'new fair-sounding [bird]', from Greek *neos*, new, with old genus name *Euphema*, named by Wagler (1832b) from Latin *bonae famae* (of good report, cf. Greek *euphēmos*, fair-sounding or auspicious). *Euphema* was a large genus later split into a variety of new ones, including *Neophema*.

Parvipsitta Mathews, 1916 [par-vi-SIT-tuh]: 'small parrot', from Latin *parvus*, little, and *psittacus*, parrot.

Psitteuteles Bonaparte, 1854 [si-TOY-te-lehz]: 'very perfect parrot', made up from Temminck's (1838) name *Psittacus euteles* (the Olive-headed Lorikeet, now *Trichoglossus euteles*, from Timor). As far as the second part of the name is concerned, two possibilities have been offered (Temminck 1838 being silent on the subject): first, that it is from Greek *eutelēs*, paltry (or cheap or frugal or simple), or second, that it is made up of Greek *eu*, good or well or very, and *telēeis*, perfect. On the whole, we think the second fits better!

Trichoglossus Stephens, 1826 [tri-ko-GLOS-sʊs]: 'hairy-tongue', from Greek *thrix/trikhos*, hair, and *glōssa*, tongue. Refers to the brush-like papillae on the end of the tongue, which help the birds feed, particularly on pollen.

Glossopsitta Bonaparte, 1854 [glos-so-SIT-tuh]: 'tongue-parrot', see *psitta* under family name, and from Greek *glōssa*, tongue. Again the brush adaptation in the tongue for feeding on pollen is highlighted.

Melopsittacus Gould, 1840 [me-lo-SI-ta-kʊs]: 'tuneful parrot', see *psitta* under family name, and from Greek *melos*, song, tune or melody. The chatter and warble of a flock of budgerigars is one of the loveliest sounds of the outback.

Cyclopsitta Reichenbach, 1850 [si-klo-SIT-tuh]: 'circle-eyed parrot', see *psitta* under family name, and from Greek *kuklōps*, round-eyed. So the Cyclops, giants of Greek myth, had one eye in the middle of their foreheads, and it was **round**. According to Robert Graves (1996), this myth may have arisen from a guild of smiths of Bronze Age Greece, who are said to have had concentric rings tattooed on their foreheads. The birds are certainly not one-eyed and may even be four-eyed – see species name *diophthalma*.

The species

Superb Parrot (breeding resident)

See family introduction. Indeed it is, but it is unclear why it was considered more superb than others. Gould didn't know this name, though he waxed lyrical: 'few species are more elegant in form or more exquisitely coloured' (Gould 1848). Hall (1899) didn't call it Superb and neither did Lucas and Le Soeuf in 1911 though in the same year Leach had it as first-choice name; perhaps he coined it and, as chair of the RAOU's committee responsible for the second checklist (RAOU 1926), he doubtless had some influence on its adoption.

Other names: Green Leek, see also Scaly-breasted Lorikeet, though it is to the Superb Parrot that the name is most widely applied; Scarlet-breasted Parrot, offered as an alternative name by Leach (1911) for no discernible reason, though his illustration shows a greatly exaggerated red breast band. Barraband or Barraband Parrot or Parrakeet, from the name *Psittacus barrabandi* given by William Swainson, unaware that the name had already been applied to a South American parrot; he was honouring Jacques Barraband, distinguished French bird and plant artist, much of whose work was commissioned by Napoleon – the name was still regarded as first choice by RAOU (1913). Swainson's Parrot, from the species name.

Polytelis swainsonii (Desmarest, 1826) [po-LI-te-lis SWAYN-suh-ni-ee]: 'Swainson's costly [bird]', see genus and for William Swainson, a talented zoological artist and unfortunately also a prime example of going a bridge too far when he came to Australia and took up botanical taxonomy. Sir William Hooker of Kew – a man noted for his tolerance and tact – wrote to von Mueller, 'in my life I think I never read such a series of trash and nonsense. There is a man who left this country with the character of a first-rate naturalist, and of a very first-rate natural history artist, and he goes to Australia and takes up the subject of Botany, of which he is as ignorant as a goose' (ANBG 2019).

Regent Parrot (breeding resident)

See family introduction. For the black and gold plumage, which it shares with Regent Bowerbird (see that species for the origin of the name). Its actual adoption is a mystery – it doesn't seem to have been of 19th-century origin, as might be expected; Morris (1898) was unaware of it, and it springs fully formed to life in RAOU (1926). John Leach, chair of the committee responsible for that checklist, was often apparently responsible for such neologisms but he didn't mention this one in his 1911 bird book.

Other names: Black-tailed Parrot or Parrakeet, used by Gould (1848) for an unusual character, based on the early species name *Palaeornis melanura* (in fact this name was applied to the male, just after the female had been given the current name) and used into the 20th century (RAOU 1913); Plaide-wing Parrot, we suppose that you could see the black, yellow and red as tartan if you chose; Smoker, we suspect for the yellow-black plumage like nicotine-stained fingers; Marlock Parrot or Parrakeet, a Western Australian name, for a habitat composed of stunted but single-stemmed eucalypts; Mountain Parrot, reported by Gould (1848) to be used by the 'Colonists of Western Australia' (see also Mountain Duck); Yellow King-Parrot, see Australian King Parrot; Royal or Regal Parrot, perhaps by extension of the previous association; Blossom-feathered Parakeet, used by Nicholas Vigors in Lear's book of parrot illustrations, from the species name. Pepplar, Pebbler, Peblar, Rock Pebbler, one of the great mysteries of Australian bird names: we will stick our necks out and suggest that it is of Aboriginal origin (probably from Western Australia) and anglicised for familiarity, with the 'rock' added later by association, but we have no proof of this. We note the similarity to the species name, but suspect that to be coincidence.

Polytelis anthopeplus (Lear, 1831) [po-LI-te-lis an-tho-PEP-lʊs]: 'splendid flower-mantled costly bird', see genus name, and from Greek *anthos*, blossom or flower, and *peplos*, a splendid robe or mantle, referring to the red or salmon pink wing feathers. Edward Lear was an accomplished young bird artist (Lear 1832) as well as author of limericks and children's much-loved poems (see also Baudin's Black-Cockatoo).

Princess Parrot (breeding resident)

See family introduction. Nothing equivocal about this one. Gould allotted both species and common name (The Princess of Wales' Parrakeet) from specimens shot by Frederick Waterhouse on one of John McDouall Stuart's expeditions, writing in 1869: 'I feel sure … all ornithologists … will readily assent to its bearing the specific name of *Alexandrae*, in honour of the princess destined, we trust, at some future time to be the queen of these realms and their dependencies, of which Australia is by no means the least important'. (She did become queen 32 years later, after her mother-in-law, Queen Victoria, died.)

Other names: Alexandra, Princess Alexandra, Queen Alexandra Parrot; Rose-throated Parrot.

Polytelis alexandrae Gould, 1863 [po-LI-te-lis a-lek-ZAN-dreh]: 'Alexandra's costly bird', see genus and common names. 1863 was the year of Alexandra's marriage to the future Edward VII, so Gould may have been carried away by royal wedding fever.

Australian King Parrot (breeding resident)

See family introduction. The origin of this name has been largely forgotten – in fact it was forgotten within a few decades of being applied. George Caley, a botanist working for Sir Joseph Banks in New South Wales from 1800, sent back to England many plant and bird specimens, among them this species which he called 'King's Parrot' for Governor Philip Gidley King. This is quoted by Nicholas Vigors and Thomas Horsfield in their address to the Linnean Society of London in 1825 (Vigors and Horsfield 1827). However, another publication at the same time showed that the apostrophe was already fading: 'It was also observed that the well known species of Parrot ... commonly called the King Parrot, was also originally called King's Parrot after the same gentleman' (Zoological Club of the Linnean Society 1826). By 1848 Gould was unaware of this, and simply called his bird King Lory (see family introduction). For reasons that are abstruse to us at least, the RAOU decreed in 1978 that the link should be further obscured by introducing a hyphen (RAOU 1978); fortunately it has now been removed again.

Other names: Australian King-Parrot, King Lory see common name; Blood Rosella, probably for the colour of the male; Queensland King Parrot, King Parrakeet; Spud Parrot, used in Victoria 'as they attacked the potatoes when dug and lying on the ground' (Ford 1918); Tabuan Parrot, based on a misconception that it was the same as the Fijian species *Prosopeia tabuensis*, which has plumage quite similar to that of the male King Parrot – White, in his 1790 journal, labelled King Parrot illustrations thus, and Latham (1801) was still confusing the two; Scarlet and Green Parrot (Latham 1821).

Alisterus scapularis (Lichtenstein MHK, 1816) [A-li-stuh-rʊs ska-pʊ-LAH-ris]: 'Alister's shouldered bird', see genus name, and from Latin *scapulae*, the shoulders or shoulderblades, for the striking light blue-green streak on the wing.

Red-winged Parrot (breeding resident)

See family introduction. The red wing against green body is most striking; Gould (1848) used Red-winged Lory, from the species name.

Other names: Red-winged or Crimson-winged Lory, see family introduction; Blood-winged, Crimson-winged, Scarlet-winged Parrot; Red-wing.

Aprosmictus erythropterus (Gmelin JF, 1788) [ap-ros-MIK-tʊs e-rith-ro-TE-rʊs]: 'red-winged misery-guts', see genus name, plus Greek *eruthros*, red, and *pteros*, winged.

Eclectus Parrot (breeding resident)

See family introduction. Direct from the genus name.

Other names: Red-sided Parrot or Red-sided Eclectus Parrot, long a preferred name (CSIRO 1969), though it's an obscure choice; Rocky River Parrot, for the river on the east coast of Cape York Peninsula, east of Coen, which also gives its name to various rainforest plant species.

Eclectus roratus (Müller PLS, 1776) [ek-LEK-tʊs ro-RAH-tʊs]: 'dewy chosen one', see genus name, and from Latin *roro*, drop dew, and *roratus*, bedewed, for the sheen on the bird's plumage. Müller (1776) called it '*der Thaukakatu*' (in Modern German, this would be *Taukakatu*), or dew cockatoo. He described it as red, with blue shoulders and flight feathers, and said that its back and belly are 'bedewed with a blue haze'. We conclude that he had access only to the female bird, and not to the brilliant green and red male.

Red-cheeked Parrot (breeding resident)

See family introduction. A feature limited to the male.

Other names: Geoffroy's Parrot, see genus (and species!) name.

Geoffroyus geoffroyi (Bechstein, 1811) [zho-FRWAH-ʊs zho-FRWAH-ee]: Geoffroy's Geoffroy, see genus name.

Red-rumped Parrot (breeding resident)

See family introduction. Used by Gould, from his species name – appropriate only for the male (if he holds his wings apart!).

Other names: Grass or Ground Parrot, commonly used names, for the ground-feeding habits and its familiarity; Green Leek, see Scaly-breasted Lorikeet; Red-backed Parrot (from a time when even bird names were bowdlerised, and 'belly' or 'rump' were not 'nice', though to be fair it was a more accurate rendering of the species name); Red-rumped or Grass Parrakeet; Grassie; Redrump.

Psephotus haematonotus (Gould, 1838) [se-FOH-tʊs hee-ma-to-NOH-tʊs]: 'bloody-backed mosaic bird', see genus name, and from Greek *haima*, blood, and *nōton*, the back.

Eastern Bluebonnet (breeding resident)

See family introduction. For the not-very-conspicuous blue forehead and face. The name seems not to have been in formal use in the 19th century (the oldest usage we can find is from 1890) though it seems likely that it was used colloquially because RAOU (1913) offered it as an alternative name. Before that, in 1911 John Leach noted that 'Its vernacular name – Yellow-vented Parrakeet – has now been altered to Blue Bonnet Parrot' (though he didn't suggest by whom). The second RAOU checklist (1926) used it (two words) as first-choice name. 'Eastern' to distinguish it from the following species after they were (again) split in recent times.

Other names: Blue Bonnet or Bluebonnet (when this species and the next were lumped); Crimson-bellied Parrot or Parrakeet, for the obvious red belly in flight, which also inspired Gould's species name – it was still in use by RAOU (1913), thus calling into question Leach's comment and suggesting that he favoured Blue Bonnet; Red-vented Parrot, for race *haematorrhous*; Oak, Pine or Bulloak Parrot, for its favoured Casuarina habitat. Yellow-vented Parrakeet for Gould's later species *Psephotus xanthorrhous*, whose introduction he explained in Gould (1865), essentially because he believed that *haematogaster* covered two species and oddly, thought that the best way around it was to abolish that published name and replace it with two more. Needless to say, that didn't work, though RAOU (1913) perpetuated it.

Northiella haematogaster (Gould, 1838) [nor-thi-EL-uh hee-ma-to-GAS-tehr]: 'North's bloody-bellied little bird', see genus name, and from Greek *haima*, blood, and *gastēr*, stomach or belly (see common name).

Naretha Bluebonnet (breeding resident)

See family introduction and previous species. Naretha is a railway siding on the Transcontinental Railway at the western end of the Nullarbor Plain, in its core habitat.

Other names. Naretha Parrot or Naretha Blue Bonnet; see also previous species from which it was recently split.

Northiella narethae (White HL, 1921) [nor-thi-EL-uh na-REE-theh]: 'North's little bird from Naretha', see genus and common names.

Mulga Parrot (breeding resident)

See family introduction. Misleading, because mulga (*Acacia aneura*) forms only part of its habitat; it could have as validly been Mallee or Gidgee Parrot. The name seems to have been introduced, with little prior use, by RAOU (1926).

Other names: Various Parrot, from the species name; Many-coloured Parrakeet or Parrot, introduced by Gould (1848) and used throughout the 19th century, from the previous name, which was *Psittacus multicolor* Kuhl 1820 – much later Austin Clark of the US noted that Kuhl's name was preoccupied by a Gmelin name for Rainbow Lorikeet, also since supplanted (Clark 1910).

Psephotellus varius Clark, 1910 [se-fo-TEL-lʊs VAH-ri-ʊs]: 'little variegated mosaic bird', see genus name, and Latin *varius*, referring to the wonderful colours, especially of the male –another way of saying *multicolor.*

Hooded Parrot (breeding resident)

See family introduction. For the black hood pulled down over the eyes; the name was already in use by RAOU (1913).

Other names: Note that, Collett's name notwithstanding, for much of the 20th century it was regarded as conspecific with the Golden-shouldered Parrot, so all of those names potentially applied here too.

Psephotellus dissimilis (Collett, 1898) [se-fo-TEL-lʊs dis-SIM-i-lis]: 'little dissimilar mosaic bird', see genus name, and from Latin *dissimilis.* Collett (1898) remarked that the sexes are 'different in coloration', but sexual dimorphism is such a common feature that we feel this is unlikely to be the reason for his choice of specific name. However, the bird is also unlike other *Psephotus* parrots: Collett said it is 'nearest to' the Golden-shouldered Parrot, *P. chrysopterygius*, 'but lacks the yellow band across the forehead'.

Golden-shouldered Parrot (breeding resident)

See family introduction. Descriptive, but only of the male; used by Gould (1869) but a loose translation of his name.

Other names: Golden-winged Parrot; Chestnut-crowned Parrot, quite misleading; Hooded Parrot, from when, until recently, it was considered the same species as the previous one; Antbed or Anthill Parrot, for its habit of nesting in termite mounds.

Psephotellus chrysopterygius (Gould, 1859) [se-fo-TEL-lʊs kri-so-te-RI-gi-ʊs]: 'little golden-winged mosaic bird', see genus name, and from Greek *khrusos/khruso-*, gold/golden, and *pterux*, wing (cf. *pterugion*, anything like a wing, part of a shoulder-blade).

Paradise Parrot (extinct)

See family introduction. Curiously, although Gould was rapturous about this gorgeous bird, calling it the Beautiful Parrakeet when he described it, the 'Paradise' appellation was a marketing term, not introduced to Australia until 1922, shortly before the parrot's tragic demise (Olsen 2007). It was applied to the bird when it was introduced to the British cage bird trade and the great Alec Chisholm (1922) brought it to Australia's attention in the lovely *Mateship with Birds.*

Other names: Beautiful Parrakeet or Parrot, used by Gould from his species name; Ground Parrakeet, Parrot or Rosella; Anthill Parrot (it too nested in termite mounds, contributing to its downfall when the mounds were levelled to make tennis courts and landing fields); Scarlet-shouldered, Red-shouldered or Red-winged Parrot (but see this last-named species); Elegant Parrot, an Australian dealers' name; Soldier Parrot or Soldier Grass-Parrot, presumably for the red shoulder-flashes (though Olsen 2007 was of the firm opinion that the name stemmed from the mistaken memory of one individual in the 1920s, and was never otherwise used).

Psephotellus pulcherrimus (Gould, 1845) [se-fo-TEL-lʊs pʊl-KE-ri-mʊs]: 'most beautiful little mosaic bird', see genus name, and from Latin *pulcher*, beautiful, with superlative ending. Most beautiful in this case unfortunately equals most likely to die out through the greed of humans.

Red-capped Parrot (breeding resident)

See family introduction. For the extensive red cap down to the eyes; Gould (1848) used (and possibly coined) Red-capped Parrakeet, doubtless influenced by the invalid name *Platycercus pileatus* that he was using.

Other names: Red-capped Parrakeet, see common name; King Parrot, Western or Western Australian King Parrot, though there is little real resemblance – it is a Western Australian endemic, though; Pileated Parrot or Parrakeet, from a former species name *pileatus*, meaning capped; Hookbill, for the distinctive upper mandible (a specialisation for extracting seeds from Marri, *Eucalyptus calophylla*, capsules); Blue Parrot, the name of the rather unimaginative 'Colonists' (Gould 1848).

Purpureicephalus spurius (Kuhl, 1820) [pʊr-pʊr-e-i-SE-fa-lʊs SPOO-ri-ʊs]: 'bastard red-head', see genus name, and Latin *spurius*, of illegitimate birth. Kuhl (1820b) said he believed, on first seeing the birds, that the young bird was so unlike its parents as to appear unrelated.

Green Rosella (breeding resident)

See family introduction. Greenish-yellow anyway … Doubtless from Green Parrot, used by the 'Colonists of Van Diemen's Land' (Gould 1848).

Other names: Van Diemen's Parrot (Latham 1823), Tasmanian Rosella or Tasman Parrot, as endemic to the island; Green Parrot, see common name; Yellow-bellied or Yellow-breasted Parrot or Parrakeet, accurate; Caledonian or New Caledonian Parrot, both used by Latham, from the unfortunate species name.

Platycercus caledonicus (Gmelin JF, 1788) [pla-ti-SEHR-kʊs ka-le-DO-ni-kʊs]: 'Scottish broad-tail' – well, that seems as reasonable as 'New Caledonian broad-tail' for a bird coming from neither region. Somehow Gmelin's Tasmanian type specimen was labelled 'nova Caledonia' (New Caledonia) in error. See also genus name.

Crimson Rosella (breeding resident)

See family introduction. For the adult body colour, though oddly the name wasn't used until into the 20th century (e.g. Leach (1911), an eminent ornithologist, referred to the 'cousin of the Rosella [i.e. Eastern Rosella] … the Crimson Parrot'. Crimson Parrot was still used by RAOU (1913) but in its second checklist it introduced Crimson Rosella (RAOU 1926).

Other names: Red Lory (see family introduction) or Parrot; Crimson Parrot; Mountain Lowry (a version of lory) for its Blue Mountains stronghold; Pennant's Parrakeet or Pennantian Parrot, from an early species name per John Latham, for Thomas Pennant, a highly influential 18th-century Welsh naturalist. Campbell Parrakeet or Northern Crimson Parrot, for the tropical Queensland race *nigrescens*, once regarded as a separate species: we may reasonably assume that it was named for Alexander James Campbell, a prominent 19th-century Australian ornithologist who co-founded the RAOU. His son Alexander George also became prominent, but given that he was only 8 years old when Edward Ramsay named the bird in 1888, it seems unlikely to have been for him.

Two subspecies are very distinctive and have been regarded as representing separate species with their own common names.

P. e. adelaidae, from the Adelaide region – Gould published this as a species name: Adelaide Rosella or Parrakeet. Pheasant Parrot or Parrakeet, apparently for the perceived similarity of the reddish scalloped back to that of the European Common Pheasant *Phasianus colchicus*; this name goes back to the earliest years of settlement in Adelaide.

P. e. flaveolus, from the central Murray valley: Yellow Rosella. Murrumbidgee Rosella or Lowry; Murray Smoker, see Regent Parrot; Swamp Lory; Yellow Parrot or Parrakeet; Yellow-rumped Rosella or Parrakeet, applied by Gould (1848), who described it as a full species (from Latin *flavus*, yellow, with diminutive ending).

Platycercus elegans (Gmelin JF, 1788) [pla-ti-SEHR-kʊs EH-le-ganz]: 'elegant broad-tail', see genus name, and from Latin *elegans*, fine, elegant or handsome.

Northern Rosella (breeding resident)

See family introduction. The only Top End rosella, the name was formalised by RAOU (1926).

Other names: Brown's Rosella, Parrot or Parrakeet for the great Scottish botanist Robert Brown who accompanied Flinders on the *Investigator* and was present when the bird was collected – the bird was originally named for him by Vigors and Horsfield, but publication delays meant the name was gazumped; Smutty Rosella, Parrot or Parrakeet, in the sense of sooty rather than its taste in dinner stories – Gould (1848) reported that the name was used by the 'Residents at Port Essington' and it was still the primary name at the end of the century (Morris 1898); hence also Sooty Parrot, preferred by RAOU (1913) –it's unclear where it evolved, but we suspect deliberate bowdlerisation.

Platycercus venustus (Kuhl, 1820) [pla-ti-SEHR-kʊs ve-NʊS-tʊs]: 'lovely broad-tail', see genus name, and Latin *venustus*, lovely, charming or graceful. Yet another word for beautiful.

Pale-headed Rosella (breeding resident)

See family introduction. For the distinctive creamy head, as recognised by Gould, see 'Other names'.

Other names: Pale-headed Parrakeet, applied by Gould (1848) from his species name *P. palliceps*, which he applied not realising that Latham had already named it; Pale-headed Parrot, still used by RAOU (1913); Moreton Bay Rosella, being the common open country rosella in the Brisbane area; Blue-cheeked Rosella or Parrakeet, for what is now race *adscitus* from Cape York Peninsula, which has small blue cheek patches, but which Gould named *P. cyanogenys* ('blue-cheeked'), believing it a new species – RAOU still recognised it in 1913; White-cheeked or White-headed Rosella; Mealy Rosella or Parrakeet, meaning spotty, especially in regard to animals; Custardhead or Custard-head, a lovely folk name, still apparently in wide use in Queensland, albeit mostly orally.

Platycercus adscitus (Latham, 1790) [pla-ti-SEHR-kʊs ads-SEE-tʊs]: 'approved broad-tail', see genus name, and from Latin *a(d)scisco*, approve or receive as true. Latham (1790) didn't say, but according to HANZAB (1990–2006, vol. 4) it refers to the brightly variegated plumage. This is something of a mystery to us – approved as having true parrot colours, perhaps?

Eastern Rosella (breeding resident)

See family introduction. Eastern by contrast with the closely related Western Rosella from Western Australia, but surprisingly the name wasn't generally used (the simple Rosella being preferred) until the RAOU's excellent rationalisation job in its second checklist (RAOU 1926).

Other names: Rosehill or Rose-hill Parrakeet, reported by Gould (1848) to be the name of the 'Colonists of New South Wales' – it must have been well established because he rarely adopted such names, but did so on this occasion; Common Rosella, Rosella, Rosy, all for its familiarity to early Sydney settlers; Red-headed, Golden Mantled, White-cheeked Rosella, all descriptive; Nonpareil Parrot, the wonderful name appended by John Latham (1802) in the *Second Supplement to the General Synopsis of Birds* and meaning 'unequalled'. Splendid Parrakeet, for Gould's species *P. splendidus*, not now recognised, from the northern Darling Downs; Yellow-mantled Parrot, for the same taxon, as used by RAOU (1913); Fiery Parrakeet, for Gould's species *P. ignitus* from the Moreton Bay area, based on a single specimen – Gregory Mathews believed it to be an abnormally coloured Eastern Rosella, while we tend slightly towards the result of hybridisation between Eastern and Crimson Rosellas.

Platycercus eximius (Shaw, 1792) [pla-ti-SEHR-kʊs ek-SI-mi-ʊs]: 'excellent broad-tail', see genus name, and Latin *eximius*, uncommon, excellent or select. We are not going to argue.

Western Rosella (breeding resident)

See family introduction. The only south-western Australian rosella; again the name was introduced for consistency in naming of rosellas by RAOU (1926).

Other names: Earl of Derby's Parrot, Stanley Rosella or Parrakeet, Lord Stanley's Parrakeet, all from the species name *stanleyii* given it by Nicholas Vigors in 1830 to commemorate Lord Stanley, later the Earl of Derby, not realising that Coenraad Temminck had beaten him to naming rights by 10 years. Stanley was an English parliamentarian, zoologist and private zoo owner who was later president of both the Linnean Society and London Zoological Society. Gould (1848) pointed out Vigors' error, but attempted to keep the honour alive by updating the name and calling the bird The Earl of Derby's Parrakeet 'as bound by justice to the first describer … I feel I shall have the acquiescence of all ornithologists'. Yellow-cheeked Parrot (preferred by RAOU 1913) or Rosella; Rose-hill, see 'Rosella'.

Platycercus icterotis (Temminck & Kuhl, 1820) [pla-ti-SEHR-kʊs ik-te-ROH-tis]: 'yellow-eared broad-tail', see genus name, and Greek *ikteros*, yellow (as in jaundice), and ōtis, eared. The only yellow-cheeked rosella (close enough to the ears, we suppose).

Australian Ringneck (breeding resident)

For the distinctive yellow collar, consistent across several races; it doubtless arose spontaneously and was known to Morris (1898) in the form Ring-neck Parrakeet. The earliest reference we can find is to 'Ring-necked Parrots' for sale in the *Sydney Empire* of 12 January 1852, but they could well have been Ring-necked (Rose-ringed) Parakeets *Psittacula krameri*, a popular imported cage bird.

Other names: Banded Parrot or Parrakeet, Yellow-naped Parrakeet.

The four races have at times been regarded as separate species, and have some race-specific names.

B. z. zonarius, from Eyre Peninsula to northern Australia: Port Lincoln Parrot, for the major town on southern Eyre Peninsula; Western Ringneck; Yellow-banded or Northern Yellow-banded Parrot, RAOU (1913); Bauer's Parrot or Parrakeet, from the species name *Platycercus baueri* of the Dutch ornithologist Coenraad Temminck, for Ferdinand Bauer, brilliant but self-effacing Austrian natural history artist who sailed with Flinders on the *Investigator*.

B. z. semitorquatus, from south-western Western Australia: Twenty-eight Parrot, evocative of the ringing call and established early, because Gould (1848) reported that it was already in use by the 'Colonists of Swan River' and it is still in wide usage; Yellow-collared Parrot or Parrakeet, used by Gould (1848) who recognised it as a separate species, and still used into the 20th century (RAOU 1913). From Latin *semi*, half, and *torquatus*, wearing a collar or neck-chain.

B. z. barnardi, from inland south-eastern Australia: Barnard's Parrot or Parrakeet, as used by Gould (1848) – see genus name; Ring-necked Parrot (RAOU 1913); Eastern or Mallee Ringneck, for one of its habitats though it is certainly not mallee-dependent; Buln Buln or Bulla Bulla, from an Indigenous language, but we don't know which one.

B. z. macgillivrayi, from north-western Queensland: Cloncurry Parrot, for a major town in its range; Northern Buln Buln, see previous race; MacGillivray's Collared Parrakeet, from the race name, for Alexander Sykes MacGillivray (1853–1907), a grazier and amateur naturalist in the Cloncurry area who recognised the different nature of the bird.

North's Parrot is also cited by both CSIRO (1969) and HANZAB (1990–2006, vol. 4) – the former as either *zonarius* or *semitorquatus*, which it lumps together, the latter unspecified – but we can neither shed any light on it nor even find another reference to it; the likely reference is to Alfred North, assistant in Ornithology at the Australian Museum from 1891 to 1917, and author of *Nests and Eggs of Birds found Breeding in Australia and Tasmania*, but this is mere supposition.

Barnardius zonarius (Shaw, 1805) [bar-NAR-di-ʊs zoh-NAH-ri-ʊs]: 'girdled Barnard's bird', see genus name, and from Latin *zona*, a girdle or belt, referring to the 'ring neck', the yellow band across the back of the neck.

Swift Parrot (breeding resident)

See family introduction. For its dramatically quick flight between feed trees; Gould used Swift Lorikeet in 1848, though he reported that the 'Colonists of Van Diemen's Land' called it Swift Parrakeet, which he apparently adopted. He was smitten with the bird, describing small flocks of them flying over Hobart 'chasing each other with the quickness of thought'.

Other names: Swift or Swift-flying Lorikeet, for its passing resemblance in flight to the smaller lorikeets, in whose subfamily it used to be included; Keet or Swift-flying Keet, see Scaly-breasted Lorikeet; Clink for the call; Red-faced or Red-shouldered Parrot or Parrakeet, both descriptive – Latham used Red-shouldered Parrakeet in 1801.

Lathamus discolor (Shaw, 1790) [LEH-tha-mʊs DIS-ko-lor] 'Latham's bird of different colours', see genus name, and from Latin *discolor*, particoloured or variegated.

Norfolk Parakeet (Norfolk Island)

See family introduction. The relationships of the *Cyanoramphus* parrots are complex and understandings of which taxa comprise species as opposed to subspecies (and if the latter, of which species!) have changed regularly; it is not at all certain that the last word has been spoken.

Other names: Green Parrot; Green or Red-fronted Parrot; Tasman Parakeet, from when Christidis and Boles (2008) regarded the Norfolk Island Parakeet and the extinct Lord Howe Parakeet as of the same species (both have until recently been generally seen as subspecies of *Cyanoramphus novaezelandiae*). They coined the name Tasman Parakeet to take into account its believed former occurrence on both islands.

Cyanoramphus cookii (Gray GR, 1859) [si-a-no-RAM-fʊs KʊK-i-ee]: 'Cook's blue-bill', see genus name, and for Captain James Cook, who landed on Norfolk Island in 1774 and remarked on the parakeets as being the same sort as in New Zealand.

Lord Howe Parakeet (Lord Howe Island, extinct)

See family introduction. See comments under Norfolk Parakeet.

Other names. Lord Howe Island Parakeet; Tasman Parakeet, see previous species.

Cyanoramphus subflavescens Salvadori, 1891 [si-a-no-RAM-fʊs sʊb-fla-VE-senz]: 'yellowish blue-bill', see genus name, and from Latin *sub*, under, and from Latin *flavus*, golden yellow, from *flagro*, burn, with *-escens*, becoming (i.e. '-ish'). Salvadori describes the 'yellowish tinge', particularly on the 'underparts' of the bird.

Macquarie Parakeet (Macquarie Island, extinct)

See family introduction, and for its former home, to which it was endemic.

Other names: Macquarie Island Parakeet; Red-crowned Parakeet (*Cyanoramphus novaezelandiae*) – until recently it was believed the Macquarie Parakeet was a race of this species.

Cyanoramphus erythrotis (Wagler, 1832) [si-a-no-RAM-fʊs e-ri-THROH-tis]: 'red-eared blue-bill', see genus name, and from Greek *eruthros*, red and ōtis, eared.

Eastern Ground Parrot (breeding resident)

See family introduction. For its near-exclusively terrestrial habits; the name was apparently first used by Latham (1801). 'Eastern' since it was split from the following species.

Other names: Ground Parrakeet, used by Gould (1848); Black-spotted Parrakeet (of Van Diemen's Land), reported by Gould to have been used by Bruni d'Entrecasteaux in his account of the *Recherche* expedition of 1791–93; Swamp or Button Grass Parrot for habitat, the latter referring to the extensive *Gymnoschoenus sphaerocephalus* boggy plains of Tasmania; Green Quail, for its dumpy demeanour; Madden's Swamp Parrot, purportedly for Madden's Plains 'forty miles south of Sydney, a once famous haunt of the species' (Hindwood 1932), though we note that near Fitzgerald River National Park in Western Australia (a stronghold of the next species, former regarded as conspecific) there is a Mount Madden named for a colonial secretary.

Pezoporus wallicus (Kerr, 1792) [pe-zo-POR-ʊs WAL-li-kʊs]: 'New South Wales pedestrian', see genus name, and Modern Latin for the type locality, *Nova Wallia Australis*, (i.e. New South Wales).

Western Ground Parrot (breeding resident)

See family introduction and previous species. Restricted to the south-west.

Other names. Madden's Swamp Parrot (possibly, see previous species).

Pezoporus flaviventris North, 1911 [pe-zo-POR-ʊs flah-vi-VEN-tris]: 'yellow-bellied pedestrian' see genus name, and from Latin *flavus*, yellow, and *venter/ventris*, belly.

Night Parrot (breeding resident)

See family introduction. It is primarily nocturnal or at least crepuscular, and the name seems to have arisen naturally.

Other names: Nocturnal Ground Parrakeet, used by Gould (1869), making the point that it was closely related to the previous species; Night Parrakeet; Western Ground Parrot (there was conjecture that they were the same species), but see also previous species; Spinifex or Porcupine Parrot for its prime habitat (*Triodia* spp. hummock grasslands, also known as Porcupine Grass); Nocturnal Ground-Parrakeet, Gould's name (1848); Midnight Cockatoo, Solitaire, both reported by HANZAB (1990–2006, vol. 4), but we can find no other reference to their usage or origin, though this could simply indicate an oral tradition.

Pezoporus occidentalis (Gould, 1861) [pe-zo-POR-ʊs ok-si-den-TAH-lis]: 'western pedestrian', see genus name, and Latin *occidentalis*, western (the type specimen was collected in Western Australia).

Bourke's Parrot (breeding resident)

See family introduction. Named by explorer and Surveyor-General Thomas Mitchell in 1838, and Gould formalised the name in Latin 3 years later. It honours Governor Richard Bourke, an unhappy man whose severe facial war injuries thwarted his parliamentary ambitions. He and his wife came to New South Wales for her health; she died here and Bourke's liberal views on society were staunchly resisted.

Other names: Bourke or Bourke's Grass Parrot or Parrakeet; Blue-vented Parrot or Parrakeet; Pink-bellied Parrot or Grass-Parrot, perplexingly the choice of RAOU (1913); Night or Sundown Parrot, for its evening drinking habits.

Neopsephotus bourkii (Gould, 1841) [ne-o-se-FOH-tʊs BER-ki-ee]: 'Bourke's new mosaic-bird', see genus and common names.

Blue-winged Parrot (breeding resident)

See family introduction. The members of the *Neophema* group of parrots are very similar to each other, so unique names were difficult to arrive at; furthermore, the sexes are dimorphic and the males tend to have the striking features. This one has more blue in the wings than the others.

Other names: Blue-winged Grass Parrot, Grass-Parrot or Parrakeet; Blue-banded Parrot, Grass Parrot or Parrakeet, for the blue eyebrow band (the name used by Gould 1848, though it is a character shared by the next two species); Hobart Ground Parrot, for its terrestrial habits and being the only group member found near Hobart (the Orange-bellied is the only other one in Tasmania).

Neophema chrysostoma (Kuhl, 1820) [ne-o-FEH-muh kri-so-STO-muh]: 'golden-mouthed new fair-sounding bird', see genus name, and from Greek *khrusos/khruso-*, gold/golden, and *stoma*, mouth, for the yellow, not quite of the mouth, but of the lores and around the eyes.

Elegant Parrot (breeding resident)

See family introduction. Avoids the need to find a usefully unique name! Taken straight from the species name.

Other names: Grass Parrot or Parrakeet (reported by Gould 1848 to be the name 'of the Colonists') – oddly, in 1913 RAOU used Grass-Parrot, without elaboration, as its recommended name; Elegant Grass Parrot or Parrakeet.

Neophema elegans (Gould, 1837) [ne-o-FEH-muh EH-le-ganz]: 'elegant new fair-sounding bird', see genus name, and Latin *elegans* – beginning to feel like a somewhat overused word, but how many ways are there to say 'stunning'?

Rock Parrot (breeding resident)

See family introduction. For its exclusively coastal habitat, including (though not restricted to) cliffs.

Other names: Rock Parrakeet or Grass-Parrakeet, used by Gould (1848), when he reported that Rock Parrakeet was the name 'of the Colonists of Swan River'; Rock Elegant Parrot or Parrakeet.

Neophema petrophila (Gould, 1841) [ne-o-FEH-muh pe-TRO-fi-luh]: 'rock-loving new fair-sounding bird', see genus name, and from Greek *petros*, rock or stone, and *-philos*, lover of. See genus and common names.

Orange-bellied Parrot (breeding resident)

See family introduction. Mostly a male-only characteristic.

Other names: Orange-bellied Grass Parrakeet, used by Gould (1848); Orange-breasted Parrot, Grass-Parrot, or Grass Parrot. Trumped-up Corella, perhaps the only Australian bird name coined by a state premier, in this case Jeff Kennett of Victoria when his proposal in 1996 to move the Coode Island Chemical Works to Port Wilson was thwarted by the realisation that the site was important feeding habitat for the highly endangered bird. The name was adopted by the press, and by the Orange-bellied Parrot Recovery Team as the name of its newsletter.

Neophema chrysogaster (Latham, 1790) [ne-o-FEH-muh kri-so-GAS-tehr]: 'golden-bellied new fair-sounding bird', see genus name, and from Greek *khrusos/khruso-*, gold/golden and *gastēr*, stomach or belly.

Turquoise Parrot (breeding resident)

See family introduction. For the male's striking face.

Other names: Beautiful Parrot or Grass Parrot; Chestnut-shouldered or Chestnut-winged or Red-shouldered Parrot, Grass Parrot, Grass-Parrot or Grass Parrakeet, again for a males-only feature; Turquoisine or Turcoisine Parrot, very Victorian era! Orange-bellied Parrot or Parrakeet, confusingly used by Latham (1801) and Shaw (1819).

Neophema pulchella (Shaw, 1792) [ne-o-FEH-muh pʊl-KEL-luh]: 'very pretty new fair-sounding bird', see genus name, and from Latin *pulchellus*, a diminutive of beautiful – more loveliness.

Scarlet-chested Parrot (breeding resident)

See family introduction. Yet another name appropriate to the male only.

Other names: Scarlet-chested Grass Parrakeet, Grass-Parrot or Grass Parrot; Scarlet-breasted or Scarlet-throated Parrot; Splendid Parrot or Grass Parrot or Grass Parrakeet, straight from the species name as applied by Gould.

Neophema splendida (Gould, 1841) [ne-o-FEH-muh SPLEN-di-duh]: 'shining new fair-sounding bird', see genus name, and Latin *splendidus*, bright or shining, for its brilliant colours.

Little Lorikeet (breeding resident)

See family introduction. The tiniest lorikeet, called Little Parrakeet by Gould, from the species name.

Other names: Small Parrakeet or Parrot (Latham 1801, 1823); Red-faced Lorikeet or Lory, for the only lorikeet with a fully red face (see 'lory'); Tiny Lorikeet; Gizzie for the zitty calls; Jerryang, recorded by Gould (1848) as a name used by the 'Aborigines of New South Wales' – we may suppose that it too was onomatopoeic; Keet, Green Keet, Green Leek, see Scaly-breasted Lorikeet.

Parvipsitta pusilla (Shaw, 1790) [par-vi-SIT-tuh pʊ-SIL-luh]: 'tiny little parrot', see genus name, and Latin *pusillus*, small (yes, very) or insignificant (no!).

Purple-crowned Lorikeet (breeding resident)

See family introduction. A most apposite name, straight from the species name.

Other names: Porphyry-crowned Lorikeet, as used by Gould (1848), directly from the species name; Blue-crowned (inaccurate) or Purple-capped Lorikeet; Purple-crowned Lory, see family introduction, or Purple-crowned Parrakeet; Zit Parrot, for the buzzy flight calls.

Parvipsitta porphyrocephala (Dietrichsen, 1837) [par-vi-SIT-tuh por-fi-ro-SE-fa-luh]: 'little purple-headed parrot', see genus name, and from Greek *porphuroeis*, purple, and *kephalē*, head.

Varied Lorikeet (breeding resident)

See family introduction. From the species name (as applied by Gould 1848), but a misleading translation – there is very little variation within the species.

Other names: Varied Lory (see 'lory'); Variegated Lorikeet, probably closer to Gould's species name intent; Red-crowned or Red-capped Lorikeet, actually a good distinguishing name within its range.

Psitteuteles versicolor (Lear, 1831) [si-TOY-te-lehz vehr-SI-ko-lor]: 'many-coloured very perfect parrot', see genus name, and from Latin *versicolor*, of varied or changeable colours. And the authority is Edward Lear, he of the limericks and 'The Owl and the Pussycat', who was also an accomplished artist whose parrot illustrations were published when he was 20 years old (see also Baudin's Black-cockatoo.)

Coconut Lorikeet (regular Torres Strait Islands visitor)

See family introduction. There is certainly an association with coconut palms, but no specialisation.

Other names. Coconut Lory.

Trichoglossus haematodus (Linnaeus, 1771) [tri-ko-GLOS-sʊs hee-ma-TOH-dʊs]: 'bloody hair-tongue', see genus name, and from Greek *haimatōdēs*, blood red (from *haima*, blood), referring to the breast rather than the more striking underwing. The name has a slightly unusual history: it appeared first as the abbreviation '*haematod*'. (Linnaeus 1767) (apparently to avoid printing the '*-us*' on the next line) and was subsequently written that way, along with several other erroneous spellings. It was eventually corrected to its present (and originally intended) form. The etymological history of this whole genus is a highly complex one, too long to be told here.

Rainbow Lorikeet (breeding resident)

See family introduction. For the truly rainbowesque plumage patterns, though the name didn't gain general acceptance until the 20th century. Morris' *Dictionary of Austral English* wasn't aware of it in 1898 but John Leach used Rainbow Lory as an alternative name in 1911, so it was probably in vernacular use by then. It is quite likely that his influence as chair of the RAOU committee responsible for preparing the second checklist of recommended names (RAOU 1926) led to Rainbow Lorikeet being used, apparently for the first time in an official ornithological publication, in that document.

Other names: Blue Mountain Lorikeet or Parrot, for its former stronghold near Sydney, a name still popular into the 20th century (Leach 1911) despite its New South Wales-centricity; Blue-bellied Parrot, Parrakeet or Lorikeet, a very early name (Gmelin 1788 and White 1790 used Blue-bellied Parrot) still favoured into the 20th century (RAOU 1913); Swainson's Lorikeet, the name used by Gould for the then species name honouring William Swainson, zoological artist and self-styled naturalist – the bird had long been named by Gmelin (and subsequently others) but, by the time that William Jardine and Prideaux Selby named it that; Bluey, Lorikeet, Lory, Rainbow Lory, see family introduction; Blue Bonnet, for the blue head, but see also Blue Bonnet species; Northern Blue-bellied Lorikeet, for the briefly recognised north Queensland species *T. septentrionalis*, described in 1900 in Britain, accepted by RAOU (1913) but dropped RAOU (1926).

Trichoglossus moluccanus (Gmelin JF, 1788) [tri-ko-GLOS-sʊs mo-lʊ-CA-nʊs]: 'Moluccan hair-tongue', see genus name, and from the Maluku or Moluccan Islands. See also comments under name of previous species.

Red-collared Lorikeet (breeding resident)

See family introduction. For the major distinction between this species and the previous one, from which it was recently split.

Other names. Orange-naped Lorikeet.

Trichoglossus rubritorquis Vigors & Horsfield, 1827 [tri-ko-GLOS-sʊs roo-bri-TOR-kwis]: see genus name, and from Latin *ruber*, red, and *torquis*, a twisted neck-chain or collar. Vigors and Horsfield describe the colour of the collar as 'scarlet', in Latin '*coccineum*'.

Scaly-breasted Lorikeet (breeding resident)

See family introduction. For the scale-like yellow scalloping on the breast, but straight from the species name.

Other names: Scaly-breasted Lory (see 'lory'); Greenie; Green-and-Yellow or Green-and-Gold or Green Lorikeet; Green Keet, an abbreviation of lorikeet, which appears in the literature from the 1870s but was doubtless used well before that. Green Leek, a slightly mysterious name used for several green Australian parrots; there is no accepted origin for it (other than that it seems to have arisen in Australia) but we note that the simile 'as green as a leek' appears as far back as Chaucer. It is quite possible that it was aided by the general similarity of 'leek' and 'lorikeet'.

Trichoglossus chlorolepidotus (Kuhl, 1820) [tri-ko-GLOS-sʊs klo-ro-le-pi-DOH-tʊs]: 'greenish-yellow-scaled hairy-tongue', see genus name, and from Greek *khlōros*, greenish-yellow, and *lepidōtos*, scaly (from *lepis*, scale). See common name.

Musk Lorikeet (breeding resident)

See family introduction. For its reputed scent! We cannot verify this, but Gould (1848) reported that it was called Musk Parrakeet in Van Diemen's Land 'from the peculiar odour of the bird'. He used Musky Parrakeet (though he did use the name Lorikeet for *Trichoglossus*).

Other names: Pacific Parrot (Latham 1801) and Pacific Parrakeet (Shaw 1819), for no very obvious reason, though Latham believed that it came from New Zealand; Crimson-fronted Parrakeet (Latham 1823); Musky Lorikeet, per Gould; Musk Lory (see 'lory'); Red-crowned or Red-eared Lorikeet; Green Leek or Green Keet, see Scaly-breasted Lorikeet; King Parrot, uncertain.

Glossopsitta concinna (Shaw, 1791) [glos-so-SI-tuh con-SIN-nuh]: 'elegant tongue-parrot', see genus name, and Latin *concinnus*, elegant or neat.

Budgerigar (breeding resident)

Gould (1848) reported that the 'Natives of Liverpool Plains' called the bird Betcherrygah (though he preferred Warbling Grass-Parrakeet!). However, we have found a usage very close to the modern form earlier than this, in a Wanted to Buy advertisement in the *Sydney Morning Herald* of 27 January 1845, for 'Budgerigor of the Aborigines; the Shell Parrot of the Colonists'. The *Australian National Dictionary* says that the word came from one of the Yuwaalaraay group of languages.

Other names: Budgie, Budgerygah, Betcherrygah; Love Bird, for the cuddly courtship displays (this name probably prompted by the African parrots of the same name); Undulated Parrot or Parrakeet, used by Latham (1823) and Shaw (1819), from the species name; Scallop, Shell or Zebra Parrot, for the strong scalloped wing pattern, again referring to the species name; Shelly, from Shell Parrot; Warbling Grass Parrot or Parrakeet, after Gould, still used by RAOU (1913); Canary Parrot, for the song – it sounds like a cage bird trade name, but in fact Gould reported that it was used by 'Colonists'.

Melopsittacus undulatus (Shaw, 1805) [me-lo-SI-ta-kʊs ʊn-doo-LAH-tʊs]: 'tuneful scalloped parrot', see genus name, and from Latin *undulatus*, wavy (see common name).

Double-eyed Fig Parrot (breeding resident)

Fig Parrot (or more usually 'Fig-Parrot') from its favoured rainforest food, though a recent usage in Australia – the group name was formalised in Australia only by RAOU (1978), though Slater (1970) had used Fig Parrot in the first 'modern' Australian field guide; 'Double-eyed', from the species name.

Other names: Fig-Parrot; Double-eyed Dwarf Parrot; Lorilet, another diminutive form of lory, see 'lory' – this was favoured well into the 20th century (RAOU 1926; Cayley 1931); Two-eyed Fig-Parrot or Lorilet.

The three races have at times been regarded as separate species, with their own names, thus:

C. d. coxeni (north-eastern New South Wales and south-eastern Queensland): Coxen's Fig-Parrot, for Charles Coxen, whose sister Elizabeth married John Gould and who greatly facilitated the Goulds' extended visit to Australia. Coxen settled in the upper Hunter Valley, was an avid naturalist and collector and, after extending his interests to Queensland, helped found the Queensland Museum and later entered the Queensland parliament. He did not collect this specimen, but Gould (1848) explained that the collector 'Mr Waller' requested that Gould name it for Coxen; it could be true, too. Southern Fig-Parrot or Lorilet, to distinguish from the next two races; Blue-browed or Red-faced Fig-Parrot or Lorilet, not very convincing, though it does lack the red foreheads of the other two races.

C. d. macleayana (wet tropics): Macleay's Fig-Parrot; when Edward Pierson Ramsay described this race in 1874, it was not clear which of the mighty dynasty of Macleays he was honouring. There was Alexander Macleay (see Forest Kingfisher), New South Wales Colonial Secretary from 1825, entomologist and key figure in founding the Australian Museum, his son William Sharp Macleay, diplomat, jurist and entomologist, later a trustee of the Museum, and William's cousin William John Macleay (see Macleay's Honeyeater), pastoralist, parliamentarian, naturalist and first president of the Linnean Society of New South Wales. McCoy's Fig-Parrot (from the scientific name allotted by Gould in 1875, unaware that Ramsay had already done so) for Sir Frederick McCoy, Irish naturalist, palaeontologist and geologist who took the inaugural chair in Natural History at Melbourne University; unfortunately, he was also an avid driver of the insidious Acclimatisation Society. Leadbeater's Fig-Parrot, for John Leadbeater, chief taxidermist at the National Museum in Melbourne (which McCoy also founded); McCoy named the bird for him in the same year that Gould called it *mccoyi*! Blue-faced or Red-browed Fig-Parrot or Lorilet, again unconvincing, though the female has a pale blue cheek.

C. d. marshalli (eastern Cape York), only differentiated from the previous race by Tom Iredale in 1947. Marshall's Fig-Parrot, for Captain A.J. Marshall who collected the bird in September 1942; we assume that this was the redoubtable zoologist Jock Marshall, who had not yet been accepted into the militia (where he was made captain) and may have been in Cape York. We cannot be certain and Iredale does not assist further (Iredale 1946). Northern or Cape York Fig-Parrot or Lorilet.

Cyclopsitta diophthalma (Hombron & Jacqinot, 1841) [si-clo-SIT-tuh di-off-THAL-muh]: 'two-eyed round-eyed parrot', see genus name, and from Greek *di-*, two, and *ophthalmos*, eye. Hombron and Jacquinot (1841) described the feathers at the lores as being of a shining soft blue, and said that from a distance they resemble eyes gleaming like precious stones.

PASSERINES

The passerines encountered by the Europeans were, in contrast to many of Australia's non-passerine families, mostly quite unlike anything they knew, so it was necessary either to start from scratch in naming them or to use existing names of entirely unrelated (and often quite dissimilar) species.

PITTIDAE (PASSERIFORMES): pittas

Pittidae Swainson, 1831 [PIT-ti-dee]: the Pitta family.

'Pitta' is from the Telugu language of south-eastern India and refers in general terms to any small bird; apparently – and not for the first or last time – English speakers misunderstood what they were being told. It seems to have been first used by Europeans when Louis Vieillot used it as a genus name in 1816; from there it gradually arrived in English, so that Gould was using it as preferred common name by 1848. Before that, Vigors and Horsfield (1827) referred to the Indian Pitta *Pitta brachyura* (with which they included the Australian Noisy Pitta) by Latham's name of Short-tailed Crow (from Linnaeus' *Corvus brachyurus*).

The genera

Erythropitta Bonaparte, 1854 [e-RI-thro-PIT-tuh]: red pitta, from Greek *eruthros*, red, and Vieillot's genus name 'pitta', see family introduction.

Pitta Vieillot, 1816 [PI-tuh]: 'pitta', see family introduction.

The species

Papuan Pitta (breeding migrant)

The species, as currently understood, lives in New Guinea and some migrate to breed on Cape York Peninsula. The species complex has recently been broken up, with *Erythropitta erythrogaster*, the species to which Australian birds were previously assigned, now being limited to the Philippines.

Other names: Red-bellied Pitta, from the former species name (see common name), formalised by RAOU (1978), and the one by which it is still known in Australia; Blue-breasted Pitta, also descriptive and in preferred use for most of the 20th century (RAOU 1926 to Slater 1974); Macklot's Pitta, from the species name, for Heinrich Macklot (1799–1832), Dutch taxidermist who worked in Timor and New Guinea in the late 1820s. Macklot died far too young in a fit of macho pique – his collection was destroyed in fires associated with riots in Java and he rashly attacked the rioters. Temminck's description includes a fulsome paean of praise for Macklot's work.

Erythropitta macklotii (Temminck, 1834) [e-RI-thro-PIT-tuh MAC-lot-i-ee]: 'Macklot's pitta', see genus and common names.

Hooded Pitta (vagrant)

See family introduction. From the striking all-black head (in most races) on a green body.

Pitta sordida (Müller, PLS, 1776) [PIT-tuh sor-DI-duh]: 'dirty pitta', see genus name and from Latin *sordidus*, meaning dirty, foul, squalid, referring here to the colours, especially the black and green.

Fairy Pitta (vagrant)

See family introduction. From the species name (sort of).

Pitta nympha Temminck & Schlegel, 1850 [PIT-tuh NIM-fuh]: 'nymph pitta', see genus name and from Latin *nympha*, a demi-goddess inhabiting forests and trees.

Blue-winged Pitta (vagrant)

True, but so are many other Pitta species, including two of the three regular Australian species.

Other names: This is a bird with a wide range through Indonesia and South-East Asia, so has many names that are not relevant to Australia. The commonest is perhaps Moluccan Pitta, from the (mistaken) species name.

Pitta moluccensis (Müller PLS, 1776) [PI-tuh mo-lʊk-KEN-sis]: 'Moluccan pitta', from the Maluku (Moluccan) Islands, part of Indonesia plus the suffix *-ensis*, usually indicating the place of origin of the type specimen. This was a mistake, based by Müller on the name Merle des Moluques in Buffon's (1765) *Planches Enluminées*, the origin being actually Malacca, in what is now Malaysia. Müller (1776) was not free of error here – he claimed the bird has a blue head, which, whether he was working from a bird in the hand or from Buffon's pretty accurate illustration, is a bit wide of the mark: the bird's head has fairly striking black and yellow-brown stripes.

Rainbow Pitta (breeding resident)

Certainly a very colourful bird (like all pittas) though given the predominance of black it must be said that 'rainbow' seems an odd choice; straight from the species name, but Gould didn't assist us in following his thinking.

Other names: Black-breasted Pitta, an apparently more straightforward descriptive name.

Pitta iris Gould, 1842 [PIT-tuh EE-ris]: 'rainbow pitta', see genus name, plus from Greek (and Latin) *iris*, the rainbow (also, coincidentally this time, the name of the Greek goddess of the rainbow). This bird is also iridescent (no surprise given the meaning of that word, from Latin *iris* and *-escens*, becoming – so, becoming a rainbow).

Noisy Pitta (breeding resident)

Another bird named for a characteristic that is common to many pittas, but again the name (used by Gould 1848) is simply a translation of the old species name *P. strepitans* ascribed by Coenraad Temminck, unaware that Swainson had beaten him to it. It was a confusing situation: Gould tried to unravel it and concluded that Temminck had probably got there first, but as Temminck had published his work 'in parts at irregular periods' Gould couldn't be sure, and had to rely on (admittedly pretty convincing) circumstantial evidence. We must assume that Temminck worked on an unusually garrulous corpse.

Other names: Buff-breasted Pitta, helpful only in the context of the three breeding Australian species; Painted Thrush, understandable given pittas' thrush-like ground foraging; Dragoon Bird or Dragoon, widely used in the past, apparently for its very erect posture; Anvil Bird, for its habit of consistently using a particular rock or log to crack the shells of the snails on which it strongly depends.

Pitta versicolor Swainson, 1825 [PIT-tuh vehr-SI-ko-lor]: 'pitta of various colours', see genus name, plus Latin *versicolor*, of various or changeable colours. This pitta's colours answer both those descriptions – not only does it have at least five colours as well as black and white, but some of them are iridescent, changing according to the light.

MENURIDAE (PASSERIFORMES): lyrebirds

Menuridae Lesson, 1828 [men-OO-ri-dee]: the Moon-tail family, see genus name *Menura*.

The lyrebirds are a good example of the newcomers' need to find new names. They were called pheasant for a while after settlement, in reference to the long tail – albeit to nothing else that is obvious. (All this refers to the Superb Lyrebird, because Albert's wasn't discovered until long afterwards.) A series of more or less remarkable names followed (see 'Other names'), but Lyre Pheasant and Lyre-tailed Pheasant appeared early; the transition to Lyre-bird came soon afterwards. Jean Quoy and Joseph Gaimard, surgeon-naturalists on Louis de Freycinet's 1817–20 *L'Uranie* expedition, reported that the bird was called 'Menura (see genus), also called lyre-bird and magnificent lyre, because it spreads its tail feathers in an elegant lyre' (de Freycinet 1824). George Bennett (1834) referred to the ''Native or Wood-pheasant' or 'Lyre bird' of the Colonists', implying that it was already widely used. The catch is that no live lyrebird can or does hold its tail erect in the lyre-shape depicted in many early illustrations – it is either folded horizontally behind or cascaded forward over the body in display. This is clearly a bird characterised and named from a skin.

The genus

Menura Latham, 1802 [men-OO-ruh]: 'moon-tail', from Greek *mene*, moon, and *oura*, tail. Latham (1801, 1802) described the inner web of the outside tail feathers as 'very broad, inclining to grey, but from the middle to the edge fine rufous, marked with sixteen curved marks seemingly of a darker colour, but on close inspection are perfectly transparent'. He later referred to these curved marks in English as the 'crescents' and in Latin as '*lunulis*'.

The species

Albert's Lyrebird (breeding resident)

For Queen Victoria's consort, in Gould's words 'as a slight token of respect for his personal virtues, and the liberal support he has rendered to my various ventures'; Albert did indeed have a genuine interest in science. (Although the description was Gould's, it was published in Charles Bonaparte's massive *Conspectus Generum Avium*, so Bonaparte gets credit for the name.)

Other names: Northern, Prince Albert or Prince Albert's Lyrebird.

Menura alberti Bonaparte, 1850 [men-OO-ruh AL-ber-tee]: 'Albert's moon-tail', see genus and common names.

Superb Lyrebird (breeding resident)

See family introduction. From the name *Menura superba*, applied by Major-General Thomas Davies, part-time ornithologist, whose genus name was pinched by the great John Latham, who then added the species name we now use. It is pretty clear that Latham back-dated to obtain precedence (Chisholm 1960a).

Other names: Lyre-bird, used by Gould (1848) and still used as a stand-alone name by RAOU (1913), perplexingly, given that Albert's was well known by then; Superb and New Holland Menura, used very early, from the genus name; Pheasant; Lyretail; Queen Victoria's or Victoria Lyrebird, used in the past for Victorian lyrebirds; Edward or Prince Edward's Lyrebird, used, probably by analogy with the previous name, for the population in inland south-east Queensland; Native, Wood, Botany Bay, Blue Mountains, Lyre and Lyre-tailed Pheasant; Peacock-wren, surely one of the most creative or desperate of the many compound names coined in Australia. Parkinsonian Paradise-bird (we shall pause while you digest that ...). This is not for Banks' draughtsman Sydney Parkinson but for London real estate agent John Parkinson, who won a well-stocked private museum in a lottery! The owner was getting rid of it because its maintenance was costing him too much, as poor Parkinson soon discovered. However, he made the best of it for 20 years before selling the collection. Before doing so he off-loaded some treasures, including to Frenchman Louis Vieillot and Parkinson's compatriot George Shaw, both of whom were grateful to the point of obsequiousness. Vieillot promptly coined the name Parkinson for the bird, which Shaw, bizarrely, adopted and turned into a scientific name (though it already had one, as Shaw would have known). Hence *Paradisea parkinsoniana* was published, translated of course as Parkinsonian Paradise-bird (for more detail on this and much more, see Alec Chisholm's wonderful *Romance of the Lyrebird* 1960a).

Menura novaehollandiae Latham, 1801 [men-OO-ruh no-veh-hol-LAN-di-eh]: 'New Holland moon-tail', see genus name, from Modern Latin *Nova Hollandia*, New Holland, the old name for Australia.

ATRICHORNITHIDAE (PASSERIFORMES): scrubbirds

Atrichornithidae Stejneger, 1885 [at-ri-kor-NI-thi-dee]: the Bristle-less Bird family, see genus name *Atrichornis*.

Gould (1848) described the first species discovered as Noisy Brush-bird; the terms scrub and brush were used interchangeably for low dense understorey, generally in a derogatory sense. For such an extraordinary bird it was a pretty uninteresting name, and the evolution to Scrub-bird 20 years later when Gould described the second species as Rufescent Scrub-bird (Gould 1869) was a very small leap; Gould didn't explain the alteration. Newton, in his 1896 *Dictionary of Birds*, commented somewhat sadly, 'the name (for want of a better, for it is not very distinctive) ...'.

The genus

Atrichornis Stejneger, 1885 [at-ri-KOR-nis]: 'bristle-less bird' from Greek *a-*, without, *thrix/trikhos*, hair, and *ornis*, bird, by contrast with Bristlebirds, a close relative. Based on Gould's (1844) name *Atrichia*, which was found to be preoccupied. We note that it seems that Gould had actually published another name, *Atricha*, some 5 weeks before this one. Although this name has not hitherto been taken up, it is still available and this genus name may possibly be reinstated one day, though the usefulness of any change to it now has been the subject of intense debate (Olson 1998; Schodde and Bock 1998).

The species

Rufous Scrubbird (breeding resident)

See family introduction; it is dark rufous and the name comes straight from the species name.

Other names: Rufous Scrub-bird, the form mostly used in Australia; Rufescent Scrub-bird, see family introduction; Eastern Scrub-bird, to distinguish from the Western Australian Noisy Scrubbird; Mocking-bird, Mouse-bird, Mystery-bird, for its legendary skulking behaviour and apparent ventriloquial skills.

Atrichornis rufescens (Ramsay, EP, 1866) [at-ri-KOR-nis roo-FE-senz]: 'reddish bristle-less bird', see genus name, plus Latin *rufus*, red and *-escens*, growing or becoming (i.e. 'ish'). It is certainly a bit more reddish-brown overall than its genus-mate below.

Noisy Scrubbird (breeding resident)

See family introduction; 'noisy' straight from the species name, used from the start.

Other names: Noisy Scrub-bird, the form mostly used in Australia; Noisy Brush-bird, see family introduction; Western Scrub-bird, to distinguish from the eastern Australian Rufous Scrubbird.

Atrichornis clamosus (Gould, 1844) [at-ri-KOR-nis kla-MOH-sʊs]: 'noisy bristle-less bird', see genus name plus Latin *clamosus*, full of noise, clamorous, from *clamo*, shout aloud, complain with a loud voice. The volume of sound that this small bird can produce is quite startling.

PTILONORHYNCHIDAE (PASSERIFORMES): bowerbirds

Ptilonorhynchidae Gray GR, 1841 [ti-li-no-RIN-ki-dee]: the Feather-bill family, see genus name *Ptilonorhynchus*.

Gould (1848) seems to have introduced the term 'bower-bird' for the group; before that, Satin-bird and Regent Bird were the general terms for Satin and Regent Bowerbirds, the only ones that would have been familiar to most colonists (Bennett 1834). The name was taken from the remarkably constructed male display arena, which Gould didn't understand; he referred to it as a 'bower-like structure … for the purpose of a playing-ground or hall of assembly'.

'Catbird' is for the extraordinary yowling calls of the two *Ailuroedus* species; Gould used 'Cat Bird', noting that this was the term 'of the Colonists of New South Wales'.

The genera

Ailuroedus Cabanis, 1851 [ie-loo-ROY-dʊs]: 'cat singer', from Greek *ailouros*, a cat, and ōdos, a singer (Latin *oedus*). Singer? Perhaps … but the voice certainly deserved acknowledgment.

Scenopoeetes Coues, 1891 [seh-no-POY-e-tehz]: 'stage-maker', from Greek *skēnē*, a stage, and *poieō*, to make. Rather than making a bower, the bird clears a space, known as a court, on the ground, which it then decorates.

Prionodura De Vis, 1883 [pri-oh-no-DU-ruh]: 'saw-tail' from Greek *prionōdēs*, saw-like, and *oura*, tail, for what De Vis described as the exsertion of the slightly spiny tail feathers.

Sericulus Swainson, 1825 [seh-RI-kʊ-lʊs]: 'little silkie', from Greek *sērikon*, silk, and *sērikos*, silken, in Latin *sericus*, with *-ulus* diminutive ending. *Sēres* was also what the Greeks (and Romans) called the Chinese, from whom silk was obtained.

Ptilonorhynchus Kuhl, 1820 [ti-lo-no-RIN-kʊs]: 'feather-bill' (not to be confused with *ptilorynchus*, species name for the Oriental Honey-buzzard). From Greek *ptilon*, feather or down, and *rhunkhos*, bill. Satin Bowerbirds have a small patch of feathers extending from the forehead down over the root of the bill; however, *Prionodura* also has a distinctly feathered bill.

Chlamydera Gould, 1837 [kla-MI-de-ruh]: 'short neck-cloak', from Greek *khlamus*, a short mantle or cloak, and *dera*, neck. In the original illustration from Gould, the nuchal crest is spread right across the nape exactly like a little cloak – not quite what you see under normal circumstances, because it's tucked away unless produced for display.

The species

Green Catbird (breeding resident)

See family introduction. Not a useful distinction from the next species, but for some time it was the only one known.

Other names: Catbird, Cat-bird or Cat Bird, because until relatively recently the two species were often regarded as one; Large-billed Catbird, an imprecise translation of the species name.

Ailuroedus crassirostris (Paykull, 1815) [ie-loo-ROY-dʊs kras-si-ROS-tris]: 'thick-billed cat singer', see genus name, and from Latin *crassus*, thick or heavy, and *rostrum*, bill.

Spotted Catbird (breeding resident)

See family introduction. Much more speckled than the Green Catbird.

Other names: As per Green Catbird; Australasian Catbird, for its range into New Guinea and associated islands; Queensland Catbird, for its far northern range, though the previous species is found in southern Queensland; Black-eared Catbird, from the species name.

Ailuroedus melanotis (Ramsay EP, 1875) [ie-loo-ROY-dʊs me-la-NOH-tis]: 'black-eared cat singer', see genus name, and from Greek *melas*, black, and *ous/ōtos*, ear, for its brownish-black ear coverts.

Tooth-billed Bowerbird (breeding resident)

See family introduction. For the distinctive serrated bill for cutting leaves (not very obvious in the field).

Other names: Toothbill or Tooth-billed Catbird (though it's not a yowler); Stagemaker, from the genus name; Leaf-turner or Queensland Gardener, for a key part of its display behaviour, involving arranging large snipped-off leaves.

Scenopoeetes dentirostris Ramsay EP, 1876 [seh-no-POY-e-tehz den-ti-ROS-tris]: 'tooth-billed stagemaker', see genus name, plus Latin *dens/dentis*, tooth, and *rostrum*, bill – see common name. Simple name, charming image!

Golden Bowerbird (breeding resident)

See family introduction. Appropriate, but for the male only.

Other names: Golden or Queensland Gardener, for its use of flowers to decorate the amazing maypole-bower; Newton's Bowerbird, from the species name; Meston's Bowerbird, for Archibald Meston, a farmer and south Queensland politician who in 1881 moved to north Queensland and led an expedition to Mount Bellenden Ker in 1889. They collected a specimen of the bird there, 6 years too late to be the type specimen; however, it is evident from De Vis' description (see species name) that his specimen was a female or immature – it seems that Meston's was the first adult male known to science (Norton 1907; Kloot 1986).

Prionodura newtoniana De Vis, 1883 [pri-oh-no-DU-ruh nyoo-to-ni-AH-nuh]: 'Newton's saw-tail', see genus name, and for Alfred Newton (1829–1907), widely travelled British zoologist who helped found the British Ornithologists' Union and wrote the monumental *Dictionary of Birds* (to which we have regularly referred) in 1896. Charles De Vis was an exact contemporary of Newton and, though he spent much of his life in Australia, would have known him. The bird (not an adult male, as already noted) seemed dull in looks to De Vis, who remarked, 'In honouring this Bower Bird with the name of Professor Newton, it is hoped that the interest attaching to it will be accepted as an equivalent for its plentiful lack of beauty' (De Vis 1883). He was, in fact, doubtful about its sex, and aware of the gaps in his observations, going so far as to invite local residents to do 'a good turn to science' by looking for the bird and studying it.

Regent Bowerbird (breeding resident)

See family introduction. Much myth and misconception surrounds this name, generally regarding a supposed association of the Prince Regent, later George IV, with black and gold colours; no such association is known or was suggested at the time. Thomas Skottowe, commander of the Newcastle penal settlement, later collaborated with the convict artist Richard Browne in an illustrated manuscript that featured birds, including this one, then known (Lewin 1808) as Golden-crowned Honeysucker. As one of the few fans of the gambling, philandering, laudanum-addicted alcoholic George, Skottowe took it upon himself to name the poor bird for the royal personage, as he claimed to have shot the specimen illustrated in the manuscript on the day that Australia received 'news of the Regency restrictions on His Royal Highness the Prince Regent having been taken off'. The name stuck, due to Browne's use of it on many paintings, and later became applied to other species with similar colouring – see also Regent Honeyeater and Regent Parrot ('A. stentoreus' 2004). Oddly, Regent Bowerbird didn't replace the universally used Regent-bird until the 20th century, long after Bowerbird was being used for all other species; indeed the first use of it seems to have been in the RAOU's second *Official Checklist* in 1926.

Other names: Golden-headed Sericulus, used by Jardine and Selby (1826–35, vol. 1); Regent Bird, as used by Gould (1848); Australian Regent Bowerbird, a nod to the closely related New Guinea species; Golden-crowned Honeysucker, see common name; King Honeysucker, used by Lewin (1822), apparently for Governor Philip Gidley King (Iredale 1933), though it was well after his death; Golden-crowned Honeyeater, Latham 1823.

Sericulus chrysocephalus (Lewin, 1808) [se-RI-kʊ-lʊs kri-so-SE-fa-lʊs]: 'Little golden-headed silkie', see genus name, and from Greek *khrusos/khruso-*, gold/golden, and *kephalē*, head. A pretty apt name (though the bird itself is not little) given its gorgeous shiny black and golden plumage. Greek *sērikon* has a subsidiary meaning of 'red pigment' (from Syria, where it is said that Phoenicians created a variety of shellfish dyes ranging from blue through purple to red or pink), which reminds us of the stain of red on this bird's forehead and the faint marking of red across the back of the neck and shoulders.

Satin Bowerbird (breeding resident)

See family introduction. From Gould (1848), who noted that the 'Colonists of New South Wales' used Satin Bird, doubtless for the sheen of the mature male's plumage.

Other names: Satin Bird, see common name; Purple Satin; Satin Grackle presumably for the very vaguely superficially similar group of American Blackbirds (Icteridae), applied by Latham (1823); Lesser Satin Bowerbird, for race *minor* of the Queensland Wet Tropics, described as a species, *P. minor*, by Campbell in 1912 and recognised as such by RAOU (1913).

Ptilonorhynchus violaceus (Vieillot, 1816) [ti-lo-no-RIN-kʊs vi-o-LAY-se-ʊs]: 'violet-coloured feather-bill', see genus name, and Latin *violaceus*, violet-coloured from *viola*, a violet. Not sure we've ever seen a violet quite as splendid as this – the sheen of the plumage in the right light is stunning. And that feather-bill is definitely there!

Western Bowerbird (breeding resident)

See family introduction. Found in central Australia and Western Australia, west of the range of the closely related Spotted Bowerbird.

Other names: In recent times (CSIRO 1969; RAOU 1978; Pizzey 1980), this species and Spotted Bowerbird were regarded (though not universally) as the same, so some names applied to both. Western Spotted Bowerbird; Guttated Bowerbird, used by Gould (1869) for the species name – he did recognise that they were different; Yellow-spotted (RAOU 1913) or Pale-spotted Bowerbird.

Chlamydera guttata Gould, 1862 [kla-MI-de-ruh gʊ-TAH-tuh]: 'short neck-cloak with spots', see genus name, and from Latin *gutta*, a droplet. Ironically, spottier than the official Spotted Bowerbird.

Great Bowerbird (breeding resident)

See family introduction. Australia's largest bowerbird, name applied by Gould (1848).

Other names: Queensland Bowerbird, which it is, though not solely and not uniquely – however, Gould in 1869 described the Queensland population as a separate species *Chlamydera orientalis*, now recognised at subspecies level; RAOU (1913) still agreed it was a species, using this name; Ruffed Ptilonorhychus, for the species name and former genus name, used by Jardine and Selby (1826–35, vol. 2); Great Grey, Lilac-naped, Great Spotted Bowerbird, all descriptive; Bone Bird, for one of its favourite bower adornments.

Chlamydera nuchalis (Jardine & Selby, 1830) [kla-MI-de-ruh noo-KAH-lis]: 'naped short neck-cloak', see genus name, and from Mediaeval Latin *nucha* (itself from Arabic), originally meaning spinal cord but later taking on the meaning of nape of the neck. This refers, of course, to the male bird's striking pink or lilac nape crest. Note that the Spotted and Western Bowerbirds have similar crests.

Spotted Bowerbird (breeding resident)

See family introduction. For the buffy-spotted back, applied by Gould (1848) from his species name.

Other names: Cabbage Bird, we assume for bowerbirds' general fondness for leaves and gardens; Mimic-bird, but all bowerbirds are good at mimicry.

Chlamydera maculata (Gould, 1837) [kla-MI-de-ruh ma-kʊ-LAH-tuh]: another 'short neck-cloak with spots', see genus name, and from Latin *macula*, a spot or stain (as in 'immaculate'). Not as spotty as the Western Bowerbird.

Fawn-breasted Bowerbird (breeding resident)

See family introduction. It's actually the belly rather than the breast that is fawn, as the species name points out, though curiously Gould (1869) is responsible for both names; bowdlerism ruled in Victorian times and 'belly' just wasn't 'nice'.

Other names: Buff-breasted Bowerbird. Meyer's Bowerbird, cited by HANZAB (1990–2006, vol. 7) and Rowland (2008), though we can find no other reference to the name, but perhaps it has had some usage in the bird's New Guinea range (though separate races are not recognised), in which case the connection could be Adolf Meyer, German ornithologist who worked in the area in the late 19th century, but this is mere surmise.

Chlamydera cerviniventris (Gould, 1850) [kla-MI-de-ruh ser-vi-ni-VEN-tris]: 'short neck-cloak with a fawn belly', see genus name, and from Latin *cervinus*, pertaining to a stag, *cervus*, a stag or deer, and *venter/ventris*, belly (see common name). The colour is a robust fawn – a buff or tawny colour.

CLIMACTERIDAE (PASSERIFORMES): Australasian treecreepers

Climacteridae Selys-Longchamps, 1839 [kli-mak-TE-ri-dee]: the Ladder-rung family, see genus name *Climacteris*.

It was not until the 1960s that it was recognised that this ancient Australian group was not just a colonial branch of the widespread Northern Hemisphere family Certhiidae, the treecreepers or creepers. Their name comes from their feeding habit – shared, quite coincidentally, with the Climacteridae – of moving across the surface of tree trunks and extracting insects from bark crevices.

The genera

Cormobates Mathews, 1922 [kor-mo-BAH-tehz]: 'tree-treader', from Greek *kormos*, tree-trunk or log, and *batēs*, a walker, from *bainō*, walk or step.

Climacteris Temminck, 1820 [kli-MAK-te-ris]: 'ladder-rung [bird]', from Greek *klimax*, a stairway or ladder, and *klimaktēr*, a rung. The allusion is to their habit of spiralling up tree trunks.

The species

White-throated Treecreeper (breeding resident)

A satisfactory distinction from all other treecreepers except perhaps Red-browed; Gould was using the name in 1848.

Other names: Woodpecker, for its ecological equivalents elsewhere in the world; New Holland Nuthatch (Latham 1823); Bar-tailed Honeyeater (Latham 1823); Dirigang Creeper (Latham 1801) and Dirigang Honeyeater (Latham 1823), this supposedly being the 'native name'; Little or Little White-throated Treecreeper, solely for race *minor*, from the Queensland Wet Tropics hinterland, described by Ramsay in 1891 but recognised as a species (*Climacteris minor*) well into the 20th century (RAOU 1926).

Cormobates leucophaea (Latham 1801) [kor-mo-BAH-tehz loo-ko-FEH-uh]: 'dusky-white tree-treader', see genus name, and from Greek *leukophaios*, whitish-grey, from *leukos*, white, and *phaios*, dusky.

Red-browed Treecreeper (breeding resident)

Usefully descriptive of a unique characteristic among treecreepers.

Other names: Red-eyebrowed Treecreeper, used by Gould (1848), in reference at least to his species name; Woodpecker (see White-browed Treecreeper). (CSIRO 1969 suggested White-browed Treecreeper, but we're sure this must be in error.)

Climacteris erythrops Gould, 1841 [kli-MAK-te-ris e-RI-throps]: 'red-eyed ladder-climber', see genus name, plus Greek *eruthros*, red, and ōps, eye (or face) – the eye itself is black but it is encircled with red.

White-browed Treecreeper (breeding resident)

The brow is more distinctively white than that of the generally similar Brown Treecreeper.

Other names: White-eyebrowed Treecreeper.

Climacteris affinis Blyth, 1864 [kli-MAK-te-ris af-FEE-nis]: 'related ladder-climber', see genus name, plus Latin *affinis*, related, in this case to the Red-browed Treecreeper (Blyth 1864).

Rufous Treecreeper (breeding resident)

The only fully rufous treecreeper, named by Gould straight from his species name.

Other names: Woodpecker (see White-browed Treecreeper); Allied Rufous Treecreeper, see Black-tailed Treecreeper; Wheelbarrow, which seems to stem solely from a comment by H.E. Hill of Guildford, Perth, on a visit to a railway survey camp between Perth and Albany. He commented that the Rufous Treecreeper 'has a peculiar cry, which reminded me of that of the Rufous Bristle-bird, and, in fact, the boys [sons of his host] immediately christened it the 'wheelbarrow'' (Hill 1903). We find no other evidence that it gained public use, though it is cited by HANZAB (1990–2006, vol. 5).

Climacteris rufa Gould, 1841 [kli-MAK-te-ris ROO-fuh]: 'reddish ladder-climber', see genus name, plus Latin *rufus*, red or reddish.

Brown Treecreeper (breeding resident)

A name that reeks of desperation, though used by Gould (1848); perhaps he was using a popular name. It is no more or less brown than most treecreeper species.

Other names: Woodpecker (see White-browed Treecreeper); Buff-winged Honeyeater, Latham 1823 – the adjective at least refers to a conspicuous character; Black or Black-backed Treecreeper, for the dark race *melanotus* of north-east Queensland (regarded by Gould as a full species *C. melanota*, and probably the last species collected by Gould's collector John Gilbert on 28 June 1845, the day of his violent death near the Gulf of Carpentaria – this death was described by Gould in 1848 as an 'untoward event') – the species was well recognised into the 20th century (RAOU 1926).

Climacteris picumnus Temminck, 1824 [kli-MAK-te-ris pi-KŬM-nŭs]: 'little woodpecker ladder-climber', see genus name, plus a diminutive of Latin *picus*, the woodpecker. Picumnus was also a Roman god: a woodpecker personified and, with his brother Pilumnus, a minor god of married people and little children. The connection between the two aspects – woodpecker and god – is far from obvious.

Black-tailed Treecreeper (breeding resident)

An accurate name, applied by Gould (1848) straight from his species name, though overall the bird is distinctively dark.

Other names: Chestnut-bellied Treecreeper, for another characteristic; Allied Treecreeper, for the Pilbara race *wellsi*, formerly considered a full species *C. wellsi* (RAOU 1926). This species, Brown and Rufous Treecreepers are often regarded as forming a superspecies, so *wellsi* was also 'allied' to them; see also Rufous Treecreeper.

Climacteris melanura Gould, 1843 [kli-MAK-te-ris me-la-NOO-ruh]: 'black-tailed ladder-climber', see genus name, plus Greek *melanouros*, black-tailed.

MALURIDAE (PASSERIFORMES): fairywrens, emuwrens, grasswrens

Maluridae Swainson, 1831 [mal-OO-ri-dee]: the Delicate-tail family, from the genus name *Malurus*.

'Wren' was used from the earliest days of the colony, notably for the ubiquitous Superb Fairywren, because it reminded homesick settlers of the utterly unrelated (and dissimilar except for having a cocked tail and living in bushes) Eurasian Wren *Troglodytes troglodytes*. It soon became tacked on, with various descriptors (including grass-, emu-, scrub-, fern-, etc.), to a wealth of more or less related groups. The word dates from Old English *wrenna*; there is the suggestion (Lockwood 1984) that the reference was to a cocked tail, but that appears to be speculative.

'Fairywren' (or 'Fairy-wren' as is still the generally preferred name in Australia) on the other hand is a relatively recent coining, which doesn't seem to have existed before the 20th century. In 1913 the RAOU was recommending Wren-Warbler (RAOU 1913) but by the next list it had reverted to the simple 'Wren' (RAOU 1926). The *Australian National Dictionary*'s first reference to 'Fairy-wren' is in the Sydney *Bulletin* in 1928, citing Daisy Bates, but without making clear whether the term was that of the writer or Mrs Bates. The sense of 'fairy' is clearly that of small or delicate, presumably to reduce confusion with the other types of wren. The influential Tom Iredale claimed credit for the term, 'Some years ago I suggested that Blue Wrens be called Fairy-wrens, and that name has been accepted by recent workers' (Iredale 1939). Nonetheless, much more recently Slater (1974) and Pizzey (1980), in the earliest of the modern field guides, while using 'fairy wren' as a group name still used just 'wren' for the individual species. CSIRO (1969) also used just 'wren'. Perhaps Rowley (1974) had a handle on it when he wrote, "fairy-wrens' has been used extensively but is too prissy for the average Australian to use in everyday speech'. It wasn't until 1978 that that notion of prissiness was challenged by the RAOU's influential *Recommended English Names* (RAOU 1978) in which it noted an 'undercurrent of tradition … which we would like to support' (though its oldest citation for this tradition was 1949).

'Emu-wren' on the other hand goes back to the earliest days of the colony; the first reference seems to be a painting by Thomas Watling (dated between 1792 and 1797). The handwritten notes on the drawing refer to 'the resemblance of the Feathers of the tail … to those of the Cassuary in New South Wales and denominated the Emue, or Cassuary Titmouse' (Watling 1792–97). These filamentous tail feathers are indeed very reminiscent of the wispy long unbarbed aftershaft feathers of an emu. Gould (1848) was using Emu Wren as a first-choice name, and didn't offer any alternatives. (And please don't ask us why in the IOC list emu-wren has a hyphen, but fairywren and grasswren don't.)

'Grasswren' is more recent; Gould didn't use it 1848 (he referred to the species simply as 'wrens') but Hall was using Grass-Wren by 1899, without any explanation. The reference is clearly to the habitat – all species use arid land grasses, particularly porcupine grasses *Triodia* spp., as shelter. They are too cryptic to have attained a general popular name by then, so it could be that Hall introduced the name. Grass-wren has continued to be commonly used, as well as grasswren.

The genera

Malurus Vieillot, 1816 [mal-OO-rʊs]: 'delicate tail', said to be from Greek *malos*, soft or delicate, and *oura*, tail, though there is a little confusion over this. There is a derivation purporting to be from Vieillot that uses Greek *mala*, very or exceedingly – this is rather tempting, since they could certainly be said to be exceedingly tailed little birds! However, Vieillot clearly stated that he derived *Malurus* from Greek *malos*, which he translated into Latin *tener*, soft or delicate, and Greek *oura* or Latin *cauda*, tail (Vieillot 1816–19).

Stipiturus Lesson, 1831 [sti-pi-TOO-rʊs]: 'stick tail', from Latin *stipes/stipitis*, a tree-trunk or branch (from Sanskrit *sthapa-jami*, to cause to stand), and Greek *oura*, tail. HANZAB (1990–2006, vol. 5) extended the meaning of *stipes* to mean twigs, likening thickened shafts of the birds' tail feathers to a bunch of these. Lesson (1831) called it the '*Queue gazée*', referring to gauze or to something being veiled, and accordingly the birds are now called '*Queue-de-gaze*' in French (*gaze* = gauze, not *gaz* = gas).

Amytornis Stejneger, 1885 [a-mi-TOR-nis]: 'Amytis bird', from the old – and, as it turned out, pre-occupied – genus name *Amytis* Lesson 1831. There are at least three known Amytises, all related. One was married to Nebuchadnezzar of Babylon, who is said to have built the Hanging Gardens for her. Next was her niece, who was married to Cyrus the Great, founder of the Persian empire. Finally there was the daughter of Xerxes I, a grandson of Cyrus from his first wife. The life of this third Amytis is full of tales of lust and horrible violence; Dinon, another Greek chronicler of ancient Persia, called her the most beautiful and licentious woman of Asia (*Encyclopædia Iranica*). What Lesson thought any of this had to do with a bird is anyone's guess: he simply said that Amytis is a 'nom mythologique'. The *ornis* part was added by Stejneger, retaining the old name but revalidating it by the addition (from Greek *ornis*, bird).

The species

Lovely Fairywren (breeding resident)

See family introduction, including note on 'Fairy-wren'. From Gould's species name, which he justifies with 'the present species yields to none of them [the other *Malurus* species] either in the elegance of its form or in the beauty of its plumage' (Gould 1869). (Curiously and quite atypically Gould erred, in the very next plate and account, in describing the distinctive and unusual female of this species as the male of a different species, which he called *M. hypoleucus*, Fawn-breasted Superb Warbler; both plates contain a 'female' which looks like a generic *Malurus* female and we suspect he was bluffing.)

Other names: Lovely Fairy-wren, the preferred Australian form; Lovely Wren, Wren-warbler or Superb Warbler; Purple-breasted Fairy-wren, for no evident reason – in fact we wonder if HANZAB (1990–2006, vol. 5, the only source of this one) hadn't meant to refer to Blue-breasted Fairywren.

Malurus amabilis Gould, 1852 [mal-OO-rʊs a-MAH-bi-lis]: 'lovely delicate-tail', see genus name, and from Latin *amabilis*, worthy of love, lovely.

Purple-backed Fairywren (breeding resident)

See family introduction, including note on 'Fairy-wren'. Perhaps as helpful as most fairywren names.

Other names. Purple-backed Fairy-wren, the preferred Australian form; Purple-backed Wren, Wren-warbler or Fairy Warbler; Bernier Island Blue Wren or Wren-warbler or Shark Bay Variegated Fairy-wren, for race *bernieri* (formerly a species, by Ogilvie-Grant 1909 and still recognised by RAOU 1913) limited to Bernier and Dornier islands in Shark Bay, Western Australia; Dulcet Wren, somewhat forcedly, for Arnhem Land race *dulcis*, described as a full species by Gregory Mathews in 1908 and still recognised by RAOU (1913); Northern Variegated Wren-warbler, for Mathews' subspecies *mastersi* from the Northern Territory (Mathews 1912a), cited as a species by RAOU (1913) in the Provisional List, not recognised thereafter; Lavender-flanked Wren, Fairy-wren or Wren-warbler, for the same race.

Malurus assimilis (North, 1901) [mal-OO-rʊs a-SIM-i-lis]: 'similar delicate-tail', see genus name; North names this species *assimilis*, stating that among the section of *Malurus* with chestnut shoulders it is 'more closely allied' to *M. lamberti*, though he also states that in the colours of the head, ear coverts and mantle, the bird 'more closely resembles *M. pulcherrimus*'.

Variegated Fairywren (breeding resident)

See family introduction, including note on 'Fairy-wren'. From John Lewin's name Variegated Warbler, for the mix of colours in the male but also a reference to the fact that it was believed to be a colour variant of Superb Fairy-wren.

Other names: Variegated Fairy-wren, the preferred Australian form; Variegated Wren or Wren-warbler; Lambert's Malurus, Wren, Fairy-wren or Superb Warbler, from the species name, and the name by which it was long known (e.g. in Gould 1848).

Malurus lamberti Vigors & Horsfield, 1827 [mal-OO-rʊs LAM-ber-tee]: 'Lambert's delicate-tail', see genus name, and for Aylmer Bourke Lambert (1761–1842), an English botanist with a strong interest in conifers (not wrens, or indeed birds at all). However, he was a founding member of the Linnean Society and a contemporary of Nicholas Vigors and Thomas Horsfield who were also active members, and they may have been friends when they named it for him.

Blue-breasted Fairywren (breeding resident)

See family introduction, including note on 'Fairywren'. Perhaps in distinction from the similar Variegated Wren, which has a black breast, but the blue is so dark that it often seems black.

Other names: Blue-breasted Fairy-wren (the preferred Australian form), Wren, Wren-warbler or Superb Warbler; Purple-breasted Fairy-wren; Beautiful Wren, from the species name, explained thus by Gould (1848), 'a more beautiful bird than the present species … can scarcely be imagined'.

Malurus pulcherrimus Gould, 1844 [mal-OO-rʊs pʊl-KE-ri-mʊs]: 'most beautiful delicate-tail', see genus name, and from Latin *pulcher*, beautiful, with superlative ending, *rimus*. Gould was clearly bowled over by the fairywrens and found it hard to decide which was the most beautiful (see Lovely Fairywren, but perhaps by that time he'd forgotten what he said about this one). We do understand.

Red-winged Fairywren (breeding resident)

See family introduction, including note on 'Fairy-wren'. A curious name, though the rusty shoulder patch is larger than that of the adjacent Blue-breasted Fairywren.

Other names: Red-winged Fairy-wren (the preferred Australian form), Wren, Wren-warbler or Blue Wren; Graceful Wren, used by Gould (1848) from his species name; Graceful Blue Wren or Superb Warbler, Elegant Wren or Elegant Blue Wren, from the species name; Marsh Wren, for its wetter habitat than the previous species.

Malurus elegans Gould, 1837 [mal-OO-rʊs EH-le-ganz]: 'elegant delicate-tail', see genus name, and Latin *elegans*, elegant, fine, tasteful.

Superb Fairywren (breeding resident)

See family introduction, including note on 'Fairywren'. Although we sympathise with anyone who wishes to wax lyrical over a bird, names such as 'Superb' are not at all helpful (and are by implication dismissive of their more conventionally named relatives!). This one, although dating back to the 18th century, was eschewed by later ornithologists (from Gould 1848 to Cayley 1931) who preferred the simpler Blue Wren. It was coined by George Shaw in White (1790); he used Superb Warbler and the name *Motacilla superba* (for a plate that showed both this species and Variegated Fairy-wren). Leach (1911) resurrected it as Superb-warbler, and Slater (1974) in the first modern field guide to Australian passerines used Superb Blue Wren. As the 'first' fairywren this was the basis for all other wren common names, so many others became 'adjective + Blue Wren' or 'adjective + Superb Warbler', influenced by Leach in the early 20th century.

Other names: As one of the most familiar Australian birds, this species has attracted a wealth of folk names. Superb Fairy-wren (the preferred Australian form), Warbler or Malurus; Long-tailed Superb Warbler, see common name; Blue Wren, the most widely used name over the decades and a first-choice name for a long time; Superb Blue Wren; Fairy Wren, see family introduction; Bluecap, Bluey, Bluebonnet, Blue-head, Blue-tit, Dark Blue Wren; Jenny-wren, an old English name for Eurasian Wren, a human name implying familiarity and affection, like Willie Wagtail; Mormon Wren, implying polygamy, largely through an assumption that the immatures and eclipse plumage males were all females; Cocktail, which Littler (1903) reported was a familiar name, along with Bluecap, in Tasmania at the time. Blue Wren-warbler (RAOU 1913); Gould's Wren, for the Tasmanian and Bass Strait populations, which Gould described as a separate species, *Malurus longicaudus* (though he named it Long-tailed Wren, from his species name) – note that this name is more familiarly used for the Mexican White-tailed

Wren *Troglodytes leucogastra*. Long-tailed Blue Wren or Wren-warbler, still in use in the early 1900s for Tasmanian wrens, *M. longicaudus* (Littler 1903; RAOU 1913); Silvery-blue Wren-warbler, for former species *M. cyanochlamys* (Sharpe 1881) of south-east Queensland (used by RAOU 1913); Dark Blue Wren-warbler, for Campbell's 1901 short-lived 'species' *M. elizabethae* of King Island (though still recognised by RAOU 1913), still now recognised at race level; Flinders Wren-warbler, for Mathews' 1913 Flinders Island subspecies *M. samueli*, recognised as a full species by RAOU (1913) in the Provisional List, now seen as a race.

Malurus cyaneus (Ellis, 1782) [mal-OO-rʊs si-A-ne-ʊs]: 'blue delicate-tail', see genus name, and from Latin *cyaneus*, dark blue or sea-blue – the word Pliny (77–79 AD) used to describe the colour of the kingfisher. The range of blues on this little bird is well covered by this name.

Splendid Fairywren (breeding resident)

See family introduction, including note on 'Fairywren'. See also Superb Fairywren. Straight from Jean René Quoy and Joseph Gaimard's species name; it's nice – and totally understandable – that they were smitten, but a bit more imagination would have been helpful.

Other names: The three races were long described as separate species, with their own names.

Race *splendens*: from a human point of view, this is the south-western equivalent of the Superb Fairywren. Splendid Fairy-wren (the preferred Australian form), Wren and Splendid Blue Wren; Splendid Warbler, Blue or Mormon Wren, see Superb Fairywren; Banded Wren (used by Gould 1848 and RAOU 1913), Blue Wren, Wren-warbler or Fairy Warbler, for the broad black breastband across the otherwise all-blue undersides. See species name.

Race *melanotus*: Black-backed Fairy-wren (the preferred Australian form), Wren, Blue Wren, Fairywren, Wren-warbler, Warbler or Superb Warbler, not a useful distinguisher from the next race. From Greek *melas*, black, and *nōton*, the back.

Race *callainus*: Turquoise Fairy-wren (the preferred Australian form), Wren, Blue Wren, Fairywren, Wren-warbler, Warbler or Superb Warbler, for the turquoise cap; Turquoisine Superb Warbler (Gould 1848); Dark Turquoise Wren-warbler, for formerly recognised species *whitei* of South Australia, described by Archibald Campbell in 1902 and recognised by RAOU (1913). From Latin *caillainus*, turquoise coloured; Latin *callaïs* or Greek *kalais*, greenish-blue precious stone, turquoise.

Malurus splendens (Quoy & Gaimard, 1830) [mal-OO-rʊs SPLEN-denz]: 'bright shining delicate-tail', see genus name, and from Latin *splendeo/splendens*, being bright or shining – astonishingly so if you are lucky enough to see it with the sun on it!

Purple-crowned Fairywren (breeding resident)

See family introduction, including note on 'Fairywren'. For its most obvious character, in use by the end of the 19th century (Morris 1898).

Other names: Purple-crowned Fairy-wren (the preferred Australian form); Purple-crowned, Lilac-crowned, Lilac-wreathed or Mauve-crowned Wren; Purple-crowned Wren-warbler; Crowned Wren (as in the species name, as used by Gould 1848) or Crowned Superb Warbler (see Superb Fairywren).

Malurus coronatus Gould, 1858 [mal-OO-rʊs ko-ro-NAH-tʊs]: 'crowned delicate-tail', see genus name, and from Latin *corona*, crown or halo, described by Gould (1857) as 'crown of the head rich lilac purple, with a triangular spot of black in the centre' – creating a little circle, extraordinary if seen from above.

Red-backed Fairywren (breeding resident)

See family introduction, including note on 'Fairywren'. The red back against black background shines like a light through the bush.

Other names: Red-backed Fairy-wren (the preferred Australian form); Scarlet-backed Warbler, used by Lewin in 1808; Crimson-backed or Scarlet-backed Superb Warbler (see Superb Fairy-wren); Red, Blood, Red-backed, Red-winged, Scarlet-backed Wren or Wren-warbler; Orange-backed Wren or Wren-warbler (RAOU 1913), used while 'Red-backed' was reserved for Gould's northern Australian species *M. cruentatus* (which he later agreed was a synonym of *M. brownii*, see below, and is now not recognised at any level) – the two species were used into the 20th century (e.g. RAOU 1913) though not thereafter; Melville Wren-warbler, for Mathews' 1912 subspecies *M. melvillensis* from Melville Island, cited by RAOU (1913) in the Provisional List as a species, not now recognised at any level; Blood-backed Wren-warbler; Black-headed Wren, as used by Gould 1848 from the species name, though why Latham wasn't more struck by the red back is anyone's guess. Brown's Wren or Malurus, from Vigors and Horsfield's 1827 *M. brownii* name commemorating the great Robert Brown, eminent Scottish botanist on the *Investigator*, who shot the specimen at Broad Sound and donated it to the Linnean Society; on the face of it, it seems odd that Vigors and Horsfield weren't aware of Latham's older name, but in fact they considered it to be a different species and as late as 1848 Gould concurred with them. Elfin Wren or Elfin-wren, a term also introduced by Tom Iredale, alongside his more successful coining Fairy-wren (see family introduction), to distinguish this species that, unlike the fairywrens, has no blue on it (Iredale 1939); it's not apparent that anyone else ever used it.

Malurus melanocephalus (Latham, 1802) [mal-OO-rʊs me-la-no-SE-fa-lʊs]: 'black-headed delicate-tail', see genus name, and from Greek *melanokephalos* from *melas*, black, and *kephalē*, head – as already stated, certainly not a feature you usually notice first.

White-winged Fairywren (breeding resident)

See family introduction, including note on 'Fairy-wren'. Its wings are startlingly white! Straight from the species name.

Other names: White-winged Fairy-wren (the preferred Australian form); Blue-and-white Wren or Fairy-wren, long the widely preferred name; White-backed Wren, Fairy-wren, Wren-warbler or Superb Warbler (see family introduction); White-winged Wren, Warbler or Superb Warbler. Pied or Black-and-white Wren or Black-and-white Wren-warbler for black and white race *leucopterus* from Dirk Hartog and Barrow islands, named by Dumont (1824) and long recognised as a separate species (RAOU 1926).

Malurus leucopterus Quoy & Gaimard, 1824 [mal-OO-rʊs loo-ko-TE-rʊs]: 'white-winged delicate-tail', see genus name, and from Greek *leukopteros*, from *leukos*, white, and *pteron*, wing.

Southern Emu-wren (breeding resident)

See family introduction. Found throughout much of the southern and south-eastern coast.

Other names: Emu Wren, as until the 20th century it was the only species recognised; Soft-tailed Flycatcher (Latham 1801); Button-grass Wren, for its Tasmanian habitat in particular, in the *Gymnoschoenus sphaerocephalus* plains; Button-grass Moth, an unlikely name which apparently had some currency in Tasmania (e.g. Lord (1927) wrote that it is 'the bushmen's name for that dainty little midget, the Emu-Wren'); Sticktail; Hartog Emu-wren, refers to the race *hartogi* of the Southern Emu-wren, restricted to Dirk Hartog Island, Shark Bay; Western Emu-wren, for race *westernensis* from Western Australia, described by Archibald Campbell in 1912 as a species that was recognised by RAOU (1913), but not thereafter.

Stipiturus malachurus (Shaw, 1798) [sti-pi-TOO-rʊs ma-la-KOO-rʊs]: 'soft-tailed stick-tail', see genus name, and from Greek *malakos*, soft, and *oura*, tail. Sounds odd, but when you see it, that's just what it is.

Mallee Emu-wren (breeding resident)

See family introduction. From the species name.

Stipiturus mallee Campbell AJ, 1908 [sti-pi-TOO-rʊs MAL-lee]: 'Mallee stick-tail', see genus name, and for its near-exclusive habitat in the semi-arid multi-stemmed eucalypt shrublands of inland southern Australia.

Rufous-crowned Emu-wren (breeding resident)

See family introduction. Straight from the species name.

Other names: Hartog Emu-wren, as cited by HANZAB (1990–2006, vol. 5), is an error, see Southern Emu-wren (Rufous-crowned does not occur on Dirk Hartog Island).

Stipiturus ruficeps Campbell AJ, 1899 [sti-pi-TOO-rʊs ROO-fi-seps]: 'red-headed stick-tail', see genus name, and from Latin *rufus*, red or reddish, and *-ceps*, -headed (*caput*, head), because both male and (especially) female have a more intense and extensive rufous crown than other emu-wrens.

Grey Grasswren (breeding resident)

See family introduction. Not really as greyish as some other grasswrens.

Amytornis barbatus Favaloro & McEvey, 1968 [a-mi-TOR-nis bar-BAH-tʊs]: 'bearded Amytis bird', see genus name, and from Latin *barba*, beard, referring to its black facial and throat markings, which are not present in other grasswrens.

Black Grasswren (breeding resident)

See family introduction. Actually black and chestnut, but a good summary.

Amytornis housei (Milligan, 1902) [a-mi-TOR-nis HOW-see]: 'House's Amytis bird', see genus name, and for Frederick Maurice House (1865–1936), a doctor who came from England in 1891, served as a medical officer in rural Western Australia and was a naturalist in Brockman's 1901 government exploring party in northern Western Australia, when he found this bird. After World War I, he retired to Gnowangerup in the south-west to indulge another of his interests: breeding merinos.

White-throated Grasswren (breeding resident)

See family introduction. The only grasswren with a white throat that is not an extension of a white breast and belly.

Other names: Large White-throated Grasswren (cf. the next species); White-chested Grasswren; Spinifex Grasswren, unhelpful.

Amytornis woodwardi Hartert, 1905 [a-mi-TOR-nis WŬD-wŭ-dee]: 'Woodward's Amytis bird', see genus name, and for Bernard Henry Woodward (1846–1916), an English geologist who came to Australia in 1889 for his health. He became Western Australian government analyst, then curator of the Perth Museum, later known as the Western Australian Museum and Art Gallery, with Woodward as its director. He was also a fruit farmer and an advocate for nature reserves.

Carpentarian Grasswren (breeding resident)

See family introduction. For its limited range on the south-western hinterland of the Gulf of Carpentaria.

Other names: Lesser White-throated Grasswren, (cf. the previous species – Carpentarian was originally described as a subspecies of White-throated); Red-winged Grasswren, for rufous tints in the wings. Dorothy Grasswren, from the species name.

Amytornis dorotheae (Mathews, 1914) [a-mi-TOR-nis DO-ro-thee-eh]: 'Dorothy's Amytis bird', see genus name, and for Dorothy White (later Dorothy Minell), daughter of Hunter Valley property owner Australian-born Henry White (uncle of the author Patrick White), who obsessively collected stamps, cricket books, dead birds and their eggs. His 8500 skins and 4200 clutches of eggs were subsequently donated to the Victorian Museum. Dorothy was 26 when the then subspecies was named; she married the following year. We will never know whether her father requested the honour.

Short-tailed Grasswren (breeding resident)

See family introduction. Recently elevated from being a subspecies of Striated Grasswren; this name was coined to reflect an evident character (or as evident as grasswren characters in the field ever are!).

Amytornis merrotsyi Mellor, 1913 [a-mi-TOR-nis mer-ROT-si-ee]: 'Merrotsy's Amytis bird', see genus name, and for A.L. Merrotsy, who collected the three adult birds that have equal syntypes, because no type was actually designated in the original description (Parker 1982). The skins of these birds were believed by some to have been lost for years, but this was later shown to be incorrect (Ford and Parker 1974).

Striated Grasswren (breeding resident)

See family introduction. It is indeed striated, like all the grasswrens; taken straight from the species name as Striated Wren by Gould (1848). But look, we wouldn't want to have to find unequivocal names for them either! And, to be fair, this would have been the first one that Gould had seen.

Other names: Striped, Black-cheeked or Spinifex Grasswren, all equally true-but-unhelpful; Rufous or Large Striated Grass-Wren, for Mathews' 1910 species *A. whitei* from north-western Australia, still recognised as a species by RAOU (1926); Chestnut-mantled Grasswren (for Mellor's 1912 species *A. merrotsyi* from Lake Torrens), recognised by RAOU (1913) in the Provisional List, but not thereafter.

Amytornis striatus (Gould, 1840) [a-mi-TOR-nis stree-AH-tŭs]: 'streaked Amytis bird', see genus name, and from Latin *striatus*, streaked or furrowed, and *stria*, a furrow or channel (see common name).

Eyrean Grasswren (breeding resident)

See family introduction. For their range, in the Lake Eyre Basin.

Other names: Bicyclebird, which is cited by HANZAB (1990–2006, vol. 5) and which might be supposed to derive from their scooting behaviour between dune-top grass clumps – but that would be pure speculation as we cannot find a single other reference to it; Goyder's Grasswren, from the species name.

Amytornis goyderi (Gould, 1875) [a-mi-TOR-nis GOY-der-ee]: 'Goyder's Amytis bird', see genus name, and for George Goyder, South Australian Surveyor General who controversially (and correctly) defined Goyder's Line of Rainfall that marked the climatic limit of reliable agricultural land in the state. In 1874 he sent James Lewis to fill in the details of the geography of Lake Eyre; Lewis later wrote, 'I sincerely hope I will never see it again', but he brought back the first specimens of this grasswren.

Western Grasswren (breeding resident)

See family introduction. To distinguish it from the following species, with which it has often been lumped (though for a large part of their ranges in South Australia they are at the same longitude).

Other names. Large-tailed Grasswren or Wren, a perplexing name, but from Gould's 1847 species name *macroura* from Western Australia – he then considered it separate from the species he knew from further east; note that Western Grasswren has also been used for Milligan's short-lived name *A. gigantura* from Mount Magnet, Western Australia. Textile Wren or Grasswren, from the species name, as used by Gould (1848).

Amytornis textilis (Quoy & Gaimard, 1824) [a-mi-TOR-nis TEKS-ti-lis]: 'plaited Amytis bird', see genus name, and from Latin *textilis*, woven or plaited. It follows from Quoy and Gaimard's common name, French *Mérion natté*, meaning plaited fairywren, referring to its heavily streaked plumage. He described it thus: 'each feather [on throat and breast] is evenly covered with little rufous and whitish marks', adding that the same goes for the head and the back, 'on which each feather has a dirty-white line in the middle'.

Thick-billed Grasswren (breeding resident)

See family introduction. For a distinctively stout bill (not that we recall anyone commenting on it in the field).

Other names: Eastern Grasswren, anticipating the current split from the previous species; Textile Wren or Grasswren (see Western Grasswren) from when they were combined.

Amytornis modestus (North, 1902) [a-mi-TOR-nis mo-DES-tʊs]: 'modest Amytis bird', see genus name, and from Latin *modestus*, meaning many things including modest, unassuming and plain. The last of these was chosen by North to indicate the paler colouring and less distinct marking of the species, which North had always believed to be the immature female of *A. textilis* until he was brought a definite adult male by G.A. Keartland.

Dusky Grasswren (breeding resident)

See family introduction. For the distinctively dusky-grey underparts.

Other names: Thin-billed Grasswren, to distinguish from the previous species, of which it was originally regarded as a subspecies; Dark, Buff-throated, Grey-bellied Grasswren; Rock Wren, for its habitat in the rocky gorges of central Australia.

Amytornis purnelli (Mathews, 1914) [a-mi-TOR-nis per-NEL-lee]: 'Purnell's Amytis bird', see genus name, and for Herbert A. Purnell, naturalist, collector and RAOU member from Geelong. See genus *Purnella* (Meliphagidae) for more detail.

Kalkadoon Grasswren (breeding resident)

See family introduction. For the Kalkadoon people within whose land in north-west Queensland lies the restricted area occupied by the species, which has relatively recently been recognised as a separate species.

Amytornis ballarae Condon, 1969 [a-mi-TOR-nis BAL-la-reh]: 'Ballara Amytis bird', see genus name, and for what used to be the thriving Ballara copper-mining community in the Selwyn Range near Mt Isa, the land belonging to the Kalkadoon people.

MELIPHAGIDAE (PASSERIFORMES): honeyeaters and chats

Meliphagidae Vigors, 1825 [me-li-FA-gi-dee]: the Honey-eater family, see genus name *Meliphaga*. 'Honeyeater' was not part of our birding lexicon from the start, despite the group being by far the largest and most ubiquitous of all Australian bird families. (We note James Cook's observation, referred to by the

Australian National Dictionary, 'There are four species that seem to belong to the trochili, or honey-suckers of Linnaeus', but this observation was made in Hawaii.) The word Honey-eater is first known from South Africa, where its original usage seemed to be for honey-guides; later it was used generally for other honey-eating animals such as ratels (honey-badgers). Note that, unlike Australian honeyeaters, these animals actually do eat honey (as opposed to nectar)!

In Australia Lewin (1808) first referred to 'honeysuckers', which he applied to four species – Warty-faced (Regent Honeyeater), Blue-cheeked (Blue-faced Honeyeater), Yellow-eared (Lewin's Honeyeater) and Golden-crowned (Regent Bowerbird). By 1823 Latham had introduced the word Honey-eater for Australian birds and was applying it to everything from bee-eaters to robins to whistlers to bowerbirds to whipbirds – plus a lot of honeyeaters and several birds whose identity is not obvious (he regarded Latin names, even by then, as optional). Gould (1848) formalised and standardised the term.

'Spinebill' is apparently straight from the genus name, introduced by Gould – however, he noted (1848) that 'the Colonists of New South Wales' already called the bird 'spine-bill', so it seems that the usual naming process was reversed. Gould aside, we can't find it in print before 1862, however.

'Miner' is sometimes disputed (e.g. HANZAB 1990–2006, vol. 5 maintained, without support, that it is based on the Bell Miners' calls reminding settlers of 'distant miners hammering at the workface') but it seems more likely that the origin is with the Asian starlings usually now called 'myna' but in the past equally well known in Australia as 'mina' or 'minah' (Henderson 1832 referred to 'noisy minas'). The black mask, yellow eye-patch, bill and legs (and stroppy disposition!) are obvious similarities between Noisy Miner and Common Myna, especially to non-birders. (Even though Common Mynas were not introduced to Australia until the 1860s, they were already familiar both via the Indian and South-East Asian trade and military connections with Australia, and as popular cage birds.) The change to the more familiar form 'miner' was a likely step. Gould (1848) noted that it was used in that form by 'Colonists of Van Diemen's Land', though he did not adopt it except, oddly, for Yellow-throated Miner from inland New South Wales. On the other hand, even in the 20th century Mathews (1913) was using still 'minah' as the name of *Manorina* species.

'Wattlebird' seems to have been first used in writing by William Charles Wentworth in 1819; clearly a man lacking in perception and observational skills, he noted, 'In the feathered tribes of the two islands [the mainland and Tasmania], there is scarcely any diversity; of this the wattle bird, which is about the size of a snipe, and considered a very great delicacy, is the only instance which I can cite.' The name is for the fleshy wattles hanging from below and behind the eye of Red and Yellow Wattlebirds. Gould (1848) used Wattled Honey-eater for both these species but noted that the colonists of both Van Diemen's Land and the mainland used Wattle Bird. It is interesting that by then the more astute 'Colonists of Swan River' were already using Little Wattle-Bird for their wattle-less species (now Western Wattlebird), suggesting that the origin of the name was already being forgotten, though they recognised the birds' relationships.

'Chat' does not seem to have arrived in formal usage in Australia until the late 19th century (Hall 1899). (In 1847 Gould used Epthianura as the common name, as was his wont.) Although the association with the large group of terrestrial African and Asian Robin-chats might seem obvious, in his very comprehensive *Dictionary of Birds* Newton (1896) recognised only four chats (European Stonechat and Whinchat and a couple of American blackbirds). Accordingly, it is something of a mystery as to where Hall came up with the name, though the superficial similarity with the ground-foraging Stonechat and Whinchat cannot be discounted. Given the existence of similar call-based names (Tang, Tin Tack) for White-fronted Chat, it may have been a common name adopted by Hall, though he was not generally so frivolous. Note that it is now acknowledged that the chats sit firmly within the main body of honeyeaters, with regard to their relationships.

'Friarbird' refers perhaps somewhat mockingly to the bald head of the Noisy Friarbird, the only friarbird species the British colonists would have encountered for some time after settlement. It clearly made an early impression; as early as 1793, Governor Hunter mentioned, in a list of Aboriginal words, the 'Wir-gan, A bird called fryar'. Five years later, Collins was more explicit in his glossary: 'Wir-gan, Bird named by us the Friar' (Collins 1798). By 1848 Gould was using Friar Bird as first-choice name for Noisy Friarbird.

The genera

Sugomel Mathews, 1922 [SOO-go-mel]: 'Latin honey-sucker', from Latin *sugo*, suck, and *mel*, honey.

Myzomela Vigors & Horsfield, 1827 [mi-ZO-me-luh]: 'Greek honey-sucker', from Greek *muzaō*, suck, and *meli*, honey (or anything sweet).

Glyciphila Swainson, 1837 [gli-SI-fi-luh]: 'sweet-lover', from Greek *glukus*, sweet to the taste (the word glucose comes from the related noun), and *-philos*, lover of.

Glycichaera Salvadori, 1878 [gli-si-KEH-ruh]: 'sweet-fancier', from Greek *glukus*, sweet to the taste, and *khairō*, to enjoy. Let us count the ways of saying these birds enjoy nectar!

Acanthorhynchus Gould, 1837 [a-kan-tho-RIN-kʊs]: 'spine-bill', from Greek *acantha*, spine, and *rhunkhos*, bill.

Certhionyx Lesson, 1830 [ser-thi-O-niks]: 'treecreeper-claw', from genus *Certhia*, the Northern Hemisphere treecreepers, and Greek *onux*, talon, claw or nail. Lesson (1831) said only that they have 'weak toes' (referring to the Pied Honeyeater).

Cissomela Bonaparte, 1854 [si-so-MEH-luh]: 'honey magpie', from Greek *kissa* or *kitta*, the jay (in Byzantine times also the magpie, which is probably what it refers to here: black and white), and *meli*, honey. Interestingly, in *Mémoires ornithologiques* (1854) Bonaparte stated that what Gould called *Myzomela nigra* (the Black Honeyeater), he believed was the type of the new genus *Cissomela*. *Cissomela* is now not applied to the Black Honeyeater (known as *Sugomel niger*); it is applied only to the Banded Honeyeater.

Lichmera Cabanis, 1851 [lik-MEH-ruh]: 'snake-like licker', from Greek *likhmērēs*, a word used of snakes, meaning 'playing with the tongue'.

Phylidonyris Lesson, 1830 [fi-li-do-NEE-ris]: 'affectionate part-sun-bird' *Phylidonyris* (*phylidonyre* in French) is the name Lesson (1831) gave to what he calls the third subgenus of *Cinnyris*, a sunbird genus. Though he didn't say, we assume his word was made up of *Philedon* (see *Philemon*) plus the end of *Cinnyris*.

Trichodere North, 1912 [tri-ko-DEH-reh]: 'hairy throat', from Greek *thrix/trikhos*, hair, and *deirē*, throat, for the bird's bristly looking throat.

Grantiella Mathews, 1911 [gran-ti-EL-luh]: 'Grant's little [bird]', for William Robert Ogilvie-Grant (1863–1924), Scottish ornithologist and curator of birds at the British Museum from 1909 to 1918. Others (including HANZAB 1990–2006, vol. 5) have suggested that Robert Grant, who worked as a taxidermist and collector for the Australian Museum, is the person for whom *Grantiella* is named, a choice which seems, on the face of it, to be reasonable. We choose Ogilvie-Grant following Richmond (1917) whom we believe to be correct for the following reasons: Charles Richmond was Assistant Curator of Birds at the US National Museum, and was doubtless in correspondence with Mathews (his very lengthy article that we have already cited deals in large part with Mathews' genus names), and at the time of publication of *Grantiella*, Mathews was working in England, including at the British Museum with Ogilvie-Grant.

Plectorhyncha Gould, 1838 [plek-to-RIN-kuh]: 'pointed-bill', from Greek *plēctron*, anything pointy to strike with (e.g. a plectrum to play your lyre with, a spear-point, a cock's spur), and Greek *rhunkhos*, bill.

Xanthotis Reichenbach, 1852 [zan-THOH-tis]: 'yellow-eared [bird]', from Greek *xanthos*, yellow (with a brown or reddish tinge), and *-ōtis*, eared. You have to look pretty hard to find these 'ears' – very tiny tufts of yellow, named for the Tawny-breasted Honeyeater *Xanthotis flaviventer*.

Philemon Vieillot, 1816 [FI-le-mon]: 'affectionate [bird]', from Greek *philēmōn*, kindly or affectionate. 'Polochion' was what Buffon and Montbeillard (1779) called these birds, based on the cry of the Black-faced Friarbird, now *Philemon moluccensis*, a repeated and incessant call apparently sounding like 'let's kiss' in the Moluccan language. This striking call led to a suggestion from Philibert Commerson (sometimes spelled Commerçon), a French doctor, naturalist and collector who accompanied Bougainville on his round-the-globe trip in 1766–69 and who collected the type specimen (Newton 1896). He wanted to call the bird 'Philemon, ou Philedon, ou deosculator' (no accents), said Buffon. All these words mean affectionate, loving or kissing, and *Philemon* was taken up by Vieillot for the genus (though he also used *Polochion* as a genus name). *Philedon* was used by Cuvier (1817), reported by Vieillot (1816–19) to refer to what he called his own genera, *Creadion* and *Polochion*. The third of Commerson's suggestions seems to have sunk without trace – we can only be grateful.

Entomyzon Swainson, 1825 [en-to-MI-zon]: 'insect-sucker', from Greek *entoma*, insects (*entomos* being that which is cut in pieces, or segmented), and from Greek *muzaō*, suck. Swainson (1825) asserted: 'The Australian Meliphagidae, with the exception of one type, derive their principal sustenance … from the nectar of flowers.' The exception, according to Swainson's footnote, is: 'Entomyzon, (Mihi). The Blue-faced Grakle [sic] of Latham, whose filamentous tongue is used for extracting small insects from between the broken bark on the stems of trees'. We note HANZAB's different etymology (1990–2006, vol. 5) but consider ours well justified.

Melithreptus Vieillot, 1816 [me-li-THREP-tʊs]: 'honey-fed', from Greek *melithreptos*, meaning exactly that, though there is a tiny misprint in Vieillot's original etymology list and the Greek reads *melitherptos* (rather than *melitherōtos*, as suggested in HANZAB (1990–2006, vol. 5), but you do need a strong magnifying glass to see this!) (Vieillot 1816–19).

Nesoptilotis Mathews, 1913 [ne-so-ti-LO-tis]: 'Tasmanian feather-ear', from Greek *nēsos*, an island; *ptilon*, feather, and -ōtis, -eared. Mathews explained that the 'Type (by original designation)' was '*Ptilotis flavigula* Gould', (Mathews 1913), and the fact that this bird is found only in Tasmania and the Bass Strait islands, appears to explain Mathews' choice to include *nēsos* in the new name. (See the various species for more information.)

Ashbyia North, 1911 [ASH-bi-uh]: 'Ashby's [bird]', for Edwin Ashby (1861–1941), who came from England in 1884, a real estate agent with deeply held and practical Quaker beliefs and a great interest in natural history. His key passion was shells, on which he was a world authority, but he also wrote more than 80 papers on birds. He was a president of the RAOU, a fellow of the Linnean Society and a committed and well-informed conservationist.

Epthianura Gould, 1838 [ep-thi-a-NOO-ruh]: 'refined-tail'. Originally, and indeed usually, spelled by Gould as *ephthianura*, (1838b, 1848 – in the introductory Volume 1 of *Birds of Australia and General Index* – and 1865), but it appears as *epthianura* in the actual entries for the birds in *Birds of Australia* in Volume 3, as well as in the *Synopsis* (1838a). The latter spelling has now been adopted, though unfortunately it appears to be meaningless. Gould's intention is obscure: he may have taken the word from Greek *ephthos*, meaning refined or improved (as in *ephthos khrusos*, meaning refined gold), and *oura* tail, perhaps referring to the fine white tips on the central tail feathers noted in his original description of the type specimen Orange Chat (Gould 1838b). He described the chats in general (based on White-fronted Chat *Epthianura albifrons*) as having 'a short and truncated tail' (Gould 1838b), hence a further possibility: Greek *ephthinaō*, to waste away, or *phthinas*, decreasing, with *oura*, tail (HANZAB 1990–2006, vol. 5; Jobling 2010). We may have to agree with HANZAB that it is 'one of the mystery names of Australian ornithology and nomenclature'.

Conopophila Reichenbach, 1852 [ko-no-PO-fi-luh]: 'gnat-lover', from Greek *kōnōps*, gnat, and *-philos*, lover of, because the birds often eat insects, spiders, and so on in addition to nectar.

Ramsayornis Mathews, 1912 [ram-zeh-ORN-is]: 'Ramsay bird', from Greek *ornis*, bird, and for 19th-century naturalist-polymath Edward Pierson Ramsay (1842–1916), born and died in New South Wales, a respected authority on birds, fish, reptiles, mammals and plants, curator at the Australian Museum for 20 years from 1874 and a corresponding colleague of John Gould and Sir Richard Owen. Unfortunately he allowed himself to be used by an apparently corrupt clique of governors of the museum to replace the unfortunate and ill-used Gerard Krefft, but Ramsay himself was well regarded.

Acanthagenys Gould, 1838 [a-kan-tha-GE-nis]: 'spiny-cheeked bird', from Greek *acantha*, spine, and *genus*, cheek or jaw. The yellow-white cheek-stripe of the bird is made up of stiff spiny feathers, extending down the sides of its neck.

Anthochaera Vigors & Horsfield, 1827 [an-tho-KEH-ruh]: 'flower-fancier', from Greek *anthos*, flower, and *khairō*, enjoy.

Bolemoreus Nyári & Joseph, 2011 [bohl-MOR-ee-ʊs]: 'Boles and Longmore's bird', honouring Australian ornithologists Walter Boles and N. Wayne Longmore. (Pronunciation courtesy author Dr Leo Joseph.)

Caligavis Iredale, 1956 [ka-li-GA-vis]: 'obscure bird', from Latin *caligo*, a thick mist or fog, hence darkness or obscurity, plus *avis*, bird. Iredale wrote: 'The name 'Obscure' is most applicable as everything about the bird in literature is unclear.'

Lichenostomus Cabanis, 1851 [li-ke-no-STOH-mʊs]: 'lichen-mouth', from Greek *leikhēn*, tree-moss or lichen (or a skin eruption looking like lichen), and *stoma*, mouth – named for the group of Purple-gaped Honeyeaters and first used for *Lichenostomus occidentalis*, now a race rather than a species.

Manorina Vieillot, 1818 [ma-no-REE-nuh]: 'large-nostrilled [bird]', from Greek *manos*, loose or open in texture, and *rhin/rhinos*, nostrils. Vieillot also used the Latin terms *rarus* (which can also mean loose, open, etc.), and *narus*, nostril. His description of the bird's nostrils in the *Nouveau Dictionnaire* (Vieillot 1816–19) was surprisingly full. He wrote that it has large nostrils (*narines amples*), half as long as the upper mandible, stretching from the ridge of the beak to its edges, broad at the base and a little bit pointed at the ends; generally pretty big, in other words. Vieillot added that the nostrils are covered by a membrane with a narrow opening, which is what you see in the field – this may be what leads to an alternative interpretation: thin-nostrilled, apparently from Greek *manos* meaning thin (sparse) (Latin *rarus* can also mean thin in this sense), referring to the 'linear, arrow-shaped nostrils' of the Bell Miner, *Manorina melanophrys* (HANZAB 1990–2006, vol. 5).

Purnella Mathews, 1914 [per-NEL-luh]: 'Purnell's', for Herbert A. Purnell, RAOU member from Geelong, coined by Gregory Mathews in 1914. In November of that year Mathews accompanied Purnell to Port Addis, near Torquay – very much in Purnell's patch. Purnell's photo of Mathews with the Rufous Bristlebird's nest they found on this outing is in the National Library (Purnell 1914).

Stomiopera Reichenbach, 1852 [STO-mi-o-PE-ruh]: 'pocket mouth'. Reichenbach described the 'fleshy little mouth pocket' of the bird's gape, and said he had invented the name from the Greek *stomion*, mouth, and *pēra*, a leather pouch or pocket.

Gavicalis Schodde & Mason, 1999 [ga-vi-CA-lis]: 'anagram bird'. The name is an anagram of *Caligavis*, Iredale's 'Obscure bird'.

Ptilotula Mathews, 1912 [ti-LO-tʊluh]: 'little feather-ear', from Swainson's genus *Ptilotis* (from Greek *ptilon*, feather, and *-ōtis*, -eared), with diminutive ending *-ulus*, *-ula*.

Meliphaga Lewin, 1808 [me-li-FA-guh]: 'honey-glutton', from Greek *meli*, honey, and *phagos*, a glutton.

The species

Black Honeyeater (breeding resident)

See family introduction. Not quite, but mostly, but only adult male and only if viewed from behind or above'. Direct from species name, and used by Gould (1848).

Other names: Charcoal Bird, for the remarkable habit (mostly by females) of visiting dead campfires to collect ashes and charcoal, for reasons unknown.

Sugomel niger (Gould, 1838) [SOO-go-mel NEE-gehr]: 'black Latin honey-sucker', see genus name, and from Latin *niger*, gleaming black.

Dusky Myzomela (Honeyeater) (breeding resident)

Note that despite the best efforts of various authorities, virtually no-one in Australia – and none of the field guides – has adopted the odd recommendation to use the genus name instead of the widespread 'honeyeater' group name for this one genus.

See genus. A version of the species name, but neither does justice to the rich coppery tones.

Other names: Dusky Honeyeater, near universally used in Australia; Obscure Honeyeater, used by Gould (1848) from his species name.

Myzomela obscura Gould, 1843 [mi-ZO-me-luh ob-SCOO-ruh]: 'dark Greek honey-sucker', see genus name, and from Latin *obscurus*, dark or dusky.

Red-headed Myzomela (Honeyeater) (breeding resident)

Unlike Scarlet Honeyeater, only the head is red (male only). Applied by Gould (1848) for his species name. See also Dusky Myzomela.

Other names: Red-headed Honeyeater, near universally used in Australia; Blood-bird; Mangrove Red-headed Honeyeater or Mangrove Red-head, for key habitat.

Myzomela erythrocephala Gould, 1840 [mi-ZO-me-luh e-rith-ro-SE-fa-luh]: 'red-headed Greek honey-sucker', see genus name, and from Greek *eruthros*, red, and *kephalē*, head.

Scarlet Myzomela (Honeyeater) (breeding resident)

Yet again for the male only, a spectacularly red bird; it seems to have been coined by RAOU (1926). See also Dusky Myzomela.

Other names: Scarlet Honeyeater, near universally used in Australia; Sanguineous Honeyeater, used by Gould (1848) and still in use into the 20th century (RAOU 1913) – coined by Latham from his species name, preceded by Sanguineous Creeper, Cochineal Creeper and Red-rumped Creeper, all from Latham (1801); Small-crested Creeper (Shaw 1819), though we can only wonder what he was seeing; Northern Sanguineous Honeyeater, for Mathews' subspecies *stephensi* from Cooktown (Mathews 1912a), cited as a species *L. stephensi* by RAOU (1913) in the provisional list; Blood-bird, reported by Gould to be used by the 'Colonists of New South Wales'; Blood, Blood-red or Crimson Honeyeater; Hummingbird, for its habit of hovering briefly at flowers; Soldier-bird, for its 'redcoat' uniform/plumage.

Myzomela sanguinolenta (Latham, 1802) [mi-ZO-me-luh san-gwi-no-LEN-tuh]: 'blood-red Greek honey-sucker', see genus name, and from Latin *sanguinolentus*, bloody, full of blood, blood-red.

Tawny-crowned Honeyeater (breeding resident)

See family introduction. A surprisingly useful description that probably arose spontaneously during the second part of the 19th century – Morris used it in 1898 in his *Dictionary of Austral English*.

Other names: Fulvous-fronted Honeyeater, fulvous being another word for tawny, from Vigors and Horsfield's 1827 species name *Meliphaga fulvifrons*, not realising that Latham had already named it; Fulvous-breasted Honeyeater, not at all obvious and we suspect that someone mistook 'frons' for meaning 'front' more generally – however, we cannot find a usage of it outside HANZAB (1990–2006, vol. 5); Tasmanian Tawny-crowned Honeyeater, for Mathews' subspecies *crassirostris* (Mathews 1912a), cited as a species *G. crassirostris* by RAOU (1913) in the provisional list.

Glyciphila melanops (Latham, 1801) [gli-SI-fi-luh MEL-an-ops]: 'black-faced sweet-lover', see genus name, and from Greek *melas*, black, and ōps, the face, for the striking black mask extending down to frame the white breast.

Green-backed Honeyeater (breeding resident)

See family introduction. The olive back is as conspicuous as anything else about this pretty inconspicuous little honeyeater.

Other names: Puff-backed Honeyeater, widely cited in Australia although it refers to New Guinea race *fallax* – in any case not readily explicable, and the name is more usually applied to another New Guinea species, *Meliphaga aruensis*; White-eyed Honeyeater, again cited in the literature (e.g. HANZAB 1990–2006, vol. 5) but it too refers to race *fallax*.

Glycichaera fallax Salvadori, 1878 [gli-si-KEH-ruh FAL-laks]: 'deceptive sweet-fancier', see genus name, and from Latin *fallax*, deceptive, deceitful, fallacious. In what sense is the bird deceptive or false? Salvadori (1878) wrote: 'I believed that this species could be attributed to the genus *Euthyrhyncus*, Schleg, instead I have verified it to be identical to my genus *Timeliopsis*'. So the bird may be *fallax* because it appeared to be *Euthyrhyncus*, but turned out to be *Timeliopsis*.

Eastern Spinebill (breeding resident)

See family introduction. An easy one – there are two species, one confined to eastern Australia, the other to the south-west corner – but the apparently obvious name-pair was introduced by the RAOU's committee responsible for the second checklist only well into the 20th century (RAOU 1926).

Other names: Slender-billed Spine-bill, by Gould (1848), a name we might think superfluous but Gould took it from Latham, who had called it Slender-billed Honey-eater, *Certhia tenuirostris* (*Certhia* being the Northern Hemisphere treecreepers); it's unclear whether the common or species name came first. Flapping Honey-eater and Slender-billed Creeper, two more names used by Latham at different times! The former is interesting, in that it immediately brings to mind the familiar 'prrrp' of the spinebill's wings – but Latham would never have seen a live one. Hooded Creeper, Shaw (1819); Hummingbird or Tasmanian Hummingbird, logical for the hovering at flowers, but it's hard to find an example of its use. Slender Spinebill, used as first-choice name to the end of the 19th century (Morris 1898); Spinebill, used alone as first-choice name by RAOU (1913); Spiney; Cobbler's Awl or Awl Bird, a lovely reference to the bill and recorded by Gould (1848) as used by the 'Colonists of Van Diemen's Land'; Victorian Spinebill, for race *tenuirostris*; Tasmanian Spinebill, for race *dubius* – Gould described this as a separate species but expressed his doubts in the very name, though RAOU (1913) was still using it; Kangaroo Island Spinebill, for Campbell's 1906 name *A. halmaturinus*, though it is found on the South Australia mainland as well – it is now regarded as a race; Cairns Spinebill for Mathews' subspecies *cairnsensis* (Mathews 1912a), cited by RAOU (1913) in the provisional list as a species.

Acanthorhynchus tenuirostris (Latham, 1802) [a-kan-tho-RIN-kʊs te-noo-i-ROS-tris]: 'slender-billed spine-bill', see genus name, and from Latin *tenuis*, slender or thin, and *rostrum*, bill.

Western Spinebill (breeding resident)

See family introduction and Eastern Spinebill.

Other names: White-browed Spinebill or White-eyebrowed Spine-bill, the latter being Gould's name (1848) straight from his species name and still used into the 20th century (RAOU 1913); Western Spine-billed Honeyeater.

Eastern Spinebill *Acanthorhynchus tenuirostris*, Western Spinebill *A. superciliosus* and Pied Honeyeater *Certhionyx variegatus*

Acanthorhynchus superciliosus Gould, 1837 [a-kan-tho-RIN-kʊs soo-pehr-si-li-OH-sʊs]: 'eye-browed spine-bill', see genus name, and from Latin *supercilius*, above the eyelid (the eyebrow), for the striking facial markings.

Pied Honeyeater (breeding resident)

See family introduction. 'Pied' is a recurring bird name, based simply on the old name Pie for the European Magpie, and signifying black and white. Gould applied it when he named the bird *Melicophila picata* (not realising that Lesson had already named it); the species name was based on *Pica*, the (Northern Hemisphere) magpie genus. (Gould's error is readily forgiven – Lesson had mistakenly reported that his specimen came from Timor, rather than Shark Bay.)

Other names: Black-and-white Honeyeater; Western Pied Honeyeater, for the short-lived name *Certhionyx occidentalis*, applied by William Ogilvie-Grant of the British Natural History Museum to a specimen from Carnarvon, Western Australia.

Certhionyx variegatus Lesson, 1830 [ser-thi-O-niks va-ri-e-GAH-tʊs]: 'variegated treecreeper-claw', see genus name, and from Latin *variegatus*, in this case only two-coloured.

Banded Honeyeater (breeding resident)

See family introduction. For the conspicuous broad black breastband dividing the white undersides; used by Gould (1848) in reference to his species name.

Other names: Banded Myzomela, from when it was placed in this genus, until quite recently (see Dusky Honeyeater).

Cissomela pectoralis (Gould, 1841) [si-so-MEH-luh pek-to-RAH-lis]: 'honey-magpie with a breast', see genus name, and from Latin *pectus/pectoris*, breast. See common name.

Brown Honeyeater (breeding resident)

See family introduction. This unimaginative (but understandable) name was apparently a name 'of the colonists' (in the form Brown Honey-sucker) according to Gould (1848), who adopted it.

Other names: Least Honeyeater, a name somewhat mystifyingly in wide use well into the 20th century (Mathews 1913; RAOU 1913); it had been briefly applied by Gould to a species he called *Glyciphila subocularis* from northern Australia, but he soon after retracted and acknowledged that the small specimen was really a 'female or young male', though the 1913 RAOU checklist mysteriously still recognised it. The use of the name was not a mystery to Dr Spencer Roberts, who wrote in 1934, 'It is a pity that the name Least Honeyeater was ever abandoned in regard to the species, for I believe it accurately describes it' (Roberts 1934); unfortunately he didn't see fit to explain his reasoning. Warbling Honeyeater, understandable given the similarity of some of this bird's calls to those of a reed warbler, but it's hard to find actual usage.

Lichmera indistincta (Vigors & Horsfield, 1827) [lik-MEH-ruh in-dis-TINK-tuh]: 'indistinct snake-like licker', see genus name, and from Latin *indistinctus*, not quite properly distinguished, confused. Vigors and Horsfield (1827) said, '[The specimen] is however in very bad condition, and scarcely admits of a description'. They also thought it might be a young bird, so one way and another it was not properly distinguished.

Crescent Honeyeater (breeding resident)

See family introduction. For the broad black line curving round each side of the male's breast. Its origin is unclear – Gould (1848) used Tasmanian Honeyeater, though he was aware of very similar mainland birds. The current name was enough in general use by the end of the 19th century to be included in Morris' *Dictionary of Austral English*, so perhaps it arose spontaneously.

Other names: Tasmanian Honeyeater, see common name; Egypt or Egypt Honeyeater, for the harsh two-note call; Horseshoe Honeyeater, for the crescent; Crested Honeyeater, surely a corruption of the name; White-browed Honeyeater (Latham 1823); Island Crescent Honeyeater, for Campbell's 1906 species *Lichmera halmaturina* from Kangaroo Island – it is still recognised as a race, though from mainland South Australia as well; Chinawing or Chingwing, doubtless for the yellow wings and apparently always regarded as too crass to commit to print, as we can find no examples of usage beyond the citations in CSIRO (1969) and HANZAB (1990–2006, vol. 5), though they have the hallmarks of oral vernacular.

Phylidonyris pyrrhopterus (Latham, 1801) [fi-li-do-NEE-ris pi-ro-TE-rʊs]: 'yellow-winged affectionate part-sunbird', see genus name, and from Greek *purrhos*, flame-coloured (almost always red, but can also be used to mean egg-yolk yellow), and *pteros*, winged.

New Holland Honeyeater (breeding resident)

See family introduction. Straight from the not-very-useful species name; it is unclear whether White's New Holland Creeper or Latham's *Certhia novaehollandiae* (meaning the same thing) came first, as both were published in 1790. White reported in his diary, on 21 July 1788, that 'we found the New Holland Creeper', but he could well have added the name before publication in 1790.

Other names: Honeysucker, reflecting its familiarity to colonists; Fuchsia-bird, for its association with 'native fuchsias' (*Correa* spp.) and native heath flowers (Epacridaceae); Yellow-winged or Golden-winged Honeyeater, Yellow Wings, the first being recorded by Gould (1848) as used by 'the

Colonists' of Western Australia – the name was still the one preferred by RAOU (1926); Bearded or White-bearded Honeyeater, the latter used by Cayley in *What Bird is That?* (1931) for reasons that are not at all obvious, though we wonder if there is some apposition implied to the closely related Moustached Honeyeater (see White-cheeked Honeyeater); Long-billed Honeyeater, used by Gould (1848) for his species name *Meliornis longirostris* for the Western Australian population, which he regarded as a separate species, still recognised into the 20th century (RAOU 1913); White-eyed Honeyeater, a descriptive name and introduced somewhat belatedly (and unsuccessfully) by McDonald (1973).

Phylidonyris novaehollandiae (Latham, 1790) [fi-li-do-NEE-ris no-veh-hol-LAN-di-eh]: 'New Holland affectionate part-sunbird', see genus name, and from Modern Latin *Nova Hollandia*, New Holland, the old name for Australia.

White-cheeked Honeyeater (breeding resident)

See family introduction. For the bold white cheek patches, larger than those of the similar New Holland Honeyeater; used by Gould (1848).

Other names: New Holland Creeper 'female', as illustrated in White (1790) – a misconception, New Holland Creeper being his name for our New Holland Honeyeater (also illustrated and labelled as 'male'); Moustached Honeyeater, again by Gould (1848) from his name *Meliphaga mystacalis* for Western Australian birds, which he regarded as a separate species; this is now race *gouldii*, and it does have much more of a moustache-streak than a full white cheek. Herberton Honeyeater, for the unexpected isolated mountain rainforest population of tropical Queensland (including the Herberton area, but also well beyond that).

Phylidonyris niger (Bechstein, 1811) [fi-li-do-NEE-ris NEE-gehr]: 'black affectionate part-sunbird', see genus name, and from Latin *niger*, gleaming black – rather odd, considering that so much of it is not black, though HANZAB (1990–2006, vol. 5) suggested it is darker than the New Holland Honeyeater, which it resembles.

White-streaked Honeyeater (breeding resident)

See family introduction. For the distinctively bristly looking streaky black and white breast; this name was abruptly introduced by RAOU (1926).

Other names: Cockerell's or Cockerell Honeyeater, from the species name. Brush-throated Honeyeater.

Trichodere cockerelli (Gould, 1869) [tri-ko-DE-reh KO-ke-ruh-lee]: 'Cockerell's hairy-throat', see genus name, and who better than Gould himself to explain the species name, 'It is but an act of justice that at least one of the birds of Australia should be named after Mr. James Cockerell, inasmuch as he is a native-born Australian, has collected very largely in the northern parts of that great country, and discovered more than one new species, among which must be enumerated the present very interesting bird' (Gould 1869).

Painted Honeyeater (breeding resident)

See family introduction. Used by Gould (1848) from his species name, and not many alternatives have been offered.

Other names: Georgie, nicely evocative of the ringing two-note call.

Grantiella picta (Gould, 1838) [gran-ti-EL-luh PIK-tuh]: 'Grant's little painted [bird]', see genus name, and from Latin *picto*, paint or embroider. Gould (1838a) gave a vivid description of the plumage: yellow, black and white, and with the 'tail black, margined externally with rich yellow'.

Striped Honeyeater (breeding resident)

See family introduction. For the obvious and striking black-on-white streaked head and neck; it came into use sometime in the second half of the 19th century, being recorded by Morris (1898).

Other names: Lanceolated Honeyeater, per Gould (1848) from his species name; Summerbird, not clear, as movements are limited and unpredictable (HANZAB 1990–2006, vol. 5) – in any case, not widely used.

Plectorhyncha lanceolata Gould, 1838 [plek-to-RIN-kuh lan-se-o-LAH-tuh]: 'spear-shaped pointed-bill', see genus name, and from Latin *lanceolatus*, armed with a little lance or point. This refers to the breast feathers, which Gould described as lanceolate and which look quite spiky.

Striped Honeyeater *Plectorhyncha lanceolata* and Regent Honeyeater *Anthochaera phrygia*

Macleay's Honeyeater (breeding resident)

See family introduction. From the species name, for William John Macleay (1820–91), Murrumbidgee pastoralist turned parliamentarian, anti-bushranger vigilante hero, first president of the Entomological Society of New South Wales (the first specialist scientific society in Australia) and of its successor, the Linnean Society of New South Wales. He left parliament to concentrate on establishing the Macleay museum at the University of Sydney. He was the nephew of Alexander Macleay (see Forest Kingfisher).

Other names: Buff-striped, Yellow-streaked, Mottled or Mottle-plumaged Honeyeater, all for the unusually motley patterning.

Xanthotis macleayanus (Ramsay EP, 1875) [zan-THOH-tis mak-leh-AH-nʊs]: 'Macleay's yellow-eared bird', see genus and common names. Ramsay referred to him as 'our distinguished President'.

Tawny-breasted Honeyeater (breeding resident)

See family introduction. For the tawny underparts; the name was introduced by RAOU (1926).

Other names: Streaked Honeyeater, by Gould (1869) for his species name *Ptilotis filigera* (he was unaware that Lesson had already named it from a New Guinea specimen); Streak-naped Honeyeater, true but not very obvious, though first-choice name of RAOU (1913); Buff-breasted or Honey-breasted Honeyeater.

Xanthotis flaviventer (Lesson, 1828) [zan-THOH-tis flah-vi-VEN-ter]: 'yellow-bellied yellow-eared bird', see genus name, and from Latin *flavus*, yellow, and *venter/ventris*, belly. Given that the bird is actually very tawny, this seems like a misnomer, except that *flavus* is from *flagro*, burn, so can indicate a reddish-yellow, which is a bit closer to the mark than just plain yellow. Indeed, HANZAB (1990–2006, vol. 5) suggested that *flavus* can be used in ornithology to describe colours from bright yellow to cream through to brown!

Little Friarbird (breeding resident)

See family introduction. The smallest friarbird; though the name was formalised by RAOU (1913, 1926), Morris reported it as the first-choice name in his 1898 *Dictionary of Austral English*.

Other names: Yellow-throated Friarbird, straight from the species name and used by Gould, who reported that the 'Colonists of New South Wales' called it Yellow-throated Friar – RAOU (1913) was still using it, but also used Little Friar-bird for the then separate species *P. sordidus* of northern Australia; Leatherhead (see Noisy Friarbird, though much less relevant here, given its feathered crown) or Little Leatherhead; Leather-neck 'of the Colonists of Port Essington' Gould (1848) – Port Essington was the precursor to Darwin, and further north.

Philemon citreogularis (Gould, 1837) [FI-le-mon si-tre-o-gʊ-LAH-ris]: 'lemon-throated affectionate bird', see genus name, and from Latin *citreus*, the citron tree, in this case lemon-coloured, and *gula*, the throat. Gould was describing the immature bird.

Helmeted Friarbird (breeding resident)

See family introduction. For the casque-like protrusion of feathers on the back of the head.

Other names: Helmeted Honey-eater (Gould 1869); Melville Island Friarbird, for race *gordoni*, previously regarded as a full species *P. gordoni* (CSIRO 1969) of the Tiwi Islands (of which Melville is the largest) and the adjacent mainland; Mangrove Friarbird, for *gordoni*, which uses coastal habitats including mangroves; Sandstone Friarbird, for race *ammitophila* of the Arnhem Land sandstone escarpment; New Guinea Friarbird, for race – often described as species – *novaeguineae*.

Philemon buceroides (Swainson, 1838) [FI-le-mon bʊ-se-ROY-dehz]: 'hornbill-like affectionate bird', see genus name, and from the genus name *Buceros* Linnaeus 1758, the hornbills (and Latin *-oides*, resembling) – for the knobbly bit on the top of the bird's bill, thought to resemble the casque of the hornbills.

Hornbill Friarbird (breeding resident)

See family introduction. It's not at all clear why the 'Hornbill' was applied to this species and not the previous one, given its species name. Moreover there is some doubt as to whether this species will continue to be recognised, or relumped with Helmeted and/or New Guinea Friarbird.

Philemon yorki (Mathews, 1912) [FI-le-mon YOR-kee]: 'Cape York affectionate bird', see genus name, and from the type locality.

Silver-crowned Friarbird (breeding resident)

See family introduction. From the species name, formalised by RAOU (1926).

Other names: Silvery-crowned Friarbird (Gould 1848; RAOU 1913).

Philemon argenticeps (Gould, 1840) [FI-le-mon ar-GEN-ti-seps]: 'silver-headed affectionate bird', see genus name, and from Latin *argentum*, silver, and *-ceps*, -headed (*caput*, head).

Noisy Friarbird (breeding resident)

See family introduction. Probably no noisier than other friarbirds but this was the one encountered by the Sydney colonists, though the name was not known to Morris (1898) and not introduced into general use until RAOU (1926). However, the name appears in a notice in the *Commonwealth Gazette* of 6 April 1923 (in a list of species which 'may be exported without restriction'; this also included Black and Grey Falcons).

Other names: Friarbird or Friar Bird, used by Gould (1848) and into the 20th century (RAOU 1913); Leatherhead, widely used to the present time, for the bare black head; Four-o'clock, Pimlico, Four-o'clock Pimlico, Poor Soldier, all renditions of the chiming/gobbling call; Knobby Nose or Knobby-nose Leatherhead, for the bill knob; Monk, an alternative to 'friar'; Knob-fronted Bee-eater or Honeyeater (Latham

1801 and 1823, respectively); Cowled Bee-eater or Honeyeater (Latham 1801 and 1823, respectively), for young birds.

Philemon corniculatus (Latham, 1790) [FI-le-mon kor-ni-kʊ-LAH-tʊs]: 'affectionate bird with a little horn', see genus name, and from Latin *corniculatus*, diminutive of *cornu*, horn, for the bill knob.

Blue-faced Honeyeater (breeding resident)

See family introduction. Couldn't really be called anything else.

Other names: Blue-faced Entomyza, used by Gould (1848) for the then genus name, in line with his convention; Blue-cheeked Honeysucker (Lewin 1808); Blue-eared Grackle (Latham 1801) – grackles comprise an entirely unrelated and not particularly similar group of 'American blackbirds', Icteridae; Graculine Honeyeater (i.e. Grackle-like) and Blue-cheeked Thrush, both used by Latham (1823); (Latham also used Pale-cheeked, White-crowned and Blue-cheeked Honeyeater, and Blue-cheeked Bee-eater at various times for young birds. This is despite the fact that young Blue-face Honeyeaters have olive-coloured cheeks. We suspect that changes in pigments in preserved skins might be responsible, though Latham's rationale for names was not always obvious); Graculine Creeper (Shaw 1819); Blue-eye; Banana-bird, for its fondness for fruit; White-quilled or White-pinioned Honeyeater, for race (formerly species,

Blue-faced Honeyeater *Entomyzon cyanotis* and *E. c.* race *albipennis*

per Gould) *E. albipennis* of the north coast, with broad white bases to its primaries that show as a big white patch in flight; Northern Blue-faced Honeyeater, for Robinson and Laverock's 1900 Cooktown subspecies *harterti*, referred to as a species *E. harterti* by RAOU (1913) in the provisional list; Pandanus-bird, for a favoured habitat; Morning-bird, probably for its ringing dawn chorus; Gympie, apparently for the call.

Entomyzon cyanotis (Latham, 1801) [en-to-MI-zon si-a-NOH-tis]: 'blue-eared insect-sucker', see genus name, and from Greek *kuanos*, dark blue or sea-blue (Latin *cyaneus*, kingfisher blue as in Pliny (77–79AD), plus Greek ōtis, eared, as in one of Latham's several common names for the bird.

Black-chinned Honeyeater (breeding resident)

See family introduction. From the species name, for the black streak down chin and throat.

Other names: Black-throated Honeyeater, as used by Gould – it really is a toss-up; Yellow-backed or Golden-backed Honeyeater, for race *laetior* of northern Australia, with beautifully golden back, and regarded as a separate species *M. laetior* until recently; Black-cap, Black-chin, Black-throat.

Melithreptus gularis (Gould, 1837) [me-li-THREP-tʊs gʊ-LAH-ris]: 'throated honey-fed bird', see genus name, and from Latin *gula*, throat (see common name).

Strong-billed Honeyeater (breeding resident)

See family introduction. By Gould (1848) for his species name, though to our eyes it's not an especially striking characteristic.

Other names: Black-capped Honeyeater or Black-cap, logically to distinguish it from the only other Tasmanian *Melithreptus*, Black-headed Honeyeater, which has the entire head black – but they share the same names!; Cherry-picker, reported by Gould to be used by 'the Colonists of Van Diemen's Land'; Bark-bird, for its trunk-gleaning habits, as commented on by Gould.

Melithreptus validirostris (Gould, 1837) [me-li-THREP-tʊs va-li-di-ROS-tris]: 'strong-billed honey-fed bird', see genus name, and from Latin *validus*, strong or robust, and *rostrum*, bill, see common name. (See also Black-headed Honeyeater).

Brown-headed Honeyeater (breeding resident)

See family introduction. The only *Melithreptus* which, as an adult, does not have a black head. The name appears to have arisen spontaneously. Interestingly, Gould (1848) didn't recognise it or name it, thinking his specimens of it were immature White-napeds – later (1865) he was starting to doubt this.

Other names: Short-billed Honeyeater, from the species name, though the bill's not very different from that of other *Melithreptus*. Large-billed or Long-billed Honeyeater, an apparently wilful perversity, which RAOU (1978) and HANZAB (1990–2006, vol. 5) have perpetuated by reporting without clarification; it goes back to Alex North's description of *M. magnirostris* from Kangaroo Island in 1905, with the confusing 'large-billed' species name intended as a comparison to distinguish it from the mainland species – it is now recognised as a race. Island Brown-headed Honeyeater, for the same race. Least Honeyeater, which we suspect has been applied in error instead of for Brown Honeyeater; Brown-head; Cobbler, recorded rarely (Chisholm 1929) – we believe it refers to the sharp chipping flight calls (like a tapping cobbler's hammer); Western Brown-headed Honeyeater, for Milligan's 1903 species (now race) *M. leucogenys* from south-western Australia.

Melithreptus brevirostris (Vigors & Horsfield, 1827) [me-li-THREP-tʊs bre-vi-ROS-tris]: 'short-billed honey-fed bird', see genus name, and from Latin *brevis*, short, and *rostrum*, bill.

White-throated Honeyeater (breeding resident)

See family introduction. Gould's name, from his species name, perhaps to distinguish it from Black-chinned – it doesn't help with White-naped, though.

Other names: White-chinned Honeyeater, equally valid; Gay or Gay-tinted Honeyeater, used in the late 19th and early 20th centuries (Morris 1898; Campbell 1901; RAOU 1913) for De Vis' 1884 species *M. vinitinctus* from the Gulf of Carpentaria, no longer recognised as a taxon – it is unclear who applied the common name (it wasn't De Vis) but we may safely assume that it referred only to bright plumage; Grey Honeyeater, which makes no sense to us at all beyond a mistranscription of Gay, and we can't find usage beyond the HANZAB (1990–2006, vol. 5) citation.

Melithreptus albogularis Gould, 1848 [me-li-THREP-tʊs al-boh-gʊ-LAH-ris]: 'white-throated honey-fed bird', see genus name, and from Latin *albus*, dull white, and *gula*, throat.

White-naped Honeyeater (breeding resident)

See family introduction. Refers to the same tapering white neck band as the species name.

Other names: Lunulated Honeyeater, used by Gould (1848) directly from the species name; Black-crowned Honey-sucker (Lewin 1822); Red-eyed Honeyeater (Latham 1823), for the distinctive red eye ring of the nominate race; Black-cap, or Black-capped Honeyeater – while true of most other *Melithreptus*, this is the most familiar mainland species; White-nape.

Melithreptus lunatus (Vieillot, 1802) [me-li-THREP-tʊs loo-NAH-tʊs]: 'crescent honey-fed bird', see genus name, and from Latin *lunatus*, crescent-shaped (*luna*, moon).

Gilbert's Honeyeater (breeding resident)

See family introduction. For Gould's highly skilled and indefatigable naturalist-collector John Gilbert; Gould cited Gilbert's description of its habits, and it seems almost certain that he collected it for Gould.

Other names. White-naped Honeyeater, from when it was lumped with the previous species, until recently; Western Lunulated Honeyeater, see previous species; Swan River Honeyeater (or Honey-eater) as used by Gould, for the Swan River colony where it lives, that being the then name for the south-west of Western Australia; White-eyed Honeyeater, for Mathews' 1909 *M. whitlocki* from south-west Western Australia, cited by RAOU (1913) in the provisional list – see also New Holland Honeyeater.

Melithreptus chloropsis (Gould, 1848) [me-li-THREP-tʊs klor-OP-sis]: 'green-eyed honey-fed bird', see genus name, and from what Gould describes as the bird 'having the bare space above the eyes of a pale green' (Greek *khlōros*, green, and *opsis*, face or appearance).

Black-headed Honeyeater (breeding resident)

See family introduction. As used by Gould from his name *melanocephalus*; the only *Melithreptus* to have an all-black head, unbroken by a white band at the rear.

Other names: Black-cap or Black-cap Honeyeater; King Island Honeyeater, for Mathews' (1913) race *alisteri*, restricted to King Island.

Melithreptus affinis (Lesson, 1839) [me-li-THREP-tʊs af-FEE-nis]: 'related honey-fed bird', see genus name, and from Latin *affinis*, related, though Lesson (1839b) didn't say to what. However, Swainson (1837) originally named this genus *Eidopsarus*, with species *E. bicinctus*, an earlier name for the Strong-billed Honeyeater *Melithreptus validirostris*; Lesson placed his bird in the same genus, so it is likely that *E. bicinctus* was the bird he believed it was related to.

White-eared Honeyeater (breeding resident)

See family introduction. From the species name, used by Latham (1823).

Other names: White-eared Thrush (Latham 1801); New Norcia or Western White-eared Honeyeater, for race *novaenorciae*, described by Alexander Milligan in 1904 from near New Norcia, north of Perth, as a full species *Ptilotis novaenorciae* but now known to represent a race found right across southern Australia.

Nesoptilotis leucotis (Latham, 1802) [ne-so-ti-LO-tis loo-KOH-tis]: 'white-eared non-Tasmanian feather-ear', see genus name, and from Greek *leukos*, white, and *-ōtis*, eared. This bird is in mainland Australia, rather than Tasmania. Mathews actually included seven races of *Nesoptilotis leucotis* in his *List of the Birds of Australia*.

Yellow-throated Honeyeater (breeding resident)

See family introduction. From the species name (and for the obvious character), used by Gould.

Other names: All other names are from early Tasmania, because this bird is a common endemic to that state; it is curious that 'green' is so recurrent: it is at best only dull olive. Cherry-picker or Green Cherry-picker, not offered with affection; Linnet or Green Linnet, though it is not especially linnet-like; Green Dick, a mark of familiarity, like Willie Wagtail.

Nesoptilotis flavicollis (Vieillot, 1817) [ne-so-ti-LO-tis flah-vi-KOL-lis]: 'yellow-necked island feather-ear', see genus name, and from Latin *flavus*, golden (or reddish) yellow (from *flagro*, burn), and *collum*, the neck. This was Gould's *Ptilotis flavigula*, Vieillot's *Melithreptus flavicollis*, Mathews' *Nesoptilotis flavicollis flavigula*.

Gibberbird (breeding resident)

Gibbers are wind-polished stones, derived from the hard, often silcrete caps of eroding mesas ('jump-ups') in inland Australia; these stones cover vast plains and form the key habitat of this bird.

Other names: Gibber or Desert Chat or Bush-chat (see Crimson Chat); Desert Bird.

Ashbyia lovensis (Ashby, 1911) [ASH-bi-uh luh-VEN-sis]: 'Ashby's [bird] from Love's place', see genus name, and for the Rev. James Robert Beattie Love (1889–1947), a teacher turned clergyman and superintendent of various missions around Australia, who was perhaps a bit more enlightened than many in his time. In World War I he was in the Camel Corps and Light Horse Regiments, and won the DCM and MC. He was an amateur anthropologist who published many papers on aspects of the lives of the Indigenous people he lived among, and while in the Kimberley he translated Worora stories into English, and parts of the Bible into Worora. 'Love's place' plus the suffix *-ensis*, usually indicating the place of origin of the type specimen, which was at the time Leigh's (now Leigh) Creek in South Australia.

Crimson Chat (breeding resident)

See family introduction. The three 'coloured' Australian chats divide neatly up by colour – though mostly only per the males. The male of this species has a gloriously crimson cap and breast.

Other names: Tricoloured Epthianura, Gould's standard *modus operandi* of simply translating the species name of a bird he'd just described; Tricoloured Chat or Bush-chat – Bush-chat or Bush Chat was used widely until the early 20th century (Campbell 1901; RAOU 1913, though by 1911 Leach had dropped the 'bush'); Red or Saltbush Canary, for its chenopod habitat, though it isn't a noted songster; Crimson-breasted Nun, see White-fronted Chat; Crimson-fronted, Red-breasted, Red-capped Chat or Bush-chat; Crimson Tang, see White-fronted Chat.

Epthianura tricolor Gould, 1841 [ep-thi-a-NOO-ruh TRI-ko-lor]: 'three-coloured refined-tail', see genus name, and from Latin *tricolor* – the three colours are red, white and dark brown, according to Gould (1840), who was mightily impressed with this bird, after the fashion of the times: 'As may be supposed, the sight of a bird of such beauty, and which, moreover, was entirely new to me, excited so strong a desire to possess it that scarcely a moment elapsed before it was dead and in my hand' (Gould 1848).

Orange Chat (breeding resident)

See family introduction and Crimson Chat. Again only the male is orange, but he is **very** orange on the breast and head.

Other names: Orange-fronted Chat or Nun or Tang or Bush-chat (from the species name – Gould used Orange-fronted Epthianura – and see Crimson Chat); Golden-fronted or Orange-breasted Chat or Bush-chat (see also Yellow Chat); Bush or Saltbush Canary, see Crimson Chat; Northern Bush-chat, from Mathews' subspecies *obsoleta* (!) from the Northern Territory (Mathews 1912a), cited by RAOU (1913) in the provisional list as a species *E. obsoleta*. Plain-fronted Tit-Warbler, an odd one resulting from an unfortunate major misidentification by Edwin Ashby (see genus *Ashbya*) in 1910; he described a specimen from the McDonnell Ranges (admittedly in a 'bad state of preservation') as *Acanthiza flaviventris* and proposed this common name (Ashby 1910). RAOU (1913) still recognised it.

Epthianura aurifrons Gould, 1838 [ep-thi-a-NOO-ruh OW-ri-fronz]: 'golden-fronted refined-tail', see genus name, and from Latin *aurum*, gold, and *frons*, forehead.

Yellow Chat (breeding resident)

See family introduction and Crimson Chat. Here the female is yellow, while the male is more orange.

Other names: Yellow-breasted Chat, Bush-chat (see Crimson Chat) or Nun (see White-fronted Chat); Orange-breasted Bush-chat, from Mathews' subspecific name *tunneyi* from the Northern Territory (Mathews 1912a), cited by RAOU (1913) in the provisional list as a species *E. tunneyi*.

Epthianura crocea Castelnau & Ramsay, EP, 1877 [ep-thi-a-NOO-ruh KRO-se-uh]: 'saffron refined-tail', see genus name, and from Latin *croceus*, saffron-coloured, for the brilliant yellow of the male's head and underparts.

White-fronted Chat (breeding resident)

See family introduction. A bird's 'front' (or frons) is its forehead – but in fact the male also has a white face and all-white underparts, broken only by a broad black breastband. The name is taken straight from the species name.

Other names: The vivid pattern and relatively tame nature made this a surprisingly familiar bird in the past; it has attracted a wealth of names, several associated with its twangy call. White-fronted

Epthianura (Gould 1848); Tintack or Banded Tintack, Gar, Tang, White-fronted or Banded Tang, all from the call; Clipper or Tripper, perhaps call-based – Clipper is cited by Leach (1911); Nun or White-fronted Nun, for the black and white 'habit'; White-fronted Bush-chat (see Crimson Chat); White-fronted Acanthiza, used by Jardine and Selby (1826–35, vol. 2), who placed it with thornbills; Tasmanian Bush-chat, for Mathews' subspecies *tasmanica* (Mathews 1912a), cited as a species *E. tasmanica* by RAOU (1913) in the provisional list but not recognised thereafter; Dotterel, for its largely terrestrial habits; Baldyhead or Ballyhead, a reminder that 'bald' originally meant white, especially in reference to animals' heads ('baldy' is also used in Australia to refer to Hereford cattle); Moonbird, Moony, Bumps – we have no idea, and it is very difficult to find examples of actual usage; Jenny Wren, an old name for Eurasian Wren in England; Ringlet, Ringneck or Single-bar, all for the black breastband, the last undoubtedly by contrast with the superficially similar Double-bar (i.e. Double-barred Finch); Thistlebird, somewhat mysterious as it rarely eats seeds, though we have been informed that it regularly lines its nest with thistledown – again there is little evidence of the name being used, though as noted elsewhere this is not proof of no oral usage.

Epthianura albifrons (Jardine & Selby, 1828) [ep-thi-a-NOO-ruh AL-bi-fronz]: 'white-fronted refined-tail', see genus name, and from Latin *albus*, dull white, and *frons*, forehead.

Rufous-banded Honeyeater (breeding resident)

See family introduction. It's really the entire breast that is rufous, but we guess that's just a broader band. The name appeared apparently out of the blue in RAOU (1926).

Other names: White-throated Honeyeater, as used by Gould from his species name – a useful distinguisher from the closely related and similar Rufous-throated Honeyeater; Red-breasted or Rufous-breasted Honeyeater; Red-throated Honeyeater, the official name in RAOU (1913) though we wonder if that wasn't a mix-up with the next species.

Conopophila albogularis (Gould, 1843) [ko-no-PO-fi-luh al-boh-gʊ-LAH-ris]: 'white-throated gnat-lover', see genus name, and from Latin *albus*, dull white, and *gula*, throat.

Rufous-throated Honeyeater (breeding resident)

See family introduction. Again from Gould's species name.

Other names: Red-throated Honeyeater (Gould 1848).

Conopophila rufogularis (Gould, 1843) [ko-no-PO-fi-luh roo-foh-gʊ-LAH-ris]: 'red-throated gnat-lover', see genus name, and from Latin *rufus*, red, and *gula*, throat.

Grey Honeyeater (breeding resident)

See family introduction. Although 'grey' doesn't quite do it justice, this really is a hard bird to characterise physically; the first significant usage of this name was by Cayley in his important 1931 field guide.

Other names: Inconspicuous Honeyeater, used early (e.g. Mathews 1913, just 3 years after its description) and reiterating our earlier point, though it was also reflecting a species name, *Lacustroica inconspicua*, that North considered but rejected in favour of *whitei* (North 1910). White's or White Honeyeater, from the species name, White's being applied by North when he formally named the bird; Alfred Honeyeater, see species name.

Conopophila whitei (North, 1910) [ko-no-PO-fi-luh WIE-tee]: 'White's gnat-lover', see genus. According to HANZAB (1990–2006, vol. 5), the White was Captain Samuel Albert White (1870–1954, a naturalist who explored remote Australia in the early 20th century, collecting bird specimens for Gregory Mathews in particular. However, North makes it quite clear that he received the birds from Henry White of Belltrees, Scone, NSW, an inveterate collector. They were obtained for White by Frederick Whitlock near Wiluna, Western Australia (Whitlock 1910). **But** the bird was not named for Henry, as explained by North: 'in response to a request from the owner of the specimens, who has done so much recently to advance Australian ornithology, I have associated it with the name of his son, Mr Alfred Henry Ebsworth White, who, although yet young in years, I am informed is worthily following in his father's footsteps' (North 1910). (A photo of young Alfred pushing a home-made wooden wheelbarrow containing a massive nest appeared in 1909 as plate 9 in *Emu* 9(3), page 113, captioned 'Master Alfred H.E. White (youngest member of the A.O.U.) Showing Nest of Crow (*Corvus coronoides*) containing 6 Eggs'.) Hence Alfred Honeyeater, used, without explanation, by Whitlock (1910) – and for more detail on him see Inland Thornbill. The name was still used by the influential RAOU second checklist (RAOU 1926).

Bar-breasted Honeyeater (breeding resident)

See family introduction. Very apt, for the broad broken black bars across the chest, but surprisingly it has only recently been adopted – McDonald (1973) and particularly Slater (1974) in his ground-breaking field guide seem to have introduced it into the mainstream literature.

Other names: Fasciated Honeyeater, per Gould (1848) for his species name – confusing, see also Mangrove Honeyeater; White-breasted Honeyeater, as used by Hall (1899) and near universally thereafter until 1974 – it seems very strange, given its clear inappropriateness.

Ramsayornis fasciatus (Gould, 1843) [ram-zeh-ORN-is fa-si-AH-tʊs]: 'streaked Ramsay bird', see genus name, and for Latin *fascia*, a band or streak. Gould (1842) described it: 'Chest crossed by several semi-circular brownish-black fasciae'.

Brown-backed Honeyeater (breeding resident)

See family introduction. We can almost hear the desperation in this name. A delightful but generally nondescript Honeyeater – we might have focused on the pink bill had we been asked to provide a name for it. Hall was using the name by 1899 and may have coined it.

Other names: Unadorned Honeyeater, well, as we said; Modest Honeyeater, from the species name; Melaleuca Honeyeater or Bird, from a favoured habitat tree (especially the paperbarked species).

Ramsayornis modestus (Gray GR, 1858) [ram-zeh-ORN-is mo-DES-tʊs]: 'modest Ramsay bird', see genus name, and from Latin *modestus*, meaning many things including modest, gentle and unassuming. Or just plain plain!

Spiny-cheeked Honeyeater (breeding resident)

See family introduction. From Gould's genus name – he also used this as a common name (Gould 1848).

Other names: Western Spiny-cheeked Honeyeater, for Campbell's 1899 species *A. flavacanthus* from north-western Australia (RAOU 1913); Northern Spiny-cheeked Honeyeater, for Mathews' Northern Territory subspecies *territori* (Mathews 1912a), referred to by RAOU (1913) as a species in its provisional list; Spring-cheeked Honeyeater, surely a perpetuated error, though there is no evidence that we can see that it has been used beyond its citation in CSIRO (1969).

Acanthagenys rufogularis Gould, 1838 [a-kan-tha-GE-nis roo-foh-gʊ-LAH-ris]: 'rufous-throated spiny-cheek', see genus name, and from Latin *rufus*, red or reddish, and *gula*, throat.

Little Wattlebird (breeding resident)

See family introduction. Obviously smaller than the co-existing Red and Yellow Wattlebirds. This introduces an internal contradiction – a wattlebird without wattles, which must have been (and remains) utterly baffling to non-birders!

Other names: Brush Wattlebird (Gould 1848); Golden-winged Bee-eater (Latham 1801) and Honeyeater (Latham 1823), which are most perplexing names – see species name; Mock Gill-bird, in reference to its lack of 'gills' (see Red Wattlebird); Biddyquock, Charcoal Jack or Cookaycock, for the raucous three-part calls; Goruck Creeper, from Shaw (1819), who cited Vieillot, surely from the call – Gould reported Goo-guar-ruck of the 'Aborigines of the Coast of New South Wales' and we suggest that Shaw and Vieillot's name also originated there; Mockbird or Mockingbird, see Western Wattlebird.

Anthochaera chrysoptera (Latham, 1801) [an-tho-KEH-ruh kri-so-TE-ruh]: 'golden-winged flower-fancier', see genus name, and from Greek *khrusos/khruso-*, gold/golden, and *pteron*, wing. Golden-winged? Latham (1823) described the wing-coverts as 'fine yellow' and the wings do also have striking rufous panels visible when the bird is in flight.

Western Wattlebird (breeding resident)

See family introduction. Endemic to south-western Australia, until relatively recently regarded as conspecific with Little Wattlebird.

Other names: Little Wattlebird; Brush Wattlebird, for its dense understorey heath habitat (oddly this was the name used by Gould for the eastern Little Wattlebird – he recognised them as separate species); Lunulated Wattlebird, straight from the species name (Gould 1848). Brush Mockbird, or Mockingbird, on the face of it an odd name (and not recorded by the meticulous dictionary maker Edward Morris, 1898), since it was usually used to imply a mimic, such as a lyrebird; however, we wonder if 'mockbird' wasn't for the call (cf. Biddyquock) and 'mockingbird' followed. The other, and perhaps more convincing, potential explanation lies in the alternative name Mock Wattlebird, presumably for its lack of wattles.

Anthochaera lunulata Gould, 1838 [an-tho-KEH-ruh loo-nʊ-LAH-tuh]: 'crescent flower-fancier', see genus name, and from Latin *lunulatus*, crescent-shaped, (*luna*, moon, with *-ulus*, diminutive ending), referring to the silvery patch under the bird's eye and down the sides of the neck ('a lunulate mark of white': Gould 1848).

Red Wattlebird (breeding resident)

See family introduction. The name that is the bane of non-birders (and of birders who have to justify the name for a bird that is neither red nor especially associated with wattle trees!). 'Red-wattle Bird' would have made it clearer.

Other names: Wattled Crow or Bee-eater, from the erratic but influential John Latham (1801, 1823); Wattled Honey-eater (Gould 1848); Gill-bird, widely used from at least the middle of the 19th century, a reference to the wattles – this is still an occasional usage of 'gill' but was commonplace then; Gilly Wattler, a euphonious tautology; What's o'Clock, Chock or Barking Bird, inspired by the raucous gobble; Mutton-bird, for a bird which appears very regularly in accounts of 19th century menus and diets (more usually used for various shearwater species).

Anthochaera carunculata (Shaw, 1790) [an-tho-KEH-ruh ka-run-koo-LAH-tuh]: 'flower-fancier with little fleshy bits', see genus name, and from Latin *caruncula* (diminutive of *carno/carnis*, flesh), meaning a little piece of flesh, and referring to the wattles.

Yellow Wattlebird (breeding resident)

See family introduction and Red Wattlebird – this one's handsome wattles are yellow. The name was presumably in general use despite Gould not using it, because it appeared in Morris (1898) as first-choice name.

Other names: Wattled Honey-eater (Gould 1848) – it was odd that Gould gave this species its own species name but didn't distinguish it from Red Wattlebird by common name; Tasmanian Wattlebird, because it is endemic there; Great-wattled Honeyeater, Long-wattle Bird, for the strikingly long wattles.

Anthochaera paradoxa (Daudin, 1800) [an-tho-KEH-ruh pa-ra-DOK-suh]: 'incredible flower-fancier', see genus name, and from Greek *paradoxos*, strange, incredible, unexpected. Daudin called it *Corvus*, a crow, and it must indeed have seemed a very strange one to him.

Regent Honeyeater (breeding resident)

See family introduction. For the black and gold plumage, which it shares with Regent Bowerbird; see that species for the origin of the name. (The formal recognition that the Regent Honeyeater is actually a wattlebird is a very recent one.)

Other names: Black and Yellow Bee-eater or Honey-eater (Latham 1801, 1823); Warty-faced Honey-eater, the name introduced by John Lewin (1808) and by which the bird was widely known throughout the 19th and into the early 20th century (Campbell 1901), by which time sensibilities decided that a nobler name was required. 'Regent Honeyeater' was used by John Leach in 1911 but the name didn't appear from nowhere – Mock Regent Bird was also in use in the 19th century (Regent-bird being Regent Bowerbird), and Gould (1848) reported that 'Colonists of New South Wales' called it thus; Embroidered Merops and Bee-eater, used by George Shaw in his *Zoology of New Holland* (1794) and his mighty *General Zoology* (1819), from the species name, and for the beautiful patchwork plumage – hence also Embroidered Honeyeater; Flying Coachman, presumably again for the handsome livery-like plumage – while Morris (1898) didn't include it in his comprehensive *Dictionary of Austral English*, Leach (1911) cited it as an alternative name; Turkey-bird, for the call – it really is a wattlebird! (This name is not often recorded, but an article in the Melbourne *Argus* used it in 1927). This plethora of names for what is now a vanishingly rare species reflects its abundance in earlier times.

Anthochaera phrygia (Shaw, 1794) [an-tho-KEH-ruh FRI-gi-uh]: 'embroidered flower-fancier', see genus name, and from Latin *Phrygius*, for the inhabitants of Phrygia in Asia Minor, who were noted for their skill in gold embroidery (also, it appears, for their supposed indolence and stupidity – but that would spoil a very charming and appropriate name!).

Bridled Honeyeater (breeding resident)

See family introduction. Directly from the species name.

Other names: Mountain Honeyeater, for its montane rainforest habitat.

Bolemoreus frenatus (Ramsay, EP, 1874) [bohl-MOR-ee-ʊs fre-NAH-tʊs]: 'bridled bird of Boles and Longmore', see genus name, and from Latin *frenum*, bridle or reins, 'on account of the markings at the base of the bill and round the face', as Ramsay wrote. The word *frenum* is linked to frenulum, the little tethering membrane that attaches and limits the movement of various parts of the body, including the upper and lower lips.

Eungella Honeyeater (breeding resident)

See family introduction. Restricted to the vicinity of the town of Eungella, inland from Mackay, Queensland.

Other names: formerly regarded as an outlying population of Bridled Honeyeater.

Bolemoreus hindwoodi (Longmore & Boles, 1983) [bohl-MOR-ee-ʊs HIEND-wʊd-ee]: 'Hindwood's, Boles' and Longmore's bird', see genus name, and for Keith Alfred Hindwood (1904–71), Australian businessman and amateur ornithologist who became a world authority on Australian birds, president of the RAOU and recipient in 1959 of the Australian Natural History Medallion.

Yellow-faced Honeyeater (breeding resident)

See family introduction. Perhaps the broad yellow stripe from bill to ear below the eye, sandwiched by black, does not strictly represent a face, but it's a clear and useful name, taken pretty much from the species name and used by Gould (1848).

Other names: Yellow-eared Flycatcher, referred to in John White's journal in May 1788, making it one of the first birds named in the new colony (White 1790); Black-cheeked Honeyeater, by John Latham (1801); Yellow-gaped Honeyeater; Lesser Yellow-faced Honeyeater, for Mathews' briefly recognised species *Ptilotis barroni* from Cairns (Mathews 1912a); Chickup or Quitchup, for the two-note call.

Caligavis chrysops (Latham, 1801) [ka-li-GA-vis KREE-sops]: 'yellow-faced obscure bird', see genus name, and from Greek *khrusos/khruso-*, gold/golden, and ōps, the face (or eye).

Yellow-tufted Honeyeater (breeding resident)

See family introduction. For the yellow tufts behind the eye, which often look larger in life than they appear in the illustrations.

Other names: Yellow-tufted Flycatcher or Honey-eater, or Mustache or Tufted-eared Honeyeater, all from Latham (1801, 1823), who favoured a scattergun approach; Black-faced Honeyeater, from the species name; Golden-tufted or Gippsland Golden-tufted Honeyeater; Yellow Whiskers or Whisky; Helmeted Honeyeater, the Victorian bird emblem, for race *cassidix*, formerly a full species *Ptilotis cassidix*, named by Gould for its feathery golden 'helmet'; Spectacled or Yellow-throated Honeyeater; Speckled Honeyeater, not a very obvious name but perhaps a mishearing of Spectacled.

Lichenostomus melanops (Latham, 1801) [li-ke-no-STOH-mʊs MEL-an-ops]: 'black-faced lichen-mouth', see genus name, and from Greek *melas*, black, and ōps, the face (or eye, but in this case the black mask).

Purple-gaped Honeyeater (breeding resident)

See family introduction. A bird's gape is the bare skin immediately behind the bill (usually particularly conspicuous in young birds to assist parents in finding the relevant food receptacle!); the purple is clear in very good conditions. The name does not appear until the 20th century (RAOU 1913).

Other names: Wattle-cheeked Honeyeater, the name used by Gould from his species name; though the gape itself is not a wattle, at very close range it has a fleshy wattle-like extension 'the lengthened wattle, of a beautiful lilac colour … but slightly pendulous' (Gould 1848). Hall was still using the name in 1899 in his *Key to the Birds of Australia*. Lilac-wattled Honeyeater, widely cited although we can't find an instance of actual usage – perhaps a pity as it more closely reflects the colour than does 'purple'; Mountain Wattle-cheeked Honeyeater, used by Mathews (1913) in his (perhaps excessively) comprehensive *List of the Birds of Australia* for his race *stirlingi* from the Stirling Ranges, Western Australia – others of his included Victorian Wattle-cheeked Honeyeater for race *howei* and Gulf Wattle-cheeked Honeyeater for race *carpentariensis* (although the bird is not found within 1500 km of that Gulf!). The only two races now recognised are what he called Kangaroo Island Wattle-cheeked Honeyeater for nominate race *cratitius* and Western Wattle-cheeked Honeyeater for race *occidentalis*, though it is found right across southern Australia.

Lichenostomus cratitius (Gould, 1841) [li-ke-no-STOH-mʊs kra-TI-ti-ʊs]: 'wattled lichen-mouth', see genus name, possibly from Latin *cratitius* (also *craticius*), which means wattle, as in wattle-and-daub or wickerwork rather than dangly bits. But there is nothing in any of Gould's descriptions of this bird to explain his choice of this word other than 'the lengthened wattle, of a beautiful lilac colour'. Perhaps he could have chosen Latin *carunculatus* instead, meaning these kinds of fleshy parts (see Red Wattlebird).

Bell Miner (breeding resident)

See family introduction. The tinkling notes of a Bell Miner colony were noted very early; Banks in the *Endeavour Journal* reported a morning chorus 'almost imitating small bells' (Banks 1770). David Collins, 30 years later, was less impressed, referring to 'the melancholy cry of the bell-bird' (Collins 1802) but telling us that the name was established by then. Gould was so convinced of the ubiquity of the name that he used Australian Bell-bird as first-choice name (Gould 1848), an unusual informality for him.

Other names: Black-browed Thrush (Latham 1801) as per his species name; Green Manorina, from Vieillot's 1818 name *M. viridis*, presumably unaware of Latham's name; Bellbird, Bell Mynah or Minah.

Manorina melanophrys (Latham, 1801) [ma-no-REE-nuh me-LA-no-fris]: 'black-browed large-nostrilled bird', see genus name, and from Greek *melas*, black, and *ophrus*, eyebrow. They are very small eyebrows looking a bit like discreet eye makeup.

Noisy Miner (breeding resident)

See family introduction. A difficult name to object to, but it seems not to have become established in the literature until close to the 20th century; the earliest we know of was by C.M. Lyons of Melbourne University who used the name Noisy Miner in 1901 (Lyons 1901) – ironically, given that as he was writing about Cooper Creek, he was actually referring to Yellow-throated Miner. (Henderson (1832) referred to 'restless and noisy minas' but as this was purely descriptive it was not the origin of Noisy Miner.)

Other names: Chattering Bee-eater or Honey-eater (Latham 1801); Garrulous Honeyeater, Gould's name, from the then species name *Myzantha garrula*; Noisy Mina or Minah; Noisy Micky, a familiar name, like Willie Wagtail; hence also just Micky or Mickie; Black-headed Minah, from the species name, as used by Mathews (1913) – he also used Southern, Queensland and Tasmanian for various races, but now only Tasmanian (*leachi*) is recognised. (We believe that Black-backed Minah cited in HANZAB 1990–2006, vol. 5 may be a misprint for this name.) Black-headed Honey-eater, used by Gould (1848) for his claimed separate species in Tasmania. Cherry-eater, for its fruit-eating; Soldierbird – a name already familiar when Richard Howitt reported in 1845 that he 'found that this creature was very appropriately named the soldier-bird. It is the very sentinel of the woods, sending far on before you intelligence of your coming' (Howitt 1845). Squeaker; Snakebird, for the habit of the entire colony following and harassing predators, including snakes.

Manorina melanocephala (Latham, 1801) [ma-no-REE-nuh me-la-no-SE-fal-uh]: 'black-headed large-nostrilled bird', see genus name, and from Greek *melanokephalos*, from *melas*, black, and *kephalē*, head. The head is not completely black but there is a broad band of black from the crown down over the eyes to the side of the bill, making the bird look rather as if it wears a bandit's mask – perhaps not an entirely unkind comment, given some of the bird's habits.

Yellow-throated Miner (breeding resident)

See family introduction. From Gould's species name, true but not very obvious in the field.

Other names: Yellow-throated Minah; White-rumped Miner, very arguably a more useful name; Micky or Mickey Miner, see Noisy Miner.

Race *lutea* of north-western Australian: Luteous Honeyeater, Gould's name for what he took to be a species; Yellow Minah or Miner. This was recognised as a separate species, *Myzantha lutea*, as recently as RAOU (1913). From Latin, *luteus*, golden or saffron yellow.

Race *obscura* of south-western Australia: Sombre Honey-eater, again from Gould who thought it a separate species, for its darker plumage; Sombre Miner, Dusky Mina or Minah. This was recognised as a separate species, *Myzantha obscura*, as recently as by Cayley (1931). From Latin *obscurus*, dark or dusky.

Manorina flavigula (Gould, 1840) [ma-no-REE-nuh fla-VI-gʊ-luh]: 'yellow-throated large-nostrilled bird', see genus name, and from Latin *flavus*, golden (or reddish) yellow, from *flagro*, burn, and *gula*, throat.

Black-eared Miner (breeding resident)

See family introduction. From the species name, though it doesn't differentiate it from other miners.

Other names: Dusky Miner.

Manorina melanotis (Wilson FE, 1911) [ma-no-REE-nuh me-la-NOH-tis]: 'black-eared large-nostrilled bird', see genus name, and from Greek *melas*, black, and *-ōtis*, eared.

White-fronted Honeyeater (breeding resident)

See family introduction. A bird's front (or frons) is its forehead; in this case it is very prominently white in contrast to the dark face. This is the name used by Gould (1848), from his unambiguous species name, and nobody since then seems to have found a reason to look further, though before that Gould reported that the 'Colonists of Swan River' called it Black-throated Honeyeater.

Purnella albifrons (Gould, 1841) [per-NEL-uh AL-bi-fronz]: 'Purnell's white-fronted bird', see genus name, and from Latin *albus*, white, and *frons*, forehead.

White-gaped Honeyeater (breeding resident)

See family introduction. A bird's gape is the area immediately behind the bill and this species has a white spot there, small but conspicuous on the uniformly dull body. The name arose somewhere in the second half of the 19th century – by 1898 Morris was using it in his *Dictionary of Austral English*.

Other names: Uniform Honeyeater, used by Gould from his own species name; River Honeyeater, for an important habitat; Erect-tailed Honeyeater, cited by various sources but it is not at all clear who used it or why.

Stomiopera unicolor (Gould, 1843) [STO-mi-o-PE-ruh oo-NI-ko-lor]: 'one-colour pocket-mouth', see genus name, and from Latin *unicolor*: *unus/uni-*, one, and *color*, colour.

Yellow Honeyeater (breeding resident)

See family introduction. An undeniably yellow honeyeater, named by Gould from his species name.

Other names: Bush Canary, for obvious reasons (visual, not aural); Broadbent's Honeyeater, on the face of it a total mystery, though Kendall Broadbent's name recurs in northern Australian birds – he was an indefatigable collector for the Queensland museum in the late 19th century – and two other species, Bridled Honeyeater and Silver-eared Honeyeater (*Lichmera alboauricularis*) of New Guinea are also sometimes called Broadbent's Honeyeater (a confusion introduced into the Australian arena by RAOU 1913, apparently believing, for obscure reasons, that *L. alboauricularis* also occurred in Queensland). There seems not to have ever been a subspecies of that name and no recorded usage of it can be found, although both HANZAB and the *Handbook of the Birds of the World* cite it. The solution lies with Gregory Mathews, a law unto himself when it came to taxonomy; he assigned the new genus *Broadbentia* to this species, though nobody else seems to have used it (Mathews 1913). Inkerman Yellow Honeyeater, again by Mathews, for his race *addenda*, from a specimen from Inkerman in Queensland.

Stomiopera flavus (Gould, 1843) [STO-mi-o-PE-ruh FLAH-vʊs]: 'yellow pocket-mouth', see genus name, and from Latin, golden (or reddish) yellow, from *flagro*, burn.

Varied Honeyeater (breeding resident)

See family introduction. From the species name, as used by Gould.

Other names: Yellow-streaked Honeyeater (though perhaps Yellow Streaked would be more appropriate).

Gavicalis versicolor (Gould, 1843) [ga-vi-CA-lis ver-SI-ko-lor]: 'various-coloured anagram bird', see genus name, and from Latin *versicolor*, of a variety of colours.

Mangrove Honeyeater (breeding resident)

See family introduction. For its core habitat, though it only came to general usage in RAOU (1926).

Other names: Fasciated Honeyeater, per Gould (1869), from his species name, presumably forgetting that he'd already applied it to Bar-breasted Honeyeater – this name was still in use into the 20th century (RAOU 1913); Scaly-throated Honeyeater, very descriptive and clearly based on the species name; Island Honeyeater, cited by CSIRO (1969), but no other usage found.

Gavicalis fasciogularis (Gould, 1854) [ga-vi-CA-lis fa-si-o-gʊ-LAH-ris]: 'streaky-throated anagram bird', see genus name, and from Latin *fascia*, a band or streak, and *gula*, throat. Gould described the throat as 'feathers of the throat brownish-black, each bordered with pale yellow, presenting a fasciated appearance'.

Singing Honeyeater (breeding resident)

See family introduction. Many of us wonder what on earth this name could be about, and reading Gould's notes (he called it *Ptilotis sonorus*, as well as Singing, not realising that Vieillot had long beaten him to the naming punch) we can't help but wonder if he was listening to the same bird that we know, which at its most melodious can about manage a 'prrip'. 'As its name implies, it possesses the power of singing, and for an Australian bird, and particularly a Honey-eater, in no ordinary degree; its notes being so clear, full and loud as to be heard at a considerable distance, and very much resembling those of the Missel Thrush (*Turdus viscivorus*)' (Gould 1848). (We shall leave aside for now his scurrilous imputations on the singing powers of other Australian birds!)

Other names: Black-faced Honeyeater; Large-striped Honeyeater, presumably for the broad dark face-neck stripe, though there is no readily available evidence for its use; Grape-eater, for obvious reasons; Grey Peter, surely a mishearing of the previous though again there's little evidence of it being used; Larger Honey-sucker, reported by Gould (1848) to be used by the 'Colonists of Swan River', we suspect in contrast to the Brown Honey-sucker (Brown Honeyeater), another common and much smaller bird; Forrest's Honeyeater or Pale Singing Honeyeater, for race *forresti* of central Australia, for a while regarded as a separate species *Ptilotis forresti*; Rottnest Honeyeater, for Milligan's 1911 short-lived species name *Ptilotis insularis* from Rottnest Island, recognised by RAOU (1913).

Gavicalis virescens (Vieillot, 1817) [ga-vi-CA-lis VI-re-senz]: 'greening anagram bird', see genus name, and from Latin *viresco*, to become green. Vieillot (1816–19) considered the upper parts of the bird to have a greenish tinge, and extended that perception of the colour to the streaks on the breast and belly.

Yellow-tinted Honeyeater (breeding resident)

See family introduction. A slightly poetic version of Gould's species name, and applied by him (Gould 1848).

Other names: Pale Yellow or Yellowish Honeyeater (the latter being closest to the species name).

Ptilotula flavescens (Gould, 1840) [ti-LO-tʊ-luh fla-VE-senz]: 'little yellowish feather-ear', see genus name, and from Latin *flavus*, golden (or reddish) yellow, from *flagro*, burn, with *-escens*, becoming (i.e. '-ish').

Fuscous Honeyeater (breeding resident)

See family introduction. Straight from Gould's species name, and used by him – in fact the most he could say was that the bird was 'not distinguished by any brilliancy in its plumage'. Nobody seems to have been inspired to improve on it, either.

Ptilotula fusca (Gould, 1837) [ti-LO-tʊ-luh FʊS-cuh]: 'little dusky feather-ear', see genus name, and from Latin *fuscus*, dark or dusky, though not as dusky as the Dusky Honeyeater *Myzomela obscura*.

Grey-headed Honeyeater (breeding resident)

See family introduction. For the grey crown; the name seems to have been first generally used by RAOU (1926).

Other names: Keartland's Honeyeater, from the species name.

Ptilotula keartlandi (North, 1895) [ti-LO-tʊ-luh KERT-lan-dee]: 'Keartland's little feather-ear', see genus, and for George Keartland (1848–1926), a Melbourne *Age* photographer and typesetter who worked primarily to support his natural history studies. These included two major collecting trips to northern Australia in the 1890s, during the first of which he collected the type of this species in the Northern Territory. He later helped found the RAOU.

Grey-fronted Honeyeater (breeding resident)

See family introduction. A bird's front (or frons) is its forehead; the grey patch above the eyes is diagnostic, but reliably visible only at **very** close range. The name was introduced in *Recommended English Names* as late as 1978, when the RAOU finally put its foot down on the hitherto-universal but utterly inaccurate Yellow-fronted Honeyeater.

Other names: Yellow-fronted Honeyeater, widely used and formerly first-choice, but totally perplexing; Plumed Ptilotis or Honeyeater (the former being Gould's name, from his species and genus names);

Yellow-necked Honeyeater, for Campbell's 1910 name *Ptilotis planasi* from north-western Australia, recognised by RAOU (1913) but not thereafter, though *planasi* is still recognised as a race across the tropical hinterlands.

Ptilotula plumula (Gould, 1841) [ti-LO-tʊ-luh PLOO-mʊ-luh]: 'small-feathered little feather-ear', see genus name, and from Latin *plumulus*, a little feather or plume. The bird is named for what Gould (1848) called 'the double tuft of black and yellow feathers situated on the sides of the neck'.

Yellow-plumed Honeyeater (breeding resident)

See family introduction. For the conspicuous yellow crescent around the sides of the neck; the name came into use sometime between Gould (1848) and its appearance in Morris' 1898 *Dictionary of Austral English*.

Yellow-plumed Honeyeater *Ptilotula ornata*, Yellow-tufted Honeyeater *Lichenostomus melanops*, Grey-fronted Honeyeater *P. plumula*, Mangrove Honeyeater *Gavicalis fasciogularis* and Fuscous Honeyeater *P. fusca*

Other names: Graceful Ptilotis (per Gould 1848, for his genus name) or Graceful Honeyeater – aside from describing it as an 'elegant little bird' Gould gives no indication of why he chose this vague common name over something related to the plume, which he emphasised in his species name; Mallee Honeyeater, for its key habitat.

Ptilotula ornata (Gould, 1838) [ti-LO-tʊ-luh or-NAH-tuh]: 'little adorned feather-ear', see genus name, and from Latin *ornatus*, adorned or ornate.

White-plumed Honeyeater (breeding resident)

See family introduction. For the conspicuous white streak across the sides of the neck; applied by Gould (1848) from his species name.

Other names: A familiar and in-your-face bird that has attracted several names. Greenie; Linnet, for no obvious reason (*Carduelis cannabina* being a finch with distinct red face and breast for the breeding male); Australian Canary – more for its yellow-olive colour than its song; Chickowee, for its strong three-note call; Ringneck or Ringeye – the latter another mystery; Carter Honeyeater, a short-lived species *Ptilotis carteri* (now race *carteri*) named by Archibald Campbell in 1899 for Thomas Carter, a Yorkshire naturalist who came to Western Australia in 1887 and bought a station near Cape Range, where he collected the type (see also Spinifexbird); Cloncurry Honeyeater, for race *laelavalensis*, collected in north-western Queensland; Pallid Honeyeater, used for a while in the 1920s for *laelavalensis*, both as a full species *Ptilotis leilavalensis* (Carter 1921) and a race (Nicholls 1924) – inland birds tend to be paler.

Ptilotula penicillatus (Gould, 1837) [ti-LO-tʊ-luh pe-ni-si-LAH-tuh]: 'little feather-ear with a little paintbrush', see genus name, and from Latin *penicillus*, paintbrush, a diminutive of *peniculus*, brush (itself a diminutive of *penis*, a tail). For the bird's neck plumes, see common name.

Graceful Honeyeater (breeding resident)

See family introduction. Directly from Gould's species name, which he explains thus: 'it differs from both [Lewin's Honeyeater, and Wattled Honeyeater of the south-west Pacific] in the greater slenderness of its form, in its diminutive size' (Gould 1866).

Other names: Lesser or Little Yellow-spotted Honeyeater (RAOU 1913) – see that very similar and coexistent species; Grey-breasted Honeyeater, presumably to distinguish from the streaky-breasted Yellow-spotted; Slender-billed Honeyeater, rarely used; Graceful Meliphaga, used in New Guinea.

Meliphaga gracilis (Gould, 1866) [me-li-FA-guh GRA-si-lis]: 'slender honey-glutton', see genus name, and from Latin *gracilis*, slender.

White-lined Honeyeater (breeding resident)

See family introduction. Henry White's description of the species specified the 'conspicuous white stripe extending below the eye from the gape to behind the ear coverts' (White 1917a).

Other names: White-striped Honeyeater.

Meliphaga albilineata (White HL, 1917) [me-li-FA-guh al-bi-li-ne-AH-tuh]: 'white-lined honey-glutton', see genus name, and from Latin *albus*, dull white, and *linea*, a line. See common name.

Kimberley Honeyeater (breeding resident)

See family introduction. Endemic to the Kimberley area; formerly included in White-lined Honeyeater, described as a race in 1989 and first recognised as a new species in Christidis and Boles (2008).

Meliphaga fordiana Schodde, 1989 [me-li-FA-guh for-di-AH-nuh]: 'Ford's honey-glutton', see genus name, and for Dr Julian Ford, Western Australian chemist turned ornithologist who died in 1987 aged only 55. He was known among many other things for recognising in 1969 that the races of Australian Magpie represent one species.

Yellow-spotted Honeyeater (breeding resident)

See family introduction. The name was used by Gould (1869) for the yellow spot behind the eye – very like that of Lewin's (next species).

Other names: Yellow-spot Honeyeater; Lesser Lewin's or Little Lewin's Honeyeater, the former a first-choice name as recently as the CSIRO index in 1969, for its infamous similarity to that species.

Meliphaga notata (Gould, 1867) [me-li-FA-guh no-TAH-tuh]: 'marked honey-glutton', see genus name, and from Latin *notatus*, marked.

Lewin's Honeyeater (breeding resident)

See family introduction. From the species name, which itself has a complex history. Lewin painted the bird and called it Yellow-eared Honeyeater *Meliphaga chrysotis* (based on Latham's earlier *Certhia chrysotis*). Swainson later deemed this name inadmissible (Latham had already used it, but it is unclear for what – not for the first time) and published his own name honouring Lewin. Gould nonetheless was still using Latham's names in 1848.

Other names: Spotted-eared Creeper (Shaw 1819); Yellow-eared Honeyeater, see common name; Lewin Honeyeater; Orange- or Banana-bird, neither necessarily applied fondly, for its fruit-loving habits; White Lug or Jug, both references to its whitish 'ears', per HANZAB (1990–2006, vol. 5), though it is impossible to verify any actual usage – however, they both sound like oral tradition and as such would not necessarily leave a written trace.

Meliphaga lewinii (Swainson, 1837) [me-li-FA-guh LOO-win-i-ee]: 'Lewin's honey-glutton', see genus name, and for John Lewin, English naturalist and engraver who came to Australia with etching paraphernalia in 1800 to collect specimens, initially insects but later birds, to produce books for English patrons. His 1813 *Birds of New South Wales* (based on his earlier *Birds of New Holland* for six English subscribers) was the first illustrated natural history book to be printed in Australia.

DASYORNITHIDAE (PASSERIFORMES): bristlebirds

Dasyornithidae Schodde 1975 [da-si-or-NI-thi-dee]: the Hairy Bird family, see genus name Dasyornis.

'Bristlebird' refers to the prominent rictal 'bristles' (stiff barbless feathers around the bill, perhaps serving a sensory function). The term seems to have been coined by Banks' somewhat irascible collector George Caley; Nicholas Vigors and Thomas Horsfield, in their lengthy paper to the Linnean Society on the birds in the Society's collection, reported that Caley 'procured [the bird] in a scrubby place north of Paramatta … calls it in his notes 'Bristle Bird'' (Vigors and Horsfield 1827).

The genus

Dasyornis Vigors & Horsfield, 1827 [da-si-OR-nis]: 'hairy bird', from Greek *dasu-*, hairy or rough, and *ornis*, bird, for 'the singular bristles that spring from the front of the forehead above the bill' (Vigors and Horsfield 1827).

The species

Eastern Bristlebird (breeding resident)

The only east coast species.

Other names: New Holland Bristle Bird, used by Jardine and Selby (1826–35, vol. 2); Bristlebird or Bristle-bird, as used by Gould (1848); Eastern and Western at times were regarded as one species.

Dasyornis brachypterus (Latham, 1801) [da-si-OR-nis bra-ki-TE-rʊs]: 'short-winged hairy bird', see genus name, and from Greek, *brachy-*, short, and *pteron*, wing. Bristlebirds are mostly ground dwellers, so don't have or need well-developed wings.

Western Bristlebird (breeding resident)

Certainly limited to the south-west, but Rufous occurs there too (or did until recently).

Other names: Long-billed Bristlebird, coined by Gould (1848) from his species name; Brown Bristlebird, to distinguish from Rufous; Western Australian Bristlebird.

Dasyornis longirostris Gould, 1841 [da-si-OR-nis lon-gi-ROS-tris]: 'long-billed hairy bird', see genus name, and from Latin *longus*, long, and *rostrum*, bill. Gould explained that it differed from the previous species in 'being of a smaller size and in having a longer bill' (Gould 1848); his observations on the bill are not supported by measurements cited in HANZAB (1990–2006, vol. 6).

Rufous Bristlebird (breeding resident)

For the conspicuously rufous cap (both other species have rufous wings).

Other names: Rufous-headed Bristlebird, used by Gould (1869); Lesser Rufous Bristlebird, for the apparently extinct smaller western race *litoralis*, described by Milligan in 1902 as a full species and recognised as such by RAOU (1913); Cartwheel-bird, for the creaking reeling song.

Dasyornis broadbenti (McCoy, 1867) [da-si-OR-nis BRAWD-ben-tee]: 'Broadbent's hairy bird', see genus name, and for Kendall Broadbent (1837–1911), who migrated from England in 1853 and spent 30 years working for the Queensland Museum as a taxidermist and collector, widely respected as an authority on Australian birds. He collected throughout eastern Australia and several times in New Guinea, where he contracted a chronic 'fever' (malaria?) (Anonymous 1911). Shortly after arriving in Australia he collected the type specimen in Victoria, but it was 9 more years before it was recognised as a separate species.

PARDALOTIDAE (PASSERIFORMES): pardalotes

Pardalotidae Strickland, 1842 [par-da-LO-ti-dee]: the Leopard-spotted family, see genus name *Pardalotus*.

The name pardalote, unsurprisingly for the time, was adopted in English straight from Louis Vieillot's 1816 genus name; more surprising is that such a common and familiar group of birds should not over time have had formally bestowed on it one of the more familiar names that they attracted – Diamond Bird was one such, and it was in near-universal use in the 19th century.

The genus

Pardalotus Vieillot, 1816 [par-da-LOH-tʊs]: 'leopard-spotted [bird]'. Vieillot gave the bird the name *pardalote* in French, as well as the genus name *Pardalotus* in Latin, ultimately from Greek *pardalōtus*, spotted like a leopard (Greek *pardalis*).

The species

Spotted Pardalote (breeding resident)

Apparently first used by Gould (1848), a direct translation of the species name.

Other names: A large number of names involved 'diamond', for the white spots – as Morris (1898) put it, 'The broken colour of the plumage suggested a sparkling jewel.' It goes back to at least the 1820s: Vigors and Horsfield (1827) reported to the Linnean Society in London, 'We are informed by Mr. Caley that this species is called diamond bird by the settlers, from the spots on its body. By them it is reckoned as valuable on account of its skin.' (For what purpose is unclear, though we might suppose the trade in stuffed birds for English cabinets.) Although the term originated with this species it was later applied, by association, to all other pardalotes. Diamondbird, Diamond-bird, Diamond Bird; Spotted Diamondbird, Diamond Sparrow; Bank Diamond or Diamond Dyke – for its habit of excavating a nesting hollow in a bank, the latter referring to 'dyke' meaning a hole in the ground (or the action of digging it); Speckled Manakin (Latham 1823) – sometimes the great English ornithologist revealed that his strength lay in dogged perseverance rather than insights, the true manakins being entirely unrelated (but small and colourful) South American birds; Headache Bird or Miss Piggy for its incessant three-note call (the latter very recent but perfect – linger over the 'Miss' then hasten through 'Piggy' and you've got it!); Pantherbird, for the classical allusion in 'pardalote' (see genus name).

For race *xanthopygus*, until recently regarded as a full species: Yellow-rumped, Yellow-tailed or Golden-rumped Pardalote; Golden-rumped Diamondbird. (It was almost Leadbeater's Pardalote *P. leadbeateri* for John Leadbeater, taxidermist at the National Museum in Melbourne; he obtained specimens from near Swan Hill and expressed the desire to have the apparent new species named for him. 'Bad taste', commented the great Edward Pierson Ramsay, who nevertheless felt obliged to do so under pressure from Gerard Krefft of the Australian Museum in Sydney, who lent him the specimens. A combination of fast political footwork and mail delays meant that Ramsay's name was pre-empted; for details see Hindwood (1949b)). (Note that this was not the Benjamin Leadbeater of Major Mitchell's Cockatoo *Lophochroa leadbeateri* fame, but Leadbeater's Possum was also John's.)

Pardalotus punctatus Shaw, 1792 [par-da-LOH-tʊs pʊnk-TAH-tʊs]: 'spotted leopard-spotted bird', see genus name, and from Latin *pungo/punctus*, prick or sting, hence a speck or spot. (Pliny (77–79 AD) wrote of '*mundi punctus*' – 'this speck of a world'.) The name is somewhat tautological, but when Vieillot (1816–19) named the genus *Pardalotus*, his reference was Latham's *Pipra punctata* so he knew what he was letting himself in for.

Forty-spotted Pardalote (breeding resident)

It's initially unclear why Gould settled on 40 as the magic number of spots possessed by this species, but in fact he uncharacteristically adopted the vernacular name of 'the Colonists of Van Diemen's Land' and once he'd also decreed it in the species name, it was fixed! A writer to the *Australasian* pondered this very question in 1896: 'only one wonders why the number 40 was pitched upon. Was it a guess? Or did the namer first shoot the bird and count?' (Morris 1898).

Pardalotus quadragintus Gould, 1838 [par-da-LOH-tʊs kwa-dra-GIN-tʊs]: 'forty leopard-spotted bird', see genus name, and from Latin *quadraginta*, forty.

Red-browed Pardalote (breeding resident)

The brow is more ginger than red, save for a spot in front of the eye, but is obvious.

Other names: Red-lored Pardalote, used by Gould (1848) and a more precise name; Red-browed or Fawn-eyebrowed Diamondbird – see Spotted Pardalote for 'diamondbird'; Cape Red-browed Pardalote, for race *yorki* of Cape York Peninsula; Pale Red-browed or Pallid Pardalote, for race *pallidus* of north-western Australia, originally described as a full species by Campbell in 1909 but no longer recognised at any level; Bellbird, for its haunting five-note call.

Pardalotus rubricatus Gould, 1838 [par-da-LOH-tʊs rʊb-ri-KAH-tʊs]: 'red-marked leopard-spotted bird', see genus name, and from Latin *rubrico*, to colour red (cf. *rubrica*, red earth or red ochre), for the bird's red eyebrow. You may be tempted to rubricate if you find errors in this book ...

Striated Pardalote (breeding resident)

For the streaked head (of some races), in contrast to that of the Spotted Pardalote. It is of course an oxymoron.

Other names: This species has one of the most impressive lists of ascribed names of any Australian bird, partly because of its ubiquity and familiarity but also because its several races have often and long been described as separate species; see also Spotted Pardalote for 'diamondbird'. Striped-crowned or Striped-headed Manakin by Latham (1823) – see Spotted Pardalote. Many names are based on the familiar and distinctive three-note (sometimes two-note) call, quite different from the Spotted Pardalote's: for example, Morris (1898) explained that they 'give forth a treble note which has secured for them the name of 'Pick-it-up' from our country boys', also Whittychu, Wittachew, Chuck-e-chuc, Pickwick, Chip Chip or Chook-chook; Allied Pardalote or Diamondbird, for Gould's *P. affinis*, when he believed that the Tasmanian birds were different from but still closely related to the mainland populations (we now know that at least some of the birds actually migrate across Bass Strait).

For race (formerly species) *striatus:* Yellow-tipped Pardalote or Diamondbird, for the yellow spot at the base of the white wing streak; Red-tipped Pardalote (RAOU 1913) (see also next two races!). From Latin *striatus*, streaked or furrowed.

For race *substriatus:* Red-tipped Pardalote or Diamondbird, for the red spot at the base of the white wing streak. From Latin *sub*, meaning in this case less, and *striatus*, streaked or furrowed.

For race (formerly species) *ornatus:* Orange-tipped Pardalote or Diamondbird, again for the red spot at the base of the white wing streak – it takes a vivid imagination to call it orange! Ramsay Pardalote, for Edward Ramsay, who named 'sub-subspecies' *assimilis*, as a subspecies of then species *ornatus*. Red-tipped Pardalote (RAOU 1926). From Latin *ornatus*, adorned.

For race (formerly species) *melanocephalus:* Black-headed Pardalote (by Gould 1848) who gave it species status (still recognised into the 20th century: RAOU 1926); Black-headed Diamondbird – it does indeed have a plain black crown. From Greek *melas*, black, and *kephalē*, head.

For race *uropygialis:* Yellow-rumped Pardalote, from Gould (1848) who named it as a species, still recognised by RAOU (1913); Chestnut-rumped Pardalote or Diamondbird, not as accurate as Gould's name. From Mediaeval Latin *uropygium*, the rump of a bird, in fact the 'parson's nose'.

Pardalotus striatus (Gmelin JF, 1789) [par-da-LOH-tʊs stri-AH-tʊs]: 'striped leopard-spotted bird', see genus name, and from Latin *striatus*, streaked or furrowed (see common name).

ACANTHIZIDAE (PASSERIFORMES): scrubwrens and allies, gerygones, thornbills, whitefaces

Acanthizidae Bonaparte, 1854 [ak-an-THI-zi-dee]: the Thorn-bush Dwelling family, see genus *Acanthiza*.

This is a group which was always going to attract European names, because the ubiquitous small relatively nondescript species inevitably reminded people of familiar Northern Hemisphere groups, especially wrens, tits and warblers. This was reinforced by the scientists, who until well into the second half of the 20th century insisted on interpreting them as belonging to Northern Hemisphere families.

'Scrubwren' was a relative latecomer, at least in the literature. The first reference appears to be in Edward Morris' 1898 *Dictionary of Austral English*, where he defined Scrub-Wren as 'any little bird of the Australian genus *Sericornis*' and listed nine species. Before that they were referred to formally as Sericornis (Gould 1848; Hall 1899). Leach (1911) used Scrub-Wren as first-choice name.

'Heathwren' was long preceded by the blandly simple 'wren' (as used by Gould 1848), followed by 'Ground-Wren', as reported by Morris (1898) (and still being used by Leach 1911; Lucas and Le Souef 1911; RAOU 1926). There are newspaper references to the word (most of which seem to refer to 'our' heathwrens) from 1926 onwards, but it seems to have been Cayley who popularised the term in *What Bird is That?* (1931).

'Fieldwren' was termed 'Reed-Lark' by Gould (1848), referring to both the old genus name *Anthus* (pipits, often then referred to as larks) and his new one. Morris (1898) was unaware of the name Fieldwren, but by the following year Hall (1899) was using it (as Field-Wren) in his influential *Key to the Birds of Australia*, so the coining could well have been his.

'Gerygone' was used from early in the 19th century as a common name, as well as a genus name (Gould 1848). However, by the end of the 19th century a more familiar name was being sought. Morris (1898), in his entry on Gerygone, noted that the 'new name for them is fly-eater'. He cited W.O. Legge of the Australasian Association for the Advancement of Science in 1895 as saying, '[The habits and habitats of the genus as] applied to *Gerygone* suggested the term Fly-*eater*, as distinguished from Fly-*catcher*, for this aberrant and peculiarly Australasian form of small Fly-catchers, which not only capture their food somewhat after the manner of Fly-catchers, but also seek for it arboreally.' It caught on – Hall used it in his 1899 massive *Key* as did Lucas and Le Soeuf (1911), Leach (1911) and RAOU (1913). A little later RAOU (1926) and Cayley (1931) had dumped it for the simple Northern Hemisphere-oriented 'warbler', which was still being used by Slater in his first passerine field guide in 1974 and by Pizzey (1980). However, by then the RAOU had proposed the more precise Gerygone in its influential *Recommended English Names* (RAOU 1978); 4 years later the first *Atlas of Australian Birds* (Blakers *et al.* 1984) used it, and it has remained.

'Thornbill', perhaps surprisingly, does not appear in the Australian literature until the beginning of the 20th century. Archibald Campbell (1901), in a footnote to an entry on the 'Brown Tit', commented that, 'Some recent authors use the term Thornbill – a name already applied to several Humming Birds – as a vernacular name for the Acanthizas'. Just 2 years earlier, the authoritative Robert Hall was still using the Northern Hemisphere term 'tit' for the thornbills; perhaps more surprisingly, in 1911 John Leach wrote, 'They have been called Thornbills by Mr. A.J. North. The name Tit-Warbler has been adopted by the 'names' sub-committee of the RAOU, pending the completion of the Australian Check-List.' (It's hard enough to get the world to take birders seriously as it is!)

Alfred North was the respected assistant in Ornithology at the Australian Museum, but his influence was insufficient, though Leach did use Thornbill as a secondary name. The RAOU tipped the balance by using thornbill as primary name in its influential second *Official Checklist* (RAOU 1926).

It is perhaps a little ironic that 'whiteface', nearly always interpreted by a neophyte birder as an adjective ('white-faced what?') started out as exactly that. Gould (1848) – who discovered the genus in the streets of Adelaide while on his way to pay his respects to Governor Gawler, who in turn obliged by having one in a cage for Gould by nightfall – called it White-faced Xerophila, for his genus name. Hall (1899) called it White-faced Titmouse; the latter is an old English name, later simplified to Tit everywhere except North America, for the family Paridae. Ironically, it was the belief that the birds belonged to this family (along with Wedgebills!) that led Hall to apparently coin the noun 'whiteface' (1907); by 1911 both Leach and Lucas and Le Soeuf were using it as first-choice name.

The genera

Oreoscopus North, 1905 [o-re-o-SCOH-pʊs]: 'watcher on the mountain', from Greek *oros*, *oreo-*, mountain, and *skopos*, a watcher or look-out stationed in a high place (*skopia*), for the bird's highland rainforest habitat.

Pycnoptilus Gould, 1851 [pik-NO-ti-lʊs]: 'firm feather [bird]', from Greek *puknos*, firm or solid, and *ptilon*, feather or down, named by Gould for 'the dense and silky character of its plumage' (Gould 1850).

Acanthornis Legge, 1887 [a-kan-THOR-nis]: 'thorn bird', based on genus *Acanthiza* (thornbills), and from Greek *acantha*, a thorn or prickle or prickly plant, plus *ornis*, bird. Gould (1855) noted a resemblance of this bird, which he called *Acanthiza magna*, to the smaller species of *Sericornis*.

Origma Gould, 1838 [o-RIG-muh]: 'excavation [bird]', from Greek *orugma*, excavation, trench or tunnel, for its tendency to choose caves or the underneath of rock ledges as sites for its hanging nest.

Calamanthus Gould, 1838 [ka-la-MAN-thʊs]: 'stubble pipit', from the genus *Anthus*, pipits, and from Greek *kalamē*, a stalk or stubble. The birds of this genus like low shrubby or scrubby vegetation, including tussocky grass, and they could be said to bear a somewhat vague resemblance to pipits.

Pyrrholaemus Gould, 1841 [pi-ro-LEH-mʊs]: 'flame-throat', from Greek *purrhos*, flame-coloured or red, and *laimos*, throat, the obvious name for the Redthroat to which it was first applied, though perhaps a bit over the top – it is really more a little rusty-coloured bib.

Sericornis Gould, 1838 [seh-ri-KOR-nis]: 'silk bird', from Greek *sērikon*, silk, *sērikos*, silken, and *ornis*, bird, for the silky plumage.

Smicrornis Gould, 1843 [smik-ROR-nis]: 'small, unimportant bird', from Greek *smikros* (or *mikros*), small or unimportant, and *ornis*, bird.

Gerygone Gould, 1841 [dje-RI-go-ne]: 'child of song', from Greek *gērugonos*, child of song (an echo), *gēruō*, sing (including to sing of or celebrate), and *gonē*, offspring or birth.

Acanthiza Vigors & Horsfield, 1827 [a-kan-THEE-zuh]: 'thorn-bush dweller' from Greek *acantheōn*, a thorn-bush, and *zaō*, to live on or off something. (cf. Greek *akanthis* or *akanthos*, probably a finch of some kind, named for its fondness for thorny plants. Arnott (2007) suggested a linnet or greenfinch perhaps, on grounds of Aristotle's (c. 330 BC) account of the bird, which described it as having 'poor colouring', thus excluding another likely candidate, the goldfinch.

Aphelocephala Oberholser, 1899 [a-fe-lo-SE-fa-luh]: 'simple head', from Greek *aphelēs*, simple or plain, and *kephalē*, head. Harry Oberholser may not have meant it this way, but we think he got it right – we've always thought (rather unkindly) that Whiteface had a somewhat dopey look about them.

The species

Fernwren (breeding resident)

For its Wet Tropics rainforest habitat. Until 1905 it was included in *Sericornis* as a scrubwren and the name Collared Scrubwren was universal; even in 1911 Lucas and Le Soeuf retained this name. The first RAOU *Official Checklist* introduced Fern-bird (RAOU 1913), which its successors amended to Fern-Wren (RAOU 1926).

Other names: White-throated or Collared Scrubwren, see common name, for the broad white throat band; Australian or Southern Fernwren, not at all clear what these distinctions refer to; Collared Fernwren.

Oreoscopus gutturalis (De Vis, 1889) [o-re-o-SCOH-pʊs gʊ-tʊ-RAH-lis]: 'throated watcher on the mountain', see genus name, and from Latin *guttur*, the throat or neck, for the white collar with its black bib below.

Pilotbird (breeding resident)

The name comes from a confusion of cause and effect. The bird follows Superb Lyrebirds and benefits from the leftovers of their vigorous scratchings; observers thought that they were guiding the bigger bird. Gould (1848) called it Downy Pycnoptilus (direct from its scientific name). The first usage of Pilotbird is cited by Morris (1898) as from the Melbourne *Argus* of 1893, where it noted that the bird was very rare; presumably it arose by local usage rather than formally. It was formalised by RAOU (1913).

Other names: Downy Pycnoptilus, see common name; Guinea-a-week, as good a rendition of a call as any we've heard.

Pycnoptilus floccosus Gould, 1851 [pik-NO-ti-lʊs flo-KOH-sʊs]: 'woolly firm feather bird', see genus name, and from Late Latin *floccosus*, full of tufts of wool (*floccus*, a tuft of wool), a further reference to how dense the plumage is.

Scrubtit (breeding resident)

'Scrub' recurs in Australian bird names, referring to any low dense vegetation; surprisingly, 'tit' was used into the 20th century for thornbills (which this bird was long defined as) – they were regarded as Northern Hemisphere Sylviid warblers, but not Parid tits. Gould (1869) called it Great Acanthiza (for the genus name of Thornbills); some time between then and 1898, when Edward Morris published his very thorough *Dictionary of Austral English*, 'Scrubtit' became established through popular usage in Tasmania.

Other names: Great Acanthiza, see common name; Tasmanian or White-breasted Scrubtit, the latter to distinguish it from the similar buff-breasted Tasmanian Scrubwren; Fern-weaver, for the beautiful woven nest, often involving treefern fibres; Mountain Wren; Great Tit, from the name of the blue and yellow European *Parus major*, though surely not for any supposed resemblance to it!

Acanthornis magna (Gould, 1855) [a-kan-THOR-nis MAG-nuh]: 'large thorn bird', see genus name, and from Latin *magnus*, large. Given that it is only 11–12 cm long, it would have been interesting to see a small one, but not only is this is now the only member of the genus, even when it was known under the previous name *Sericornis magnus* it was smaller than almost all others in that genus. We have to look back to Gould's name *Acanthiza magna* (1855) to make any sense of it – it would have been a comparative giant among the thornbills (*Acanthiza*), most of which are well under 11 cm.

Rockwarbler (breeding resident)

Long assumed, like many other Australian passerines, to be another Northern Hemisphere warbler, and for the obligate Sydney sandstone habitat. Lewin (1808) described it as *Sylvia solitaria* and translated it directly as Solitary Warbler, based on Latham's Solitary Flycatcher. By 1822 he had changed the name to Rock Warbler, which Gould entrenched in 1848 as Rock-Warbler.

Other names: Origma, often used, from the genus name; Ruddy Warbler (Latham 1801); Cataract-bird, Cave-bird, Rock Robin, Sandstone Robin or Warbler, all for the definitive habitat; Hanging Dick, another 'Willie Wagtail' type familiar name.

Origma solitaria (Lewin, 1808) [o-RIG-muh so-li-TAH-ri-uh]: 'solitary excavation bird', see genus name, and from Latin *solitarius*, lonely, solitary. Although it can be seen in family parties, it often appears alone or just with a mate. And as if to prove the aptness of its name, its repertoire of calls includes a rather mournful whistle.

Chestnut-rumped Heathwren (breeding resident)

See family introduction. Straight from the species name, but no help at all in separating it from the next species.

Other names: Heath Wren, as the relatively more familiar of the two species; Red-rumped Wren, Gould's name (1848); Ground-wren, see family introduction; Chestnut-rumped or Chestnut-tailed Ground-wren; Chestnut-tailed Heathwren; Scrub-warbler or Red-rumped Scrub-warbler, cited as an alternative name by Leach (1911) but nowhere else that we can find (until CSIRO 1969 and HANZAB 1990–2006, vol. 6) – another reminder that oral usage is quite possible without a name finding its way into the literature; Geelong Groundwren, perhaps for a near location for Melbourne birders and another example of the previous principle. Northern (RAOU 1913) or Charleville (RAOU 1926) Scrubwren, based on an intriguing story that began with a Queensland Museum specimen from Charleville, described as *Sericornis tyrannula* in 1905 by Charles De Vis, inaugural director of that museum; the specimen apparently disappeared and it took some impressive detective work by Shane Parker of the South Australian Museum to track it down and identify it as an immature Chestnut-rumped Heathwren (Parker 1984).

Calamanthus pyrrhopygius (Vigors & Horsfield, 1827) [ka-la-MAN-thʊs pi-ro-PI-gi-ʊs]: 'red-rumped stubble-pipit', see genus name, and from Greek *purrhos*, flame-coloured or red, and *pugē*, rump – see common name.

Shy Heathwren (breeding resident)

See family introduction. From the species name; Gould (1848) went to some trouble to emphasise its extreme timidity. Indeed, unsportingly, it was 'so excessively shy that I obtained a single specimen only during my stay in the district'. Shyness is a (sensible) characteristic of many mallee birds.

Other names: Mallee Heathwren or Wren, for the core habitat; Cautious Wren or Ground-wren, from the species name (the former used by Gould 1848); Shy Ground-wren or Scrub-warbler (see previous species); Red-rumped or Rufous-rumped Ground-wren (and we'd like to hear the RAOU 1913 committee explain the difference between Rufous-rumped and Chestnut-rumped, which it used for the previous species); Western Ground-wren; Island Ground-Wren, for Mathews' *Hylacola halmaturina* from Kangaroo Island, described in 1912 and used by RAOU (1913), then promptly dropped.

Calamanthus cautus Gould, 1843 [ka-la-MAN-thʊs COW-tʊs]: 'shy stubble-pipit', see genus name, and from Latin *cautus*, cautious or heedful, see common name.

Striated Fieldwren (breeding resident)

See family introduction. Heavily streaked, prompting Gould (1848) to call it Striated Reed-lark. It is more heavily streaked than Rufous Fieldwren and, although the two species were combined until recently (Blakers *et al.* 1984 and field guides), Gould did recognise them as different and may well have had that in mind.

Other names: Because only one fieldwren species was recognised until recently, it is not possible to allocate many of the names recorded to just one of the species. Some general names used right across the range of both species include Field Lark (see family introduction), Reed-lark (Gould's 1848 name), Calamanthus (from the genus name), Rush Warbler (perhaps another reference to the genus name), Cocktail; Mock Quail, presumably a shooters' name.

Names that can safely be limited to this species include Stinkbird, apparently a Tasmanian name – the bird's scent was enough to attract dogs' attentions; Striated Reedlark or Calamanthus or Fieldlark; Eastern or Streaked Fieldwren or Fieldlark; White-lored Fieldwren, Fieldlark or Reedlark, from North's 1902 species *C. albiloris* from Victoria, subsumed by 1926.

Calamanthus fuliginosus (Vigors & Horsfield, 1827) [ka-la-MAN-thʊs fʊ-li-gi-NOH-sʊs]: 'Sooty stubble-pipit', see genus name, and from Latin *fuligo*, soot, for the fact that the bird is streaked with black and has brownish-black on tail and some wing feathers, though the overall impression is hardly all that 'sooty'.

Western Fieldwren (breeding resident)

See family introduction. A recent split, referring to former race *montanellus* of Rufous Fieldwren in south-western Australia, though Rufous Fieldwren occurs as far west, but further north.

Other names. See previous and next species. There are only two alternative names that can be definitely ascribed to this species; Rock Fieldwren or Fieldlark, for Milligan's 1903 species *C. montanellus* from the Stirling Ranges; Warrenbird, a name from the Nullarbor Plain, where Brooker *et al.* (1979) reported that it 'Takes refuge in rabbit burrows when pursued … and is known to some local rabbit-trappers as the 'warrenbird''.

Calamanthus montanellus (Milligan, 1903) [ka-la-MAN-thʊs mon-ta-NEL-lʊs]: 'little mountain stubble-pipit', see genus and other names.

Rufous Fieldwren (breeding resident)

See family introduction. Slightly more rufous (and less heavily streaked) than the other two fieldwrens.

Other names: See names common to all three species, under Striated Fieldwren. Field Reed-lark, used by Gould (1848), though it has been applied to the previous species also. Western Fieldwren; Rufous Calamanthus; Rusty or Desert Wren, for North's 1896 *C. isabellinus* from Central Australia, still recognised by RAOU (1926) and Cayley (1931); Sandplain Fieldwren, also from the habitat; Rusty-red Fieldwren for Campbell's 1899 *C. rubiginosus* from north-western Australia, still recognised by RAOU (1913).

Calamanthus campestris (Gould, 1841) [ka-la-MAN-thʊs kam-PES-tris]: 'stubble-pipit of flat open land', see genus name, and from Latin *campus*, a field or flat open land – a pretty good description of the habitat, especially the Nullarbor!

Redthroat (breeding resident)

Coined by Gould, straight from the genus name (also his). It was a little out of character, in that he tended to more formally use the genus name as a common name for new groups of birds (e.g. Acanthiza for thornbills, Sericornis for scrubwrens, Chthonicola for Speckled Warbler).

Other names: Brown Redthroat, used by Gould (1848), for his full species name; Red-throated Scrubwren – later writers decided it was a *Sericornis*.

Pyrrholaemus brunneus Gould, 1841 [pi-ro-LEH-mʊs BRʊN-ne-ʊs]: 'brown flame-throat', see genus name, and from Mediaeval Latin *brunneus*, brown (6th century *brunus*, 12th century *bruneus*).

Speckled Warbler (breeding resident)

A relatively recent name. Gould (1848) used the awful 'Little Chthonicola' from his name *Chthonicola minima* (not realising that Latham had already called it *Sylvia sagittata* in 1802); in 1899 Hall called it Little Field-wren, Lucas and Le Souef (1911) used Little Field Lark but in the same year Leach used Speckled Warbler, apparently for the first time. Putting aside the 'warbler' part of it, it is a descriptive name, with obvious reference to its species name.

Other names: Streaked Warbler; Fieldwren, Little or Speckled Fieldwren or Fieldlark, all references to its resemblance to a small Striated Fieldwren; Jenny Wren, Speckled Jack, testaments to its former commonness and generally tame disposition (like Willie Wagtail); Chocolate Bird, Blood Tit, both for the beautiful 'dark reddish-brown … lustrous' eggs (Beruldsen 1980).

Pyrrholaemus sagittatus (Latham, 1801) [pi-ro-LEH-mʊs sa-gi-TAH-tʊs]: 'arrowed flame-throat', see genus name (unfortunate in this case, since there is no hint of a red throat), and from Latin *sagitta*, arrow (like Sagittarius, the astrological archer), referring to the bird's markings, which do resemble arrowheads (a bit) – Latham (1802) thought so, anyway, as he referred to them as 'sagittal black streaks'.

Atherton Scrubwren (breeding resident)

See family introduction. Limited to the Atherton Tableland, north Queensland.

Other names: Bellenden Ker Scrubwren, from the species name.

Sericornis keri Mathews, 1920 [seh-ri-KOR-nis KER-ee]: 'Ker silk bird', see genus, and for Mt Bellenden Ker, the highest mountain in Queensland, named for John Bellenden Ker, born John Gawler in 1764 and changed his name at age 40 to Ker Bellenden by royal permission – but then called himself Bellenden Ker! He was an author for the *Botanical Register* in London; Phillip Parker King on the *Mermaid* expedition named the mountain for him at the request of on-board botanist Allan Cunningham.

White-browed Scrubwren (breeding resident)

See family introduction. For another conspicuous character; common usage for a common and familiar bird had changed Gould's White-fronted to White-browed by the end of the 19th century (Morris 1898).

Other names: Scrubwren, for the commonness and familiarity of this species in eastern Australia; White-fronted Scrubwren or Sericornis (the latter per Gould 1848, based on the species name); Cartwheel-bird, presumably for the reeling call, though we can find no reference to it other than the CSIRO (1969) and HANZAB (1990–2006, vol. 6) citations – it was probably essentially an oral name; Sambird, we have no idea – HANZAB is the only reference we have for it. Spotted Sericornis (Gould 1848), Spotted or Striated Scrubwren for South Australian and Western Australian races *mellori* and *maculatus*, together formerly regarded as a full species, the latter described by Gould; Allied Sericornis, applied by Gould (1848) to his species *S. osculans* from South Australia, which he described as 'most nearly allied to the *S. frontalis*'; hence Allied Scrubwren (RAOU 1913); Bernier Island Scrubwren, for race *balstoni*, also given species status in the past – despite the name, it is found along the Western Australia coast from Jurien to Shark Bay; Buff-breasted Scrubwren or Sericornis, applied by Gould to race *laevigaster* of the north-east coastal region, which he also described as a separate species; Kangaroo Island Scrubwren, for Gregory Mathews' short-lived species name *S. ashbyi*, recognised by RAOU (1913). Tasmanian Scrubwren was previously included in this species, so those names have also been formerly applied.

Sericornis frontalis (Vigors & Horsfield, 1827) [seh-ri-KOR-nis fron-TAH-lis]: 'browed silk bird', see genus name, and from Latin *frons*, the forehead or brow. See common name.

Tasmanian Scrubwren (breeding resident)

See family introduction. The only Tasmanian scrubwren, an endemic.

Other names: Brown or Sombre Scrubwren, Chocolate Wren, somewhat duller than the White-browed; Sombre or Sombre-coloured Sericornis, the latter applied by Gould (1848), who described it; Kent Scrubwren for *S. gularis*, described by Legge in 1896 for specimens from Kent Island in Bass Strait, recognised by RAOU (1913) but reduced to synonymy by RAOU (1926); similarly Flinders Scrubwren for White and Mellors' 1913 *S. flindersi*, whose taxonomic fate was identical.

Sericornis humilis Gould, 1838 [seh-ri-KOR-nis HOO-mi-lis]: 'lowly silk bird', see genus name, and from Latin *humilis*, humble or lowly. More about the bird's habitat than its status, given the relationship of *humilis* with humus (Latin for the ground).

Yellow-throated Scrubwren (breeding resident)

See family introduction. For a character unique among scrubwrens, applied by Gould (1848) from his species name.

Other names: Black-nest Bird, purportedly for the 'black and conspicuous' nests, coined by Mrs Hilda Curtis of Tamborine Mountain – for an interesting discussion of this, see Curtis (2011); Devil-bird, possibly also coined by Mrs Curtis, in this case for the shock of experiencing a small bat flying out of an abandoned nest or perhaps just for the bird itself 'darting out of gloomy spots in rainforests'.

Sericornis citreogularis Gould, 1838 [seh-ri-KOR-nis si-tre-o-gʊ-LAH-ris]: 'lemon-throated silk bird', see genus name, and from Latin *citreus*, of the citrus tree, (in this case, lemon-coloured), and *gula*, the throat or gullet. See common name.

Large-billed Scrubwren (breeding resident)

See family introduction. Does indeed have an evidently large (and crookedly stuck on) bill; taken directly from the species name.

Other names: Except for Gould's Large-billed Sericornis, it seems that no other name has ever been needed.

Sericornis magnirostra (Gould, 1838) [seh-ri-KOR-nis mag-ni-ROS-truh]: 'large-billed silk bird', see genus name, and from Latin *magnus*, large, and *rostrum*, bill, see common name. The illustration in Gould (1848) shows this admirably!

Tropical Scrubwren (breeding resident)

See family introduction. Restricted in Australia to Cape York Peninsula, and the name was coined by RAOU (1978) to supplant the unsatisfactory Little or Beccari's Scrubwren that prevailed until then.

Other names: Little Scrubwren, from Gould's 1875 name *minimus*, which was still recognised well into the 20th century (RAOU 1926) – Count Adelaro Tommaso Salvadori had only beaten Gould to naming rights by a year; Beccari's Scrubwren, from the species name; Perplexing Scrubwren, a delightful name for race *virgatus* of New Guinea (more usually now regarded as a full species), whose exact relationship to related taxa has perplexed many ornithologists!

Sericornis beccarii Salvadori, 1874 [seh-ri-KOR-nis bek-KAH-ri-ee]: 'Beccari's silk bird', see genus name, and for Odoardo Beccari (1843–1920), Italian botanist who collected birds on the side, including the type specimen of this one in the Aru Islands off south-west New Guinea in the early 1870s – he gave it to fellow Italian Tommaso Salvadori of Turin University to describe.

Weebill (breeding resident)

This is a quite recent name, apparently plucked from the air by RAOU (1926); however, a preview of the list in the Melbourne *Herald* of 30 May 1923 included Wee-Bill in a list of 'pleasing 'popular' names' to be introduced, perhaps implying it was already in wide informal use. Moreover, for what it's worth, a racehorse with the unlikely name of Weebill was reported in 1906. Until then 'Tree-tit' (and to a much lesser extent 'Scrub-tit') had been formally used universally, after Gould's use of Smicrornis as common name did not take (Morris 1898; Hall 1899; Lucas and Le Soeuf 1911; Mathews 1913). The name obviously refers to the short stubby bill – the same feature that prompted the older name.

Other names: Short-billed Smicrornis, used by Gould (1848) in describing it; Tree-tit, Short-billed Treetit; Scrub-tit, Short-billed Scrub-tit; Brown Weebill, Southern Treetit or Weebill, for southern race *brevirostris*; Yellow-tinted Smicrornis, used by Gould for northern inland race *flavescens*, which he described as a separate species (still recognised into the 20th century: RAOU 1926; Cayley 1931), thus also Yellow-tinted scrub-tit or Treetit, Yellow Weebill or Treetit, Central Australian Tree-tit; Western Scrub-tit for race *occidentalis*.

Smicrornis brevirostris (Gould, 1838) [smik-ROR-nis bre-vi-ROS-tris]: 'small short-billed bird', see genus name, and from Latin *brevis*, short, and *rostrum*, bill. Can't say fairer than that!

Brown Gerygone (breeding resident)

See family introduction. It is brown – exactly like half the other Australian gerygones.

Other names: Brown Warbler, Bush-warbler, Flyeater; Northern Warbler or Queensland Flyeater, for the tropical race *mouki*; Fuscous Gerygone, Gould's name for it (but he lumped it with Western Gerygone); Citron Bird, for which there are several citations but no explanations, and no evidence of its actual usage; Pale Brown Fly-eater, for North's *G. pallida* of north Queensland, cited on the Provisional List of RAOU (1913).

Gerygone mouki Mathews, 1912 [dje-RI-go-ne MOO-kee]: 'mouki child of song', see genus name, and named by Mathews for reasons best known to himself – none of his lists elaborated on the name. There are few clues to the meaning of this word, though there is a suggestion in HANZAB (1990–2006, vol. 6) that it may be from an Aboriginal language (unspecified). We wonder if there is some connection with the Mooki River in northern New South Wales, but as the type specimen was from Cairns this is drawing a long bow. Nonetheless, perhaps a now forgotten locality name is the most likely explanation.

Norfolk Gerygone (Norfolk Island)

See family introduction. Endemic to Norfolk Island.

Other names: Norfolk Island Flyeater or Gerygone; Grey Gerygone, used on Norfolk Island, though this is more properly assigned to the New Zealand species *G. igata;* Hummingbird, commonly used on Norfolk for the hovering behaviour; Ashy-fronted Gerygone.

Gerygone modesta Pelzeln, 1860 [dje-RI-go-ne mo-DES-tuh]: 'plain child of song', see genus name, and from Latin *modestus*, meaning many things in the moderate, gentle, temperate, unassuming line – but ultimately named for its plain plumage.

Lord Howe Gerygone (Lord Howe Island, extinct)

See family introduction. Formerly endemic to Lord Howe Island.

Other names: Lord Howe or Lord Howe Island Flyeater or Warbler; Pop-goes-the-Weasel for the song – and we can certainly hear this tune in the Western Gerygone's song; Rainbird, reportedly for its activity after rain (as for many insect-eating species).

Gerygone insularis Ramsay EP, 1879 [dje-RI-go-ne in-sʊ-LAH-ris]: 'island child of song', see genus name, and from Latin *insula*, island, and *insularis*, belonging to an island, which it did.

Mangrove Gerygone (breeding resident)

See family introduction. For its near-obligate habitat; however, the name was only formalised by RAOU (1978), though Slater had already used it in 1974.

Other names: Mangrove Warbler or Flyeater; Buff-breasted Gerygone, used by Gould (1848), though unusually it doesn't really reflect his species name; hence also Buff-breasted Warbler or Flyeater, the latter still used by RAOU (1926) and Cayley (1931) – this does distinguish it from Large-billed and Dusky Gerygones in the same habitat (though it doesn't help with Brown Gerygone, which can also be found in mangroves); Singing Flyeater, Queensland Canary, 'canary' for the song, and used also for White-throated Gerygone (also found widely in Queensland) – it is not clear that this is a notably better singer than some other gerygones, but the reference is both to and from the species name *cantator*, applied in 1908 to birds from south-east Queensland, recognised by RAOU (1913) but not much thereafter.

Gerygone levigaster Gould, 1843 [dje-RI-go-ne le-vi-GAS-tehr]: 'lightly [yellow] bellied child of song', see genus name, and from Latin *leviter*, lightly, and *gaster*, belly. Gould (1842) said, '*corpore ... inferiore albo, leviter flavido tincto*', in English, 'all the under surface white, slightly washed with yellow'. (Jobling 2010 has it coming from *laevum*, favourable).

Western Gerygone (breeding resident)

See family introduction. Found throughout much of the country, but the initial reference was Gould's who described birds from Western Australia, which he named *G. culicivorus*; as noted earlier, he considered eastern populations of this species to be the same species as Brown Gerygone.

Other names: Fuscous Gerygone, Flyeater or Warbler, per Gould – see Brown Gerygone; Inland, Western or Southern Warbler or Flyeater; Desert Flyeater; White-tailed Warbler, Flyeater or Bush-warbler, for the distinctive white-based tail; Reddish-crowned Fly-eater, for Campbell's western New South Wales *Gerygone (Pseudogerygone) jacksoni*, cited by RAOU (1913) on the Provisional List; Sleepy Dick, for the soporific sweet call, the 'Dick' being a familiar name like 'Willie Wagtail'; Psalmist, an apparently spontaneous name from Western Australia in the 1920s: 'The Psalmist, a local bird name mentioned recently in 'Denizens of the Bush', seems the most fitting name to adopt for the bird which earns it. *Gerygone fusca*, a small grey-brown bird hitherto only known by such cumbersome appellations as 'white tailed flyeater', 'Western warbler' and 'Southern warbler ...'' (*Western Australian*, 9 October 1928 – 'Denizens of the Bush' was a regular column by 'Mo'Poke'). The name was used in an *Emu* article 2 years later (Pollard 1930), but not apparently in print thereafter.

Gerygone fusca (Gould, 1838) [dje-RI-go-ne FʊS-kuh]: 'dusky child of song', see genus name, and from Latin *fuscus*, dark or dusky (but see next species).

Dusky Gerygone (breeding resident)

See family introduction. From the species name.

Other names: Dusky Warbler or Flyeater. Was previously included in the next species, so doubtless some of those names applied too.

Gerygone tenebrosa (R. Hall, 1901) [dje-RI-go-ne te-ne-BROH-suh]: 'dark and gloomy child of song', see genus name, plus Latin *tenebrosus*, from *tenebrae*, darkness or shadows or places to lurk – not that this bird is particularly given to lurking, but it is dark in colour.

Large-billed Gerygone (breeding resident)

See family introduction. From the species name (Gould, who described it, used Great-billed).

Other names: Great-billed Gerygone, see common name; Large-billed Warbler, Flycatcher, Bush-warbler, Flyeater; Brown-breasted Flyeater, for the buffy sides to the breast; Swamp Warbler, for one of its habitats; Flood-bird, for the resemblance of its hanging nest to flood debris (Campbell 1901).

Gerygone magnirostris Gould, 1843 [dje-RI-go-ne mag-ni-ROS-tris]: 'large-billed child of song', see genus name, and from Latin *magnus*, large, and *rostrum*, bill – just like the common name – a heavier bill than any of the others.

Green-backed Gerygone (breeding resident)

See family introduction. Distinctively olive-green above, and the name was taken by Gould straight from his species name.

Other names: Green-backed Flyeater or Warbler; Hornet-nest Bird, for a propensity to nest near wasp nests for protection, though the name was more often applied to Fairy Gerygone – Campbell (1901) used it for that species, but not this one.

Gerygone chloronota Gould, 1843 [dje-RI-go-ne kloh-ro-NOH-tuh]: 'green-backed child of song', see genus name, and from Greek *khlōros*, greenish-yellow, and *nōton*, the back – Gould described it in English as olive-green.

White-throated Gerygone (breeding resident)

See family introduction. Used by Gould (1848) from his species name, for an obvious characteristic.

Other names: White-throated Warbler, Flyeater or Bush-warbler; Bush or Native Canary, for the glorious song (more beautiful than that of any canary!); Grey Flyeater, for Sharpe's 1878 species name *G. cinerascens* from north-western Australia, still recognised into the early 20th century (RAOU 1913).

Gerygone olivacea (Gould, 1838) [dje-RI-go-ne o-li-VA-ce-uh]: 'olive green child of song', see genus name, and from Latin *oliva*, an olive, and Modern Latin *olivaceus*, olive green. (Gould mistakenly had *G. albogularis* – as this species was known until very recently – and *G. olivacea* as separate birds in the *Synopsis of the Birds of Australia* with *G. olivacea* as No. 3, and *G. albogularis* as No. 4, so the less descriptive term *olivacea* has priority as the bird's correct name.)

Fairy Gerygone (breeding resident)

See family introduction. It is not clear why this name was adopted, though it has generally been applied to the plainer southern race *flavida* (and three races – that one, plus *personata* from far north Queensland and *palpebrosa* of New Guinea – have often been regarded as separate species). It wasn't used by Mathews (1913) (who used Herbert River Flyeater for *flavida*) or by RAOU (1913), which used Black-throated Flyeater, but was first – and only – choice in RAOU (1926).

Other names: See common name. For race *personata*: Black-headed, Black-throated or Masked (from the subspecies name, originally Gould's species name) Gerygone, Warbler or Flyeater – Black-throated Flyeater was the name of choice by RAOU (1913); Hornet-nest Bird, for a noted preference for nesting near wasp nests (North 1909).

For race *flavida*: Fairy or Yellow Flyeater or Warbler; Yellow Gerygone; Cardwell Gerygone; Herbert River Flyeater (see common name).

Gerygone palpebrosa Wallace, 1865 [dje-RI-go-ne pal-pe-BROH-suh]: 'child of song with prominent eyelids', see genus name, and from Latin *palpebra*, eyelid, for the small white spot above each eye.

Mountain Thornbill (breeding resident)

See family introduction. Restricted to the hinterland ranges of the Queensland wet tropics.

Other names: Mountain Tit-warbler.

Acanthiza katherina De Vis, 1905 [a-kan-THEE-zuh ka-the-REE-nuh]: 'Katherine's thornbush-dweller', see genus name, and for Katherine De Vis, second wife of English naturalist Charles De Vis, who named the species. He came from England in 1870, settling in Queensland. He was curator then director of, and consultant to, the Queensland Museum from 1882 to 1912 and was an indefatigable naturalist, writing numerous papers on many aspects of natural history.

Brown Thornbill (breeding resident)

See family introduction. We might think this a notably bland and unhelpful name (which it is) but it dates back to Shaw's description (and was one of the first of the new colony's birds to be described) as Brown Warbler (*Motacilla pusilla*).

Other names: Little Brown Acanthiza (Gould 1848); Dwarf Warbler (Latham 1823), Dusky Warbler, Tit-warbler; Brown Tit or Tit-warbler; Large-billed Tit, from the former name *A. magnirostris*, assigned to birds from King Island (Campbell 1903b); Gregory Mathews later changed *A. magnirostris* to *A. archibaldi* for reasons too obscure for us (Mathews 1910b), presumably to honour Archibald Campbell, but RAOU (1913) listed Large-billed Tit-Warbler under Mathews' name; Western Port Tit-warbler, cited by RAOU (1913) as a species name for Quoy and Gaimard's 1830 *Saxicola macularia*, which was accepted as a race until relatively recently (Boles 1983); Brown-rumped Tit-Warbler, used by RAOU (1913) for Gould's species (now race) *diemensis* of Tasmania; Kangaroo Island Tit-Warbler, for race *zietzi*, formerly described as a species by Alfred North in 1904 and supported by RAOU (1913); Broad-tailed Thornbill, a name more usually, though not much more aptly, applied to Inland Thornbill; Brown-tail; Tit-bat, cited by CSIRO (1969) and HANZAB (1990–2006, vol. 6) but not to be found anywhere in the literature; Cotton-legs, cited by HANZAB but nowhere else.

Acanthiza pusilla (Shaw, 1790) [a-kan-THEE-zuh pʊ-SIL-luh]: 'tiny thornbush-dweller', see genus name, and from Latin *pusillus*, very small or insignificant. We accept the first meaning but not the second – their widespread presence alone ensures their significance to the birdwatcher ('It's just a Brown Thornbill' …).

Inland Thornbill (breeding resident)

See family introduction. As distinct from the very similar Brown Thornbill, which replaces it along the east coast; the name did not become established until recommended by RAOU (1978), though 50 years earlier its predecessors (RAOU 1926) used Inland Thornbill for *A. albiventris* (see 'Other names').

Other names: There are many races and formerly recognised species, which makes complex any examination of this species. Western Thornbill, by Gould (1848) who believed it to be found only in Western Australia; Red-tailed or Red-rumped Thornbill, Acanthiza, Tit or Tit-Warbler, for Gould's name *A. pyrrhopygia* (later replaced by Mathews' *A. hamiltoni*), which was recognised well into the 20th century (RAOU 1926); Broad-tailed Thornbill, Tit or Tit-Warbler, a reference to the species name, widely used in recent times; Western, Inland or Rufous-rumped Brown Thornbill. Whitlock's Thornbill or Tit or Tit-Warbler, for race *whitlocki*, collected at Lake Way in Western Australia by Frederick Whitlock and described by Alfred North in 1909; Whitlock was an Englishman who arrived in Perth in 1901 at the age of 41 and was soon taken on as a collector for the Western Australian Museum (Whittell 1939). Lake Way Tit, for the same race and reason; Tanami Thornbill or Tit or Tit-Warbler, for race *tanami* (not now recognised as separate from race *whitlocki*), described by Mathews (1911) from a single specimen from the Tanami Desert – it was sporadically recognised even as a species until well into the century (RAOU 1913; Mathews 1946). White-vented Tit-Warbler, for Alfred North's species *A. albiventris* (now regarded as a race, from central New South Wales and southern Queensland), recognised by RAOU (1913) and RAOU (1926), which called it Inland Thornbill.

Acanthiza apicalis Gould, 1847 [a-kan-THEE-zuh a-pi-KAH-lis]: 'tipped thornbush-dweller', see genus name, and from Latin *apex* (plural *apices*), point or top or tip (Mediaeval Latin *apicalis*), referring to the tail, which Gould (1847) described as 'largely tipped with white'.

Tasmanian Thornbill (breeding resident)

See family introduction. A Tasmanian endemic. (Gould described also another Tasmanian species, which he called *A. diemensis* ('of Van Diemen's Land') and named Tasmanian Thornbill – not now recognised.)

Other names: Ewing's Acanthiza (Gould's name), Thornbill, Tit or Tit-warbler, for the 'Reverend Thomas James Ewing, a gentleman ardently attached to the study of Natural History, and a sincere friend to all who have the advantage of his acquaintance' (Gould 1848) – Gould stayed with him when visiting Tasmania. Browntail, perhaps to distinguish from the co-existing and similar Brown Thornbill, with a brighter rufous rump – Gould recorded this as a name of the 'Colonists of Van Diemen's Land'; Creek Tit.

Acanthiza ewingii Gould, 1844 [a-kan-THEE-zuh yoo-WING-i-ee]: 'Ewing's thornbush-dweller', see genus and common names.

Chestnut-rumped Thornbill (breeding resident)

See family introduction. A descriptive name, in the format of the next species and Yellow-rumped, though not uniquely.

Other names: Chestnut-rumped Acanthiza (Gould 1848), from his species name. Chestnut-rumped Tit or Tit-warbler; Chestnut-tailed or Chestnut-backed Thornbill, Tit or Tit-warbler.

Acanthiza uropygialis Gould, 1838 [a-kan-THEE-zuh oo-ro-pi-gi-AH-lis]: 'rumped thornbush-dweller', see genus name, and from Mediaeval Latin *uropygium*, the rump of a bird (from Greek *ouropugion*, a combination of the words for tail and rump), meaning where the tail feathers grow from (i.e. the 'parson's nose' – not that other birds don't have one, but that this bird's is so striking).

Buff-rumped Thornbill (breeding resident)

See family introduction. Another in the 'colour-rumped' series, but this one is unique.

Other names: Regulus-like Acanthiza (Gould 1848) – Gould had something of a gift for infelicitous common names but surpassed himself here; this gift was based on uncritical translation of scientific names (often his own, but not on this occasion). *Regulus regulus* is the European Goldcrest – the recognition of whose resemblance to this species is something of a winningly original mental gymnastic move; however, see species name. Buff-rumped Tit or Tit-warbler; Buff-tailed Thornbill, Tit or Tit-warbler; Bark Tit, for the trunk-gleaning behaviour; Scaly-breasted Tit-Warbler, Varied Thornbill, Tit or Tit-warbler, for Queensland coastal race *squamata*, previously regarded as a full species (as recently as RAOU 1926); South Australian Tit, for race *australis*.

Acanthiza reguloides Vigors & Horsfield, 1827 [a-kan-THEE-zuh re-gʊ-LOY-dehz]: 'thornbush-dweller like a Goldcrest', see genus name, and from *Regulus regulus*, the Eurasian Goldcrest (Latin *regulus*, little king), and from Latin *-oides*, resembling. As mentioned, this is a mystery, in that there is only the most superficial of resemblances – in particular the Goldcrest's most striking feature, its black-margined gold crown, is entirely absent in this thornbill. We have followed the somewhat laybrinthine and nebulous path and these are, we believe, the pertinent markers of the route: Latham (1823) seems to have described the Buff-rumped Thornbill twice; once is almost certainly species 161, Dwarf Warbler variety B (as noted by Vigors and Horsfield 1827), but we believe that his species 164, Buff-headed Warbler, also refers to the same species. (His descriptions are often very brief and his *laissez faire* attitude to scientific names makes interpretation difficult.) Crucially, we think, he commented that species 164 is the 'size of the Gold-crowned Wren [the Goldcrest]', though he made no other comparison with it. Although 4 years later, in their publication of the name *reguloides*, Vigors and Horsfield (1827) made no reference to the Buff-headed Warbler, they made much of supposed resemblances to the Goldcrest in introductory comments before the brief descriptions of their two new species (this one and *A. nana*). These resemblances appear spurious at best: 'a general similarity in the disposition of their colours ... having the webs of their feathers, particularly about the head and neck, more than usually loose and decomposed ... inhabitants of bushes and low scrubs'. Although they do not further explain their name *reguloides*, we propose that they were influenced, perhaps unconsciously, by Latham's comment and sought to justify it, albeit somewhat unconvincingly.

Western Thornbill (breeding resident)

See family introduction. Endemic to south-western Western Australia.

Other names: Bark-tit, see previous species; Plain-coloured Acanthiza (Gould 1848) or Tit, from the species name – this is a very Little Brown Bird; Western Tit-warbler; Master's Tit or Tit-warbler, from *A. mastersi*, named by Alfred North in 1901, apparently for the collector G. Masters who worked for Sir William MacLeay in the Albany area in the 1860s – it was still recognised as a species into the 20th century (RAOU 1913). (Note that Gould used Western Thornbill for *A. apicalis*.)

Acanthiza inornata Gould, 1841 [a-kan-THEE-zuh in-or-NAH-tuh]: 'unadorned thornbush-dweller', see genus name, and from Latin *inornatus*, unadorned, without ornament.

Slender-billed Thornbill (breeding resident)

See family introduction. From the old species name *A. tenuirostris* (Zeitz 1900), which was soon afterwards deemed to be unavailable; it is small-billed, even for a thornbill, and the contrast with *A. robustirostris* was surely not coincidental.

Other names: Slender-billed Tit; Samphire Thornbill, for a key habitat; Small-billed Thornbill, Tit or Tit-warbler (RAOU 1913) or Slender Thornbill, used by RAOU (1926) and Cayley (1931) for Mathews' *A.*

morgani, since absorbed into this species; Dark Thornbill, for Mathews' *A. hedleyi* from Meningie in South Australia, still recognised by RAOU (1926) and Cayley (1931); Small-tailed Thornbill, which we wonder about – there seems no usage other than the citation in HANZAB (1990–2006, vol. 6) and it doesn't sound like an oral name. Perhaps in error for Small-billed?

Acanthiza iredalei Mathews, 1911 [a-kan-THEE-zuh IER-deh-lee]: 'Iredale's thornbush-dweller', see genus name, and for Tom Iredale, born 1880, an Englishman who came to Australia as Gregory Mathews' amanuensis in 1923 and became a most significant ornithologist, conchologist and zoological polymath in his own right, working at the Australian Museum until his death in 1972.

Yellow-rumped Thornbill (breeding resident)

See family introduction. Perhaps the most distinctive characteristic possessed by any thornbill.

Other names: Yellow-tailed Acanthiza (Gould 1848), or Thornbill; Yellow-rumped or Yellow-tailed Tit or Tit-warbler; Yellow-tail or Tomtit, both acknowledgments of how common and familiar it is; Chigaree, we venture to suggest for the flight call though we can find no evidence of its use – however, an essentially verbal folk-name doesn't necessarily leave written traces. Leigh's Thornbill, from the name *A. leighi* (now recognised as a race) coined in 1909 by Grant, Scottish curator of Ornithology at the British Natural History Museum, from a museum skin – we know that the skin was presented to the museum by a Lord Leigh, but there the trail ends (Mathews 1909). Pallid Tit or Tit-Warbler, meaning pale, from old name *A. pallida* for Western Australian birds (by Alexander Milligan, honorary consulting ornithologist to the Western Australian Museum, in 1903), still used by RAOU (1913); Tasmanian Yellow Tit-warbler, for Mathews' subspecies *A. leachi* (Mathews 1912a, still recognised), cited by RAOU (1913) in the Provisional List as a species.

Acanthiza chrysorrhoa (Quoy & Gaimard, 1830) [a-kan-THEE-zuh kri-so-ROH-uh]: 'golden-rumped thornbush-dweller', see genus name, and from Greek *khrusos/khruso-*, gold/golden, and *orrhos*, rump. See common name.

Yellow Thornbill (breeding resident)

See family introduction. A variably coloured thornbill throughout its range, but only consistently and evidently yellow in the south-east.

Other names: Long known as Little Thornbill from its species name, though the basis of that is far from evident now – it arose from Latham's name Dwarf Warbler, when he was almost certainly not comparing it with other thornbills; Little Acanthiza (Gould 1848); Little Tit or Tit-warbler; Yellow Dicky, another familiar personal name (like Willie Wagtail), but we can find no evidence of it in use anywhere beyond the citations in CSIRO (1969) and HANZAB (1990–2006, vol. 6); Yellow-breasted Thornbill or Tomtit. Mathews' (not Mathew's, as per HANZAB) Thornbill, from subspecies *mathewsi*, supposedly found coastally from Sydney to south-eastern South Australia (Condon 1951); the race was identified and named by German ornithologist Ernst Hartert in 1910 from skins in the Tring Museum in England, for the then Australian ornithological giant (only partly self-styled) Gregory Mathews. Oddly, in the same year Mathews himself referred to it as a full species, *A. mathewsi* (Mathews 1910a); the taxon is not now recognised at any level. (For more on Mathews, see Rufous Songlark.) Victorian Tit-warbler, for *A. mathewsi*, as recognised by RAOU (1913); Northern Striated Tit-warbler, for Charles De Vis' *A. modesta* from Charleville, recognised by RAOU (1913) but not for long after that, though it is still a race, widespread in inland south-east Australia; Fairy Tit-warbler, for Milligan's 1913 species *A. pygmaea* from north-western Victoria, recognised by RAOU (1913) in the Provisional List, but not thereafter.

Acanthiza nana Vigors & Horsfield, 1827 [a-kan-THEE-zuh NAH-nuh]: 'dwarf thornbush-dweller', see genus name, and from Latin *nanus*, a dwarf, see common name.

Striated Thornbill (breeding resident)

See family introduction. The streakiest thornbill, especially round the head, taken from the species name.

Other names: Striated Acanthiza (Gould 1848), straight from his species name; Striped-crowned Thornbill or Tit-warbler; Striped or Striated Tit or Tit-warbler; Green Thornbill, for the tinge on the back (in the right light).

Acanthiza lineata Gould, 1838 [a-kan-THEE-zuh lin-e-AH-tuh]: 'lined thornbush-dweller', see genus name, and from Mediaeval Latin *lineatus*, lined (from Latin *linea*, a linen thread, string or line, Latin *linum*, flax).

Slaty-backed Thornbill (breeding resident)

See family introduction. It does have a greyer back than the co-existing similar Inland Thornbill (again, in the right light). The name seems to have been introduced by McDonald (1973), followed closely by Slater (1974) (both using Slate-backed), and confirmed in the current form by RAOU (1978).

Other names: Slate-backed or Grey-backed Thornbill; Large-billed, Robust-billed or Thick-billed Thornbill, from the species name; Thick-billed Tit or Tit-warbler; Robust Thornbill, for no obvious reason, though perhaps influenced by (part of) the species name.

Acanthiza robustirostris Milligan, 1903 [a-kan-THEE-zuh ro-bʊs-ti-ROS-tris]: 'robust-billed thorn-bush-dweller', see genus name, and from Latin *robustus*, originally 'made of oak' (*robur*) but here meaning strong or robust, and *rostrum*, bill.

Southern Whiteface (breeding resident)

See family introduction. Found across southern Australia.

Other names: White-faced Xerophila, applied by Gould (1848) from his genus and species names; Whiteface, for the only species familiar to most people; Eastern or Central Australian Whiteface for race *leucopsis*; Western Whiteface for race *castaneiventris* of Western Australia; Chestnut-bellied Whiteface, from the same race – this was formerly regarded as a full species, described by Alexander Milligan in 1903 and still recognised as such by RAOU (1926); Squeaker or White-faced Squeaker, used as a first-choice name by Alfred North of the Australian Museum (North 1916), but not much elsewhere in written form; White-faced Titmouse, see family introduction.

Aphelocephala leucopsis (Gould, 1841) [a-fe-lo-SE-fa-luh loo-KOP-sis]: 'white-faced simple head', see genus name, and from Greek *leukos*, white, and *opsis*, face or appearance.

Chestnut-breasted Whiteface (breeding resident)

See family introduction. For the broad chestnut breastband, unique among whitefaces.

Other names: Chestnut-breasted Tit.

Aphelocephala pectoralis (Gould, 1871) [a-fe-lo-SE-fa-luh pek-to-RAH-lis]: 'simple head with a chest', see genus name, and Latin *pectus/pectoris*, the chest, for the breastband.

Banded Whiteface (breeding resident)

See family introduction. For the unique black breastband.

Other names: Black-banded Whiteface or Squeaker, directly from the species name.

Aphelocephala nigricincta (North, 1895) [a-fe-lo-SE-fa-luh ni-gri-SINK-ta]: 'black girdled simple head', see genus name, and from Latin *niger*, gleaming black, and *cinctus*, a band or girdle for the toga, for the breastband.

POMATOSTOMIDAE (PASSERIFORMES): Australian babblers

Pomatostomidae Schodde, 1975 [po-ma-to-STOH-mi-dee]: the Covered-mouth family, see genus name *Pomatostomus*.

The word 'babbler', indicating a foolish chattering person, goes back to the 16th century in England. The application to birds seems to date only to the 1830s, to the family Timaliidae (formerly Timeliidae), long regarded as notoriously an odds and ends basket. Newton's influential 1896 *Dictionary of Birds* commented, 'many have not unjustifiably regarded it as 'a refuge for the destitute' – thrusting into it a great number of forms … [which] cannot, in the opinion of some, be conveniently stowed elsewhere'. The Australian babblers were long stowed thus (along with a surprising number of other Australian groups), but the name babbler did not come into general usage until the second half of the 19th century. Gould did not use it in 1848 or 1869, but by 1898 Morris' *Dictionary of Austral English* regarded the name as standard.

The genus

Pomatostomus Cabanis, 1850 [po-ma-to-STOH-mʊs]: 'covered mouth', from Greek *pōma*, a lid or cover, and *stoma*, mouth. Not a reference to the birds' silence … Cabanis based the name on Horsfield's *Pomatorhinus*, which are Old World babblers. Horsfield chose the name for what he described as 'the corneus covering of the nares', which he considered the key distinguishing feature; until renamed by Cabanis, Australian babblers were included in that genus.

The species

Grey-crowned Babbler (breeding resident)

See family introduction. For this very distinctive feature, between broad white eyebrows; it was formalised (and perhaps coined) by RAOU (1926).

Other names: One of the most-monikered Australian birds. Dusky Bee-eater (Latham 1823) – he had quite a fixation on bee-eaters, seeing them in some surprising guises; Temporal Pomatorhinus, from Gould (1848) at his direct-translating best from the species name (and *Pomatorhinus* is the Asian genus of scimitar-babblers, Timaliidae, to which Australian babblers were then ascribed); Australian Babbler (RAOU 1913), an impressively unhelpful name; Three-banded Pomatorhinus, from an invalid name; Temporal Babbler; Apostlebird, Twelve Apostles, Happy Family, for the group living which is an integral part of their lifestyle; Barker, Cackler, Quackie, Cur-Cur, Catbird, Dog Bird, Chatterer, Grey-crowned Chatterer, Red-breasted Babbler, Rufous-breasted Chatterer (both referring to rusty-bellied northern race *rubeculus*, still recognised as a species *P. rubeculus* well into the 20th century: RAOU 1926), Yahoo, all for the constant and sometimes crazy-sounding vocalisations – and it really does yell 'ya-hoo'!; Pine Bird, for its liking for *Callitris* woodlands; Happy Jack, an affectionate personalisation, like Willie Wagtail; Fussy, Hopper, Hopping Dick or Jumper, for its relentless energy. Codlin-Moth-Eater, apparently assigned by Archibald Campbell (1901), who reported in the columns of the *Australasian* on 27 June 1896, 'Mr. G.E. Shepherd, Somerville, brought under notice the valuable work performed by Babblers, which he has seen persistently destroying the larvae of the pestilent codlin moth'. He added sadly, 'However, it is feared these wild and restless birds will need much encouragement to come about orchards or the habitations of man.' Parson Bird, somewhat of a mystery as we can't find any usage of it – Morris (1898) didn't know it, though he discussed at length its use for the New Zealand Tui. Avachat, a widely used bush colloquialism (as in 'always happy to have a chat') for all our babbler species.

Pomatostomus temporalis (Vigors & Horsfield, 1827) [po-ma-to-STOH-mʊs tem-po-RAH-lis]: 'covered-mouth with temples', see genus name, and from Latin *tempora*, the temples of the head, rather than *temporalis*, temporary. The broad white eyebrow described earlier extends down past the temples.

Hall's Babbler (breeding resident)

See family introduction. For Harold Wesley Hall (1888–1964), Melbourne born, English educated, a wealthy man who used his money to fund British Museum research, culminating in the five somewhat controversial Harold Hall Australian Expeditions during the 1960s, whose unabashed aim was to shoot lots of Australian birds to replenish the museum stocks that had been depleted when it flogged its Rothschild and Mathews bird collections to the US. The babbler was collected – and, more to the point, recognised for the first time as being different from White-browed Babbler – in southern Queensland on the first of the expeditions, in 1963; Graham Cowles of the British Museum, who described it, may have felt he had little choice in the naming. (J.D. McDonald of the museum, who led the first expedition, wrote *Birds of Australia* in 1973).

Other names: Black-bellied, Dark-bellied, White-breasted or White-throated Babbler – all cited by HANZAB (1990–2006, vol. 6) and perplexing to us because we can find no other example of their usage, nor would we expect to for a species so recently discovered. Avachat, see Grey-crowned Babbler.

Pomatostomus halli Cowles, 1964 [po-ma-to-STOH-mʊs HAWL-lee]: 'Hall's covered-mouth', see genus and common names.

White-browed Babbler (breeding resident)

See family introduction. White eyebrows are common to all babblers; this one was named simultaneously with Grey-crowned and named very similarly to it – either could have got the 'white-browed' label.

Other names: White-eyebrowed Pomatorhinus (Gould 1848), see Grey-crowned Babbler; White-eyebrowed Babbler; Cackler or White-eyebrowed Chatterer for its constant vocalisations; Apostlebird or Happy Family, for the cooperative living; Catbird, Go-away for its yowling calls; Hopping Dick or Hopping Jenny, affectionate familiar names, like Willie Wagtail; Jumper or Kangaroo-bird, for its constant hopping around; Stickbird, cited without explanation by Leach (1911) but not otherwise evident – we venture to suggest it was for the big stick roosting nests, but without total conviction; Western Babbler, for race *ashbyi* of south-western Australia, described by Mathews in 1911 (*Pomatorhinus ashbyi*) and cited as a species by RAOU (1913) in the provisional list. Avachat, see Grey-crowned Babbler.

Pomatostomus superciliosus (Vigors & Horsfield, 1827) [po-ma-to-STOH-mʊs soo-pehr-si-li-OH-sʊs]: 'eyebrowed covered-mouth', see genus name, and from Latin *supercilius*, above the eyelid (i.e. the eyebrow) (and what you do with your eyebrows when you're feeling supercilious).

Chestnut-crowned Babbler (breeding resident)

See family introduction. From the species name.

Other names: Chestnut-crowned Pomatorhinus (Gould 1869), see Grey-crowned Babbler; Red-capped Babbler; Chatterer, Chestnut-crowned Chatterer, for its constant commentary. Avachat, see Grey-crowned Babbler.

Pomatostomus ruficeps (Hartlaub, 1852) [po-ma-to-STOH-mʊs ROO-fi-seps]: 'red-headed covered-mouth', see genus name, and from Latin *rufus*, red, and *-ceps*, -headed (*caput*, head).

ORTHONYCHIDAE (PASSERIFORMES): logrunners

Orthonychidae Gray GR, 1840 [or-tho-NI-ki-dee]: the Straight-claw family, see genus name *Orthonyx*.

'Logrunner' is for key behaviour of the group, foraging for invertebrates on the forest floor and along rotting logs, but its actual origin is not clear. Gould (1848, 1869) used the genus name Orthonyx as a common name, as was his wont. Robert Hall used Log-runner in his *Key to the Birds of Australia* in 1899 and Morris' *Dictionary of Austral English* recorded the name in 1898, so it was presumably generally accepted by then.

The genus

Orthonyx Temminck, 1820 [or-THO-niks]: 'straight-claw', from Greek *orthos*, straight, and *onux*, claw or talon. As well as noting that the claws were not very curved, Temminck (1820) stressed their length and strength – longer than the toes themselves, he said.

The species

Australian Logrunner (breeding resident)

See family account. 'Australian' is a recent addition, with the recognition that the New Guinea populations represent a separate species.

Other names: Spine-tailed Orthonyx, used by Gould (1848), from Temminck's then species name *spinicaudus*, referring to the stiff barbless tail feather tips; Log-runner or Logrunner, until recently, when Australian was added; Spinetail, Jungle Spinetail, Spine-tailed Logrunner (RAOU 1913); Southern or Spine-tailed Chowchilla, see Chowchilla; Scrub-quail or Scrub-hen for a widespread (and somewhat dismissive) term for rainforest, and the terrestrial habitat; Brown Logrunner, hmm.

Orthonyx temminckii Ranzani, 1822 [or-THO-niks tem-MIN-ki-ee]: 'Temminck's straight-claw', see genus name, and for Coenraad Temminck (1778–1858), Dutch ornithologist who directed the Dutch National Natural History Museum in Leiden for the last 38 years of his life. Temminck had published the genus name *Orthonyx* in 1820, describing the type species (clearly a Logrunner, though Temminck got the male and female mixed up, asserting that the male has the red throat) but somehow failing to give it a species name! Camillo Ranzani (Catholic priest, primicerius of the Metropolitan Cathedral of Bologna and professor of Mineralogy and Zoology at the University of Bologna, as well as director of its Museum of Natural History) must have realised this, and when he put out Volume 3 of his *Elementi di Zoologia* in 1822 he took the opportunity to name the species. Ranzani's account of the bird included a word-for-word translation of Temminck's original descriptions of the genus and type species (including Temminck's error of taking the more colourful bird to be the male), though he did shuffle a couple of the sentences around. He added speculative comments on what some of the physical aspects of the bird might indicate about its behaviour, and criticised Temminck sharply for lack of detail on the bird's habits (Ranzani 1822). We don't know what Temminck thought about this name but he did redescribe the bird as *Orthonyx spinicaudus* in 1827, apparently because he erroneously believed the specimens had come from New Zealand, so assumed it was a new species. In the end Ranzani's name prevailed.

Chowchilla (breeding resident)

A pretty fair rendition of the ringing contact call; of Aboriginal origin, from the Barron River Valley in north Queensland according to Sidney Jackson (1908), which seems to be the first publication of the name. It caught on quickly – by 1926 the RAOU had adopted the name (for both species) in the *Official*

Checklist of the Birds of Australia. Much later Alec Chisholm (1950) wrote that he had been told the name by 'a man from Atherton' and that he had passed it on to John Leach of the RAOU checklist committee, who got it accepted.

Other names: Spalding's Orthonyx, Gould's usual direct translation (1869); Spalding's Spinetail (see Australian Logrunner); Northern Chowchilla (from the time when Australian Logrunner was Southern Chowchilla); Black-headed Logrunner, again by contrast with brown-headed southern species; Auctioneer-bird, reported by Chisholm (1950) from the same man from Atherton (see common name) 'because it was forever chattering'.

Orthonyx spaldingi Ramsay EP, 1868 [or-THO-niks SPAWL-ding-ee]: 'Spalding's straight-claw', see genus, and for Edward Spalding, an entomologist who did several collecting trips in north Queensland, including on behalf of Sir William MacLeay, before being employed as a taxidermist by the Queensland Museum; he collected the type specimen on an earlier trip near Cardwell. (Ramsay also named the Top End race of Black Butcherbird as a species for him.)

PSOPHODIDAE (PASSERIFORMES): quail-thrushes, whipbirds, wedgebills

Psophodidae Bonaparte, 1854 [so-FO-di-dee]: the Noisy family, see genus name *Psophodes*.

'Quail-thrush' seems to have been invented by the RAOU expert committee tasked with preparing the second *Official Checklist of the Birds of Australia* (RAOU 1926). In its words, 'Some indefinite names like Ground-bird have been replaced by more appropriate names, such as Quail-thrush'. Until then 'ground-bird' was pretty universal (Leach 1911; Mathews 1913; RAOU 1913). Why 'quail-thrush', a force-blend of names of utterly unrelated birds not even belonging to the same Order, should be 'more appropriate' is beyond us but we suspect that the port decanter played a role. Well, we suppose that they (the quail-thrushes, not the committee) spend most of their time on the ground, like a quail, and sort of sing, sort of like a thrush ...

'Whipbird', on the other hand, was a name that invented itself. The *Australian National Dictionary* recorded 'whip-bird' from 1843 but it was long pre-dated by 'coach-whip'; for example, Watkin Tench, in one of the earliest accounts of the colony, wrote, 'To one of them, not bigger than a tomtit, we have given the name of coach-whip, from its note exactly resembling the smack of a whip' (Tench 1793) – though his estimate of the size is a bit curious, leading us to wonder if there was also some confusion with a distinctive call of the Golden Whistler. Latham (1801) used Coachwhip-bird, the redoubtable Barron Field in his collection of others' writings (Field 1825) recorded Coachman's Whip-bird, while Lieutenant William Breton (1833), in reporting on his meanderings in NSW, used both Coachman and Coachwhip, clarifying that both refer to the startling call. The RAOU was still using Coachwhip-Bird as its recommended name in 1913, but recommended Whipbird in 1926. All these referred to the Eastern Whipbird – the two western species, with totally unwhiplike calls, were found much later and would never have entered the consciousness of most citizens.

'Wedgebill' was also used from the start of European familiarity with the birds, though in this case by scientists. John Gould used the genus name *Sphenostomus* ('wedge-mouth') for the short sharp broad-based bill, translated as 'Wedge-bill', and the name has been used ever since, though in recent times two species have been recognised and the genus has been subsumed into that of the whipbirds, *Psophodes*.

The genera

Psophodes Vigors & Horsfield, 1827 [so-FOH-dehz]: 'noisy [bird]', from Greek *psophōdēs*, full of noise.

Cinclosoma Vigors & Horsfield, 1827 [sin-klo-SOH-muh]: 'thrush body', from Greek *kikhlē*, thrush, and *soma*, body. Vigors and Horsfield derived the first of these from *kinkhlos*, which according to Arnott (2007) was probably a wagtail whereas *kikhlē* was a thrush, but *kinkhlos* later came to be used for a variety of birds including thrushes and Latin *cinclus* is used in ornithology for many thrush-like birds. In any case, they gave the Latin *turdus*, about which there is no doubt – they meant a thrush.

The species

Eastern Whipbird (breeding resident)

See family introduction. The three species are widely geographically separated – this one is all along the east coast.

Other names: See family introduction – Whip-bird, Coachwhip-bird, Coachman's Whip-bird, Coachman and Coachwhip; Coach-whip Flycatcher or Honeyeater (Latham 1801 and 1823, respectively); Stockwhip Bird; Lesser Coachwhip-Bird for North's 1897 north Queensland species (now race) *P. lateralis*.

Psophodes olivaceus (Latham 1802) [so-FOH-dehz o-li-VA-se-ʊs]: 'olive-green noisy-bird', see genus name, and from Latin *oliva*, an olive, and Modern Latin *olivaceus*, olive green. Latham also called the bird *Muscicapa crepitans* (from Latin *crepito*, creak, clatter or chatter), hence it was also known for a considerable time as *Psophodes crepitans* – a rather more evocative name than the present one, we think.

Black-throated Whipbird (breeding resident)

See family introduction. From the specific name.

Other names: Western Whipbird, until recently, when this species (from the far south-west) and the next (from South Australia and Victoria) were regarded as one; Black-throated Psophodes, by Gould (1848), straight from his species name; Whipbird or Coachwhip-bird – even though it doesn't 'whip'; Black-throated Whipbird or Coachwhip-bird; Mallee, Mallee Black-throated or Mallee Western Whipbird, for its main habitat.

Psophodes nigrogularis Gould, 1844 [so-FOH-dehz ni-gro-gʊ-LAH-ris]: 'black-throated noisy-bird', see genus name, and from Latin *niger*, black, and *gula*, throat. Though this bird has a less extensive black throat than *P. olivaceus*, it seems Gould was impressed by it, describing it as having a 'very black' throat (*gula nigerrima*).

White-bellied Whipbird (breeding resident)

See family introduction. From the specific name.

Other names. See previous species; it is not possible to say if any of those names were exclusively ascribed to one or other of the two current species, though this one was less well known.

Psophodes leucogaster (Howe & Ross, 1933) [so-FOH-dehz loo-ko-GAS-tehr]: 'white-bellied noisy-bird', see genus name, plus Greek *leukos*, white, and *gastēr*, stomach.

Chirruping Wedgebill (breeding resident)

See family introduction. It was only in the 1970s that one wedgebill species was divided in two, by elevating race *occidentalis* to species status. Apart from range, the calls are the only realistic way to distinguish them; 'chirrup' doesn't really do justice to the remarkable song, but it gives the idea.

Other names: Wedgebill; Crested Wedgebill (from the species name) and Daylight-bird for its dawn calling – both these referred to both species.

Psophodes cristatus (Gould, 1838) [so-FOH-dehz kris-TAH-tʊs]: 'crested noisy-bird', see genus name, and from Latin *cristatus*, tufted or crested.

Chiming Wedgebill (breeding resident)

See family introduction and Chirruping Wedgebill. This call has a beautiful hollow-chiming quality.

Other names: See Chirruping Wedgebill names; Western Wedgebill; Chimes Bird, Chime, Kitty Lintol, Sweet Kitty Lintol, Wheelbarrow Bird or Wagon Bird, all for aspects of the call.

Psophodes occidentalis (Mathews, 1912) [so-FOH-dehz ok-si-den-TAH-lis]: 'western noisy-bird', see genus name, and from Latin *occidentalis*, western, for its range (see Chirruping Wedgebill).

Spotted Quail-thrush (breeding resident)

See family introduction. From the species name, for the heavily speckled undersides of both sexes.

Other names: The array of awkward names emphasises the difficulty faced by both scientists and laypeople in defining the genus. Spotted-shouldered Thrush, used by George Shaw when he named the species, one of the first birds in the colony to be described; Spotted Ground-thrush (Gould 1848); Babbling-thrush or Spotted Babbling-thrush, still being used in a formal way in Australia into the 20th century (Hall 1907); Ground-dove or Spotted Ground-dove (Hall 1907); Ground-bird or Spotted Ground-bird, the name of choice of many respected professionals until 1926 (see family introduction); Tasmanian Ground-bird, cited by RAOU (1913) in the provisional list as a species *C. dovei*, for Mathews' subspecies *dovei* (Mathews 1912a), still recognised.

Cinclosoma punctatum (Shaw, 1794) [sin-klo-SOH-muh pʊnk-TAH-tʊm]: 'spotted thrush-body', see genus name, and from Latin *pungo/punctus*, prick or sting, hence a speck or spot.

Chestnut Quail-thrush (breeding resident)

See family introduction. This is a seriously odd name; Gould used *castanotum* to draw attention to the patch of chestnut on the lower back, but Chestnut Quail-thrush implies all-over chestnut – it is not, and Gould had it spot on.

Other names: Chestnut-backed Quail-thrush (see common name), Ground-thrush (Gould 1848), Ground-bird; Chestnut Ground-bird; Black-breasted Quail-thrush or Ground-bird, for the male's all-black throat and breast, unique among quail-thrushes except for the following species, with which it was long lumped; Copperbird (perhaps for the population that became the next species). Nullarbor Quail-thrush, for former sub-species *C. alisteri* – but the bizarre thing is that people couldn't really agree which species it was sub- to, this one or Cinnamon! (This would seem to raise some questions on why it wasn't accepted as a species at the time; it has now been reinstated as such – see Copperback Quail-thrush.)

Cinclosoma castanotum Gould, 1840 [sin-klo-SOH-muh kas-ta-NOH-tʊm]: 'chestnut-backed thrush-body', see genus name, and from Greek *kastanon*, chestnut, and *nōton*, the back.

Copperback Quail-thrush (breeding resident)

See family introduction. From the larger and more intense copper saddle than the previous species, in which it was long included.

Other names. Copper-backed Quail-thrush; Copperbird (perhaps, see previous species).

Cinclosoma clarum (Morgan, 1926) [sin-klo-SOH-muh KLA-rʊm]: 'bright thrush-body', see genus name, and from Latin *clarus, -a, -um,* clear, bright, shining. The name indicates Morgan's observation that in his specimen, from south of Lake Gairdner, SA, 'the chestnut colour of the back is brighter and of a lighter shade' than that of the Chestnut Quail-thrush found in the Murray scrub.

Cinnamon Quail-thrush (breeding resident)

See family introduction. Straight from Gould's species name.

Other names: From here on things get very complicated, because for a time during the 20th century this species and Chestnut-breasted were lumped together again (CSIRO 1969; RAOU 1978). Ironically, for much of that time Nullarbor Quail-thrush was regarded as a separate species before being demoted again (see Chestnut Quail-thrush). We have tried to keep the names to the taxa as they now are, but bear this in mind if you're looking at material from the mid 20th century.

Cinnamon-coloured Cinclosoma (Gould 1848) – the interesting question here is why Gould didn't use Ground-thrush, as he had done for the previous two species and would do later for Chestnut-breasted; Cinnamon Babbling-thrush, Ground-bird, Ground-thrush; Cinnamon-chested Ground-thrush.

Cinclosoma cinnamomeum Gould, 1846 [sin-klo-SOH-muh sin-na-MOH-me-ʊm]: 'cinnamon thrush-body', see genus name, and from Latin *cinnamomum.*

Nullarbor Quail-thrush (breeding resident)

See family introduction. From the arid treeless (i.e. Nullarbor) plain, which is its home.

Other names. Black-breasted or Black-throated Quail-thrush or Ground-bird, for the male's black breast band and vertical throat stripe.

Cinclosoma alisteri (Mathews, 1910) [sin-klo-SOH-muh A-li-stuh-ree]: 'Alister's thrush-body', see genus name, and for Mathews' son Alister, then about 3 years of age.

Western Quail-thrush (breeding resident)

See family introduction. Relative to the next species, from which it was recently split.

Other names. From when it was regarded as a race of the next species: Black-vented Ground-bird (dismissed by RAOU 1926 as 'inelegant'), Quail-thrush, Ground-thrush or Babbling-thrush; Western, Chestnut-backed, Chestnut-bellied or Chestnut Quail-thrush, Ground-bird, Ground-thrush or Babbling-thrush – all those colour features that distinguished it then from race *castaneothorax*, now Chestnut-breasted Quail-thrush, whose underparts are a darker and more muted chestnut.

Cinclosoma marginatum (Sharpe, 1883) [sin-klo-SOH-muh mar-dji-NA-tʊm]: 'thrush-body with a border', see genus name and from Latin *marginatus –a –um*, enclosed with a border, furnished with a margin. Sharpe wrote that this species 'is distinguished by the colour of its under tail-coverts, which are black margined with white'. He made no mention of the narrow white line between the black lower breast band and the rufous upper breast, but admitted that the skin he was looking at was 'in bad condition'.

Chestnut-breasted Quail-thrush (breeding resident)

See family introduction. Straight from the species name.

Other names: Chestnut-breasted Ground-thrush (Gould 1869), Babbling-thrush or Ground-bird.

Cinclosoma castaneothorax Gould, 1849 [sin-klo-SOH-muh kas-ta-ne-o-THOR-aks]: 'chestnut-breasted thrush-body', see genus name, and from Latin *castanea*, a chestnut or chestnut-tree, hence chestnut-coloured, and *thorax*, the breast.

MACHAERIRHYNCHIDAE (PASSERIFORMES): boatbills

Machaerirhynchidae Schodde & Mason, 1999 [ma-keh-ri-RIN-ki-dee]: the Knife-bill family, from the genus name.

The genus

Machaerirhynchus Gould, 1851 [ma-keh-ri-RIN-kʊs]: 'knife-bill', from Greek *makhaira*, a large knife or dagger, for the birds' enormous flat bill. A bit like the expression, '**that's** a knife'!

The species

Yellow-breasted Boatbill (breeding resident)

This (Boatbill as a 'group' name for the two species in the family) is a coined name, courtesy of the committee for the RAOU's third list of recommended names (RAOU 1978); RAOU preferred it to the previous Boatbilled or Flatbill Flycatcher, all referring to what Pizzey and Knight (2007) described as its 'grotesquely large, flat bill'. It was fortuitous, because then it was believed they were monarch flycatchers, to which they are not in fact closely related. Yellow-breasted is by contrast with the only other boatbill, the Black-breasted of New Guinea.

Other names: Yellow-breasted Flycatcher, per Gould from his species name (albeit bowdlerised from bellied to breasted, though either is accurate); Boat-billed Flycatcher, which seems to have been introduced (and possibly invented) by the authors of the first *Official Checklist* (RAOU 1913); Yellow-breasted Flatbill; Wherrybill, an alternative to boatbill (a wherry is a light rowed boat or barge), but it's hard to find written usage of it.

Machaerirhynchus flaviventer Gould, 1851 [ma-keh-ri-RIN-kus flah-vi-VEN-ter]: 'yellow-bellied knife-bill', see genus name, and from Latin *flavus*, golden or reddish yellow (from *flagro*, burn), and *venter*, belly.

ARTAMIDAE (PASSERIFORMES): woodswallows, magpies, currawongs, butcherbirds

Artamidae Vigors, 1825 [ar-TA-mi-dee]: the Greek Butcher family, see genus name *Artamus* (and cf. Laniidae, the Latin Butcher family).

'Woodswallow' was a name popularly applied to Dusky Woodswallow (as Wood Swallow) from quite early in the history of the Sydney colony. Vigors and Horsfield (1827) noted that Banks' collector George Caley reported 'Wood Swallow' as a familiar name; bear in mind that Caley was withdrawn from New South Wales by the ageing Banks in 1810. Caley (again reported by Vigors and Horsfield) further commented that 'Their resting places were on the stumps of trees which had been felled', suggesting the significance of the 'wood'. Many Australians felt the distinction was unnecessary, so the meticulous Edward Morris (1898) in his *Dictionary of Austral English* lumped the woodswallows in with 'other' swallows, hence the widespread use of 'swallow' as an alternative.

The same combination of informants – Vigors and Horsfield (1827) quoting Caley – showed that 'Butcher-bird' was also used by colonists in the first decade of the 19th century. The term did not arise in Australia – it had been in used in England since the 17th century for the Shrikes (Laniidae) for their habit of storing prey on thorns and in branch forks, a trait shared by Australian butcherbirds. 'Crow-shrike' was also used widely, but understandably little distinction was made between butcherbirds, magpies and currawongs, and group names were used indiscriminately.

'Magpie' is one of the archetypal Australian names based on a superficial similarity to an unrelated European species that was familiar to settlers. In this case the similarity of Australian Magpies to the European Magpie (*Pica pica*) is pretty sketchy, beyond being largish and black and white. Nonetheless, the latter character was the originator of the name, originally just Pie, from French *pie* and ultimately Latin *pica*, the name of the bird. Like other familiar birds it was 'rewarded' with a person's name, originally Magot Pie, again straight from French *Margot la Pie* from the 16th century; it is first recorded in written English in *Macbeth* in 1605 but was doubtless in use before that. This evolved to Magget Pie and Mag Pie to Magpie (though the original Pie long survived alongside it; Newton had a long entry under that name in his comprehensive 1896 dictionary).

'Currawong' is clearly of Aboriginal origin – from the Jagawa language of south-east Queensland according to the *Australian National Dictionary*, though the Australian Oxford suggested it could have been from Dharuk, in the Sydney area – and surely also onomatopoeic. It does not seem to have appeared in English until the 20th century, the awkward 'crow-shrike' or 'bell-magpie' having been prevalent until then. The *Australian National Dictionary*'s earliest version is 'churwung' from Brisbane in 1905; certainly none of the bird books used it before then, nor did it appear for the next 20 years in several landmark bird books. It arrived, without fanfare or explanation (there is no apparent use of it in *Emu* before that), in the RAOU (1926) *Official Checklist of the Birds of Australia* (without having been mentioned in the first list, in 1913) and became established thereafter. (A property named Currawong 'beyond Yass' or 'Lake George' features heavily in advertisements from 1837 onwards, but this seems pure coincidence.)

The genera

Artamus Vieillot, 1816 [AR-ta-mʊs]: 'butcher', from Greek *artamos*, butcher (and, metaphorically, murderer). Vieillot said that *Artamus* refers to 'Langraien' from Buffon (1770) (who is citing Brisson 1760, though Brisson himself said the inhabitants of Manila call that bird *Langni-Langnaien* – 'Chinese whispers', anyone?). Buffon expressed some doubts but settled on the fact that the bird 'Langraien' looks a little bit like a shrike, and agreed to call it '*pie-grièche*' (shrike) in French (literally meaning Greek magpie, but that's another story). He did this largely because others seem to do it, he said. So, though Vieillot (1816) is separating this genus, along with many others, from *Lanius* (the shrikes), it seems likely that the general belief that this bird was a shrike, rather than any behaviour of the woodswallows themselves, influenced his choice of the name *Artamus*. Although the woodswallows do eat insects they certainly don't eat small birds, as the shrikes and butcherbirds do.

Melloria Mathews, 1912 [mel-LOR-ee-uh]: 'Mellor's bird', in honour of John White Mellor (1868–1931), Australian ornithologist, and a founder member and President of the RAOU as well as President of the South Australian Ornithological Association.

Gymnorhina Gray GR, 1840 [djim-no-RY-nuh]: 'naked nose', from Greek *gumnos*, naked and *rhis*, nose. Gray did not explain, but Vieillot, whose *Cracticus* Gray was commenting on, described the base of the bill as '*glabre*', meaning 'hairless' or 'beardless'.

Cracticus Vieillot, 1816 [KRAK-ti-kʊs]: 'Greek noisy [bird]', from Greek *kraktikos*, noisy. Not quite how we would have described the fluting tones of the butcherbirds.

Strepera Lesson, 1830 [STRE-pe-ruh]: 'Latin noisy [bird]', from Latin *strepo/strepere*, to make a noise. We can't help being reminded that the birds are also fairly obstreperous, from Latin *obstrepo* – to be annoying and disruptive, to other birds at least.

The species

White-breasted Woodswallow (breeding resident)

See family introduction. Far from the only white-breasted woodswallow in its range (from western Indonesia east to New Caledonia), but the vernacular name seems to be related to an alternative name, *A. leucogaster*, still in use in Australia into the 20th century (Lucas and Le Soeuf 1911; RAOU 1913), though that name meant literally 'white belly'.

Other names: White-rumped Swallow or Woodswallow, the latter widely used, including by Gould (1848) who based it on his own species name *A. leucopygialis* – he believed the Australian birds to be of a different species from the 'Indian' ones (they do not occur that far west); Swallow-shrike, another widely used term – it may have been influenced by the genus name.

Artamus leucorynchus (Linnaeus, 1771) [AR-ta-mus loo-ko-RIN-kʊs]: 'pale-billed butcher', see genus name, and from Greek *leukos* (usually white, but can range through to pale grey) and *rhunkos*, bill. Doubt is sometimes expressed about this species name, as Linnaeus actually only wrote '*leucoryn.*', and you might expect 'rh' rather than just 'r' for the Greek. However, his original description has the bird as 'black, with the bill, chest, belly and rump white', so the full name was eventually accepted.

Masked Woodswallow (breeding resident)

See family introduction. From the species name, as applied by Gould (1848).

Other names: Blue-bird, Blue Martin, Bluey, Skimmer, all used for other woodswallows too (see White-browed and Dusky), originating with Dusky Woodswallow to which the 'blue' is slightly more appropriate; Wood or Bush Martin; Blue Jay, more commonly applied to Black-faced Cuckoo-shrike and quite likely a case of mistaken identity; Summerbird, also used for Black-faced Cuckoo-shrike, and the use was doubtless reinforced by the similarity of the birds but it also reflects the appearance of flocks in settled southern areas over summer – Leach (1911) rhapsodised of woodswallows in general that 'they are the Blue-birds, Summerbirds or Martins of our youth'; Cherry-bird, quite likely from the name Cherry-hawk for the somewhat similar Black-faced Cuckoo-shrike, but it's hard to find an example of its usage outside the HANZAB (1990–2006, vol. 7a) citation.

Artamus personatus (Gould, 1841) [AR-ta-mus pehr-soh-NAH-tʊs]: 'masked butcher', see genus name, and from Latin *persona*, a mask or false face or character in a play – a word perhaps more familiar to us nowadays as an aspect of human behaviour (thanks to psychiatrist Carl Jung): 'adopting another persona' (i.e. wearing a different mask).

White-browed Woodswallow (breeding resident)

See family introduction. A unique character in the genus, from Gould's name (see 'Other names').

Other names: White-eyebrowed Woodswallow, used by Gould (1848) directly from his species name; Martin, Blue Martin, Bluey, Cherry-bird, Skimmer, Summerbird (see Masked); Sky Summerbird or Sky, evocative, but not easy to find an example of its usage. Four-year-bird, for its sporadic appearances. F.A. D'Ombrain, writing from the Hunter Valley in 1934, said, 'During school days, the White-browed Wood-swallow was known to us by many names, but perhaps the most outstanding name was that of 'Four-Year-Bird'. That name was given to it by reason of the fact that the birds were reputed to appear every four years only.'

Artamus superciliosus (Gould, 1837) [AR-ta-mus soo-pehr-si-li-OH-sʊs]: 'eyebrowed butcher', see genus name, and from Latin *supercilium*, above the eyelid (i.e. the eyebrow). And so to 'supercilious' again – what you do with your eyebrows when you're feeling like that.

Black-faced Woodswallow (breeding resident)

See family introduction. True, but still one of the great mysteries; why was Masked Woodswallow, with a fully black face, called thus, while this species has a little black mask? The answer lies with Gould, who described this as *A. melanops* (i.e. black-faced), unaware of Vieillot's earlier name (he can be forgiven, as Vieillot mistakenly cited his type as being from Timor). Black-faced was not formally introduced until RAOU (1926).

Other names: Grey-breasted Woodswallow, the name used by Gould (1848), for an apparently unlikely characteristic but one that actually distinguishes it from most other species – this name was still in use into the 20th century (RAOU 1913); White-bellied or White-vented Woodswallow, for race *albiventris* of tropical Queensland – Gould described it as a separate species *A. albiventris* and used White-vented as his common name; RAOU (1913) also applied White-bellied to Sharpe's perplexing 1890 name *A. hypoleucos*, which he proposed to replace *albiventris* – it was not used much after that; RAOU (1913), in the provisional list, also used White-vented for Sharpe's 1878 species *A. venustus* from the Top End, to add to the confusion; Grey Woodswallow, straight from the species name.

Artamus cinereus Vieillot, 1817 [AR-ta-mus si-NE-re-ʊs]: 'ashy-grey butcher', see genus name, and from Latin *cinereus*, like ashes.

Dusky Woodswallow (breeding resident)

See family introduction. Perhaps influenced by Latham's alternative species name *sordidus*, which was still being used, as *A. sordidus*, by RAOU (1913); Dusky Woodswallow seems to have been coined by the compilers of RAOU (1926).

Other names: Sordid Thrush, from Latham (1801), a direct translation of his species name; Wood-swallow, used by Gould (1848) and still being used by Mathews and the RAOU in 1913 for the only woodswallow common in the south-east; Blue-bill, a descriptive early settlers' name, reported by Caley per Vigors and Horsfield (see family introduction); Bee-bird, perhaps for its reputation as a bee-eater or for its formation of semi-torpid swarms in cold weather (see Wigan 1931 for a discussion – though we think we can dismiss her speculation that 'the birds could possibly have learnt this swarming habit from the bees themselves'). Martin, Blue Martin, Bluey, Skimmer, Summerbird, Sky – see White-breasted and White-browed Woodswallow, though all such names originated with Dusky Woodswallow; Jacky Martin, interesting as 'Martin' itself for the bird was originally coined (in England) as a human name; Sordid Woodswallow.

Artamus cyanopterus (Latham 1802) [AR-ta-mus si-a-no-TE-rʊs]: 'blue-winged butcher', see genus name, and from Greek *kuanos/kuaneos*, dark blue – the colour of blue steel, a porpoise's skin or the deep sea. It is also the colour of the enamel on Achilles' breast-plate, of lapis lazuli and of a swallow's (or even a woodswallow's) wing (Greek *pteron*). The word Latham used in his Latin description is *caeruleus*, which is actually dark-coloured, dark blue and dark green, as well as azure. In the current case, darkish blue-grey might be nearer the mark.

Little Woodswallow (breeding resident)

See family introduction. Used by Gould (1848), straight from the species name.

Artamus minor Vieillot, 1817 [AR-ta-mus MEE-nor]: 'smaller butcher', see genus name, and from Latin *minor* (comparative of *parvus*, small). By far the smallest of the woodswallows.

Black Butcherbird (breeding resident)

See family introduction. The only all-black butcherbird.

Other names: Quoy's Crow Shrike, as used by Gould (1848) or Quoy Butcherbird, from the species name. Black Crow-shrike; Rufous Crow-shrike or Brown Butcherbird, both from the unusual situation of the rufous morph of race *rufescens*, regarded as a separate species *C. rufescens* into the 20th century (RAOU 1913), which expresses itself only in juveniles; Spalding's Butcherbird, for Edward Pierson Ramsay's 1878 Top End species *C. spaldingi* (still recognised as a race) cited by Morris (1898) – for Edward Spalding, an entomologist who did several collecting trips to north Queensland, including on behalf of Ramsay, but there seems to be no suggestion that he collected this specimen (unlike the Chowchilla, which was 'his').

Melloria quoyi (Lesson & Garnot, 1827) [mel-LOR-ee-uh KWA-ee]: 'Quoy's and Mellor's bird', see genus name, and for Jean René Constant Quoy, correspondent with Charles Darwin and surgeon-naturalist to Louis de Freycinet in *L'Uranie* on his 1817–20 expedition of exploration to the Pacific and Australia; Quoy made at least two further journeys of exploration, and collected the type specimen of this species in New Guinea.

Australian Magpie (breeding resident)

See family introduction. The 'Australian' is both to distinguish it from the European Magpie and to unify the three former species – *tibicen*, *dorsalis* and *hypoleuca* – which were recognised until recently (Slater 1974).

Other names: Former *Gymnorhina tibicen* (when three species were recognised), across much of Australia – including current races *tibicen*, *longirostris*, *terraereginae* and *eylandtensis*; Black-backed Magpie, this being the key distinction from the next two taxa; Piping Roller, per Latham (1823) from his species name; Piping Crow-shrike (Gould 1848); Piper, Organ Bird, from the species name and the rich song; Singing White Crow, cited only by Leach (1911), indicating oral use; Long-billed Magpie, from the race name *longirostris* of the north-west, described as another species *G. longirostris* by Milligan in 1903, given brief credence by RAOU (1913); Little Magpie, for Mathews' subspecies *terrae-reginae* from near Cairns (Mathews 1912a), described as a species *G. terrae-reginae* by RAOU (1913) in the provisional list; Varied Crow-shrike, cited by CSIRO (1969) as applying to this 'species' but we are sure it should refer to *dorsalis*.

Former *G. dorsalis*, now race *dorsalis*: Western Magpie, limited to south-western Western Australia; Varied-backed Magpie, for the mottled back of the female (to which *dorsalis* refers), or Varied Magpie (RAOU 1913).

Former *G. hypoleuca*, of south-eastern Australia and central South Australia – current races *hypoleuca*, *tyrannica* and *telonocua*: White-backed Magpie or Crow-shrike, in direct contrast to Black-backed; Lesser White-backed, Tasmanian or Island Magpie, for Tasmanian race *hypoleuca*, formerly separated from mainland White-backed Magpies; Piping Shrike, as formally described on the South Australian flag and coat of arms.

And for all forms, Maggie – surely the 'real' name for this bird!

Gymnorhina tibicen (Latham 1802) [djim-no-RY-nuh tee-BEE-sen]: 'piping naked-nose', see genus name, and from Latin *tibicen*, a piper or flute-player – very apt.

Grey Butcherbird (breeding resident)

See family introduction. By contrast with Pied and Black Butcherbirds; the name is based on Gould's *C. cinereus* (ashy-grey) which he applied to Tasmanian birds, regarding them as a separate species. He didn't use Grey Butcherbird, even in his 1865 *Handbook*. The name became established in general use between then and the end of the century – Edward Morris included it in his *Dictionary of Austral English* (1898) – but it didn't appear 'officially' until RAOU (1926). (We suspect that an 1878 advertisement appearing in various Melbourne newspapers for a lost 'small grey BUTCHERBIRD' refers to a horse, especially as the owner offered the significant sum of ten shillings for its return.)

Other names: Butcher Bird, as used simply by Gould (1848); by 1865 he had dropped the 'butcher bird' and amended it to Collared Crow-shrike (from the species name); hence Collared Butcherbird, widely used into the early 20th century; Derwent, Tasmanian or Whistling Jackass, all for race *cinereus* of Tasmania (formerly seen as a species, see common name); Whistling Dick or Jack; Grey Shrike; White-winged Butcherbird, Grey-butcherbird or Crow-shrike, from the race name *leucopteris* – this race is found over most of southern Australia, though Gould regarded it as a Western Australian species *C. leucopteris*, as did RAOU (1913); Durbaner, an unexplained name cited only by Leach (1911); Grey-backed Butcherbird, used sparingly but apparently in contrast to Black-backed; Koolide, Coolady or Coolidie, presumably onomatopoeic, recorded by McGilp (1934) and Cohn (1934) as being used on southern Eyre Peninsula, and Cohn also cited an unnamed correspondent as claiming that Koolide was used 'in Queensland'.

Cracticus torquatus (Latham 1801) [KRAK-ti-kus tor-KWAH-tʊs]: 'noisy bird with a collar', see genus name, and from Latin *torquis*, a twisted neck-chain or collar. Its white partial collar is narrower and perhaps better defined than that of other *Cracticus* species.

Silver-backed Butcherbird (breeding resident)

See family introduction. From the species name; the back is a much paler grey than that of the Grey Butcherbird. Gould referred to it as 'silvery grey'.

Other names. Silvery-backed Butcherbird, as used by Gould himself; it was lumped with the previous species for a long time, until reinstated by Schodde and Mason (1997).

Cracticus argenteus (Gould, 1841) [KRAK-ti-kʊs ar-GEN-tee-ʊs]: 'silvery noisy-bird', see genus name and from Latin *argentum*, made of silver.

Black-backed Butcherbird (breeding resident)

See family introduction. Resembles a black-backed Grey Butcherbird.

Other names: White-throated Butcherbird, to distinguish from the similarly black and white Pied Butcherbird, which has also been known as Black-throated Butcherbird.

Cracticus mentalis Salvadori & D'Albertis, 1876 [KRAK-ti-kus men-TAH-lis]: 'mental noisy bird', see genus name, and no, not its brain – this has to do with its chin (from Latin *mentum*, chin, rather than *mens/mentis*, mind, intellect, soul, etc.). It refers to the minute black patch on the bird's chin, quite hard to see in the field and not even shown in some field guides. There is a secondary meaning of 'mental' in English, pertaining to the chin; the 'mentalis' muscle (*levator menti*), between the lower lip and chin, wrinkles the skin of the chin.

Pied Butcherbird (breeding resident)

See family introduction. From Gould's species name *picatus*, now a race, from northern and western Australia.

Other names: Pied Crow-shrike, from Gould (1848), for his species *C. picatus* from northern Australia; Black-throated Butcherbird or Crow-shrike, the latter from Gould again, directly from his species name, used by RAOU (1913), which still recognised both *picatus* and *nigrogularis*; Crow-shrike; Break-o'-day Boy, Organ Bird, both for the gloriously rich song, which starts early; Jackeroo, though the only usage we can find is McGilp (1935), in an account of the Musgrave Ranges, who stated baldly, 'It is known as the 'Jackeroo'', without explaining why, or whether he was referring only to local usage.

Cracticus nigrogularis (Gould, 1837) [KRAK-ti-kus ni-gro-gʊ-LAH-ris]: 'black-throated noisy bird', see genus name, and from Latin *niger*, gleaming black, and *gula*, throat.

Pied Butcherbird *Cracticus nigrogularis* and Grey Butcherbird *C. torquatus*

Pied Currawong (breeding resident)

See family introduction. Gould (1848) seems to have introduced the 'pied', meaning 'like a magpie' (black and white).

Other names: White-vented Crow, from John White (1790), thus making it among the first Australian birds to acquire an English name; Noisy Roller (Latham 1802); Pied Crow-shrike, from Gould (1848) – oddly the same name he used for Pied Butcherbird, and furthermore he retained the name for both species in 1865; Bell, Black, Scrub, Port Macquarie, Talking or Mountain Magpie – the similarity with magpie is obvious, 'bell' for the ringing call and 'mountain' for its preferred habitat; North (1904) regarded Black Magpie as the name by which it was most commonly known; Black, Port Macquarie, Talking or Mountain Scrub-magpie; Pied Bell-magpie or Crow-shrike; Chillawong, Currawang,

Tullawong, Tallawong, Charawack, Kurrawack, all variants on the Indigenous word (or perhaps different words); Mutton-bird, and there are many records of currawongs appearing on plates but North (1904) recorded this usage from the Albury area; Hircine Magpie, reported by George Caley to be used 'by the Colonists', as cited by Vigors and Horsfield (1827), 'It is very good eating, except the hinder parts, which have a strong goatish smell', from Latin *hircus*, a billy-goat; Lord Howe Island Pied Currawong or Crow-shrike, for race *crissalis*; Otway Forester and Pied Afternoon-tea Bird are both cited by HANZAB (1990–2006, vol. 7a), but we can find no occurrence of either term that pre-dates that.

Strepera graculina (Latham 1802) [STRE-pe-ruh grah-kʊ-LEE-nuh]: 'jackdaw-like noisy bird', see genus name, and from Latin *graculus*, very likely a jackdaw, with *-inus*, of or pertaining to. Onomatopoeic ('gra gra'?), said Quintilian (c. 95 AD).

Black Currawong (breeding resident)

See family introduction. Exactly what it says.

Other names: Sooty Crow-shrike, by Gould (1848) from his species name, Sooty Currawong; Bell or Mountain Magpie; Jay; Black Bell-magpie, Crow-shrike, Magpie or Jay; Muttonbird (see Pied Currawong).

Strepera fuliginosa (Gould, 1837) [STRE-pe-ruh fʊ-li-gi-NOH-suh]: 'sooty noisy bird', see genus name, and from Latin *fuliginosus*, sooty, from *fuligo*, soot.

Grey Currawong (breeding resident)

See family introduction. Again, exactly what it is.

Other names: Grey Crow-shrike (Gould 1848). Variable Crow, applied by Latham and from which he later coined his species name; Grey Magpie, Bell-magpie or Crow-shrike; Squeaker, used in Western Australia, for what Pizzey described as its 'conversational notes'; Rainbird, cited by Leach (1911), who claimed that its ringing call says 'It's-going-to-rain' – unlike other 'rainbirds', this one is not a migrant whose arrival coincides with the rains.

For race (formerly species: RAOU 1926) *melanoptera* from the South Australian–Victorian mallee, a very dark bird: Black-winged Currawong, Bell-magpie or Crow-shrike, straight from the race name. (Greek *melas*, black, and *pteron*, wing.)

For race (formerly species: RAOU 1926) *arguta* of Tasmania, another very dark race: Clinking Currawong or Bell-magpie, for the metallic chiming call; Black Jay or Magpie (probably by confusion with Black Currawong); Hill Bell-magpie (RAOU 1913) or Crow-shrike. (Latin *argutus*, quick, expressive).

For race (formerly species: RAOU 1926) *intermedia* (Latin *intermedius, -a*, intermediate) from Eyre Peninsula: Brown Bell-Magpie, Crow-shrike or Currawong.

For race (formerly species: RAOU 1913) *plumbea* from south-western Australia: Leaden Crow-shrike or Bell-magpie, from Gould's species name; Western Grey Bell-magpie. (Latin *plumbeus*, leaden).

Strepera versicolor (Latham 1802) [STRE-pe-ruh ver-SI-ko-lor]: 'noisy bird of various colours', see genus name, and from Latin *versicolor*, of changeable or various colours. What are the colours? In his English version Latham (1801), working from a drawing and calling it the Variable Crow, described the bird, 'the plumage dusky brown, with reflections of blue and reddish in different lights'. In his official Latin version, calling it *Corvus versicolor*, he described it in two ways: first under the species description, 'darkish-blue varying with red', then under general notes, 'darkish, shining with red and blue' (Latham 1802). What can we say – except perhaps to wonder about the artist?

CAMPEPHAGIDAE (PASSERIFORMES): cuckoo-shrikes and trillers

Campephagidae Vigors, 1825 [kam-pe-FA-gi-dee]: the Caterpillar-eater family, from Greek *kampē*, caterpillar (or silk-worm), and *phagein*, eat.

'Cuckoo-shrike' is something of a mystery, not least because it is another awful combination of names of birds of entirely different orders, which tells us very little about the group in question. Presumably it took 'cuckoo' from the fact that many species are predominantly grey and have a dipping flight, and 'shrike' from the strong insect-eating bills, but really! It gained currency in Australia some time during the second half of the 19th century; in 1898 it was included in Edward Morris' *Dictionary of Austral English*. The group is found throughout southern Africa, south and south-east Asia and the western Pacific so there is no reason to suppose that the name arose in Australia, but evidence for another origin is lacking. In fact the Englishman Alfred Newton (1896), in his very comprehensive *A Dictionary of*

Birds – published only 2 years before Morris' work – seemed unaware of the term, offering only 'caterpillar-eater', which suggests that 'cuckoo-shrike' may well have arisen in Australia. Unfortunately we cannot offer any further support to that suggestion. The earliest reference we can find is in the *Sydney Morning Herald* of 18 August 1887, but as that is part of a very lengthy report on the 'Malay Archipelago', it's not clear if the term was local or if the Australian author brought it with him. Another mystery is the decision by IOC to ignore its own rule about retaining a hyphen in group names were both names are those of birds, and thus offering us cuckooshrike; it is probably due to an apparently conflicting rule relating to compound names.

'Triller' did not arise until well into the 20th century, and appears to have been coined by the RAOU (1926) – before that no bird book (or the first *Official Checklist*, RAOU 1913) offered it as an alternative to the universally used caterpillar-eater.

The genera

Coracina Vieillot, 1816 [ko-ra-SEE-nuh]: 'young raven', from Greek *korakinos*, young raven (cf. *korax*, raven). Vieillot found that some of these birds shared a characteristic with ravens: their nostrils were hidden under forward-pointing feathers.

Edolisoma Jacquinot & Pucheran, 1853 [eh-doh-lee-SOH-muh]: 'drongo body', from Edolius, Cuvier's specific name for the Drongo, *Cuculus edolius* (from the Greek name of a somewhat uncertain type of bird mentioned by Hesychius, and possibly Aristophanes: see Arnott (2007) for more detail), and from Greek *soma*, body. The type genus was found off the west coast of New Guinea during the return voyage of L'Astrolabe and La Zélée, on the expedition led by Jules Dumont d'Urville.

Lalage Boie, 1826 [LA-la-ge]: 'chirruper', from Greek *lalax*, a babbler or croaker (cf. *lalageō*, to babble or chirrup). Arnott (2007), in his comprehensive study of Ancient Greek bird-names, discussed an unidentified bird named *lalax*, mentioned by Hesychius, and felt that on balance this was likely to be a small green bird, on the grounds that it is also a name for a frog (which is small and green), so it is not a good model for the genus under discussion!

The species

Ground Cuckooshrike (breeding resident)

See family introduction. For its primarily terrestrial habits, unique in the family.

Other names: Ground Cuckoo-shrike, the preferred form in Australia; Ground Graucalus, used by Gould (1848) – *Graucalus* was the genus name used for a long time (RAOU 1913) until it was realised that Louis Vieillot had just pipped Georges Cuvier by a matter of months with his name *Coracina*; Ground or Long-tailed Jay – 'jay' has long been used for cuckooshrikes in Australia, presumably for the bluish-grey and predatory nature of the familiar Black-faced Cuckoo-shrike (but at least it didn't get added into the name!); Eastern, Western and Northern Ground Cuckoo-shrike, for the three subspecies almost inevitably recognised by Gregory Mathews – almost equally inevitably, they are not recognised now.

Coracina maxima (Rüppell, 1839) [ko-ra-SEE-nuh MAK-si-muh]: 'biggest young raven', see genus name, and from Latin *maximus*, superlative of *magnus*, large, great, tall, and so on.

Barred Cuckooshrike (breeding resident)

See family introduction. From the species name, for the strikingly black-barred belly.

Other names: Barred Cuckoo-shrike, the preferred form in Australia; Swainson's Graucalus (see Ground Cuckoo-shrike) or Cuckoo-shrike, for William Swainson, who named it. Gould renamed it for Swainson and it's best to let him explain himself: 'This species … was originally named by Mr Swainson under the appellation of *lineatus*; but that term having been previously applied to another species of the group, it became necessary to change it; and in substituting that of [*Graucalus*] *Swainsoni*, I was desirous of paying just tribute to the talents of a gentleman who has laboured most zealously in the cause of natural science, and whose researches and writings are so well known to all ornithologists.' It is now unclear why Gould thought *lineatus* was unavailable, but he was atypically wrong on this occasion. Not everyone shared his high opinion of Swainson, though it was mostly only when Swainson dabbled in botany that he got into trouble (see Superb Parrot). Yellow-eyed Cuckoo-shrike, for another striking characteristic.

Coracina lineata (Swainson, 1825) [ko-ra-SEE-nuh lin-e-AH-tuh]: 'lined young raven', see genus name, and from Mediaeval Latin *lineatus*, lined, from Latin *linea*, a linen thread, string or line (from Latin *linum*, flax.)

Black-faced Cuckooshrike (breeding resident)

See family introduction. For its most obvious characteristic, which extends from frons to ears to throat and upper breast; it also refers to Gould's name *Graucalus melanops*, which he applied, not realising that Johann Gmelin had long pre-empted him.

Other names: Black-faced Cuckoo-shrike, the preferred form in Australia; for Graucalus and jay which follow, see Ground Cuckoo-shrike; this is a very familiar and conspicuous species that has attracted many common names. Black-faced Crow (Latham 1823); Black-faced Graucalus (Gould 1848), see common name; Black-faced Greybird, the latter a name used for *Coracina* in Africa; Bluebird, Blue Peter, Blue Jay, Grey Jay; Blue or Mountain Pigeon, probably using 'pigeon' as a general term for 'bird'; Cuckoo-shrike; Small-billed Cuckoo-shrike, for the species name *Graucalus parvirostris* used briefly by Gould for Tasmanian birds (though RAOU 1913 resurrected it) and used by Mathews (1913) for his race *novaehollandiae*, also limited by him to Tasmania; Shuffle-wing or Lapwing, for the engagingly fussy habit of rearranging its wings every time it alights; Leatherhead, probably for a passing resemblance (in the dusk with the light behind it) to a Noisy Friarbird; Summer-bird, Rainbird, Stormbird for its partly migratory habits – it arrived in eastern Australia at the start of the rainy season; Cherry Hawk, bestowed unlovingly by orchardists – it is quite omnivorous; Banded Thickhead, listed by HANZAB (1990–2006, vol. 7a) but we are sure they did so in error – it was a name used by Gould for Golden Whistler and there is no evidence of it being used for this species as well.

Coracina novaehollandiae (Gmelin JF, 1789) [ko-ra-SEE-nuh no-veh-hol-LAN-di-eh]: 'New Holland young raven', see genus name, and from Modern Latin *Nova Hollandia*, New Holland, the old name for Australia.

White-bellied Cuckooshrike (breeding resident)

A puzzling name, because this is one character it actually has in common with the previous species; it was used by Gould (1865) though it doesn't match his species name *Graucalus hypoleucos* – again, he didn't realise that Gmelin had already named it.

Other names: White-bellied Cuckoo-shrike, the preferred form in Australia; For Graucalus and jay which follow, see Ground Cuckoo-shrike. Papuan Cuckoo-shrike, from the species name; White-breasted Cuckoo-shrike; Little Cuckoo-shrike, used widely into recent times (Pizzey 1980); Varied Graucalus (Gould 1838a); White-bellied Graucalus (Gould 1848); Lesser Blue Jay; Greybird or Papuan Greybird.

Coracina papuensis (Gmelin JF, 1789) [ko-ra-SEE-nuh pah-pʊ-EN-sis]: 'young raven from Papua', see genus name, and from Papua, plus the suffix *-ensis*, indicating the place of origin (the type specimen was from Cendrawasi, West Papua).

Common Cicadabird (breeding resident)

For the extraordinarily cicada-like extended call of the male (except in north Queensland); the first recorded use that we can find is as an alternative name in the RAOU's second *Official Checklist of the Birds of Australia* (RAOU 1926). Its pioneering use as first-choice name seems to have been in *What Bird is That?* (Cayley 1931). The group name is now used for several species in the genus, from islands north of Australia; this is the most widespread.

Other names: Cicadabird, widely still used in Australia; Jardine Caterpillar-eater (still used by RAOU 1926 as first-choice name) or Triller, for Sir William Jardine, the 19th-century Scottish ornithologist who described the species and whose 40-volume *The Naturalist's Library* brought natural history to the masses – in what seems an odd coincidence Wilhelm Rüppell named it *Ceblepyris jardinii* in 1839; as late as 1926 the RAOU was still promulgating this vernacular name as the one of choice, but just 5 years later Cayley (1931) had relegated it to an alternative. Jardine's Campephaga (Gould 1848), for the then genus name – that genus is now applied only to a few African species. Most campephagids spend a lot of time gleaning for caterpillars, hence Caterpillar-catcher, the preferred name of Gregory Mathews (1913); Cricketbird, still referring to the call; Slender-billed Cicadabird, for the species name; Great Caterpillar-eater (RAOU 1913), presumably by comparison with the trillers.

Edolisoma tenuirostre (Jardine, 1831) [eh-doh-lee-SOH-muh te-noo-i-ROS-trey]: 'slender-billed drongo-body', see genus name, and from Latin *tenuis*, slender or thin, and *rostrum*, bill. Jardine gave the bird this specific name on the suggestion of Swainson, saying, 'The name is characteristic of the slender and more than usually attenuated bill, being actually as high as it is broad, in which it differs from its congeners' (Jardine 1831). The congeners in question are in *Graucalus*, the name replaced by *Coracina*.

Long-tailed Triller (Norfolk Island, extinct – though extant elsewhere in the south-west Pacific)

Has a slightly longer tail than other trillers.

Other names: Norfolk Island or Black and White Sparrow.

Lalage leucopyga (Gould, 1838) [LA-la-ge loo-ko-PI-guh]: 'white-rumped chirruper', see genus name, and from Greek *leukos*, white, and *pugē*, rump.

White-winged Triller (breeding resident)

See family introduction. For the white wings, but actually it seems to have evolved from Gould's 'White-shouldered', based on his species name *Ceblepyris humeralis* (he didn't know that Louis Vieillot had already named it, albeit from Timor). The RAOU's checklist committee (RAOU 1926) fortunately opted for the current name over the cumbersome White-shouldered Caterpillar-eater.

Other names: White-shouldered Caterpillar-eater, regarded as the formal name well into the 20th century, see common name; White-shouldered Campephaga (Gould 1848), see Cicadabird; White-shouldered Triller, White-winged Caterpillar-eater; Peewee-lark or Peewit-lark, for its resemblance to a small Magpie-lark; Jardine's Caterpillar-eater or Triller, as cited by HANZAB (1990–2006, vol. 7a), though we wonder if there isn't some confusion with Common Cicadabird.

Lalage tricolor (Swainson, 1825) [LA-la-ge TRI-ko-lor]: 'three-coloured chirruper', see genus name, and for the black, white and grey colouring of the male. Swainson describes the black as 'glossed with a metallic lustre of dull greenish-blue'.

Varied Triller (breeding resident)

See family introduction. At least 15 races are recognised, mostly from New Guinea and offshore islands.

Other names: Black and White Campephaga, per Gould (1848), from the species name – see Cicadabird; Northern Campephaga, also from Gould, who believed that northern Australian specimens were of a different species; Pied, White-browed or White-eyebrowed Caterpillar-eater (see White-winged Triller) or Triller, the white eyebrows being a good distinction from White-winged Triller.

Lalage leucomela (Vigors & Horsfield, 1827) [LA-la-ge loo-ko-ME-luh]: 'white and black chirruper', see genus name, and from Greek *leukos*, white, and *melas*, black, with no mention of the tawny vent described by Vigors and Horsfield.

NEOSITTIDAE (PASSERIFORMES): sittellas

Neosittidae Ridgway, 1904 [ne-o-SIT-ti-dee]: the New Nuthatch family, from Greek *neos*, young or new, and genus *Sitta* (see genus). The family name is a taxonomic oddity. Not everyone was so apparently myopic as to lump them with the Northern Hemisphere family Sittidae. Neosittidae was based on Hellmayr's 1901 genus *Neositta* for the current species. However, De Vis had in 1897 described the New Guinea Black Sittella as *Daphoenositta miranda*; when eventually it was agreed that the Australian and New Guinea birds were congeneric his genus took precedence. **But**, Neosittidae had been validly published, so it stood, despite the disappearance of its genus base!

'Sittella' comes directly from a genus name erected by William Swainson in 1837, recognising that these Australian ecological counterparts of the European nuthatches (family Sittidae, genus *Sitta*) warranted their own genus – the word was just a diminutive of *Sitta*. Astonishingly (in retrospect it's easy to be astonished, of course) it was only in the 1960s that they were removed from Sittidae and universally recognised as an unrelated Australia–New Guinea family. (Swainson's genus stood throughout the 19th century, until someone noticed that the almost identical name *Sitella* had previously been published by Constantine Rafinesque in 1815, applied to *Sitta*; that name was immediately suppressed when it was realised that the genus had already been named *Sitta*, but its brief stint in the literature was enough to make its subsequent usage by Swainson (even with an additional 't') invalid.)

The genus

Daphoenositta DeVis, 1897 [da-foy-no-SIT-tuh]: 'bloody nuthatch' from Greek *daphoinos*, tawny, and, less commonly, 'reeking with blood'. The genus was named for the Black Sittella *Daphoenositta miranda* of New Guinea, which has a pinkish-red forehead, chin and tail-tip. Also after genus *Sitta*, nuthatches (from Greek *sittē*, which Arnott's (2007) convincing evidence confirmed as the ancient name for a nuthatch).

The species

Varied Sittella (breeding resident)

For the five strikingly different races, which until the 1970s were generally regarded as representing five species. At that time all were in the genus *Neositta* (see family name).

Other names: Sittella, Nuthatch, Bark-pecker, Tree-runner, Woodpecker.

For race *chrysoptera*, south-eastern Australia: Orange-winged Sittella, Nuthatch, Bark-pecker or Tree-runner, for broad orange wing band (see species name).

For race *leucocephala*, south and central coastal Queensland: White-headed Sittella, Nuthatch, Bark-pecker or Tree-runner; Pied Sittella or Tree-runner, from the species name, for white head and dark body. From Greek *leukos*, white, and *kephalē*, head.

For race *pileata*, most of southern Australia: Black-capped Sittella, Nuthatch, Bark-pecker or Tree-runner, for a character shared by most of the races; Slender-billed Tree-runner, from an old name *Sittella tenuirostris*, by Gould. From Latin *pileus*, the felt cap or 'cap of liberty' worn by Roman slaves who had been granted their freedom.

For race *leucoptera*, tropical Australia except Cape York Peninsula: White-winged Sittella, Nuthatch, Bark-pecker or Tree-runner, for the spectacular white wing bar in flight. From Greek *leukos*, white, and *pteron*, wing.

For race *striata*, Cape York Peninsula: Striated Sittella or Tree-runner. From Latin *striatus*, streaked or furrowed.

Daphoenositta chrysoptera (Latham 1802) [da-foy-no-SIT-tuh kri-so-TE-ruh]: 'golden-winged bloody nuthatch', see genus name, and from Greek *khrusos/khruso-*, gold/golden, and *pteron*, wing, for the orange wing band. So this species is not bloody at all.

OREOICIDAE (PASSERIFORMES): bellbirds

Oreoicidae Schodde & Christidis, 2014 [o-re-OH-i-si-dee]: the Mountain-dwelling family, from genus name.

The genus

Oreoica Gould, 1838 [o-re-OH-i-kuh]: 'mountain-dweller', from Greek *oreioikos*, mountain dwelling (*oros*, *oreo-*, mountain, and *oikos*, a house or dwelling-place). Its application to this bird of the inland plains seems odd, though Gould (1848) commented that 'it appeared to give a decided preference to the naked sterile crowns of hills and open bare glades in the forests'. However, that was with the benefit of hindsight: he may originally have been swayed by Vigors and Horsfield's (1827) erroneous description of the type locality for the bird they called *Falcunculus gutturalis*, the Kent Group of rocky islands in Bass Strait. Vigors and Horsfield's only real error was that they believed the Mr Brown who had presented the specimen to the Linnean Society when he told them this.

The species

Crested Bellbird (breeding resident)

For the resonant hollow chiming call, perhaps the most archetypal inland call of all, and for the male's little erectile crest. Gould (1848) reported that Bell-bird was already being used by the 'Colonists of Swan River'.

Other names: Crested Oreoica, per Gould (1848) from the genus name; Crested Thrush (Lewin 1808) – though the specimen's origin gives us pause for thought, given that the inland had not yet been penetrated; Dick-Dick-the-Devil, Pan-pan-panella, Wack-to-the-rottle (from western Victoria, Sullivan 1911), Bunbundalui (from north-central New South Wales, Sullivan 1931), all renditions of the male's territorial call.

Oreoica gutturalis (Vigors & Horsfield, 1827) [o-re-OH-i-kuh gʊt-tʊ-RAH-lis]: 'throaty mountain-dweller', see genus name, and from Latin *guttur*, the throat or neck. Even if the genus name doesn't quite reflect the habitat as we think of it today, the species name is certainly apt for the male's large black bib.

Crested Bellbird *Oreoica gutturalis* and Spangled Drongo *Dicrurus bracteatus*

PACHYCEPHALIDAE (PASSERIFORMES): shriketit, whistlers, shrikethrushes

Pachycephalidae Mayr, 1941 [pak-i-se-FA-li-dee]: the Thick-head family, see genus name *Pachycephala*.

'Whistler' didn't appear in the literature until the 20th century – throughout the 19th century the bird suffered under the unfortunate name 'thickhead' (a direct translation of the genus name *Pachycephala*). It seems to have been the influential John Leach who introduced the word whistler in *An Australian Bird Book* (1911), though his comment 'the Thickheads (now called Whistlers)' doesn't make it clear if he coined it. It was certainly new – Lucas and Le Soeuf in the same year didn't use the new name – and it is likely that Leach's usage encouraged the RAOU to adopt 'whistler' in the first *Official Checklist of the Birds of Australia* (RAOU 1913).

'Shrike-thrush' (IOC removes the hyphen), despite its awkwardness (though at least both the names are of passerines, albeit unrelated to each other or to these birds), was used a little earlier than 'whistler' and is based on the genus name. 'Thrush' was used for Grey Shrikethrush in particular from the earliest days of the colony, for its superb song, while the 'shrike' addition refers to the strong predatory bill. Colluricincla (the genus name) was used as a group name for much of the 19th century. The earliest use of shrike-thrush known to Edward Morris and cited in his *Dictionary of Austral English* (1898) was in the *Melburnian* in 1896, rather than in a bird book, but we can take the use back another 20 years, to a reference to Harmonious Shrike-thrush in the *Melbourne Weekly Times* of 12 October 1878. This was in a report of a bill in the Victorian Assembly 'to protect insectivorous and other birds', so the name carried official gravitas. The fact that Hall used shrike-thrush in his major key to Australian birds in 1899 strongly suggests that the term was already in general use in the birding community.

The genera

Falcunculus Vieillot, 1816 [fal-KʊN-kʊ-lʊs]: 'little falcon', from Latin *falco/falconis*, a falcon, with diminutive ending *-ulus*. Vieillot's common name for this genus was *Falconelle*, from *falconella* (said by Belon 1555 to be the Italian word for shrike) and he seems to have based his Latin term on this. This usage reflects the fact that shrikes are birds of prey, and Vieillot's description focused on the Shriketit's hooked bill (Vieillot 1816–19).

Pachycephala Vigors, 1825 [pa-ki-SE-fa-luh]: 'thick-head', from Greek *pakhus*, thick, and *kephalē*, head. This is descriptive, not intended to be insulting: the birds do have rather large round heads.

Colluricincla Vigors & Horsfield, 1827 [col-lʊ-ri-SINK-luh]: 'shrike-thrush', from Greek *kollyriōn*, said by Belon to be a shrike, and chosen by Linnaeus to use as species name for *Lanius collurio*, the Red-backed Shrike. It was later used as a transitory genus name *Collurio* by Rafinesque. The Greek word was used by Aristotle and may have been one of the shrikes – even Arnott (2007) didn't commit himself on this – but it is also sometimes rendered as thrush (in which case our translation could be 'thrush-thrush'). The latter part of the genus name, *-cincla*, is from Greek (Vigors and Horsfield 1827), this time spelled *khinkhlos* (cf. *Cinclosoma*, quail-thrushes). According to Arnott (2007), *Kinkhlos* was probably a wagtail whereas *kikhlē* was a thrush (there are many variant spellings of both words), but Vigors and Horsfield held that this genus 'has also some general resemblance to the neighbouring family of the thrushes', so they definitely meant thrush!

The species

Crested Shriketit (breeding resident)

Another hybrid name. Latham referred to it as Frontal Shrike in 1801 (Shrike doubtless because of the powerful bill), while Gould (who used 'shrike-tit' and may have invented it – he didn't record the name as being used by 'the colonists', as was his wont) commented, 'in many of its habits it bears a striking resemblance to the Tits, particularly in the manner in which it clings to and climbs among the branches in search of food'. What is also striking is the resemblance of the yellow, white and black plumage to that of the European Great Tit (*Parus major*). 'Crested' is self-evident, and also refers to the species name.

Other names: Crested Shrike-tit, the form used in Australia; Frontal Shrike, see common name; Frontal Shrike-tit, straight from the species name; Barktit, Crested Tit, Falcon-shrike (in reference to the genus name), Yellow-hammer, presumably in reference to its vague resemblance to the Northern Hemisphere bunting *Emberiza citrinella*; Northern Shrike-tit or Yellow Shrike-tit, for race *whitei*, which has been regarded as a separate species (RAOU 1926; Cayley 1931); Eastern, Yellow-bellied or Yellow-breasted Shrike-tit, for race *frontatus* of eastern and south-eastern Australia, with yellow undersides in contrast with race *leucogaster*; Western or White-bellied Shrike-tit, for race *leucogaster* of the south-west, named by Gould (1848) – this race and the previous one were seen as separate species for nearly 100 years (RAOU 1926).

Falcunculus frontatus (Latham 1802) [fal-KʊN-kʊ-lus fron-TAH-tʊs]: 'little falcon with a forehead', see genus name, and from Latin *frons*, forehead. Latham described what he called a white line going from the frons and down both sides of the bill, as well as white bands down the sides of the head.

Olive Whistler (breeding resident)

See family introduction. From the species name.

Other names: Olivaceous Pachycephala, being from Gould (1848) at his orotund best; Olivaceous or Olive Thickhead; Olivaceous Whistler; Whipbird, for the call that Pizzey and Knight (2007) described as 'a long, ringing 'peeee''; Mystery Bird, a name that anyone who has tried to track the thin ventriloquial calls through dense vegetation will relate to; Eel-bird, a complete mystery – we cannot add anything to

HANZAB's citation of it (1990–2006, vol. 6) nor find a single other usage, though the implication of elusiveness is beguiling.

Pachycephala olivacea Vigors & Horsfield, 1827 [pa-ki-SE-fa-luh o-li-VAH-se-uh]: 'olive-green thick-head', see genus name, and from Latin *oliva*, an olive, and Modern Latin *olivaceus*, olive green.

Red-lored Whistler (breeding resident)

See family introduction. For the rufous lores (the area between eye and bill).

Other names: Red-throated Pachycephala (Gould 1848), Thickhead or Whistler, straight from the species name; Rufous-throated Pachycephala, Thickhead or Whistler; Buff-breasted Whistler or Thick-head, hmm – the throat and belly are rufous-buffy, the breast is grey.

Pachycephala rufogularis Gould, 1841 [pa-ki-SE-fa-luh roo-foh-gʊ-LAH-ris]: 'rufous-throated thick-head', see genus name, and from Latin *rufus*, red, and *gula*, the throat.

Gilbert's Whistler (breeding resident)

See family introduction. From Gould's published species name *gilbertii*, for the remarkable John Gilbert, a taxidermist who came to Australia at 26 to work for Gould, taught himself bushcraft to an impressive degree and for the remaining 6 years of his life travelled in some very remote areas collecting specimens and recording information. By our count he was responsible for a startling 8% of all Australian bird and mammal type specimens, before being killed in an Aboriginal attack on Leichhardt's camp near the Gulf of Carpentaria in 1845. To quote Gould, 'Although the practice of naming species after individuals is a means by which the names of men eminent for their scientific attainments may be perpetuated for after-ages, I have ever questioned its propriety, and have rarely resorted to it; but in assigning the name *Gilbertii* to this new and interesting species, I feel that I am only paying a just compliment to one who has most assiduously assisted me in the laborious investigations required for the production of the present work, and who was the discoverer of the birds forming the subject of this paper' (Gould 1848). Unfortunately, he seemed to overlook the fact that he had already named the species *inornata* some years previously!

Other names: Gilbert's Pachycephala (Gould) or Thickhead; Red-throated Whistler or Thickhead (as valid for this as for the previous species); Black-lored Whistler, in direct contrast to the previous species.

Pachycephala inornata Gould, 1841 [pa-ki-SE-fa-luh in-or-NAH-tuh]: 'unadorned thick-head', see genus name, and from Latin *inornatus*, unadorned, without ornament. Pity about the loss of *gilbertii* (see common name) – Gilbert clearly deserved the honour. And as for *inornata*, although male birds are not completely unadorned, Gould was dealing with a female or an immature male – his description lacks any mention of the mature male's rufous throat (Gould 1840).

Grey Whistler (breeding resident)

See family introduction. An uninspired name for a fairly unexceptional-looking bird – which isn't really grey.

Other names: Plain-coloured Pachycephala (Gould 1848); Grey Thickhead; York Thickhead, for race *peninsulae* of Cape York Peninsula, recognised as a separate species into the 20th century; Grey-headed Whistler or Thickhead, for the same race, perhaps a more pertinent name and based on the species name *P. griseiceps*, published by Gray in 1858 and still used by RAOU (1926); Brown Whistler or Thickhead, for race (species well into the 20th century: RAOU 1926) *simplex* of the Top End, perhaps more un-grey than *peninsulae*; Plain-coloured Shrike-Robin, from Ramsay's unfortunate misidentification of a specimen from north Queensland, which he called *Eopsaltria inornata*.

Pachycephala simplex Gould, 1843 [pa-ki-SE-fa-luh SIM-pleks]: 'plain thick-head', see genus name, and Latin *simplex*, simple or plain.

Australian Golden Whistler (breeding resident)

See family introduction. An obvious name (albeit only for the male) but it doesn't seem to have appeared until the 20th century. White-throated was popular coming into the century, Yellow-breasted was used for the RAOU's first *Official Checklist* (1913), with Leach coming closest in 1911 with Golden-breasted. Perhaps it was his convenorship of the second checklist committee (RAOU 1926) that helped persuade it to adopt Golden Whistler. 'Australian' is a recent addition, to reflect the fact that a plethora of former races spread through the islands to the north have now been described as separate species, with this species now restricted to Australia.

Other names: Golden Whistler (see previous sentence); Guttural Thrush, by Latham from his name *Turdus gutturalis* in Supplement II to his mighty *General Synopsis of Birds* in 1801 (not his *Index Ornithologicus* as suggested by Mathews 1912c) – and thereby hangs a tale. Unfortunately the name had already been used by Philipp Ludwig Statius Müller for another species 25 years previously but nobody seemed to notice that for the next 110 years; fortuitously, albeit unwittingly, Latham himself provided the solution by redescribing the bird with a different name (*Muscicapa pectoralis*, Black-breasted Flycatcher) 10 pages further on in the same book! In 1911 both Leach, and Lucas and Le Soeuf, in their separate books, used *Pachycephala pectoralis*, apparently for the first time. Black-crowned Thrush, by Lewin (1808) in his pioneering *Birds of New Holland*; Guttural Pachycephala, by Gould (1848) for the prevailing species name – hence the previous story; Thunder-bird, for its propensity to sing in response to loud noises (including thunder) – Gould (1848) reported that it was already in use by the 'Colonists of New South Wales'; Whipbird, Coachwhip Bird, Ringcoachie, all for its whip-like single sharp contact call – the 'ring' for the black breastband; Golden-breasted Whistler or Thickhead; Cut-throat, perhaps also for this call; White-throated Whistler or Thickhead, popular in the late 19th and early 20th centuries, though we might think it odd that the throat attracted more attention than the gloriously golden breast; Yellow-breasted Whistler (RAOU 1913); Grey-tailed Pachycephala, used by Gould (1848) for his species *P. glaucura* of Tasmania, now recognised as a race, though RAOU (1913) still saw it as a species and called it Grey-tailed Whistler; Southern Whistler, for *P. meridionalis* from South Australia, described in 1904 and recognised by RAOU (1913) but forgotten soon afterwards; similarly Queensland Whistler, for *P. queenslandica* described by Reichenow in 1899 from near Cairns and listed by RAOU (1913) on the provisional list; Robust Whistler or Thickhead, in confusion with *P. robusta* (Mathews 1913) – see Mangrove Golden Whistler; Lord Howe Island Whistler, Robin or Yellow Robin for race *contempta*, formerly described by Hartert in 1898 as a species and recognised as such by RAOU (1913); Norfolk Island Thickhead or Whistler, for race *xanthoprocta*, formerly recognised as a full species (RAOU 1913); Tamey, a Norfolk Island name.

Pachycephala pectoralis (Latham 1802) [pa-ki-SE-fa-luh pek-to-RAH-lis]: 'thick-head with a breast', see genus name, and from Latin *pectus/pectoris*, the chest (see common name).

Western Whistler (breeding resident)

See family introduction. An obvious name for this recently resurrected species (long subsumed into Golden Whistler), restricted to south-western Australia.

Other names. Western Thickhead.

Pachycephala occidentalis (Ramsay, 1878) [pa-ki-SE-fa-luh ok-si-den-TAH-lis]: 'western thick-head', see genus name and from Latin *occidentalis*, western, see common name.

Mangrove Golden Whistler (breeding resident)

See family introduction. A very similar bird to the Golden Whistler but restricted to mangroves; the name was apparently formally introduced by Slater (1974), though it was doubtless in use before that.

Other names: Black-tailed Thickhead or Whistler, straight from the species name, used by Gould (1848) and for over 100 years thereafter; Robust Whistler or Thickhead or Big-billed Whistler (RAOU 1913) for former species *P. robusta* of Cape York, described by Masters in 1875 – it is now recognised as a race.

Pachycephala melanura Gould, 1843 [pa-ki-SE-fa-luh me-la-NOO-ruh]: 'black-tailed thick-head', see genus name, and from Greek *melas*, black, and *oura*, tail – its tail is indeed very black, and slightly shorter than that of the Golden Whistler.

Rufous Whistler (breeding resident)

See family introduction. For the male's rufous breast, a direct contrast with Golden Whistler.

Other names: Orange-breasted Thrush (Lewin 1808); Rufous-vented Honeyeater (Latham 1823); Banded Thickhead, per Gould (1848) who confusingly used the name *pectoralis* (see Golden Whistler), following Vigors and Horsfield (1827) – and their description ('with a rusty belly') makes it clear that they really were referring to Rufous Whistler; Lunated Pachycephala for Gould's species (now race) *P. falcata* of northern Australia – he refers to 'the black crescent which bounds the white throat of the male not extending upwards to the ear-coverts'; Northern Thickhead or Whistler, for the same race, still recognised at species level by RAOU (1913); Echong, Joey-joey, Mock Whipbird, Ring Coachman, Coach-whip-bird, all for the whip-like contact call (it has a legendary repertoire of calls) – see also Golden

Whistler; Little or Small Thrush, for the glorious song; Pale-breasted Thickhead or Whistler for Cape York Peninsula race (formerly species: RAOU 1913) *pallida* – though inland birds are also often pale; Rufous-breasted Thickhead or Whistler; White-throated Whistler, from the true-but-unhelpful department; Thunderbird or Thundering Joey, see Golden Whistler.

Pachycephala rufiventris (Latham 1802) [pa-ki-SE-fa-luh roo-fi-VEN-tris]: 'rufous-bellied thickhead', see genus name, and from Latin *rufus*, red, reddish (of all shades), and *venter*, belly.

White-breasted Whistler (breeding resident)

See family introduction. For the obvious and unique character, albeit again of males only.

Other names: Shrike-like Pachycephala, per Gould (1848) from his species name, or Shrike-like Thickhead or Thrush; White-bellied Whistler or Thickhead; Torres Strait Whistler or Thickhead (Mathews 1913), for short-lived species *P. fretorum* (De Vis, 1889), now recognised as a race – but it isn't found as far east as the strait and in fact the type specimen was from far to the west, near Wyndham in Western Australia; Large-billed Whistler, used for no obvious reason by RAOU (1913) for the same species.

Pachycephala lanioides Gould, 1840 [pa-ki-SE-fa-luh la-ni-OY-dehz]: 'thick-head like a shrike', see genus name, and from *Lanius*, a genus in the shrike family, with Latin *-oides*, resembling, for the very strong and slightly hooked black bill.

Bower's Shrikethrush (breeding resident)

See family introduction, and from species name.

Other names: Bower's Shrike-thrush, the preferred form in Australia; Bower Thrush or Shrike-thrush; Stripe-breasted Thrush or Shrike-thrush, for an obvious character.

Colluricincla boweri Ramsay EP, 1885 [col-lʊ-ri-SINK-luh BOW-er-ee]: 'Bower's shrike-thrush', see genus name, and for Thomas Henry Bowyer-Bower, an English collector who came twice to Australia to acquire skins for the British Museum, was curator of birds at the Western Australian museum and died of typhoid fever at Port Darwin 1886, aged only 24; the bird was named while he was still alive.

Little Shrikethrush (breeding resident)

See family introduction. The smallest shrikethrush, named Little Colluricincla by Gould (1848) for his species *parvula*, now a race, from the Top End.

Other names: Little Shrike-thrush, the preferred form in Australia; Little Colluricincla (Gould 1848); Little Thrush; Red, Red-breasted, Rufous, Rufous-breasted, Fawn-breasted or Rusty-breasted Shrike-thrush, especially race *rufogaster* (formerly species, named by Gould in 1845, still recognised by RAOU 1913) from central east Queensland, though all races have some rustiness; Allied or Allied Rufous Shrike-thrush (RAOU 1913) for formerly recognised species *C. parvissima* (described by Gould in 1872) from north Queensland, presumably for its close relationship with *megarhyncha* – good call, because it's not now even regarded as a race.

Colluricincla megarhyncha (Quoy & Gaimard, 1830) [col-lʊ-ri-SINK-luh me-ga-RIN-kuh]: 'large-billed shrike-thrush', see genus name, and from Greek *megas*, large, and *rhunkhos*, bill.

Grey Shrikethrush (breeding resident)

See family introduction. Not a very satisfying name for a magnificent bird; Morris (1898) reported that Colonel Legge had recently applied the name. (William Legge was an internationally respected Australian ornithologist who wrote a massive treatise on the birds of Ceylon and was shortly afterwards the founding president of the RAOU, all while a serving officer of the British Army.) However, Legge was far pre-empted by Jardine and Selby (1826–35, vol. 2).

Other names: Grey Shrike-thrush, the preferred form in Australia; Port Jackson Thrush (White 1790), still used as first-choice name by Morris (1898) in his *Dictionary of Austral English*; Harmonic, Dilute (?), Austral or Grey-headed Thrush (Latham 1801, 1823); Harmonious Colluricincla, per Gould (1848), straight from the species name; Harmonious or Harmonic Thrush or Shrike-thrush; Native Thrush; Duke-Wellington, Jock Whitty, Joe Wickie, Wick, Whistling Dick (reported by Gould 1848 as used 'by the Colonists of Van Diemen's Land'), all references to elements of the beautiful and complex song; Pluff, a curious one but in use in the early years of the 20th century (Leach 1911) – we cautiously suggest an association with the strong single-note winter call; Brown Colluricincla, Thrush or Shrike-thrush, for race *brunnea* (named as a species by Gould and still recognised well into the second half of the 20th century) of northern Australia; Buff-bellied Shrike-thrush, by Gould for his race *rufiventris*, likewise named as a species and still recognised well into the second half of the 20th century, more generally as Western Thrush or Shrike-thrush; an article in the *South Australian Chronicle and Weekly Mail* of 9 June 1877 refers to this race as Buff-Bellied Crow-Shrike Thrush(!), though it's quite possible this was self-coined by the author; Selby's Colluricincla, for Jardine's Tasmanian species *C. selbii*, not now recognised, for English ornithologist and bird artist Prideaux Selby; Whistling Shrike-thrush, also for *C. selbii* (RAOU 1913); Pallid Grey Shrike-thrush, for Mathews' race *oblita* from south-east Queensland; White-browed Shrike-thrush, for Masters' 1875 northern Cape York species *C. superciliosa*, still recognised by RAOU (1913) and now regarded as a race; Grey-backed Shrike-thrush for Mathews' 1912 subspecies *C. pallescens* from north Queensland, cited as a species by RAOU (1913) in the provisional list.

Colluricincla harmonica (Latham 1802) [col-lʊ-ri-SINK-luh har-MO-ni-kuh]: 'musical shrike-thrush', see genus name, and from Greek *harmonikos*, skilled in music, and Latin *harmonicus*, harmonious. That's better – much more satisfying than the common name!

Sandstone Shrikethrush (breeding resident)

See family introduction. For its obligate habitat in the escarpments of the Top End and Kimberley, formalised by CSIRO (1969).

Other names: Sandstone Shrike-thrush, the preferred form in Australia; Cliff or Rock or Sandstone Thrush; Brown-breasted Thrush or Shrike-thrush, a preferred name until quite recently (CSIRO 1969); Red-bellied Shrike-thrush. Woodward's Shrike-thrush, from the species name.

Colluricincla woodwardi E. Hartert, 1905 [col-lʊ-ri-SINK-luh WʊD-wʊd-ee]: 'Woodward's shrike-thrush', see genus name, and for Bernard Henry Woodward (1846–1916), an English geologist who came to Australia in 1889 for his health and became Western Australian Government analyst, then inaugural director of the Western Australian Museum and Art Gallery. He founded the Western Australia Natural History Society and was an advocate for reserved lands for conservation. The German ornithologist Ernst Hartert worked for Lord Rothschild at his private museum at Tring in England. Woodward arranged for John Tunney, Western Australian Museum collector, to extend a major expedition to northern Western Australia in 1901 to include the Northern Territory; sets of skins were provided to both museums, so we may assume that Rothschild paid for at least some of it. This species was among them.

LANIIDAE (PASSERIFORMES): shrikes

Laniidae Rafinesque, 1815 [la-NEE-i-dee]: the Latin Butcher family, see genus name *Lanius* (cf. Artamidae, the Greek Butcher family).

'Shrike' as a word is shrouded in mystery, though it appears in English in the 16th century. It is suggested that it was onomatopoeic, there being an obsolete meaning of 'shriek' for the word, but this is not a characteristic of shrikes; it is complicated by also being a dialect word for missel-thrush. Sadly, we cannot shed any further light on this.

The genus

Lanius Linnaeus, 1758 [LA-ni-ʊs]: 'butcher', from Latin *lanius*, a butcher (*lanio*, tear in pieces), from the bird's habit of hanging up the bodies of small birds or insects on thorns in order to feed from them or keep them for later.

The species

Tiger Shrike (vagrant to Christmas Island; also a dead bird from Fremantle, suspected of being a ship stowaway)

For dark scalloping across the chestnut back.

Lanius tigrinus Drapiez, 1828 [LA-ni-ʊs ti-GREE-nʊs]: 'shrike striped like a tiger', see genus name, and from Latin *tigrinus*, meaning just that.

Brown Shrike (vagrant to Christmas Island)

A sort of 'average' name, because races range from grey to chestnut.

Lanius cristatus Linnaeus, 1758 [LA-ni-ʊs kris-TAH-tʊs]: 'crested shrike', see genus name, and from Latin *cristatus*, tufted or crested. Not, actually – Linnaeus said that he had this from George Edwards'

(1747) *Natural History of Birds* ('the Crested Red, or Russit Butcher-bird'). Edwards' illustration is from a dead bird which 'has something of a Crest' but which to us just looks a bit ruffled. And it could well have been, because Edwards reported that it had been sent from Bengal to London (and not on a plane).

ORIOLIDAE (PASSERIFORMES): orioles and figbirds

Oriolidae Vigors, 1825 [o-ri-OH-li-dee]: the Golden Bird family, see genus name *Oriolus*.

'Oriole' is somewhat unusual in that it derives from the genus name; Thomas Pennant (1776), influential Welsh naturalist, coined the word in his monumental *British Zoology*, specifically from Linnaeus' genus.

'Figbird' is known from the 19th century as a name used in Sussex for the Chiffchaff (*Phylloscopus collybita*), though that species is an insect-eater so the reference is obscure. This is unlikely to have influenced its use for *Sphecotheres*, however, because 'figbird' also seems to have arisen in the 19th century in Australia. Gould (1848) inevitably used Sphecotheres as a group name; while he would say that anyway, he didn't record a colonists' name for it, and it had probably been rarely encountered by Europeans before that. The first-named species was from Timor in 1816 and it is unlikely to have attracted an English name by Gould's time. As settlement increased in the bird's subtropical and tropical east and north coast habitats, its passion for figs means that the choice of name was logical and almost certainly spontaneous. Edward Morris' 1898 *Dictionary of Austral English* included the name but didn't give any references.

The genera

Sphecotheres Vieillot, 1816 [sfeh-co-THEH-rehz]: 'wasp-hunter', from Greek *sphēx*, wasp, and *-thēras*, hunter. Perhaps the fact that the bird is an avid fruit-eater caused confusion, giving the impression it was hunting wasps that might be on the fruit? Though it does occasionally take insects, they are not a significant part of its diet. Vieillot (1816) is ambivalent about the spelling; the name appears in *Analyse* once spelt as shown and twice spelt *Sphecotera*.

Oriolus Linnaeus, 1766 [o-ri-O-lʊs]: 'golden [bird]', from Mediaeval Latin *oryolus* from Latin *aureolus*, golden (diminutive of *aureus*, golden from *aurum*, gold). It also appears in Old French: in the 12th century as *orïoel*, in the 14th as *lorio* and by the early 17th century the French is *lorion* or *loriot*, also *loriol* in various dialects; for some reason it didn't enter English until the late 18th century (see family).

The species

Australasian Figbird (breeding resident)

The current trend is to recognise three species, with this as the only Australian one (the others being from Timor and nearby Wetar).

Other names: Until relatively recently (Slater 1974) two Australian species were recognised, which are now regarded as races.

Race *vieilloti*, south from the Queensland wet tropics: Australian Sphecotheres (Gould 1848); Figbird, Southern Figbird; Green, Grey, Grey-breasted Figbird, in contrast to race *flaviventris*, with yellow undersides; Green Sphecotheres; Banana-bird or Mulberry-bird, for various of its tastes – Morris (1898) cited Alfred North of the Australian Museum, 'on the Tweed River, where it is locally known as the 'Mulberry-bird', from the decided preference it evinces for that species of fruit amongst many others attacked by this bird'; Shrike – well, nearly everything else was, why not this one too?

Race *flaviventris*, across the far north, with bright yellow underparts: Yellow-bellied Sphecotheres, per Gould (1869), from his species (now race) name; Northern, Yellow, Yellow-throated, Yellow-breasted Figbird. From Latin *flavus*, yellow, and *venter*, belly; Grey-throated Figbird, for Ingram's 1908 *S. stalkeri* from Mt Elliott, north Queensland, recognised by RAOU (1913) and doubtfully by RAOU (1926); Stalker Figbird, cited by HANZAB (1990–2006, vol. 7), we can find no other reference to it, but surely it refers to *S. stalkeri*.

Sphecotheres vieilloti Vigors & Horsfield, 1827 [sfeh-co-THEH-rehz vi-eh-YOH-tee]: 'Vieillot's wasp-hunter', see genus name, and for French ornithologist Louis Jean Pierre Vieillot (1748–1831). Vigors and Horsfield wrote that if this bird should prove to be a different species from *S. viridis*, 'it may appropriately receive the specific name *Vieilloti* after the founder of the genus'. Vieillot was one of the first ornithologists to really interest himself in plumage changes over the lifetime of a bird and in studying birds in the field, rather than dead in the hand. He made a major contribution to ornithology, but (and?) is said to have died in extreme poverty.

Olive-backed Oriole (breeding resident)

Actually this is a most unusual character among orioles, but the name appeared only at the behest of the RAOU's second official checklist committee (RAOU 1926).

Other names: Green Grackle (Latham 1802) – see also Blue-faced Honeyeater; Oriole, used alone somewhat improbably by Campbell (1901) and Lucas and Le Soeuf (1911); New South Wales Oriole (Gould 1848); Australian Oriole (Leach 1911; RAOU 1913); Green Thrush for its song; Greenback; Cedar Bird or Cedar Pigeon, for its fondness for cedar berries – 'pigeon' was not infrequently used as a general term for 'bird' (e.g. see Black-faced Cuckoo-shrike); Striated and Streaked Roller (Latham 1801, 1823); Green-backed or White-bellied Oriole.

Oriolus sagittatus (Latham 1802) [o-ri-O-lus sa-git-TAH-tʊs]: 'arrowed golden bird', see genus name, and from Latin *sagitta*, an arrow, for the actually quite arrow-like streaks on the bird's breast and belly.

Green Oriole (breeding resident)

Sort of greenish-yellow, though younger birds are more yellow beneath.

Other names: Yellow Oriole, widely used in Australia; Crescent-marked Oriole (Gould 1848, 1865), for the same curved wingbars that prompted the species name; Yellow-bellied Oriole, perhaps for young birds; Green Mulberry, perhaps by contrast with the figbird (see profile); Northern Oriole, for Gould's Top End species *O. affinis* (1865), still given species status by RAOU (1913) but now regarded as a race. Bull-voiced Oriole, though we can't for the life of us hear it – however, Walter Petrie could, because C.H.H. Jerrard informed us that 'Mr. Petrie gives 'Bull-voiced Oriole' (from its deep, rich notes) as a local name' (Jerrard 1931). Sadly, he omitted to say where 'local' was.

Oriolus flavocinctus (King PP, 1826) [o-ri-O-lus flah-vo-SINK-tʊs]: 'yellow-girdled golden bird', see genus name, and from Latin *flavus*, yellow, and *cinctus*, a girdle or belt. This refers to 'the yellow fascia which is formed on the wing, when closed by the junction of the apical spots on the quill coverts', a feature that King considered to contribute to the resemblance of this bird to 'the true orioles' (King 1827).

DICRURIDAE (PASSERIFORMES): drongoes

Dicruridae Vigors, 1825 [di-KROO-ri-dee]: the Forked-tail family, see genus name *Dicrurus*. 'Drongo' is generally accepted to be of Malagasy origin (i.e. from Madagascar). A widely accepted route is via the French zoologist François Le Vaillant, as *drongeur*, applied to a South African *Dicrurus*. He was in South Africa in the 1780s (though he didn't necessarily name it then) but 'drongo' didn't apparently appear in English until well into the 19th century; on the other hand, Gould's use of 'drongo shrike' (1838a) gives the lie to the *Oxford Dictionary*'s earliest date of 1841.

(It is pretty well established that use of the word as an Australian term of abuse is only indirectly associated with the bird; the owners of the 1920s Australian racehorse named Drongo presumably hoped that it would be as nippy as the bird they named it for. It wasn't quite, but neither was it devoid of talent and even managed the odd gallant second, contrary to modern presumptions. It never won though, so the original connotation was more of a trier who couldn't quite crack it.)

The genus

Dicrurus Vieillot, 1816 [di-KROO-rʊs]: 'forked-tail', from Greek *dikroos*, forked, and *oura*, tail.

The species

Crow-billed Drongo (vagrant)

See family introduction. From the 'thick bill with equal width and depth at nostril level' (Rocamora and Yeatman-Berthelot 2018).

Dicrurus annectens (Hodgson, 1836) [di-KROO-rʊs an-NEK-tenz]: 'connected forked-tail', see genus name, and from Latin *annecto*, to tie or bind to, connect. In Hodgson's description of this bird (species 'Annectans'), he appears to have seen it as connected with two other birds he was reporting on. In his detailed description of this fifth species in the list he was describing, he mentioned various elements, including 'the distinctly forked tail' and 'the distinctly hooked bill' as 'attest[ing] its intimate affinity' with his fourth species 'Albirictus'. However, in his introductory paragraph to 'Annectans', he described it as 'A singular species, returning, both by its form and habits, toward the forest-haunting birds first described, through the 3rd, or Tectirostris, which [the current bird] very closely resembles through the form of its bill'. In spite of *tectirostris* lacking one distinctive feature shared by Hodgson's *annectans* and *albirictus*, the hooked bill, other features led to his considering this bird to be the link between numbers 1 and 2 of his list (*casia* and *malabaroides*) and numbers 4 and 5 (*annectans* and *albirictus*). The spelling of Hodgson's '*annectans*' was altered to *annectens* to reflect correct Latin (*annecto, annectere*).

Spangled Drongo (breeding resident)

Used by Gould (1848), from his species name.

Other names: Drongo (there's only one in Australia); Drongo Shrike (Gould 1838a), Spangled Drongo-shrike; Fishtail, from the genus name; King-crow, an Anglo-Indian term (Newton 1896); Hair-crested Drongo, which is now applied to *D. hottentottus* (an Asian species despite the name), formerly included in *D. bracteatus*; Satinbird, for the glossy plumage, but more generally applied to Satin Bowerbird.

Dicrurus bracteatus Gould, 1843 [di-KROO-rʊs brak-te-AH-tʊs]: 'gold-plated forked-tail', see genus name, and from Latin *bracteatus*, gold-plated or shining like gold (Latin *bractea*, gold-leaf).

RHIPIDURIDAE (PASSERIFORMES): fantails

Rhipiduridae Sundevall, 1872 [ri-pi-DOO-ri-dee]: the Fantail family, see genus name *Rhipidura*.

'Fantail' for rhipidurids seems first to have been used in Captain Cook's journals in 1873 in the form Fan-tail, for New Zealand birds (Hay *et al.* 2008), though the term was already in use in England for a breed of domestic pigeon. It didn't catch on for a while in Australia; Lewin (1822) still used 'flycatcher'. On the other hand Vigors and Horsfield (1827) in their description of the genus commented that George Caley – Banks' botanical collector who left the colony in 1810 – used 'Fan-tail' to refer to Grey Fantail. The fact that Gould (1848) used Fantail as a common name instead of Rhipidura, his default position, suggests that it was well established by then.

The genus

Rhipidura Vigors & Horsfield, 1827 [ri-pi-DOO-ruh]: 'fantail', from Greek *rhipis*, fan, and *oura*, tail.

The species

Willie Wagtail (breeding resident)

It would be a little ironic if it turned out that this most archetypal of Australian bird names actually arose in Britain. The *Australian National Dictionary* certainly believes this to be so, asserting it to be 'transferred use of *willy-wagtail* the water wagtail *Motacilla lugubris*' (the latter being a synonym of Pied Wagtail *M. alba*). The term was apparently used only on the Isle of Man and in Northern Ireland – it is interesting that the enormously comprehensive *Oxford Dictionary of British Bird Names* (Lockwood 1984) was unaware of the name. The earliest usage offered by the *Australian National Dictionary* is from 1885; we can only improve 3 years on that, from near Gympie in the *Logan Witness* of 26 August 1882. Gould (1865) didn't report it (he made a point of commenting on names 'of the Colonists' but he had been away from Australia for some time by then), so this might be close to the start of its use here, though oral use is likely to have long preceded written occurrences. We are unlikely to know whether it arose in Australia uninfluenced by the British use, but that would seem to be a big coincidence.

Other names: Black Fantailed Flycatcher (Gould 1848); Wagtail Flycatcher, per Gould again, 'of the Colonists of Swan River'; Willy Wagtail; Black-and-White Fantail or Flycatcher, cited by Leach (1911) and recommended by RAOU (1913); Shepherd's Companion, also per Leach, though the attraction for the bird is more the sheep (stirring up insects) than the shepherd; Morning Bird, for its dawn singing; Frog Bird, which we think is by confusion with Leaden Flycatcher, though Leach (1911) cited it; Australian Nightingale, cited by Baker (1945), presumably for its nocturnal song; Pied Fantail, cited by HANZAB (1990–2006, vol. 7a), though this is a name applied to *R. javanica* of Indonesia and South-East Asia and it's not clear where it has been used for Willie Wagtail; White-browed Fantail, from the species name, though again it's more generally used for an Asian species.

Rhipidura leucophrys (Latham, 1802) [ri-pi-DOO-ruh LOO-ko-fris]: 'white-eyebrowed fantail', see genus name, and from Greek *leukophrys*, white-browed (*leukos*, white, and *ophrus*, eyebrow).

Northern Fantail (breeding resident)

Found right around the tropical Australian coast and hinterland. The name was used by Gould (1848), playing an uncharacteristically straight bat, though perhaps he just couldn't bring himself to use the inappropriate translation of Vieillot's species name.

Other names: Northern Flycatcher; Red-vented Fantail, taken uncritically (and incorrectly) from the species name; White-throated Fantail, comprehensively true-but-unhelpful in distinguishing it from similar species; Queensland Fantail, described as species *R. superciliosa* by Ramsay from Cape York Peninsula, cited by RAOU (1913) in the provisional list.

Rhipidura rufiventris (Vieillot, 1818) [ri-pi-DOO-ruh roo-fi-VEN-tris]: 'red-bellied fantail', see genus name, and from Latin *rufus*, red or reddish (of all shades, but perhaps a bit of a stretch as the belly is at best pale buff), and *venter*, belly.

Grey Fantail (breeding resident)

Perhaps surprisingly, this is a name that seems to have been coined in the 20th century (RAOU 1926) to replace the universally used White-shafted Fantail; it seems to have coincided with dropping the name *albiscapa* (ironic in that we're now using that name again) in favour of the New Zealand name *flabellifera* (which itself was later superseded by *fuliginosa*).

Other names: White-shafted Fantail, by Gould (1848) from his species name, still used by RAOU (1913); White-shafted Flysnapper, reported by Leach (1911), also Snapper; White-tailed Fantail, for race *albicauda* of central Australia, described by Alfred North in 1895 as a separate species *R. albicauda* – and there are renewed suspicions that it might be (e.g. Christidis and Boles 2008); Western Fantail, for race *preissi* of south-western Western Australia, regarded as a separate species *R. preissi* for the second half of the 19th century and until at least 1913 (RAOU 1913); Dusky Fantail, for Sharpe's 1879 *R. diemensis* from Tasmania, recognised until RAOU (1913) but not much beyond – note that the name is now more properly applied to *R. tenebrosa* of the Solomon Islands; Pelzeln Fantail, for race (formerly species) *pelzelni* of Norfolk Island – for Austrian ornithologist Auguste von Pelzeln of the Viennese Imperial Museum, named by his contemporary George Gray of the British Museum, to whom Pelzeln returned the compliment with at least one South American species; Cranky or Mad Fan, for its antics around walkers; Fanny; Devil-bird, reported by Leach (1911), also Fanny Devilbird, sadly not explained, and apparently an essentially oral tradition; the only earlier reference we can find to its usage is in a series on birds in the *Australian Town and Country Journal* in which the article of 5 December 1896 on this species is headed *White-Shafted Fantail, 'Devil-bird'*, with no further explanation! Land Wagtail or Land-wagtail, apparently from oral tradition also; a brief anonymous column in the Sydney *Sun* of 18 April 1932 on this bird entitled *Cranky Little Fanny*, tells us that it has six different names, including Devil Bird and Land Wagtail – but that the latter (only!) is 'erroneous'. There is no reference that we can find to either name after 1953.

Rhipidura albiscapa Gould, 1840 [ri-pi-DOO-ruh al-bi-SCA-puh]: 'white-shafted fantail', see genus name, and from Latin *albus*, dull white, and *scapus*, shaft. The shafts of most of the grey tail-feathers are white.

New Zealand Fantail (formerly Lord Howe Island, extinct)

Limited to, and found throughout, New Zealand, other than the former outlying Lord Howe Island population. (The separation of this species from Grey Fantail is recent – Schodde and Mason 1999 – and still controversial, though Gould was quite sure they represented different species.)

Other names: Lord Howe Island or Fawn-breasted Fantail, the latter also cited in RAOU (1913) under Sharpe's 1881 name *R. macgillivrayi*. (Names used only in New Zealand are not considered here.)

Rhipidura fuliginosa (Sparrman, 1787) [ri-pi-DOO-ruh fʊ-li-gi-NOH-suh]: 'sooty fantail', see genus name, and from Latin *fuligo*, soot. This species comes in two colour forms (phases or morphs): 'pied', which is brown, grey and yellowish, and 'black' (mainly in the South Island of New Zealand), which is definitely sooty.

Mangrove Grey Fantail (breeding resident)

For its obligate habitat.

Other names: Mangrove Fantail; Pheasant Fantail from the species name, though that doesn't excuse it, even though the RAOU (1913) recommended its use.

Rhipidura phasiana De Vis, 1884 [ri-pi-DOO-ruh fa-zi-AH-nuh]: 'pheasant fantail', see genus name, and from Latin *phasianus* (Greek *phasianos*), pheasant, apparently describing what HANZAB (1990–2006, vol. 7a) called 'the long expressive tail of all fantails'.

Rufous Fantail (breeding resident)

For the gloriously rich rusty rump and tail base; apparently first used in its simplified modern form by RAOU (1926).

Other names: Rufous-fronted Flycatcher, used by Latham (1801), the same as his species name; Orange-rumped Flycatcher (Lewin 1808); Rufous-fronted Fantail (Gould 1848); Red Fantail; Rufous Flycatcher; Allied Fantail, for North's 1902 species *R. intermedia* from north Queensland, still recognised by RAOU (1913) but dropped soon afterwards; Redstart, for a passing (just) resemblance to either Old World (Muscicapidae) or New World (Parulidae) redstarts – there seems to have been only one usage recorded, that of William Swainson in 1831 in an advertisement in the *Sydney Gazette* for suppliers of bird skins, 'immediately wanted, in half dozens or more' (Hindwood 1949a); he would have been aware of Lewin's and Latham's names but presumably was reflecting a popular usage.

Rhipidura rufifrons (Latham 1802) [ri-pi-DOO-ruh ROO-fi-fronz]: 'fantail with a red forehead', see genus name, and from Latin *rufus*, red, and *frons*, forehead. Why focus on this tiny (albeit lovely) feature? Latham described the bird, 'with forehead, back, and base of the tail rufous', so we idly wonder if the forehead made it into the species name simply because it had first place in the list?

Rufous Fantail *Rhipidura rufifrons,* Grey Fantail *R. albiscapa* and Willie Wagtail *R. leucophrys*

Arafura Fantail (breeding resident)

Lives around the Arafura Sea, in northern Australia, southern New Guinea and the Lesser Sundas. Gould described this as a separate species and it was recognised well into the 20th century (RAOU 1913) but was subsequently subsumed into Rufous Fantail (RAOU 1926) until Schodde and Mason (1999) separated them out again (these authors seem to have coined the current common name too).

Other names: Wood Fantail, used especially in Western Australia until very recently but it's an old name, probably introduced by Gould and in wide use until the species was merged with Rufous Fantail; Scaly-breasted Fantail, for the short-lived species *R. mayi* from the Northern Territory (described in 1911, used by RAOU 1913, subsumed soon afterwards).

Rhipidura dryas Gould, 1843 [ri-pi-DOO-ruh DRI-as]: 'Dryad fantail', see genus name, and from Greek *dryad*, a tree-nymph.

MONARCHIDAE (PASSERIFORMES): monarch-flycatchers

Monarchidae Bonaparte, 1854 [mo-NAR-ki-dee]: the Monarch family, see genus name *Monarcha*.

'Flycatcher', with the obvious implication, was introduced to English by the 'father of English natural history' John Ray in 1678, a translation of the Latin *muscicapa* as previously used (French *gobe-mouche*, fly-eater, also appeared late, dated as a bird-name in print only from 1611). Thomas Pennant, the influential 18th-century Welsh naturalist, applied it to the familiar (to him) Spotted Flycatcher (*Muscicapa grisola*) and from there it was 'used in a general and very vague way for a great many small birds from all parts of the world, which have the habit of catching flies on the wing' (Newton 1896). Australian birds were no exception.

One attempt to reduce the plethora of 'flycatchers' resulted in the introduction of 'monarch' for the genus *Monarcha* (and obviously enough, straight from that name) in the latter part of the 20th century. RAOU (1978) specifically recommended it but noted that it wasn't new, citing a New Guinea handbook (Rand and Gilliard 1967) as precedent.

The genera

Symposiachrus Bonaparte, 1854 [sim-po-si-AK-rʊs]: 'colourless drinking-party', from Greek *symposion*, a regulated drinking-party (from *sum*, in company with, and *posis*, a drink), and *akhrōs*, colourless or pallid. Something of a puzzle, this one. Bonaparte said it was impossible to leave the species type *trivirgata* among the *Monarcha* as it is a genuine '*Myiagrien*' – but gave no further explanation of his choice of name. Sundevall (1872), on the other hand, spelled it *Symposiarchus*, while referring to the original quote in Bonaparte and nowhere suggesting that he was correcting Bonaparte's spelling ... Since '*arkhos*' means 'leader', this seemed to make sense: the birds often forage with other small birds and would be perhaps seen as the 'leader' because their active, fluttering behaviour makes them conspicuous. Of course, Bonaparte has to take the priority guernsey, but what on earth does his symposi*achrus* mean? A colourless symposium ... not unfamiliar, but perhaps a bit hard to equate with the very colourful example of this genus.

Monarcha Vigors & Horsfield, 1827 [mo-NAR-kuh]: 'monarch', from Greek *monarkhos*, a sole ruler. Although Vigors and Horsfield don't elaborate on the choice of name, it seems likely to reflect their opinion that the bird's very strong bill could place it among the Tyrant flycatchers (Tyrannidae being a New World family, from Greek *turannos*, an absolute ruler), though its other features seemed to place it squarely with the Muscapidae (Old World flycatchers, from Latin *musca*, a fly, and *capesso*, seize eagerly). Even Vigors and Horsfield stated that they believed more research into the bird's habits was needed to shed light on this.

Carterornis Mathews, 1912 [car-ter-OR-nis]: 'Carter's bird', for Thomas Carter (1863–1931), a Yorkshire naturalist who explored various parts of Australia and collected birds, when he wasn't being a pastoralist.

Arses Lesson, 1830 [AR-sayz]: 'Arses'. No, it's the name of a somewhat weak king of the Ancient Persians (also known as Artaxerxes IV) who was poisoned by his close adviser, whom Arses was himself planning to murder. The name is connected with Greek *arsēn*, male, manly, mighty. Lesson's variant on the monarch theme.

Grallina Vieillot, 1816 [gral-LEE-nuh]: 'stilt-walker', from Modern Latin *grallinus*, with stilts (Latin *grallae*, stilts), for what Vieillot (1816) called the bird's 'elongated tarsi'.

Myiagra Vigors & Horsfield, 1827 [mi-i-AG-ruh]: 'flycatcher', from Greek *muiagros* (*muia*, fly, and *agreō*, seize). Myiagros was a minor Greek god whose function was to chase away flies (and mice), which might have been a nuisance during the ritual sacrifices to Zeus and Athena. These sacrifices could include offerings of fruit and vegetables or little cakes, and sometimes the slaughter of large animals.

The species

Spectacled Monarch (breeding resident)

See family introduction. For the black mask going well behind the eyes; the name seems to have arisen spontaneously and was used by Morris (1898) in his *Dictionary of Austral English*.

Other names: Spectacled Flycatcher or Monarch-flycatcher; Black-fronted Flycatcher (Gould 1848), again recalling that the frons is a bird's forehead – though this does nothing to distinguish it from many other monarchs; White-bellied Flycatcher, for race *albiventris* from Cape York Peninsula, in the past regarded as a separate species *Monarcha albiventris* (RAOU 1913), as described by Gould.

Symposiachrus trivirgatus (Temminck, 1826) [sim-po-si-AK-rʊs tri-veer-GAH-tʊs]: 'three-striped colourless drinking-party', see genus name, and from Latin *tri-*, three, and *virga*, stripe, for the 'tripod' of black marks, one down the throat and one down past lores and eye on each side, connected at the frons or forehead, all of which Temminck (1826–27) described in detail.

Island Monarch (vagrant)

See family introduction. Scattered across dozens of islands all around New Guinea.

Monarcha cinerascens (Temminck, 1827) [mo-NAR-kuh si-ne-RA-senz]: 'ashy-greyish monarch', see genus name, and from Latin *cinereus*, like ashes, with *-ascens*, growing or becoming '-ish'.

Black-faced Monarch (breeding resident)

See family introduction. From the species name.

Other names: Black-faced Flycatcher; Carinated Flycatcher, by Gould from Swainson's name *Muscipeta carinata*, assigned in ignorance of Louis Vieillot's name – it means 'keeled', as in breastbone, which of course any flying bird has; Pearly Flycatcher, from Italian ornithologist Adelardo Salvadori's name *Monarcha canescens* (1876), for an apparently atypical bird from Cape York Peninsula – the name persisted into the 20th century; Grey-winged Monarch or Monarch-flycatcher, used more in New Guinea, as a distinction from the otherwise similar next species.

Monarcha melanopsis (Vieillot, 1818) [mo-NAR-kuh mel-a-NOP-sis]: 'black-faced monarch', see genus name, and from Greek *melas*, black, and ōps, face, which it has, though not much more than the next bird, its 'brother'.

Black-winged Monarch (breeding resident)

See family introduction. This is the key distinguishing factor from the previous species, with which it has often been combined.

Other names: Black-winged Flycatcher.

Monarcha frater Sclater PL, 1874 [mo-NAR-kuh FRAH-tehr]: 'black-faced's brother monarch', see genus name, and from Latin *frater*, brother (see common name).

White-eared Monarch (breeding resident)

See family introduction. Used by Gould (1869) from his species name.

Other names: White-eared Flycatcher or Monarch-flycatcher.

Carterornis leucotis (Gould, 1851) [car-ter-OR-nis loo-KOH-tis]: 'Carter's white-eared bird', see genus name, and from Greek *leukos*, white, and *-ōtis*, eared – the bird has several white markings on its face and head, one of which may count as the ear.

Frilled Monarch (vagrant to Torres Strait islands)

See family introduction. For the male's lovely erectable neck ruff, as with Frill-necked Monarch.

Arses telescopthalmus (Lesson RP and Garnot 1827) [AR-sayz te-les-kop-THAL-mʊs]: 'conspicuous-eyed Persian king', see genus name and from Greek *tēleskopos*, far-seeing (or in this case especially) far-seen (i.e. conspicuous), and *ophthalmos*, eye, referring to the blue skin round the eyes. Lesson and Garnot spelt the name *telescopthalmus* rather than the more correct *telescophthalmus*, meaning that those who now spell it correctly are, in this case at any rate, wrong. They also gave it the French name of *Gobe-mouche à lunettes* (flycatcher with spectacles), not to be confused with the English name Spectacled Monarch, *Symposiachrus trivirgatus*.

MONARCHA

M. TRIVIRGATA
Black fronted Fly-catcher

M. LEUCOTIS
White eared Fly-catcher

M. CARINATA
Carinated Fly catcher

Spectacled Monarch *Symposiachrus trivirgatus*, White-eared Monarch *Carterornis leucotis* and Black-faced Monarch *Monarcha melanopsis*

Frill-necked Monarch (breeding resident)

See family introduction. For the male's neck ruff, as with Frilled Monarch.

Other names: Frilled Monarch, until the Australian and New Guinea populations were split into separate species; Frill-necked or Frilled Flycatcher; Frilled Monarch-flycatcher.

Arses lorealis (De Vis 1895) [AR-sayz lo-reh-AH-lis]: 'lored Persian king', see genus name, and from Modern Latin *lorealis*, concerning the lores. De Vis writes 'lore white, with the bases of its feathers black' – not very distinctive or enlightening.

Pied Monarch (breeding resident)

See family introduction. For the usual 'black and white' implication of pied, referring to a (European) magpie; Hall (1899) was among the first to drop Kaup's Flycatcher for this name and Pied has been in use since.

Other names: Kaup's Flycatcher, Gould's name from his species name; Pied Flycatcher or Monarch-flycatcher; Australian Pied Flycatcher, to distinguish from the European Pied Flycatcher (*Ficedula hypoleuca*); Black-breasted Flycatcher, in contrast to the previous species.

Arses kaupi Gould, 1851 [AR-sayz KOW-pee]: 'Kaup's Persian king', see genus name, and for Johann von Kaup; in Gould's words, 'I am happy to have the opportunity of paying a just compliment to my friend Dr Kaup of Darmstadt, an ornithologist of vast acumen and research, and whose philosophical labours are well known to all naturalists; the compliment is the more appropriate, as he is at this time engaged in preparing a monograph of the Muscicapidae, to which family the present bird belongs' (Gould 1850). Modern taxonomists would disagree with the last comment. He wisely avoided mentioning Dr Kaup's association with the briefly popular, but frankly batty, quinary classification system of animals; Kaup was also an anti-Darwinian, with which Gould would have empathised.

Magpie-lark (breeding resident)

A true folk name, recorded by Gould (1848) as being used by the 'Colonists of New South Wales'; the magpie part is obvious enough (and recurs in other folk names), but the lark connection seems pretty tenuous.

Other names: Pied Grallina, a true Gouldism (Gould 1848) from the genus name, surprisingly still used by RAOU (1913); Little Magpie, Murray Magpie (in South Australia); Mudlark, for its need for mud to build its nest; Pugwall, another reference to the mud nest, pug being brick clay; Peewee, Peewit, Tilyabit, all for its piercing (and duetting) song; we suspect that Soldiers (Leach 1911) has the same origin but cannot confirm that.

Grallina cyanoleuca (Latham, 1802) [gral-LEE-nuh si-a-no-LOO-kuh]: 'blue and white stilt-walker', see genus name, and from Greek *kuanos/kuaneos*, dark blue, and *leukos*, white. Even dark blue at its darkest does not sound black enough, but in fact when you look very closely (and in the right light!) there is a hint of blue sheen to the bird's back – just like the colour of a swallow's wing, which is said to be one of the blues described by *kuanos*.

Leaden Flycatcher (breeding resident)

See family introduction. Named thus by Gould (1848) from the name *M. plumbea*, applied by Vigors and Horsfield well after Latham had named it – and they were describing the male only, which is something of a balancer given that the species name is pertinent only to the female.

Other names: Leaden-coloured Flycatcher or Flysnapper; Frog-bird, for the unquestionably frog-like croaky calls (or at least like a frog of popular imagination); Pretty Flycatcher, used by Gould (1848) for his *M. concinnna* (now a race) of northern Australia – he didn't explain what moved him; Blue Flycatcher, used by RAOU (1913) for the same race (which it still saw as a species) – in the right light the male is strikingly blue; Leaden Monarch, mostly a New Guinea appellation.

Myiagra rubecula (Latham, 1802) [mi-i-AG-ruh roo-BE-kʊ-luh]: 'redbreast flycatcher', see genus name, and from Latin *ruber*, red, possibly with a diminutive ending (though we feel Latin *gula*, throat, is probably the origin). *Rubecula* was used in Mediaeval Latin for the familiar Robin Redbreast (now officially known as the European Robin): Aldrovandus (1600) remarked that Theodorus Gaza (early 15th-century translator of Aristotle into Latin) used *rubecula* for Greek *eritakos*, the robin, 'doubtless from the red chest'. Both Greek and Mediaeval Latin names are reflected in the Robin's scientific name *Erithacus rubecula*. So this refers to the red throat, one of the few cases where a feature of the female bird is chosen: as Latham (1802) put it, 'with throat and chest orange', though it is open to speculation whether he recognised that it was the female.

Broad-billed Flycatcher (breeding resident)

See family introduction. Applied by Gould (1848) from his own name *M. latirostris*, not realising that Louis Vieillot had already named it from a specimen from Timor. Especially noticeable from beneath, the wide bill.

Other names: Broad-billed Monarch, more used outside Australia, particularly in New Guinea.

Myiagra ruficollis (Vieillot, 1818) [mi-i-AG-ruh roo-fi-KOL-lis]: 'red-necked flycatcher', see genus name, and from Latin *rufus*, red, and *collum*, neck. Red-throated would have been more apt and indeed, in French, Vieillot (1816–19) called it just that.

Shining Flycatcher (breeding resident)

See family introduction. Again for the glossy deep blue male. Confusingly, Gould (1848) used this name for both this bird and the next, in this case from his species name *Piezorhynchus nitidus*.

Other names: Glossy Flycatcher; Shining Monarch or Monarch-flycatcher, basically New Guinea usage; Satinbird, like Satin Bowerbird, though despite the HANZAB (1990–2006, vol. 7b) citation it doesn't seem to have been used much in Australia; Long-billed Shining Flycatcher, for Mathews' very short-lived species *Piezorhynchus wardelli* of north Queensland – he named it in 1911, RAOU (1913) used it and it vanished thereafter; Papuan Shining Flycatcher, used by RAOU (1913) to accommodate the fact that it recognised both Gould's *nitidus* and *alecto* (it used this for *alecto*).

Myiagra alecto (Temminck, 1827) [mi-i-AG-ruh a-LEK-toh]: 'Alecto flycatcher', see genus name, and said to be for the implacable Alecto, one of the Furies of Roman mythology (corresponding to the Greek *Erinnyes*). These were 'crones, with snakes for hair, dogs' heads, coal-black bodies, bats' wings and blood-shot eyes' (Graves 1996). What has all that to do with these lovely birds? Temminck didn't say, but because he also gave that species name to the Black Flying Fox, *Pteropus alecto*, we take it to come from the 'coal-black bodies'. (Linnaeus also used that species name, for a mostly grey and black butterfly.) The choice of a word for the colour (black with violet-blue sheen) of the male of a sexually dimorphic pair of birds (the female is mostly a brilliant orange-chestnut) should not surprise us – not a lot of attention was usually given to female birds. In fact no attention at all was given in this case, because Temminck claimed there was no difference in the plumage of the two sexes (Temminck 1826–27). We conclude that he must have had access only to male birds.

Satin Flycatcher (breeding resident)

See family introduction. Another male-only name, for the deep rich blue-black sheen (as for Satin Bowerbird); seems to arisen in Tasmania, where a 19th-century name for it was Satin-robin (Morris 1898). It's not clear when it became established, though Morris used it as first-choice name.

Other names: Shining Flycatcher, per Gould (1848), from his own name *M. nitida* (again he'd been gazumped by Louis Vieillot, but since Vieillot listed his bird as being from Timor – it actually came from Sydney – Gould could be excused); see also previous species. Satin Sparrow, which, according to an article in the Brisbane *Courier Mail* of 1879 was used in Tasmania ('A Bush Naturalist' 1879) – see Satin Flycatcher; Glossy Flycatcher, although this is an alternative name for the previous species (mostly in New Guinea) and we suspect that this is a confusion stemming from Gould's use of Shining Flycatcher – it does not appear to have been used for this species; Satin Monarch, cited by HANZAB (1990–2006, vol. 7b), but the only usage we can find is in an undated NSW Roads and Traffic Authority environmental report, where both Satin Flycatcher and Satin Monarch are included in a short list of 'Migratory Species' (Fork-tailed Swift is also considered to have 'potential habitat' present, so the report could not be regarded as authoritative).

Myiagra cyanoleuca (Vieillot, 1818) [mi-i-AG-ruh si-a-no-LOO-kuh]: 'blue and white flycatcher', see genus name, and from Greek *kuanos/kuaneos*, dark blue, the colour of blue steel, the deep sea or a swallow's wing, so it well expresses the satiny sheen of this bird's upper plumage. Also Greek *leukos*, for the white underparts.

Paperbark Flycatcher (breeding resident)

See family introduction. From its tropical woodland habitat, usually near water, including in paperbarks (*Melaleuca* spp.).

Other names. It was until recently (Schodde and Mason, 1999) regarded as conspecific with Restless Flycatcher, so any of those names could have applied too; Little Flycatcher, from Gould's species name.

Myiagra nana (Gould, 1870) [mi-i-AG-ruh NA-nuh]: 'dwarf flycatcher', see genus name and from Latin *nanus*, dwarf.

Restless Flycatcher (breeding resident)

See family introduction. From the species name; Latham called it Restless Thrush from his species name *Turdus inquietus* (Latham 1802).

Other names: Volatile Thrush, also one of Latham's, two pages after his previous name! Dishwasher was recorded by Vigors and Horsfield (1827) in their report to Linnean Society, 'This bird is called by the colonists Dishwasher. It is very curious in its actions. In alighting on the stump of a tree it makes several semi-circular motions, spreading out its tail, and making a loud noise somewhat like that caused by a razor-grinder when at work.' None of this explains the name. We must go back to England for that, where the Pied Wagtail was called Washtail because the vertical movement of the tail suggested the operation of a battledore, the wooden paddle for beating washing. The 'wash' was later interpreted to mean washing up, and the connection was lost. Dishlick presumably follows from this. From the extraordinary grinding call, often given while hovering, also comes Grinder (already used by 'the Colonists' as reported by Gould in 1848), Razor Grinder or Scissors Grinder; Who-are-you, traced to a report in the *Sydney Morning Herald* in 1913, which described the bird 'intermittently inquiring of the intruder, 'Who are you?'' (Sydney Morning Herald 1913); Willie Wagtail, by confusion. Crested Wagtail, of which we can find only two instances before the HANZAB (1990–2006, vol. 7b) citation; one is from the Perth *Sunday Times* of 30 December 1934, explained as a 'habit ... of raising the feathers on the head on various occasions', the other in a letter to the *Canberra Times* of 13 August 1963, both as alternative names. The separation of these examples in space and time suggests a greater oral usage than we've been aware of. Fascinating Thrush (Latham 1801), Fascinating or Greater Fascinating Bird, a fascinating yarn indeed. Latham (1823) reported that it was 'observed to hover, frequently in company with a bird of the Grosbeak Genus [the Jacky Winter], about two feet from the ground, making sudden darts on something, which, by attention, was found to be a sort of worm, which this bird, by a chirping note, and tremulous motion of the wings, with the tail widely expanded, seemed to fascinate, or entice out of its hole'.

Myiagra inquieta (Latham 1802) [mi-i-AG-ruh in-kwi-EH-tuh]: 'restless flycatcher', see genus name, and from Latin *inquietus*, restless or disturbed (see common name).

CORVIDAE (PASSERIFORMES): ravens and crows

Corvidae Leach, 1820 [KOR-vi-dee]: the Raven/Crow family, see genus name *Corvus*.

'Raven' has a long lineage, as might be expected from a bird (in this case the Common Raven *Corvus corax*) that has such a long association with human society and mythology. We might reasonably assume it to be onomatopoeic, from the growling call, but we can't be sure. We know of Old English *hraefn*, evolving from its first written appearance in the 9th century via *rauen*, *rauon*, *rauyn* and *ravyne*, to Christopher Marlowe's use of the modern form in the 1590s. The Australian Raven is virtually identical in appearance (and very closely related – although the ancestral corvids seem to have arisen in 'Australia' and dispersed very long ago, modern Australian crows and ravens 'came home' only very recently) so the use of the term here was inevitable and appropriate. The allocation of Australian corvids to 'ravens' and 'crows' is pretty arbitrary, though the size, call and throat hackles of the Australian Raven (for a long time the only Australian species recognised) made its naming pretty obvious.

'Crow' has a similar history, and presumably its name too was inspired by its call – it is easy to hear in the Old English *crawe*. *Crauuae* appears in 8th-century literature, followed by *crowe* in the 12th century, with crow appearing in the 14th century.

For 'magpie' see family Artamidae.

The genera

Pica Brisson, 1760 [PEE-kuh]: 'magpie', 'painted or spotted bird', from Latin *pica*, probably from *pingo*, paint or embroider (Lewis and Short 1879).

Corvus Linnaeus, 1758 [KOR-vʊs]: 'raven', Latin *corvus*. Aristotle (c. 330 BC) stated that while the partridge and the 'barn-door cock' are salacious, 'the whole tribe of crows is inclined to chastity, and indulge but rarely in sexual intercourse'. Perhaps it was just that he noticed that they tended to form long-lasting relationships: many of the Corvidae pair for life. Ovid (c. 8 AD), less down-to-earth, held that this bird was as white as a dove, a goose or a swan until it reported the infidelity of Coronis, daughter of Phlegyas, to her lover Apollo. Apollo became intensely jealous and shot Coronis with an arrow, then turned on the raven and, in a classic case of shooting the messenger, banished it from the ranks of white birds. Other versions of the myth report Apollo's wrath as being fiery enough to singe the bird's feathers and turn them black.

The species

Eurasian Magpie (ship-borne vagrant)

See family introduction.

Pica pica (Linnaeus, 1758) [PEE-kuh PEE-kuh]: 'magpie-magpie', from Latin *pica*, magpie. This is the Magpie of Europe, from whose common name the Australian 'Magpie' was copied (see family Artamidae).

House Crow (ship-borne vagrant)

See family introduction. Throughout its natural Asian range it is closely associated with human habitations.

Corvus splendens (Vieillot, 1817) [KOR-vus SPLEN-denz]: 'shining raven', see genus name, and from Latin *splendeo/splendens*, bright or shining. Vieillot (1816–19) described all the plumage, both black and what he called 'dark pearl-grey', as 'very lustrous'.

Torresian Crow (breeding resident)

See family introduction. 'Torresian' is the name for the Australian tropical biogeographical zone.

Other names: Crow; Kelly, generally agreed to be an allusion to the Kelly gang, for this and other corvids' alleged thievery, but it doesn't seem to have appeared until the 1920s; Large-billed Crow, to distinguish it from Little Crow; Western Crow; Intermediate Crow – presumably between Australian Raven and Little Crow, though we can find no usage outside HANZAB (1990–2006, vol. 7a); Papuan Crow, because it is also found in New Guinea.

Corvus orru Bonaparte, 1850 [KOR-vus OR-roo]: 'Orru raven', see genus name, and, though Bonaparte gave no details, perhaps from one of the many West Papuan languages – the type specimen was from Triton Bay, West Papua.

Little Crow (breeding resident)

See family introduction. The smallest Australian corvid, especially relative to Torresian Crow, from which it was split.

Other names: Kelly, see Torresian Crow; Small-billed or Short-billed Crow, for its noticeably slenderer bill than Torresian Crow (RAOU 1913); Bennett's or Bennett's Small-billed Crow, from the species name; Queensland or Scrub Crow.

Corvus bennetti North, 1901 [KOR-vus BEN-net-tee]: 'Bennett's raven', see genus, and for Kenric Bennett (1835–91), bush naturalist, self-taught ornithologist, botanist and collector of Aboriginal artefacts, who collected the specimens from Moolah near Ivanhoe in 1883 and gave them to Alfred North who named the species well after Bennett's death. He referred to Bennett as 'my esteemed friend … to whom I am deeply indebted for information on this, and many other species' (North 1904).

Forest Raven (breeding resident)

See family introduction. On the mainland lives mostly in wet eucalypt and rainforests; in Tasmania, with no competition, it spreads out.

Other names: Crow or Raven; New England or Relict Raven, for the isolated New England race *boreus*, which may yet be deemed a separate species; Tasmanian Raven.

Corvus tasmanicus Mathews, 1912 [KOR-vus taz-MA-ni-kʊs]: 'Tasmanian raven', see genus and common names.

Little Raven (breeding resident)

See family introduction. Marginally shorter, but lighter, than Australian Raven from which it was split.

Other names: See Australian Raven.

Corvus mellori Mathews, 1912 [KOR-vus MEL-lor-ee]: 'Mellor's raven', see genus name, and (we assume, not having found any explanation by Mathews– his initial naming of the species focused entirely on its features, not on his choice of name) for John White Mellor (1868–1931), well known Australian ornithologist and founding member of the RAOU.

Australian Raven (breeding resident)

See family introduction. At the time it was deemed the only Australian species, so it was then logical.

Other names: Having said that, Gould (1848) called it White-eyed Crow – he anguished over whether it was closer to the English Carrion Crow or Raven, deeming it 'so intermediate in size, in the development of the feathers of the throat, and in many parts of its economy', but eventually came down on the side of the 'true Crows'. For the record, Carrion Crow (*C. corone*) and most other *Corvus* have dark eyes. Raven or Southern Raven (southern in the world, rather than in Australia); Crow, Southern Crow, Australian Crow (RAOU 1913, but there was a lot of confusion in that it recognised both Gould's *C. australis* and *C. coronoides*, though Gould himself didn't); Kelly, see Torresian Crow.

Corvus coronoides Vigors & Horsfield, 1827 [KOR-vus ko-ro-NOY-dehz]: 'crow like a Carrion Crow', see genus name, and from Greek *korōnē*, the word for crow (and the species name of the Carrion Crow – see common name), with Greek *-oeidēs*, resembling. Ovid (c. 8 AD) told two Coronis stories: one about the raven (see genus name *Corvus*) and the other about the daughter of King Coroneus of Phocis, who was turned into a crow by Athene to save her from Poseidon, who was about to rape her – a fortunate transformation, but Athene's favour was, sadly, not long-lived. Fickle gods.

Australian Raven *Corvus coronoides*

CORCORACIDAE (PASSERIFORMES): chough and apostlebird

Corcoracidae Mathews, 1927 [kor-kor-A-si-dee]: the Chough family, see genus name *Corcorax*.

'Chough' is an old English word traceable to the guttural form *choghen* (doubtless onomatopoeic) from the very early 14th century, applied to the ubiquitous little crow the Jackdaw (*Corvus monedula*). Later the cliff-dwelling crow found, among other places, in Cornwall (*Pyrrhocorax pyrrhocorax*) became Cornish Chough (i.e. 'Cornish Jackdaw'); as the Jackdaw became known by its current name and the origin of 'chough' was forgotten, the 'Cornish' began to be dropped from the chough's name.

The genera

Corcorax Lesson, 1830 [kor-KOR-aks]: 'chough-raven', from a combination of Greek *korakias*, mentioned by Aristotle (c. 330 BC) and from his description probably a Chough *Pyrrhocorax pyrrhocorax*, or possibly an Alpine Chough *P. graculus* (Arnott 2007), and *korax*, raven. Lesson coined the French term *corbicrave* to describe the Australian bird (from *corbeau*, crow, and *crave*, chough) then adapted it for his genus name. (Hesychius also mentioned a bird he said is called *korkora* but Arnott (2007) believed this is probably a domestic fowl, so irrelevant here.)

Struthidea Gould, 1837 [stroo-THI-de-uh]: 'sparrow-type bird', from Greek *strouthos*, a sparrow or other smallish bird (when it's not being an ostrich – see family Struthionidae).

The species

White-winged Chough (breeding resident)

See family introduction. This is the name used by Gould, and it was doubtless well established by then. We say this because there was little or no doubt in anyone's mind at the time that the bird was a true chough; Vigors and Horsfield (1827) put it into the genus *Fregilus* (then the Cornish Chough genus) apparently based on little more than a general resemblance, including the down-curved bill. Even by the end of the century Newton (1896) could still write, 'Commonly placed by systematists next to *Pyrrhocorax* is the Australian genus *Corcorax*', though he conceded that more skeletal studies were required before this association could be 'admitted without hesitation'. By 1913 Leach was saying only that the Cornish Chough was 'possibly' its nearest relation, but the name was by then too well established to be lost – especially as most Australians hadn't heard of the 'Cornish' bird. Gould was still using Temminck's species name (*Pyrrhocorax leucopterus*, 'white-winged').

Other names: Jay, Black Jay or Black Magpie; Apostle-bird, for its group living (but see next species); Mutton-bird, perhaps because it was eaten but more likely because of confusion with Pied Currawong (see profile); Chatterer, well merited.

Corcorax melanorhamphos (Vieillot, 1817) [kor-KOR-aks mel-an-o-RAM-fos]: 'black-billed chough-raven', see genus name, and from Greek *melas*, black, and *rhamphos*, bill, to emphasise that it's not the red-billed Chough, which he named next: *Coracia erythroramphos*, the Chough – now *Pyrrhocorax pyrrhocorax*. Vieillot's (1816–19) lengthy description of this also encompassed the bird now known as *Pyrrhocorax pyrrhocorax graculus*, the Alpine Chough; these two species overlap in some of the areas Vieillot described, such as the mountains of France and Switzerland, which explains why he considered them to be the same bird. The suggestion (in HANZAB 1990–2006, vol. 7a) that Vieillot intended *Coracia erythroramphos* to be the Dollarbird *Eurystomus orientalis* is not supported by Vieillot's text but is not surprising, given that the very similar *Coracias* was the genus name of the Rollers and used to include the Dollarbird (and that Vieillot's name for these birds in French is *Coracias*: indeed he has two entries under the name in his dictionary, one for his own French *Coracias*, genus name *Coracia*; the second for Linnaeus' genus name *Coracias*). There is a further minor point about Vieillot's original text: his spelling of *rhamphos* in both the species names lacked the first 'h' and was subsequently corrected.

Apostlebird (breeding resident)

Named for its supposed group size of 12 – it's actually a bit more varied than that – but apparently a name that arose spontaneously during the 19th century, though the earliest reference to it known to the *Australian National Dictionary* was in 1894, in the form Apostle's Bird. We can go a bit further back than that, in a trip account south of Bourke in *The Sydney Mail and New South Wales Advertiser* of 10 November 1883, though it's not possible to say if the writer was referring to this species or the previous one. Edward Morris' very thorough *Dictionary of Austral English* (1898) made no mention of it, referring instead to the generally used Grey Jumper. This is the only name mentioned by Hall (1899) and Mathews (1913). However, in 1902, H.E. Hill, writing in the *Emu*, saw no need to explain his use of the term

Apostlebird so it was clearly known reasonably widely; Leach (1911) and RAOU (1913) used Grey Jumper as first choice, with Apostlebird as an alternative. RAOU (1926) reversed this order, and the name was established.

Other names: Grey Struthidea, from Gould (1865) in direct translation of his species name, as was his wont; Grey Jumper, for its constant movement, the name of choice until into the 20th century; C.W.A Bird, in reference to the estimable Country Women's Association, which has branches in virtually every Australian town – it is fair to say that the social aspects of their meetings are not neglected; Happy Family, Twelve Apostles. Lousy Jack, allegedly because it warns prey of a hunter's progress (this is oft-cited, though generally by way of 'cut and paste' but we wonder if it isn't more literal; Sullivan (1931) noted that, 'I am told they are very verminous'). It may also be relevant that Campbell (1901) described the closely related choughs as 'very repulsive in this respect'. Grey Bullfinch-jay, surely a candidate for most bizarrely creative hyphenated name of all time, introduced by Major S.S. Flower for no obvious reason (Flower 1925); fortunately nobody else seems to have shared his enthusiasm for it.

Struthidea cinerea Gould, 1837 [stroo-THI-de-uh si-NE-re-uh]: 'ashy-grey sparrowish bird', see genus name, and from Latin *cinereus*, like ashes.

PARADISAEIDAE (PASSERIFORMES): birds of paradise, riflebirds, manucodes

Paradisaeidae Vigors, 1825 [pa-ra-di-ZEH-i-dee]: the Paradise family, from Greek *paradeisos*, enclosed royal park (of Persian kings) or garden. According to the *Oxford Dictionary*, this is from Avestan (the language of Zoroastrian scripture) *pairidaēza*, an enclosure, from *pairi*, around, and *diz*, to form. This is also the family of the birds of paradise of Papua New Guinea and nearby islands, many of which are somewhat more spectacular than the Australian members.

'Manucode', according to Newton (1896), has a somewhat convoluted history, beginning with a Malay word *Manukdewata*; HANZAB (1990–2006, vol. 7a) preferred 'Old Javanese' and *manok dewata* but both agreed that it means 'bird of the gods' and referred to birds of paradise in general. This was Latinised to *Manucodiata*, thence Boddaert's 1783 genus name *Manucodia* (still used by many authorities), then via French to the current form as a common name for the particular group of birds.

'Rifle Bird' was well enough established in 1848 for Gould to be using it without comment. The first reference that we are aware of is by Judge Barron Field in 1825; he simply listed the name Rifle-bird in a short appendix of Australian animals. The following year it appears in a *Glossary of the Most Common Productions in the Natural History* of the colonies, in the *Sydney Gazette* of 28 January 1826, with no need for explanation. The *Australian National Dictionary* recorded the alternative form 'rifleman' from 1833. Edward Morris (1898) neatly summed up the alternative explanations much later in his *Dictionary of Austral English*, 'The male is of a general velvety black, something like the uniform of the Rifle Brigade. This peculiarity, no doubt, gave the bird its name, but, on the other hand, settlers and local naturalists sometimes ascribe the name to the resemblance they hear in the bird's cry to the noise of a rifle being fired and its bullet striking the target.' With no further evidence available or likely to be, we prefer the former explanation, if for no other reason than that we have never heard the bird make the call that Morris described. Newton (1896) cited René Lesson and Prosper Garnot, who visited the colony in the early 1820s on the *Coquille*, as reporting that it was named 'to remind us that it was a soldier of the garrison who first killed one'. We think Newton's observation that this 'seems to be insufficient reason' is restraint personified.

The genera

Phonygammus Lesson & Garnot, 1826 [fo-ni-GAM-mʊs]: 'wedded to sound', the rather nice translation offered by Jobling (2010), from Greek *phōnē*, sound, and *gamos*, marriage. The name was originally used by Lesson and Garnot for a new subspecies of what was then called *Garita keraudrenii* (see Trumpet Manucode). They based their identification of the subspecies on what they call 'a significant modification of the trachea'. They describe this extraordinary structure – and it must be said that both the description and the trachea are very long and convoluted – which enables the bird to make sounds like a horn. Exactly what kind of horn would be hard to say: Morcombe (2000) described the call as 'a rasping, abrupt, gurgling and inwards gulping 'owwgk'', but see also the gentler Olsen and Joseph (2011) interpretation (see Trumpet Manucode).

Ptiloris Swainson, 1825 [TI-lo-ris]: 'feather-nose', from Greek *ptilon*, feather or down, and *rhis*, nostril. Swainson (1825) said, 'the nostrils are completely hid by the frontal feathers'.

The species

Trumpet Manucode (breeding resident)

See family introduction. RAOU (1978) introduced the name as a 'compromise' between the two traditional names, Manucode and Trumpet-bird.

Other names: Keraudren's Crow-shrike, as used by Gould (1869), from the species name; the use of crow-shrike was interesting, because it implies a relationship with Australian magpies, though Gould went to some trouble to point out its New Guinea affinities. Australian Bird-of-Paradise; Trumpeter Manucode, Trumpet Bird or Australian Trumpet Bird, for the male's extraordinarily loud 'deep mellow bugling call' (Olsen and Joseph 2011), still used as first-choice name by RAOU (1913), though it changed to Manucode for the second checklist (RAOU 1926).

Phonygammus keraudrenii (Lesson & Garnot, 1826) [fo-ni-GAM-mʊs ke-roh-DRE-ni-ee]: 'Keraudren's sound-wedded bird', see genus name, and for Pierre Keraudren, French surgeon-naturalist on the *Géographe* under Nicolas Baudin on the mighty, but ill-fated, exploring expedition from 1801; he may have collected the type specimen, which was from New Guinea, and the authors (also doctors on the expedition) expressed fulsome praise for him, noting his 'quite paternal benevolence' towards them.

Paradise Riflebird (breeding resident)

See family introduction. From the species name.

Other names: New Holland Ptiloris, used by Jardine and Selby (1826–35, vol. 1); Rifle Bird or Riflebird; Rifle Bird of Paradise or New South Wales Rifle Bird of Paradise; Velvet Bird, for the male's glossy black plumage.

Ptiloris paradiseus Swainson, 1825 [TI-lo-ris pa-ra-DEE-se-ʊs]: 'paradise feather-nose', see genus and family names.

Victoria's Riflebird (breeding resident)

See family introduction. Direct from the species name, for the reigning British queen (who was not a noted ornithologist).

Other names: Subsequently and until recently seen as the same species as the previous one (CSIRO 1969), so those names applied; Queen Victoria's Riflebird (and variations on the apostrophe theme); Victoria Rifle Bird of Paradise; Lesser Riflebird, because it is noticeably smaller than Paradise; Barnard Island Riflebird, for the type location near Innisfail.

Ptiloris victoriae Gould, 1849 [TI-lo-ris vic-TOR-ee-eh]: 'Victoria's feather-nose', see genus and common names.

Magnificent Riflebird (breeding resident)

See family introduction. From the species name.

Other names: Albert's or Prince Albert's (also without apostrophes) Riflebird – the Australian birds represent race *alberti* of an essentially New Guinea species, though this race has at times been seen as a full species *P. alberti*; Cape York Riflebird, for its Australian range; Scale-breasted Paradise-bird, an old name used in New Guinea.

Ptiloris magnificus (Vieillot, 1819) [TI-lo-ris mag-NI-fi-cʊs]: 'magnificent feather-nose', see genus name, and from Latin *magnificus*, magnificent or splendid (also meaning a bit of a show-off, which could refer, rather unkindly, to the spectacular display of the male).

PETROICIDAE (PASSERIFORMES): Australian robins

Petroicidae Mathews, 1920 [pet-ROY-si-dee]: the Rock-dweller family, see genus name *Petroica*.

'Robin' had its origin in England, from an unrelated English bird, the European Robin *Erithacus rubecula* (a Muscicapid flycatcher). Sometime in the late 14th century the descriptive term 'redbreast' arose and it gradually overtook the older word Ruddock (which was based in Old English). The bird, familiar and popular, eventually acquired a human 'first name' and ~150 years later was known as Robin Redbreast, Robin being an affectionate form of Robert. Later the 'surname' was increasingly dropped, leaving Robin as the single name, though still associated with a red-breasted bird. As such, when red-breasted birds were encountered in the new colony they too were labelled Robin – even when species then appeared that were yellow, or black and white! (The same thing happened in North America, where a red-breasted thrush, *Turdus migratorius*, became the American Robin.) By 1848 Gould was using 'robin' for most of the same birds that we do, noting that the 'Colonists' used simple 'robin' for Scarlet Robin. Vigors and Horsfield (1827) cited George Caley (who left the colony in 1810) as reporting that colonists were already using the term 'yellow Robin'.

'Flyrobin' for the two yellow *Microeca* robins is somewhat perplexing. It might be tempting to suggest that it's a nod to the fact that here some of the robin species have traditionally been called 'flycatchers', but in fact 'flyrobin' has no history in Australia (except for Grey-headed Robin and two *Poecilodryas* species, where it was hyphenated) and is still rarely used, including in the field guides. It seems to have arisen in Papua New Guinea and is applied there to several species; perhaps it relates to the Fly River, but that feels like a long bow to draw.

'Scrub Robin' is a term introduced by Gould (1848), who commented that 'I discovered this singular bird in the great Murray Scrub [mallee] of South Australia.' (It's worth noting that well over 100 years later some Australian ornithologists were still doubting his perceptive analysis of the bird's relationships.)

The genera

Heteromyias Sharpe, 1879 [he-te-ro-MEE-i-as]: 'the other flycatcher', from Greek *heteros*, other or different, and *myia*, a fly, the one it is 'other' to being *Poecilodryas*.

Poecilodryas Gould, 1865 [poy-si-lo-DREE-as]: 'dappled tree-nymph', from Greek *poikilos*, pied or dappled, and *dryad*, a tree-nymph.

Peneoenanthe Mathews, 1920 [peh-neh-ee-NAN-theh]: 'almost a wheatear', from Latin *paene* or *pēne*, almost, and *Oenanthe*, the genus name for wheatears. The spelling of this one has varied, in that Christidis and Boles (2008) spelt it without the third 'e', but other authorities including IOC spell it as here, which makes more sense in terms of *Oenanthe*, wheatear. Mathews himself spelt it *Peneœnanthe*.

Tregellasia Mathews, 1912 [tre-gel-LA-si-uh]: 'Tregellas' [bird]', for Thomas Henry Tregellas, 1864–1938, a foundry-worker who spent much of his spare time camping out with lyrebirds to study their habits. (And could he be the man of the same name and dates, and apparently diverse interests, who made tin violins in his spare time?) Chisholm (1960b) gave his rather acerbic opinion of the name, '*Tregellasia*, it may be noted, is one of the many clumsy and 'empty' names coined by or for Mathews when he was busily engaged in splitting genera.' The 'or for Mathews' was a particularly biting reference to a belief that not everything published over Mathews' name was his own work.

Eopsaltria Swainson, 1832 [e-o-SAL-tri-uh]: 'dawn singer', from Greek *eōos* or ēōs, dawn, and *psaltria*, a female singer or harpist. It is clear from what Swainson (1831b) wrote in Richardson's *Fauna boreali-americana* that he took the name to mean a singer.

Melanodryas Gould, 1865 [me-la-no-DRI-as]: 'black tree-nymph', from Greek *melas*, black, and *dryad*, a tree-nymph.

Microeca Gould, 1841 [mi-KROH-e-kuh]: 'dweller in a small house', from Greek *mikros*, small, and *oikos*, house or dwelling place. Gould (1840) used this name for a new genus that would encompass both the species then known as *Muscicapa macroptera* – the Jacky Winter – and a new species (now just a race) from Western Australia. He called the new species *Microeca assimilis*, stating that it was 'nearly allied to but much less in size than *Muscicapa macroptera*'. We believe it is this sentence that has led to confusion about the meaning of the genus name and the belief, shared by us until recently, that it was derived from Greek *mikros*, small, and *eoika/oika*, to be like. Gould did not state his thinking on this, and nothing in his description of the characteristics of the genus helps us either. However, if we take it that '*assimilis*' takes care of the 'nearly allied to' issue, and that Gould would name the genus to reflect all members, not just a characteristic of one species (i.e. the smallness of the new bird), then Pizzey and Knight's 2007 explanation of the genus name as meaning 'small house' is as good as any other. Although Gould does not mention the nest under the characteristics of the genus, in his entry on the Jacky Winter in *The Birds of Australia*, he gives one of his full and delightful descriptions of it, including its dimensions: 'two inches and a half in diameter by half an inch in depth'. A small house indeed!

Petroica Swainson, 1829 [pet-ROY-kuh]: 'rock-dweller', from Greek *petros*, rock, and *oikos*, a house or dwelling-place, reflecting the habitat on Norfolk Island, whence the type specimen originated.

Drymodes Gould, 1841 [dri-MOH-dehz]: 'lover of woodland places', as Gould (1840) put it, from Greek *drumōdēs*, woody.

The species

Grey-headed Robin (breeding resident)

See family introduction. Directly (or nearly so) from the species name; until very recently this species was regarded as conspecific with the New Guinea species *H. albispecularis*.

Other names: Ashy-fronted Robin, a more precise translation of the species name; Fly-robin or Ashy-fronted Fly-robin, used regularly in Australia in the late 19th and early 20th centuries, without explanation – see family introduction; Ground Thicket-flycatcher and Black-cheeked Robin, cited in HANZAB (1990–2006, vol. 7a) but usage restricted to Papua New Guinea – in particular the latter name belongs to *H. armiti*, which until recently was lumped with the current species.

Heteromyias cinereifrons Ramsay EP, 1876 [he-te-ro-MEE-i-as si-ne-RE-i-fronz]: 'ashy-fronted other flycatcher', see genus name, and from Latin *cinereus*, like ashes, and *frons*, forehead – the light grey forms a striking contrast with the brown and rufous back.

White-browed Robin (breeding resident)

See family introduction. From the species name.

Other names: White-eyebrowed Robin, by Gould (1848) directly from his species name; White-browed Fly-robin or Shrike-robin (see Eastern Yellow Robin and Grey-headed Robin).

Poecilodryas superciliosa (Gould, 1847) [poy-si-lo-DREE-as soo-pehr-si-li-OH-suh]: 'eye-browed dappled tree-nymph', see genus name, and from Latin *supercilium*, above the eyelid (i.e. eyebrow) – a very obvious feature.

Buff-sided Robin (breeding resident)

See family introduction. By Gould (1869), using his species name, as was his normal approach. Note that for a long time, and until very recently, this species and the previous one were lumped together.

Other names: Buff-sided Fly-robin or Shrike-robin.

Poecilodryas cerviniventris (Gould, 1858) [poy-si-lo-DREE-as sehr-vi-ni-VEN-tris]: 'stag-bellied dappled tree-nymph', see genus name, and from Latin *cervinus*, pertaining to a stag, *cervus*, a stag or deer, and *venter/ventris*, belly (see common name). The colour is a robust fawn – a rufous buff or tawny colour, rather than the *Oxford Dictionary*'s 'light yellowish brown' or a light greyish brown. Gould described it as 'a deep fawn colour' (Gould 1865).

Mangrove Robin (breeding resident)

See family introduction. For its essential habitat.

Other names: White-tailed Robin, per Gould (1869) from his own species name *Eopsaltria leucura*, not realising that Charles Lucien Bonaparte had already assigned a name to a specimen from New Guinea – ironically this name (White-tailed) was still being used into the 20th century (RAOU 1913) after the current species name had been adopted. Mangrove or White-tailed Robin or Shrike-robin, see Eastern Yellow Robin; Ashy Robin, from the current species name, though this is a name more usually associated with a New Guinea species, *Heteromyias albispecularis*; Grey-headed Robin, for race *cinereiceps* of north-west Australia, but see that species.

Peneoenanthe pulverulenta (Bonaparte, 1850) [peh-neh-ee-NAN-theh pʊl-ve-rʊ-LEN-tuh]: 'dusty almost-wheatear', see genus name, and from Latin *pulverulentus*, covered in dust (cf. *pulvis*, dust, powder or ashes), originally for its slate grey back, but we would add that the light touch of pale grey on the white breast also suggests this.

White-faced Robin (breeding resident)

See family introduction. From the species name.

Other names: White-faced Yellow Robin; White-throated Robin or Yellow Robin or Shrike-robin, as distinct from the previous species; Little Robin or Yellow Robin; White-faced Flycatcher.

Tregellasia leucops (Salvadori, 1876) [tre-gel-LA-si-uh LOO-kops]: 'Tregellas' white-face', see genus name, and from Greek *leukos*, white, and ōps, face (or eye, but not in this case).

Pale-yellow Robin (breeding resident)

See family introduction. On the face of it, this name seems to have been introduced without explanation by the RAOU's committee responsible for the second *Official Checklist* of names (RAOU 1926), but this is one of the few situations where we know exactly how and when a name arose. The great Alec Chisholm, writing long after the event, noted that 'when I first became acquainted with *capito* its 'official' vernacular was 'Large-headed Robin', a term no more appealing than the old name of 'Thickhead', as applied to the bird now known as Whistlers ... Casting about for something better, I hit upon Pale-yellow (with a hyphen) as the adjective, and this was adopted by the Checklist Committee' (Chisholm 1960b). We note, however, that there is nothing particularly pallid about its yellow undersides.

Other names: Large-headed Robin, by Gould (1869) from his species name, a name that persisted well into the 20th century; Buff-faced or Rufous-lored Robin, both for race *nana* of the Queensland wet tropics – the southern nominate race has white lores; Half Robin, a curious one to which Chisholm (1960b) again provides the key: 'On Tambourine [Mt Tambourine, near Brisbane] the local name was 'Half Robin', a quaint title coined to distinguish the species from its brighter coloured relative in the same area, the Yellow-rump [Eastern Yellow Robin]'. Noting our objection to the fact that it's not really paler than the Eastern Yellow, we wonder if the locals weren't being more literal, and referring to the obvious size difference?

Tregellasia capito (Gould, 1854) [tre-gel-LA-si-uh KA-pi-to]: 'Tregellas' big-head', see genus name, and from Latin *capito*, having a large head – see common name, but actually its head **does** look biggish, or maybe it's just the rather thick neck ...

Eastern Yellow Robin (breeding resident)

See family introduction. Straightforward – it is yellow, and there is also a western species.

Other names: Southern Motacilla (a pipit) by John Shaw, for his own name *Motacilla australis*; Southern Flycatcher (Latham 1801), homing in on the nature of the bird; Yellow-breasted Thrush (Lewin 1822) – getting cold again; Yellow-breasted Robin (Gould 1848); Southern Yellow Robin, from the species name, long used but not very helpful. Shrike-robin or Yellow-breasted Shrike-robin, the latter being a popular term in the late 19th and early 20th centuries (RAOU 1913; Chisholm 1922) for *Eopsaltria* and some apparently associated genera, in part reflecting an uncertainty that they were real robins. For example, Leach placed them in the same 'subfamily' as whistlers; later, when discussing 'red' robins, he explained that 'the Shrike Robins belong to the Shrike family, so they need not be mentioned here' (Leach 1911). The second (RAOU (1926) checklist dropped the 'shrike-' overboard, and it didn't resurface. Golden-rumped or Golden-tailed Yellow Robin, Yellow-rump, apparently mainly used where the species overlapped with Pale-yellow Robin, this being a useful distinguishing feature (see Pale-yellow Robin); Yellowbob; Bark Robin, for its habit of clinging sideways on trees; Creek Robin, presumably for a preferred (but not obligate) habitat, though usage outside HANZAB (1990–2006, vol. 7a) is hard to find; Wild Canary.

Eopsaltria australis (Shaw, 1790) [e-o-SAL-tri-uh ows-TRAH-lis]: 'southern dawn-singer', see genus name, and from Latin *australis*, southern.

Western Yellow Robin (breeding resident)

See family introduction. The western equivalent of the previous species.

Other names: Grey-breasted Robin, by Gould (1848), partly representing his species name – this is the key distinguishing factor from Eastern Yellow Robin; Grey-breasted Yellow Robin or Shrike-robin, see Eastern Yellow Robin.

Eopsaltria griseogularis Gould, 1838 [e-o-SAL-tri-uh gree-se-o-gʊ-LAH-ris]: 'grey-throated dawn-singer', see genus name, and from Mediaeval Latin *grisus/griseus*, grey, and Latin *gula*, throat (see common name).

White-breasted Robin (breeding resident)

See family introduction. Named for the obvious reason, but also as a distinction from the previous species, which has a similar range.

Other names: White-bellied Robin, by Gould (1848) for his species name *E. leucogaster* – again he'd missed the Quoy and Gaimard name; White-breasted Shrike-robin, see Eastern Yellow Robin.

Eopsaltria georgiana (Quoy & Gaimard, 1830) [e-o-SAL-tri-uh ge-or-gi-AH-nuh]: 'dawn-singer from King George Sound', see genus name, and for the type locality in Western Australia.

Hooded Robin (breeding resident)

See family introduction. From Latham's species name (he used Hooded Flycatcher – as often with Latham, his common name came first).

Other names: Hooded Flycatcher (see common name); Pied Robin, per Gould (1848) from the species name *Grallina bicolor* (of Vigors and Horsfield) he used; Black Robin, reported by Gould to be used by 'the Colonists'; Black-and-White Robin.

Melanodryas cucullata (Latham, 1802) [me-la-no-DRI-as kʊ-kʊ-LAH-tuh]: 'hooded black tree-nymph', see genus name, and from Latin *cucullus*, a hood, Mediaeval Latin *cucullatus*, wearing a hood.

Dusky Robin (breeding resident)

See family introduction. Named by Gould (1848) from his own species name *Petroica fusca*, not having realised that the French had beaten him to it.

Other names: This bird has attracted some wry names from its fellow Tasmanians. Dozey or Dozey Robin, Stump Robin, Sleepy Robin, Wood Robin, for its hunting technique of being a very still and watchful 'perch and pouncer'; Sad Robin or Sad Bird – 'from the sad note of its call', as noted by Miss J.A. Fletcher (1924); Tasmanian Robin, because it is endemic.

Melanodryas vittata (Quoy & Gaimard, 1830) [me-la-no-DRI-as vi-TAH-tuh]: 'banded black tree-nymph', see genus name, and from Latin *vittatus*, bound with a ribbon or band, named for the pale wing bar (Quoy and Gaimard 1830).

Yellow-legged Flyrobin (breeding resident)

See family introduction, and named for the obvious character.

Other names: Yellow-legged Flycatcher, near universally used in Australia; Little Yellow Robin or Flycatcher; White-lored Robin or Shrike-robin, shrike-robin being a commonly used term for non-red robins into the 20th century (RAOU 1913) – the lores are the areas between eyes and bill; Grey-headed Flycatcher or Robin-flycatcher, from the species name.

Microeca griseoceps De Vis, 1894 [mi-KROH-e-kuh GREE-se-o-seps]: 'grey-headed dweller in a small house', see genus name, and from Mediaeval Latin *grisus/griseus*, grey, and Latin suffix *-ceps*, -headed (*caput*, head). The grey head stands out a touch oddly against the lemon and brown body.

Lemon-bellied Flyrobin (breeding resident)

See family introduction, and named from the species name.

Other names: Lemon-bellied Flycatcher, near universally used in Australia; Lemon-breasted Flycatcher, from more genteel times; Yellow-bellied or Yellow-breasted Flycatcher; Brown-tailed Flycatcher, from *M. brunneicauda*, described by Archibald Campbell (1902) from a specimen 'from the Northern Territory' – Shane Parker (1973) convincingly argues that it was based on a damaged Brown (Grey) Whistler (*Pachycephala simplex*) skin; Kimberley Flycatcher, for race *tormenti*.

Microeca flavigaster Gould, 1843 [mi-KROH-e-kuh flah-vi-GAS-tehr]: 'yellow-bellied dweller in a small house', see genus name, and from Latin *flavus*, golden or reddish yellow (from *flagro*, burn), and *gaster*, belly.

Jacky Winter (breeding resident)

As a spontaneously derived folk name, its precise origin is unlikely to be resolved but Edward Morris' assertion (1898) that it was 'ascribed to the fact that it is a resident species, very common, and that it sings all through the winter, when nearly every other species is silent' is unlikely to be far from the mark. Reported movements in some areas in winter into more open area are probably pertinent too. The 'winter' part goes back to the early 19th century; Vigors and Horsfield (1827) wrote that 'Mr. Caley [George Caley, Banks' botanical collector in New South Wales, who left the colony in 1810] informs us that the boys of the colony used to call it Winter, the reason of which he does not give.' (It has occurred to us, however, that the oft-cited 'Peter Peter' rendition of the call could also be heard as 'Winter Winter'.) The first written reference we have found to the name Jacky Winter is in the *Australian Town and Country Journal* of 15 December 1883, in inverted commas, suggesting that the author regarded it as an informal name, but it was almost certainly in general use before that, given that less than a decade later Morris defined it as 'the vernacular name in New South Wales of the Brown Flycatcher'. Jacky is an affectionate moniker (like Willie Wagtail); perhaps we have lost touch with the bird since the early days when Caley reported 'the bird has all the actions of the British Robin Redbreast, except coming inside houses. When a piece of ground was fresh dug, it was always a constant attendant'.

Other names: Lesser Fascinating Bird, from the species name as applied by Latham, who had previously noted, regarding his 'Fascinating Grosbeak', that 'at Port Jackson [it] is known as the smaller fascinating bird, having the manners exactly of the Fascinating Thrush' (Latham 1801). (The latter bird, *Turdus volitans*, we now know as Restless Flycatcher; see that profile for an account of its supposed hypnotic powers.) Great-winged Microeca, used by Gould (1848), from the genus name and the species name *M. macroptera* that he was using (assigned by Vigors and Horsfield, not realising that Latham already had named it – scarcely their fault, because it was not always easy to be sure what Latham was referring to); Brown Flycatcher, reported by Gould to be a name 'of the Colonists' and used (usually with 'Australian') well into the 20th century (RAOU 1913); Allied Flycatcher, for Gould's *M. assimilis* of Western Australia (now a race), 'so nearly allied to' the current species that he saw no need to illustrate it separately; Post-boy, Post-sitter, Stump-bird, for its 'perch and pounce' hunting technique; Peter-Peter for its repetitive call; Spinks (Leach 1911), perhaps also reflecting the call; White-tail, for the conspicuous outer tail shafts.

Microeca fascinans (Latham, 1802) [mi-KROH-e-kuh FA-si-nans]: 'bewitching dweller in a small house', see genus name, and from Latin *fascino*, enchant, bewitch, fascinate (see common name).

Rose Robin (breeding resident)

See family introduction. Used by Gould (1848) as Rose-breasted, from his species name; for male only.

Other names: Rose-breasted Wood Robin, Gould's full name – he used Wood Robin (and Wood-Robin) for this and the next species only, presumably based on his observation of habitat for both: 'They prefer the most secluded and remote parts of the forest.' The term did not live to see the end of the 19th century. Rose-breasted Robin.

Petroica rosea (Gould, 1840) [pet-ROY-kuh RO-se-uh]: 'rosy rock-dweller', see genus name, and from Latin *roseus*, rose-coloured, for the beautiful breast.

Pink Robin (breeding resident)

See family introduction. A simplified version of Pink-breasted Wood-Robin (Gould 1848), from the species name – see also Rose Robin. Again, for male only.

Other names: Pink-breasted or Magenta-breasted Robin, the latter very rarely used.

Petroica rodinogaster (Drapiez, 1819) [pet-ROY-kuh ro-di-no-GAS-tehr]: 'pink-bellied rock-dweller', see genus name, and from Greek *rhodinos*, pink (or made of roses!) and *gastēr*, belly.

Flame Robin (breeding resident)

See family introduction. Used by Gould (in the form Flame-breasted), indirectly at least from his species name; for male only.

Other names: Robin Redbreast (from Britain); Flame-breasted Robin, see common name; Bank Robin, a mystery both in origin and intent, though cited by CSIRO (1969) – the fact that it is known to nest in bank hollows may be relevant.

Petroica phoenicea (Gould, 1837) [pet-ROY-kuh foy-NEE-se-uh]: 'scarlet rock-dweller', see genus name, and from Latin *phoenicius*, scarlet or purple-red – Gould chose his colour word more accurately in his common name, unless he was thinking of a phoenix rising from flames?

Norfolk Robin (Norfolk Island)

See family introduction. Endemic to Norfolk Island. It has a complex history; until recently this species and the next were regarded as one, but coincidentally the type specimen was from Norfolk Island so the older name was retained for this species. More recently still it was lumped for a while with Pacific Robin, so it too needed a name change when split.

Other names: Norfolk Island Robin; Scarlet and Pacific Robin, see common name.

Petroica multicolor (Gmelin JF, 1789) [pet-ROY-kuh mʊl-TI-ko-lor]: 'many-coloured rock-dweller', see genus name, and from Latin *multicolor*. Well, it's red, white and black, at least – and Gould (1865) emphasised the beauty of the colours: 'scarlet', 'jet-black' and with the forehead 'snowy white'.

Scarlet Robin (breeding resident)

See family introduction. Another name referring only to the male (and only his breast and upper belly at that, though undoubtedly his most striking feature). It's not clear exactly when it arose but it was early and probably spontaneous – Gould (1848) used Scarlet-breasted (and logically, used '-breasted' with a colour to distinguish all of the common mid-east coast robins).

Other names: Crimson-breasted Warbler, by Lewin, the painter, in 1808; 14 years later he retreated to Red Breast Warbler (Lewin 1822); Robin, Robin Redbreast, directly from Britain; Scarlet-breasted Robin, see common name; Western Scarlet-breasted Robin, for race *campbelli*; White-capped Robin, by distinction from Red-capped Robin, though it is hard to find an example of its use.

Petroica boodang (Lesson, 1838) [pet-ROY-kuh BOO-dang]: 'boodang rock-dweller', see genus name, and from an Aboriginal word (language unspecified): Lesson (1837) stated that the natives of the country around 'Sidney' call the bird 'bouddang' (though he spelt it 'booddang' in his French common name). The first known usage was on a painting by 'Port Jackson Painter', number 282 in the Watling Paintings, created 'between 1788 and 1797' and entitled 'Crimson-breasted Warbler, native name Bood-dang'.

Red-capped Robin (breeding resident)

See family introduction. Again for a most obvious character of the adult male, used by Gould (1848) and as used by 'the Colonists', so evidently a spontaneously generated name.

Other names: Redhead, Redcap; Goodenough's Flycatcher, used in Jardine and Selby (1826–35, vol. 1) for the species name; Red-throated Robin, a seemingly contradictory name, but was used for the briefly recognised *P. ramsayi* of the Northern Territory, which was apparently based on an anomalous specimen.

Petroica goodenovii (Vigors & Horsfield, 1827) [pet-ROY-kuh gŭd-e-NO-vi-ee]: 'Goodenough's rock-dweller', see genus name, and for Samuel Goodenough (1743–1827), Bishop of Carlisle and an amateur naturalist who held office in both the Linnean Society and the Royal Society, of which he eventually became a Fellow. Vigors and Horsfield (1827) named the bird 'in honour of this most reverend and most erudite man'.

Northern Scrubrobin (breeding resident)

See family introduction. There are two widely separated species of scrubrobins – this one from the rainforests of Cape York, and the next from the mallee-heaths of southern Australia.

Other names: Northern Scrub-robin, the preferred form in Australia; Papuan Scrub-robin, from when this species and New Guinea *D. beccarii* were regarded as the same.

Drymodes superciliaris Gould, 1850 [dri-MOH-dehz soo-pehr-si-li-AH-ris]: 'eyebrowed lover of woodland places', see genus name, and from Latin *supercilium*, above the eyelid (i.e. eyebrow). There is a sort of eyebrow, but the eye is notable rather for the very bold thick black line that runs diagonally through it. The 'woodland places' are what MacGillivray, who collected the bird, described as 'a thin open scrub of small saplings growing in a stony ground thickly covered with dead leaves, about five or six miles inland from Cape York' (Gould 1865).

Southern Scrubrobin (breeding resident)

See family introduction and Northern Scrub-robin.

Other names: Southern Scrub-robin, the preferred form in Australia; Scrub-robin, used by Gould (1848); Mallee Scrub-robin, for the habitat; Pale Scrub-robin, for Mathews' Western Australia race *pallida*, no longer recognised.

Drymodes brunneopygia Gould, 1841 [dri-MOH-dehz brŭn-ne-o-PI-gi-uh]: 'brown-rumped lover of woodland places', see genus name, and from Mediaeval Latin *brunneus*, brown (6th century *brunus*, 12th century *bruneus*), with Latin *pyga*, rump (cf. Greek *pugē*). Gould's 'woodland places' are what he calls 'the great Murray scrub' – the tough and beautiful mallee.

ALAUDIDAE (PASSERIFORMES): larks

Alaudidae Vigors, 1825 [a-LOW-di-dee]: the Lark family, see genus name *Alauda*.

'Lark' comes from Old English, with the form *lauerce* (and others) being known; Chaucer used the form 'lark' in c. 1366. The original meaning is lost – Lockwood (1984) postulated something like 'little song', but there seems little need to look for anything beyond a name for a familiar bird (see Eurasian Skylark).

'Skylark' was unnecessary in England, which had only the one lark, but was introduced anyway by John Ray (the 'father of English natural history') in the 17th century, translated from the German.

The genera

Mirafra Horsfield, 1821 [mi-RAF-ruh]: 'mirafra', meaning – well, pretty much a mystery, seemingly. Horsfield gave no explanation, giving only the Javanese name *Branjangan*. Most of his successors used

Mirafra without comment, though Agassiz unhelpfully said it is '*vox barbar*' (foreign), and variants of this are repeated in subsequent syntheses of information. HANZAB (1990–2006, vol. 7a) cited Gotch as claiming in 1981 that the name derives from Latin *mirus*, wonderful and *afra*, African, but objects that the type species *M. javanica*, 'does not occur in the Afrotropics'. (We note, though, that most of the many species currently included in the genus are African birds.) But it is hard to see why you would name a bird for Africa in one part of its name and for Java in the other. Not to be outdone, we will have a go ourselves: what if it were from Greek *mura-*, scented or sweet, and *phrasis*, speech, giving myrafra? It's a bit of a stretch, we know, but our bird does have a lovely song, or sweet speech, and in the absence of any other more convincing suggestion …

Alauda Linnaeus, 1758 [a-LOW-duh]: 'lark', from *alauda*, possibly a Gallic word. Pliny (77–79 AD) explained that the Lark (probably the Crested Lark *Galerida cristata*) used to be called *galerita*, for the appearance of its tuft, but that this has been replaced by the Gallic name *alauda*, and mentioned that this name has transferred to a Roman legion (the Fifth). It seems that in 52 BC Julius Caesar created a legion of Gauls called *Alaudae*, possibly because of the crests or the pairs of wings embossed on the forehead and sides of their helmets (not the 20th-century Gauloises cigarettes/Asterix cartoon mythical winged version, though). There have been many attempts over the centuries to explain the derivation of the word *alauda*, most of them complicated by the fact that much has been made of the similarity to the word 'laud' and the related folk belief that the lark sings songs of praise to God, and so they are largely unconvincing. The French word *alouette* is related, but is said to come from the Latin *alauda* via Old French *aloe*, rather than taking us back to Gallic. In the absence of any certainty, we could just trust Pliny (though given some of **his** tall tales …).

The species

Horsfield's Bush Lark (breeding resident)

Named by Gould (1848) for his species name *horsfieldii* 'in honour of the founder of the genus'. The founder was Dr Thomas Horsfield, a US doctor and naturalist who worked for the British East India Company in Java, and later at its London museum. Horsfield had described the genus for the Javanese species, which he predictably called *M. javanica*; until 1911 (when Leach was still using *horsfieldii* for the Australian birds) Gould's name prevailed here, but 2 years later Mathews (1913) had demoted *horsfieldii* to a subspecies of *javanica*, where it remains.

Other names: Horsfield's Mirafra (Gould 1848); Bushlark; Singing Bushlark, a name long used – this is another tangled story. At the same time that *horsfieldii* was subsumed into *javanica*, the widespread species *cantillans* (from southern Asia to Africa) was as well – it is called 'Singing' from the species name and this name became informally adopted for the whole complex until the RAOU formalised it in *Recommended English Names* (RAOU 1978). The logic was that this was the 'usual international name' (though we might point out that this was for *cantillans* rather than *javanica*). Soon afterwards *cantillans* was removed from the complex, taking with it the rationale for this name (see Dabb 2008 for more discussion of this). Horsfield's or Australasian Bushlark, and the latter might usefully be reconsidered one day; Brown Bushlark or Fieldlark; Cinnamon Bushlark, for central Western Australian race *woodwardi*, a distinctly warm cinnamon colour. Woodward's Bushlark, for the same race (also called Onslow Bushlark by Mathews 1913), for Bernard Henry Woodward (1846–1916), an English geologist who came to Australia in 1889 for his health and became Western Australian government analyst, then inaugural director of the Western Australian Museum and Art Gallery. He founded the Western Australia Natural History Society and was an advocate for reserved lands for conservation. Rufous Bushlark, for race *rufescens* of northern inland Australia; Rufous-winged Bushlark, for a useful field character in flight, but not often used, because it is more widely applied to *M. assamica* of India and South-East Asia; Rufous Pipit. Fieldlark, Croplark, Lucernelark, though some of these names are also used for songlarks, and the application was probably somewhat indiscriminate; Australian Skylark. The incorrect form Horsefield's also occurs from time to time.

Mirafra javanica Horsfield, 1821 [mi-RAF-ruh jah-VA-ni-kuh]: 'javanese mirafra', see genus and common names.

Eurasian Skylark (introduced breeding resident)

Found right across Europe and Asia.

Other names: Skylark, English or Northern Skylark.

Alauda arvensis Linnaeus, 1758 [a-LOW-duh ar-VEN-sis]: 'field lark', see genus name, and from Latin *arvensis*, from *arvum*, field or cultivated land, with the suffix *-ensis* this time meaning 'pertaining to'. Farmland or moorland are the bird's favourite habitats.

PYCNONOTIDAE (PASSERIFORMES): bulbuls

Pycnonotidae Gray GR, 1840 [pik-no-NOH-ti-dee]: the Firm-backed family, see genus name Pycnonotus.

'Bulbul' is of Arabic origin, coming into English in the late 18th century via Persian, apparently from poetry translations though there seems to have been some confusion with the word for Nightingale. The family is a huge one, spread across Africa and south and east Asia.

The genus

Pycnonotus Boie, 1826 [pik-no-NOH-tʊs]: 'dense-backed [bird]', from Greek *puknos*, firm or dense, and *nōton*, the back. The genus was named for the Cape Bulbul *Pycnonotus capensis*, and while neither Boie, nor Linnaeus (1758) who named the species now found in Australia, nor Brisson (1760) who gave the most detailed description of the bird, mentioned it, HANZAB (1990–2006, vol. 7b) pointed out that *capensis* has a thickly feathered back, a feature it shares with some others in the genus.

The species

Red-whiskered Bulbul (introduced breeding resident)

See family introduction. For the red patch below and behind the eye.

Other name: Red-eared, Red-eyed or Indian Bulbul – its natural range is from eastern India through South-East Asia.

Pycnonotus jocosus (Linnaeus, 1758) [pik-no-NOH-tʊs yo-KOH-sʊs]: 'jokey firm-back', see genus name, and from Latin *iocosus*, full of jokes, humorous, droll. Linnaeus (1758) said only that the bird has the lower eyelid dark red ('*purpurea*'), and the vent ('*anus*') blood-red. We're not quite sure what the joke was, but HANZAB (1990–2006, vol. 7b) claimed *jocosus* 'alludes to the merry red cheeks'. The pleasant liquid whistling call must be in there with a chance too, surely?

HIRUNDINIDAE (PASSERIFORMES): swallows and martins

Hirundinidae Rafinesque, 1815 [hi-rʊn-DI-ni-dee]: the Swallow family, see genus name *Hirundo*.

'Swallow' comes from Old English *swealwe*; Chaucer used the form *swalwe* in the 14th century and the modern spelling appeared in the 16th century. Lockwood (1984) suggested it meant 'cleft stick', from the forked tail, but this appears to be speculative.

'Martin' is a roundabout one: it was a French (human) name, which came to England with the Normans and became established there from the 12th century. It was applied, in various forms (including *martoune*, *marten* and *martinet*) to swallows, eventually being associated with the commensal House Martin (*Delichon urbicum*). Meanwhile, back in France, *martinet* (opinions differ on whether this was from the human name) was becoming the name of swifts (Apodidae) rather than swallows, and Kingfishers (river kingfishers first as *martinet pescheur*, now *martin-pêcheur*, and tree kingfishers, by analogy, as *martin-chasseur*). And *martin* on its own in French (origin also unclear), just to round things off with even more confusion, is now the common name for genus *Acridotheres*, mynas.

The genera

Cheramoeca Cabanis, 1850 [keh-ra-MOY-kuh]: 'hole-dweller', from Greek *khēramos*, hole or hollow (in the earth), and *oikeō*, to live or be housed.

Hirundo Linnaeus, 1758 [hi-RʊN-do]: 'swallow', from Latin *hirundo*, meaning just that.

Delichon Horsfield & Moore, 1854 [DE-li-kon]: 'anagram swallow', from Greek *khelidōn*, swallow – the use of anagrams for names being entertaining for some but described by Strickland *et al.* (1842) as 'verbal trifling' and 'peculiarly annoying' (see also *Dacelo*, under Alcedinidae).

Cecropis Boie, 1826 [SE-kro-pis]: 'tribal Athenian'. Cecrops was a mythical king of Athens and one of the Athenian tribes was called after him. *Cecropis* really means a female of this tribe, but what Boie took it to mean is unclear.

Petrochelidon Cabanis, 1850 [pet-ro-KE-li-dohn]: 'rock swallow', from Greek *petros*, a stone or boulder, and *khelidōn*, swallow.

The species

White-backed Swallow (breeding resident)

For the large white saddle; it seems to have first been used by John Leach (1911), whose influence doubtless led to it being adopted by the RAOU (1913).

Other names: Black and White Swallow, reported by Gould (1848) as being a name 'of the Colonists' but still used as first-choice name by Hall (1899) and Lucas and Le Soeuf (1911) – we might be forgiven for thinking this a pretty banal name for a very beautiful bird; White-breasted Swallow, used by Gould (1848) from his species name; White-capped Swallow, for another striking feature; Sand Martin, presumably for its habit of nesting in sand banks, but doubtless influenced by the European bird of the same name.

Cheramoeca leucosterna (Gould, 1841) [ke-ra-MOY-kuh loo-ko-STEHR-nuh]: 'white-chested hole-dweller', see genus name, and from Greek *leukosternos*, white-chested.

Barn Swallow (uncommon summer migrant to northern Australia)

For a typical nesting site across its huge Northern Hemisphere range.

Other names: In Britain it is just 'the swallow'; it also has other names there, but none of them are relevant in Australia.

Hirundo rustica Linnaeus, 1758 [hi-RUN-do RUS-ti-kuh]: 'rustic swallow', see genus name, and from Latin *rusticus*, of the country, rural, rustic.

Welcome Swallow (breeding resident)

It seems from his own account that Gould coined the species name first, and thence the common name.

Other names: Swallow; House Swallow (it commonly nests under eaves); Australian or Australian Welcome Swallow; Pacific Swallow, from times when it has been considered conspecific with *H. tahitica*.

Hirundo neoxena Gould, 1842 [hi-RUN-do ne-o-ZE-nuh]: 'new visiting swallow', see genus name, and from Greek *neos*, new, and *xenos*, a guest, visitor or stranger. Gould said: '*neoxena* has suggested itself as appropriate, from the circumstance of its appearance throughout the whole of the southern portions of Australia being hailed as a welcome indication of the approach of spring' (Gould 1848).

Asian House Martin (vagrant)

The three *Delichon* species are referred to as house martins, from the widespread habit of the European (and Asian) Northern House Martin (*D. urbicum*) of nesting in or on buildings. This species is widespread in East and South-East Asia.

Delichon dasypus (Bonaparte, 1850) [DE-li-kon DA-si-pus 'shaggy-footed anagram swallow', see genus name, and from Greek *dasu-*, rough, shaggy or hairy, and pous, *foot*. Bonaparte (1850) described the feet as 'densely feathered'.

Red-rumped Swallow (scarce migrant to northern Australia)

For its distinctive rusty rump.

Other names: Daurican Swallow, from the species name.

Cecropis daurica Laxman, 1769 [SE-kro-pis DOW-ri-kuh]: 'tribal Athenian from Dauria', see genus name, and from Dauria or Transbaikal, a mountain region to the east of Lake Baikal in the south of the Russian Federation and part of the huge region of Siberia.

Fairy Martin (breeding resident)

Seems to have been coined by Gould, for his species name.

Other names: Bottle or Retort Swallow, for the elegant bottle-shaped mud nest ('retort' in this sense being more familiar to habitués of chemistry laboratories, as a narrow-necked glass flask); Land Swallow, does not seem to have been much used, though cited by Leach (1911), and rather obscure.

Petrochelidon ariel (Gould, 1842) [pet-ro-KE-li-dohn AH-ri-el]: 'Ariel rock-swallow', see genus name, and probably from Shakespeare's 'aryie spirit' in *The Tempest* given the common name Gould chose, though there are other 17th-century literary uses. The name Ariel and its connected name Riel (and other variants) date from possibly a thousand years before that. It is used as the name of an angel or spirit or hero or even of hell, in translations of the Bible. (See also Lesser Frigatebird, under Fregatidae).

Tree Martin (breeding resident)

Called by Gould (1848) from his species name *Collocalia arborea*, referring to its nesting habitat in hollow trees (as Gould inimitably said, it 'invariably selects the holes of trees for the purpose of nidification'). He was aware of Vieillot's name, and it is not clear why he felt the need to coin a new one.

Other names: Tree Swallow; Australian Tree-martin; Western Tree-martin, for race *neglecta*.

Petrochelidon nigricans Vieillot, 1817 [pet-ro-KE-li-dohn NI-gri-kanz]: 'blackening rock-swallow', see genus name, and from Latin *nigricans*, becoming blackish (*nigrico*, to be blackish, *niger*, black). Vieillot (1816-19) described the bird as having blackish-brown upper parts.

CETTIIDAE (PASSERIFORMES): Cettia bush-warblers and allies

Cettiidae Alström, Ericson, Olsson, & Sundberg, 2006 [se-TEE-i-dee (or tsheh-TEE-i-dee)]: Cetti's family, from the genus name *Cettia*, itself from *Sylvia cetti* Temminck, 1820, named for Francesco Cetti, priest and zoologist based in Sardinia.

The genus

Urosphena Swinhoe, 1877 [yoo-ro-SFEH-nuh]: the wedge-tail family, and from Greek *oura*, tail, and *sphēn/sphēnos*, a wedge.

The species

Asian Stubtail (vagrant)

These warblers do indeed have very stubby tails; 'Asian' here implies mainland Asia, to distinguish from species in Borneo and Timor.

Urosphena squameiceps (Swinhoe, 1863) [yoo-ro-SFEH-nuh skwa-MEH-i-seps] 'scaly-headed wedge-tail', see genus name, and from Latin *squama*, scale and *-ceps*, -headed (*caput*, head).

PHYLLOSCOPIDAE (PASSERIFORMES): leaf-warblers and allies

Phylloscopidae Jerdon, 1873 [fil-lo-SKO-pi-dee]: the Leaf Inspector family, see genus name *Phylloscopus*.

The genus

Phylloscopus Boie, 1826 [fy-LO-sko-pʊs]: 'leaf inspector', from Greek *phullon*, leaf, and *skopeō*, to watch, contemplate, inspect. The birds are very active among tree leaves, looking for insects.

The species

Yellow-browed Warbler (vagrant)

The whitish eyebrow does sort of have a yellowish tinge, though less so than other members of the genus, which tend to look dauntingly similar.

Phylloscopus inornatus (Blyth, 1842) [fy-LO-sko-pʊs in-or-NAH-tʊs]: 'unadorned leaf inspector', see genus name, and from Latin *inornatus*, unadorned, without ornament.

Dusky Warbler (vagrant)

Slightly duskier than a cohort of other dusky warblers in the genus. From the species name.

Phylloscopus fuscatus (Blyth, 1842) [fy-LO-sko-pʊs fʊs-KAH-tʊs]: 'dusky leaf inspector', see genus name, and from Latin *fuscus*, dark or dusky.

Willow Warbler (vagrant)

It seems as though Birch Warbler could be more accurate, but it certainly visits willows too!

Phylloscopus trochilus (Linnaeus, 1758) [fy-LO-sko-pʊs tro-KI-lʊs]: 'very small leaf inspector', see genus name, and from Greek *trochilos*, Latin *trochilus*, a small bird, identity much pondered upon (see Arnott 2007). Options include wren, sanderling, and others much larger.

Eastern Crowned Warbler (vagrant)

Refers to its east Asian origins and the pale crown stripe, from the species name.

Phylloscopus coronatus (Temminck & Schlegel, 1847) [fy-LO-sko-pʊs ko-ro-NAH-tʊs]: 'crowned leaf inspector', see genus name, and from Latin *corona*, crown or halo.

Kamchatka Leaf Warbler (vagrant)

For the Kamchatka Peninsula in eastern Siberia, though it also breeds in Japan.

Phylloscopus examinandus Stresemann, 1913 [fy-LO-sko-pʊs ek-sa-mi-NAN-dʊs]: 'careful leaf inspector', see genus name and from Latin *examinare*, to weigh, consider or examine.

Arctic Warbler (vagrant)

In part from the species name; it has a vast northern range, including within the Arctic Circle, from Scandinavia east to Alaska.

Phylloscopus borealis (Blasius, 1858) [fy-LO-sko-pʊs bo-re-AH-lis]: 'northern leaf inspector', see genus name, and from Latin (and Greek) *Boreas*, the north wind, hence the North (cf. *aurora borealis*, the northern lights).

ACROCEPHALIDAE (PASSERIFORMES): reed-warblers

Acrocephalidae Salvin, 1882 [ak-ro-se-FA-li-dee]: the Peak-head family, see genus name *Acrocephalus*.

'Reed-Warbler' arose as a name in England (for the Eurasian Reed Warbler *A. scirpaceus*), though it has only been known since the early 19th century.

The genus

Acrocephalus Naumann & Naumann, 1811 [a-kro-SE-fa-lʊs]: 'peak-head', perhaps from Greek *akron*, a peak or top (cf. *akros*, highest or furthest), and *kephalē*, head. There has been speculation (BOU 1915) that the Naumanns mistakenly thought Greek *akros* meant Latin *acutus* (pointy), but if *akron* was what they had in mind as the origin, it seems reasonable to us as a way to describe the slight raising of crown feathers into a tiny crest.

The species

Oriental Reed Warbler (vagrant)

From the species name.

Other names: Great or Eastern Reed-Warbler.

Acrocephalus orientalis (Temminck & Schlegel, 1847) [a-kro-SE-fa-lʊs o-ri-en-TAH-lis]: 'eastern peak-head', see genus name, and from Latin *orientalis*, eastern.

Australian Reed Warbler (breeding resident)

Endemic to Australia, plus nearby parts of New Guinea (though it has only recently been recognised as such).

Other names: Australian Reed-Warbler, the preferred form in Australia; Reed Warbler, by Gould (1848), who considered it to be a separate species; Long-billed Marsh Warbler, by Gould for his Western Australian 'species' *Calamoherpe longirostris*, not now recognised even as a race; hence also Long-billed Reed-Warbler; Clamorous Reed-Warbler, long used when it was believed to be a race of the Asian species *A. stentoreus*; Great Reed-Warbler, a strange one, given that this is an alternative name for Oriental Reed-Warbler (see previous species); Reed-lark, Reed-bird, Swamp Tit, Nightingale, Water Sparrow.

Acrocephalus australis (Gould, 1838) [a-kro-SE-fa-lʊs ows-TRAH-lis]: 'southern peak-head', see genus name, and from Latin *australis*, southern.

LOCUSTELLIDAE (PASSERIFORMES): grassbirds, songlarks and allies

Locustellidae Bonaparte 1854 [lo-kʊs-TEL-li-dee]: the Grasshopper family, from the genus name *Locustella* Kaup, 1829, from Latin *locusta*, locust or grasshopper (or lobster – perhaps not relevant here), for the sound the bird may make.

'Grasshopper warbler', from the fact that the genus *Helopsaltes* is noted for grasshopper-like trilling calls, though not all species conform.

'Grassbird' was coined by Gould during the mid 19th century, relating to the birds' strong correlation with long grass habitats; in 1848 he referred to Little Grassbird as Grass-loving Sphenœacus (the latter being his genus name) but in 1865 was using Grass-bird for both members of his genus (they are now in separate genera). It was sufficiently established by the end of the century for Edward Morris (1898) to include it in his *Dictionary of Austral English*. Although both Australian species and four other genus members are found well to the north of Australia, Newton (1896) in his great *Dictionary of Birds* only referred to Gould's usage. Today the name is used for all species recently or currently in *Megalurus*.

'Songlark' was apparently also a home-grown name (though clearly tightly based on the familiar Skylark of Europe). Although Gould (1848, 1865) considered it a little too imprecise (using, as was his wont, the genus name as group common name), he noted that 'Colonists' used 'Singing Lark' for Rufous Songlark. Doubtless that evolved in common usage to 'songlark', and Morris recorded this without further comment in 1898.

The genera

Helopsaltes Alström *et al*., 2018 [heh-lo-SAL-tes]: 'marsh musician', from Greek *helos*, marshy ground, and *psaltēs*, a harper, or musician playing a stringed instrument. The song elements of these birds were found by Alström *et al*. to be more complex and varied than those of others they were being compared with.

Poodytes Cabanis, 1850 [poh-oh-DY-tes]: 'grass plunger' from Greek *poa*, grass, and *duō*, get into or plunge. Cabanis uses the German *Grasschlüpfer*, meaning one who slips through the grass.

Cincloramphus Gould, 1838 [sin-klo-RAM-fʊs]: 'thrush-bill', from Greek *kinkhlos* or *kikhlē*, thrush, and *rhamphos*, bill. See genus *Cinclosoma* under Psophodidae.

The species

Gray's Grasshopper Warbler (vagrant)

See family introduction. George Gray headed the ornithology department of the British Museum for many years from 1831, and published widely, especially with his brother John; he described this species among many others.

Helopsaltes fasciolatus (Gray GR, 1861) [heh-lo-SAL-tes fa-shi-o-LAH-tʊs]: 'banded marsh musician', see genus name, and from Latin *fasciola*, a small bandage, from *fascia*, a band or bandage. Gray described the bird's 'throat and breast yellowish-white, banded with dusky'. 'Banded' is a bit of an exaggeration: the bird appears only lightly streaked.

Pallas' Grasshopper Warbler (vagrant)

See family introduction, though this species is not very grasshopper-sounding; German ornithologist Peter Pallas named the species and, given that the type locality was in the Transbaikal region of eastern Russia, which Pallas pioneered, it is probable that he collected it too.

Helopsaltes certhiola Pallas, 1811 [heh-lo-SAL-tes sehr-thi-OH-luh]: 'little creeping marsh musician', see genus name, and from a diminutive of genus name *Certhia*, treecreeper (from Greek *kerthios*, a treecreeper). It was its appearance rather than its behaviour that led Pallas to name it thus. A small mystery attaches to this one: *Zoographia Rosso-Asiatica* vol. 1 p. 509 is the reference given, but the year of publication appears to be 1831 rather than 1811. This was debated for many years and, though the decision was made by the International Commission on Zoological Nomenclature in 1954 to go with 1811, it still raises questions (Banks and Browning 1995).

Middendorff's Grasshopper Warbler (vagrant)

See family introduction. For Alexander Middendorf, a Russian zoologist who worked widely across the Russian Arctic from west to east, and who named this species. Given the species name, it is probable that he also collected it.

Helopsaltes ochotensis (Middendorff, 1853) [heh-lo-SAL-tes o-koh-TEN-sis]: 'Okhotsk marsh musician', see genus name, and from the Sea of Okhotsk (between Eastern Russia and the Kamchatka Peninisula) plus the suffix *-ensis*, usually indicating the place of origin of the type specimen.

Spinifexbird (breeding resident)

The name was coined some time between the RAOU's first *Official Checklist of the Birds of Australia* in 1913 and the second in 1926, when the name apparently first appeared. It refers to the bird's near-exclusive habitat in the *Triodia* hummock grasslands of inland Australia.

Other names: Desert-bird, from the former genus name *Eremiornis*; Carter's Desert-bird, from the specific name. Both were in general use until 1926.

Poodytes carteri (North, 1900) [poh-oh-DY-tes CAR-ter-ee]: 'Carter's grass-plunger', see genus and for Thomas Carter, a one-eyed Yorkshire adventurer who came to Western Australia from England in 1887 and bought a station near Cape Range, virtually commuting between Yorkshire and Australia (including a trip to England to get married) and doing much collecting, including this species in 1899. He contributed many valuable observations of Australian bird life and helped found the RAOU.

Little Grassbird (breeding resident)

See family introduction. Much smaller than the Tawny Grassbird, which was the obvious basis of comparison, though now placed in a separate genus.

Other names: Grassbird; Little Marshbird (see Tawny Grassbird); Little Reed Bird; Marsh Warbler; Striated Grassbird – unlike the Tawny Grassbird it is heavily streaked below, but this was a name particularly applied by Mathews (1913) to one of his subspecies, *striatus*, not now recognised; other such names of his, now sunk in history, were Flinders Island, Victorian, Southern and Kangaroo Island Grassbird; he also coined Tasmanian, for race *gramineus*, Allied, for race *goulburni* and Dark, for race *thomasi*, which are all still recognised. Regatta Bird, cited by HANZAB (1990–2006, vol. 7b) but not otherwise evident, and not at all obvious.

Poodytes gramineus (Gould, 1845) [poh-oh-DY-tes gra-MI-ne-ʊs]: 'grassy grass-plunger', see genus name, and from Latin *gramineus*, of grass or grassy, which can refer to colour (green) but here refers to the habitat. This bird clearly favours grass!

Brown Songlark (breeding resident)

See family introduction. Used by Gould (1848), in conjunction with the genus name. A name sufficiently uninspired as to be unarguable.

Other names: Brown Cinclorhamphus, used by Gould (1848); Black-breasted Lark or Songlark, more à propos though only in regard to the male – Gould used the name (with Cinclorhamphus), for his species *cantillans* from the south, west and north of the continent, though it is not now recognised at any level; Cock-tailed Lark or Songlark, for the distinctive stance of the male; Australian Skylark; Brown Singing-lark (see family introduction); Harvest Bird, for its habit of nesting and foraging in crops; Western, Northern and Southern Brown Songlark, names for races named by Mathews (1913) but no longer in use; Skit-scot-a-wheeler or Cock-chit-a wheeler, a nice rendition of the aerial courtship call – though it's not easy to find an example of the latter usage outside the HANZAB (1990–2006, vol. 7b) citation, that's to be expected for an essentially oral tradition.

Cincloramphus cruralis (Vigors & Horsfield, 1827) [sin-klo-RAM-fʊs kroo-RAH-lis]: 'leggy thrush-bill', see genus name, and from Latin *crus*, leg, shank or shin. The long dangling legs are most noticeable during the 'parachute' descent of the bird from its high climb into the sky.

Rufous Songlark (breeding resident)

See family introduction. Although the name isn't immediately obvious, it refers to the bright rufous rump, partly covered by the wings. It's a useful distinguisher from the Brown Songlark. More specifically it was taken directly from the original species name, *rufescens* (see species name).

Other names: Rufous-tinted Cinclorhamphus, used by Gould (1848), straight from the then species name *rufescens*; Rufous-tinted or Red-rumped Songlark; Rufous-rumped Singing-lark (see 'songlark'); Eastern, Western, North-western and Northern Rufous Songlark, all for subspecies coined by Mathews but no longer recognised.

Cincloramphus mathewsi Iredale, 1911 [sin-klo-RAM-fʊs MATH-yooz-ee]: 'Mathews' thrush-bill', see genus name, and for Gregory Mathews, ubiquitous Australian ornithologist. Arriving at this name was perhaps a more than usually complicated business: Vigors and Horsfield described the bird in 1827 as *Anthus rufescens*. Gould realised that it wasn't a pipit (*Anthus*) and coined the new genus, calling it *Cincloramphus rufescens*. In 1911 Mathews' amanuensis Tom Iredale described the Western Australian form (not now recognised) as *C. r. mathewsi*. The following year Mathews fortuitously realised that *Anthus rufescens* had been a preoccupied name in 1827 (Temminck had used the name in 1820) and so it hadn't been available for Gould's use, rendering *Cincloramphus rufescens* invalid too. No subspecies had been suggested until Iredale's, so his subspecies name *mathewsi* was the next name in line for the species. As we say, a remarkably lucky outcome for Mathews ... and imagine his surprise! (Van Gasse 1999).

Tawny Grassbird (breeding resident)

See family introduction. This was the name used by Gould (1848), though he seems not to have been making comparisons, because the only other grassbird he knew was Little.

Other names: Tawny Sphenœacus, used by Gould (1848), from the genus to which he ascribed it (*Sphenoeacus* is now recognised as containing just one South African species); Grassbird; Tawny or Rufous-capped Grass-warbler, the last being also the name used for the South African *Sphenoeacus*; Tawny Marshbird, for the habitat; Rufous-capped Grassbird – this name has rarely been used in Australia, being more applied to races to the north.

Cincloramphus timoriensis Wallace, 1864 [sin-klo-RAM-fʊs ti-mor-i-EN-sis]: 'Timorese thrush-bill', see genus name, and from Timor plus the suffix *-ensis*, usually indicating the place of origin of the type specimen.

CISTICOLIDAE (PASSERIFORMES): cisticolas and allies

Cisticolidae Sundevall, 1872 [sis-ti-KO-li-dee]: the Basket-dweller family, see genus name *Cisticola*.

'Cisticola' is taken directly from the genus name, but was only adopted in Australia relatively recently; Slater (1974) was one of the first, and the RAOU formalised it 4 years later (RAOU 1978). This is only peripherally an Australian group, with some 150 species right across southern Asia and Europe and throughout Africa; both Australian species have ranges far beyond our shores.

The genus

Cisticola Kaup, 1829 [sis-TI-ko-luh]: 'basket-dweller', from Greek *kistē*, a basket, and *-cola*, dweller, for the striking nest. The genus name is taken from Temminck's (1820) name for the Zitting Cisticola *Sylvia cisticola*. Temminck stressed that the difference from similar species is in the nest, and Kaup commented on its funnel shape. Gould (1837) gave a wonderful if somewhat hyperbolic description of this (then *Salicaria cisticola*), saying that the bird, though 'incapable, from its small size, of entwining the larger reeds … avails itself of the tall blades and stalks of grass, among which it places its nest'. It makes the nest 'by piercing each blade, and drawing the whole together by means of cottony threads, secured at each perforation by a knot so ingeniously executed as to appear the work of reason'. Well! (We note that others have put forward different derivations for 'cisticola', but consider ours well supported both linguistically and in terms of the citations used.)

The species

Zitting Cisticola (breeding resident)

For the male's courtship call; almost certainly named in Africa where the species is very extensively found. There are scores of African species, many of them named for calls (Winding, Rattling, Wailing, Trilling, Bubbling, Chattering, Croaking, Whistling, etc.).

Other names: Lineated Warbler, used by Gould (1848) who called it *C. lineocapilla*; Fantail-warbler, a widely used name in Britain (the species is common in southern Europe – the type specimen is from Sicily), Common Fantail-warbler, cited in CSIRO (1969) and HANZAB (1990–2006, vol. 7b) but not much used in Australia (where it is far the less common of the two species); Grass-warbler (RAOU 1913); Streaked Grass-warbler, not much used, but appears in the Australian literature (Lavery and Seton 1967); Streaked Cisticola or Fantail-warbler.

Cisticola juncidus (Rafinesque, 1810) [sis-TI-ko-luh YʊN-si-dʊs]: 'reedy basket-dweller', see genus name, and from Latin *juncus*, a rush: wetlands are the bird's habitat, and *Juncus* is the name of a large genus of rushes.

Golden-headed Cisticola (breeding resident)

Again for a male-only and breeding-only feature. Although the species is found throughout South-East Asia, according to HANZAB (1990–2006, vol. 7b) this name (used beyond Australia as well) is based on *C. ruficeps* that Gould published – he believed that the breeding birds were of a different species from the non-breeding ones that he recognised as *C. exilis*. However, this is not a close translation of *ruficeps* – Gould used the more accurate Rufous-capped – so we are not totally convinced.

Other names: Exile Warbler, by Gould (1848) from the species name; Rufous-capped Warbler, see common name; Square-tailed Warbler, again by Gould for another 'false species' based on *C. exilis*, which he called *C. isura*; Golden-headed Fantail-warbler (preferred by RAOU 1926 and Cayley 1931),

Warbler or Grass-warbler; Barley-bird, Cornbird, Grassbird, Grass-warbler, all for its fondness for crops; Tailorbird, for its beautifully crafted nest (see genus name); Bright-headed or Bright-capped Cisticola, mostly Asian names.

Cisticola exilis (Vigors & Horsfield, 1827) [sis-TI-ko-luh EK-si-lis]: 'slender basket-dweller', see genus name, and from Latin *exilis*, small, slender or feeble. Note that there is an English word 'exile' meaning the same, now obsolete but, according to the *Oxford Dictionary*, still in use in the late 18th century. We venture to think that Gould was still using the word in this sense in his common name.

ZOSTEROPIDAE (PASSERIFORMES): silvereyes, white-eyes

Zosteropidae Bonaparte, 1853 [zos-teh-RO-pi-dee] the Girdle-eye family, see genus name *Zosterops*. The authority for this family name, Bonaparte, 1853, is often mentioned as 'disputed', and the basis for this authority is not clear. The earliest use of 'Silver-eye' reported by the *Australian National Dictionary* is from 1862 in South Australia, though Gould in 1865 was still using Zosterops as the common name. Newton (1896), however, reported that the name arose in New Zealand, from when the Australian Silver-eye was first reported there in substantial numbers in 1856. Some 80 other species of *Zosterops* from Africa, Asia and the Pacific are known as white-eyes; the allusion in each case is to the ring of white feathers around the eyes.

The genus

Zosterops Vigors & Horsfield 1827 [ZOS-teh-rops]: 'girdle-eye', from Greek *zōstēr*, a girdle or warrior's belt, and ōps, eye.

The species

Ashy-bellied White-eye (breeding resident, Torres Strait)

A recent and needed name amendment; this name clearly distinguishes it from the closely related Yellow-bellied White-eye (*Z. chloris*), found scattered on islands further north and west, with which it was often lumped and shared the comprehensively unhelpful moniker Pale.

Other names: Pale White-eye, see common name. Pale-bellied White-eye, refers to Australian (Torres Strait Islands) race *albiventris* – a more descriptive name; Pale or Pale-bellied Silvereye.

Zosterops citrinella Bonaparte, 1850 [ZOS-teh-rops si-tri-NEL-luh]: 'yellow finch girdle-eye', see genus name, and *citrinella*, which is what Mediaeval Italians called a yellow finch (Aldrovandus 1600). Ultimately from Greek *kitrea/kitron*, the citron tree and its fruit (*Citrus medica*), and Latin *citrus or citreus*, which Pliny said refers to an African tree with fragrant wood (possibly the cedar *Cedrus atlantica*) and to the citron tree. From the second (and Mediaeval Latin *citrinus*) comes our own word 'citrus', French *citron* meaning lemon (and related words). (In case you're wondering, *citrinella*, as used here, is a noun in apposition, so not agreeing with the masculine *Zosterops*.)

Christmas White-eye (endemic to Christmas Island; introduced to Cocos (Keeling) Islands)

Restricted (other than through translocation) to Christmas Island.

Other names: Christmas Island White-eye or Silvereye.

Zosterops natalis Lister, 1889 [ZOS-teh-rops na-TAH-lis]: 'birthday girdle-eye', see genus name, and from Latin *natalis*, birthday, for the birth of Christ, and hence for the place-name.

Canary White-eye (breeding resident)

For the colour.

Other names: Yellow White-eye (the preferred name in Australia) or Silvereye; Yellow Zosterops, used by Gould (1848, 1865); Carnarvon White-eye, for north-western populations; Golden White-eye; Mangrove White-eye, for its preferred habitat. Gulliver's White-eye, from the name *Z. gulliveri*, described by Ramsay and Castelnau (François Louis Nompar de Caumont Laporte, Comte de Castelnau to his friends) in 1877 from the Gulf of Carpentaria. It was later demoted to a subspecies but is now not recognised as distinct from *Z. l. luteus*. It was named for Thomas Gulliver who 'discovered' it (we assume this is a euphemism for 'collected') while working in the remote telegraph station at Normanton. He had come to Australia to work as a botanical collector for Ferdinand von Mueller and later established some noted gardens around Townsville; he died aged 84 when he fell into a goldfish pond in his garden at night.

Zosterops luteus Gould, 1843 [ZOS-teh-rops LOO-te-ʊs]: 'yellow girdle-eye', see genus name, and from Latin *luteus*, golden or saffron yellow. The bird's underparts are completely yellow.

Silvereye (breeding resident)

See family introduction.

Other names: Grey-backed Zosterops (used by Gould 1848, 1865), relating to the species name *dorsalis* that he was using; Eastern, Grey or Grey-backed Silvereye, formerly used to distinguish eastern birds from western ones with green backs, which were regarded as a separate species; Grape-eater, for one of its less universally admired characteristics; Blight-bird, Blightie or Spectacled Blight-bird, which might be supposed a variation of the previous but in fact refers to its habit of eating aphids, scale insects and other plant pests – it seems to have arisen in New Zealand; Grinnell or Little Grinnell, a name widely used on Norfolk and Lord Howe islands, but its origin is a mystery – we wonder if it derives from 'green'; Silve, a long-used affectionate abbreviation, cited for instance in Leach (1911); Yellow-vented White-eye, for coastal and island Queensland races *ramsayi* and *chlorocephala*; Ring-eye, Button-eye, Girdle-eye, Waxeye; Rusty-sided Warbler, Caerulean or Bluish-breasted Creeper, all names used by Latham (1801, 1823), the last two from his species name *Z. caerulescens*, though the basis of that is obscure. Until recently race *gouldi* of south-western Australia was regarded as a separate species *Z. gouldi*, Western Silvereye, also known as Green-backed Silvereye and White-eye, Greenie and Rabbit Island White-eye – the last refers to the small rocky outcrop Rabbit Island near Albany, where in 1904 Guy Shortridge (a collector more associated with South Africa) shot a silvereye that supposedly differed from mainland birds in minor plumage characters and which was named by Grant (William Ogilvie-Grant of the British Museum?) *Z. shortridgei* in 1909. It is not recognised now as a taxon at any level.

Zosterops lateralis (Latham, 1802) [ZOS-teh-rops la-te-RAH-lis]: 'sided girdle-eye', see genus name, and from Latin *latus/lateralis*, side, named by Latham for his Rusty-sided Warbler (see 'Other names'). The birds' flanks vary in colour according to which of several races they belong to.

Slender-billed White-eye (Norfolk Island endemic)

From Gould's species name; the bill is long rather than particularly slender, but the effect is the same.

Other names: Slender-billed or Long-billed Silvereye; Long-billed White-eye; Grinnell (see Silvereye).

Zosterops tenuirostris Gould, 1837 [ZOS-teh-rops te-noo-i-ROS-tris]: 'slender-billed girdle-eye', see genus name, and from Latin *tenuis*, slender or thin, and *rostrum*, bill. See common name.

Robust White-eye (Lord Howe Island, extinct)

From the species name. This was a big white-eye – although data is limited, the wing was some 10 mm longer than that of the Silvereye.

Other names: Big Grinnell (see Silvereye); Big, Robust or Powerful Silvereye.

Zosterops strenuus Gould, 1855 [ZOS-teh-rops STREH-nʊ-ʊs]: 'strong girdle-eye', see genus name, and from Latin *strenuus*, which has many meanings including vigorous, active, quick and prompt. Gould (1855) stressed that it is 'by far the largest species of the genus yet discovered'.

White-chested White-eye (Norfolk Island endemic, may be extinct)

Perhaps not the most striking characteristic of this relatively huge silvereye (and odd in that it doesn't quite reflect the species name), but accurate enough.

Other names: White-chested Silvereye.

Zosterops albogularis Gould, 1837 [ZOS-teh-rops al-boh-gʊ-LAH-ris]: 'white-throated girdle-eye', see genus name, and from Latin *albus*, dull white, and *gula*, the throat or gullet.

STURNIDAE (PASSERIFORMES): starlings

Sturnidae Rafinesque, 1815 [STOOR-ni-dee]: the Starling family, see genus name *Sturnus*.

'Starling' comes from the original name stare, derived from Old English *stær*; the Latin *sturnus* was probably the ultimate root. Starling was first used as a diminutive (as in darling or duckling), referring to young birds; somehow it came to supplant the primary word. It appears as the first-choice name in the 16th century, but wasn't unchallenged until the 19th century.

'Myna' (also mynah, mina, mino, miner, etc.) derives from the Hindi name for the birds, *maina*. According to the *Oxford Dictionary* it first appeared in English (as mino) in 1769; it was a popular cage bird because of its mimicking abilities and of course trade with India was strong.

The genera

Aplonis Gould, 1836 [ap-LOH-nis]: 'plain bird'. There has been a long-running debate about *Aplornis* and *Aplonis*. Our summary must needs be simplistic: *Aplornis* gets votes because it was published 2 weeks before *Aplonis* (albeit with deficiencies) in a report in *The Analyst* about Gould's initial presentation of the name to a meeting of the Zoological Society of London – a fact unearthed by the indefatigable Gregory Mathews in 1938. *Aplonis* gets votes because it is the only spelling Gould ever used, and it has been used by almost everyone else for the last 175 years (Schodde and Bock 2009). Either could make some sense: perhaps from Greek *haploos*, simple or plain, +/– *ornis*, bird. However, since the first edition of this book most of the heat has left the argument and *Aplonis* seems to have carried the day.

Acridotheres Vieillot, 1816 [ak-rid-DO-the-reez]: 'grasshopper-hunter', from Greek *akris*, grasshopper, locust or cricket, and *-theras*, hunter.

Agropsar Oates, 1889 [ag-ro-SAR]: 'field starling', from Greek *agros*, field and *psar*, starling. Like Mynahs, Oates (1889) remarked, it likes feeding on the ground and in trees.

Pastor Temminck, 1815 [PAS-tor]: 'herdsman', from Latin *pastor*, herdsman or shepherd. Named thus by Temminck because he remarks that this genus is particularly keen on following cattle, feeding on the insects on their skin and in their droppings.

Sturnus Linnaeus, 1758 [STOOR-nʊs]: 'starling', from Latin *sturnus*, meaning just that.

The species

Metallic Starling (breeding resident)

See family introduction. From the species name.

Other names: Shining Starling, the name used by Gould (1865) and still the preferred name until recently for the iridescent plumage; RAOU (1978) decreed the change, pointing out that the Asian Glossy Starling *A. panayensis* is also referred to as Shining Starling (though we can find very little evidence of this usage, and the ranges nowhere come close to each other). Australian Shining Starling; Colonial Starling, for its big nesting colonies; Weaver Bird, apparently a local usage, for its hanging grass nests; Whirlwind-bird, cited by Alec Chisholm in a column in the *Melbourne Herald* of 16 January 1943, where he comments that the starlings are 'called 'Whirlwind-birds' by Queenslanders' but doesn't explain why.

Aplonis metallica (Temminck, 1824) [ap-LOH-nis me-TAL-li-kuh]: 'metallic plain bird', see genus name (the meaning of which seems a bit silly in relation to this bird, which is anything but plain), and from Latin *metallicus*, made of metal, metallic (cf. Latin *metallum*, metal).

Singing Starling (Torres Strait Islands)

See family introduction. From the species name, though possibly a misunderstanding of it.

Aplonis cantoroides (Gray GR, 1862) [ap-LOH-nis kan-to-ROY-dehz]: 'simple bird like *cantor*', see genus name, and from species name *cantor*, once applied to the Asian Glossy Starling, *Aplonis panayensis* (from Panay Island in the Philippines), with Latin *-oides*, resembling.

Tasman Starling (extinct; formerly Lord Howe and Norfolk islands)

See family introduction. For its Tasman Sea range, being found (until the 1920s) only on Norfolk and Lord Howe islands.

Other names: Lord Howe, Norfolk Island or Sooty Starling; Red-eye.

Aplonis fusca Gould, 1836 [ap-LOH-nis FʊS-kuh]: 'dusky simple bird', see genus name, and from Latin *fuscus*, dark, dusky.

Common Myna (introduced breeding resident)

A ubiquitous and often abundant species in its native range (not to mention its numerous overseas ranges!).

Other names: Indian Myna, for an important part of its range, and the origin of Australian birds; Calcutta Myna; Common or Indian Mynah. Chocolate Dollarbird – we can readily conceive how this might have arisen, as in flight the myna's white wing spots can quite resemble those of the Dollarbird, and the bird is often described as chocolate brown; the problem is that no actual usage outside the HANZAB (1990–2006, vol. 7b) citation is evident but this could simply mean that it has been restricted to oral use, and there is no doubt that there's a huge oral tradition associated with this species. For instance, Rat-with-Wings or Canetoad-with-Wings are often used disparagingly to associate the bird with love-to-hate terrestrial feral pests. The remainder of the names cited all derive from a single article that outlined a brief history of the birds on the Darling Downs (Walker 1952). Walker asserted that they were 'common names used in this area'. Chocolate Bird, presumably for the colour – an association with Chocolate Dollar-bird cannot be ruled out; Thynne's Bird, for the Hon. A.J. Thynne, Queensland Agriculture Minister, who was purported to have introduced the birds locally; Tick-bird, for its alleged propensity for eating ticks; White-wings, for the obvious wing flashes in flight; Tasmanian Starling, a particularly curious one because Tasmania is the only eastern state where it has not become established – perhaps it was a general term for 'somewhere down south'!

Acridotheres tristis (Linnaeus, 1766) [ak-rid-DO-the-reez TRIS-tis]: 'sad grass-hopper-hunter, see genus name, and from Latin *tristis*, sad, gloomy, downcast – because of its colour? Or maybe because the bird thinks nobody likes it? (See Rat-with-Wings under common name). The bird may be more popular in parts of Asia, where it helps control pest insects on farming land (HANZAB 1990–2006, vol. 7b).

Daurian Starling (vagrant)

Other names: Purple-backed Starling, because of a most inconspicuous purple nape patch.

Agropsar sturninus (Pallas, 1776) [ag-ro-SAR STOOR-ni-nʊs]: 'starling-like field starling', see genus name, and from Latin *sturninus*, like a starling, speckled.

Chestnut-cheeked Starling (vagrant)

See family introduction. It is indeed, and they are quite striking on the white head.

Agropsar philippensis (Forster JR, 1781) [ag-ro-SAR fi-li-PEN-sis]: 'Philippines field starling', see genus name, and for the Philippines plus the suffix *-ensis*, usually indicating the place of origin of the type specimen. Forster included the bird in his *Indische Zoologie* in a list of birds authored by Thomas Pennant. In that list, against the name of *Motacilla philippensis,* is a reference to Brisson III 446. This bird is what Brisson called '*le Grand Traquet des Philippines*' or '*Rubetra philippensis major*'.

Rosy Starling (vagrant)

For the subtle pink wash on the undersides and back.

Other names: Rose-coloured Starling.

Pastor roseus (Linnaeus, 1758) [PAS-tor RO-se-ʊs]: 'rosy herdsman', see genus name, and from Latin *roseus*, rose-coloured.

Common Starling (introduced breeding resident)

Certainly the commonest starling in Britain, whence comes the name. (It's also the commonest in Australia, but that's coincidental.)

Other names: Starling or European, Northern or English Starling.

Sturnus vulgaris Linnaeus, 1758 [STOOR-nʊs vʊl-GAH-ris]: 'common starling', see genus name, and from Latin *vulgaris*, of the people, common.

TURDIDAE (PASSERIFORMES): thrushes

Turdidae Rafinesque, 1815 [TOOR-di-dee]: the Thrush family, see genus name *Turdus*.

'Thrush' seems to go back to Old English, with a form like *þrysce* or *þræsca*; Lockwood (1984) suggested a pre-Germanic origin in the Welsh *tresglen*. Although he didn't say so, this form would have suggested a plural (e.g. oxen; see also 'gull' under Laridae) to the neighbours across the border and a false singular might have been created by dropping the 'en'. Thrush appears in its current form (give or take an 's' or so) by the 15th century.

'Blackbird' appears in print in the late 15th century, for the obvious reason, and was used alongside the older 'Ouzel' for another 200 years, when it began to take over. It is interesting to note that until the beginning of the 14th century 'bird' referred to chicks (nestlings), after which it began to be applied to any small bird. It was not until the late 18th century that it became a general term for birds; until then 'fowl' was used for larger birds.

The genera

Geokichla Müller, S, 1836 [dje-oh-KIK-luh]: 'ground thrush', from Greek *geo*, earth, and *kichlē*, thrush.

Zoothera Vigors, 1832 [zoh-o-THEH-ruh]: 'animal-hunter', from Greek *zōon*, a living being or animal, and *thēra*, hunting.

Turdus Linnaeus, 1758 [TOOR-dʊs]: 'thrush', from Latin *turdus*, thrush or fieldfare.

The species

Siberian Thrush (vagrant)

See family introduction. From the species name, and its huge breeding range across central Siberia.

Geokichla sibirica (Pallas, 1776) [dje-oh-KIK-luh si-BEE-ri-kuh]: 'Siberian ground-thrush', see genus and common names.

Russet-tailed Thrush (breeding resident)

See family introduction. Until Julian Ford investigated the situation (Ford 1983), it was assumed that Australian *Zoothera* thrushes were part of the huge *Z. dauma* complex, White's or Scaly Thrush, stretching across north-central and South-East Asia. He separated this species and the next (both as full species) from *Z. dauma*, and coined the currently used common names for each. Mathews (1913) used Russet-tailed Ground-thrush for race *heinei*, which Ford raised to species status. It is not a very helpful field character, though the rufous rump is somewhat less scalloped than that of the Bassian Thrush.

Other names: Russet-tailed Ground-thrush.

Zoothera heinei (Cabanis, 1850) [zoh-o-THEH-ruh HIE-nuh-ee]: 'Heine's animal-hunter', see genus name, and for Ferdinand Heine (1809–94), a German ornithologist whose large private collection of birds was catalogued and documented by Cabanis (Museum Heineanum).

Bassian Thrush (breeding resident)

See family introduction and previous species. Bassian is the biogeographical term for the temperate zone of south-eastern and south-western Australia; it derives from Bass Strait, named for George Bass, navigator and coastal explorer, colleague of Matthew Flinders; Bassian Thrush is found well south of Russet-tailed Thrush (though they also overlap).

Other names: Mountain Thrush, used by Gould (1848), as used by the 'Colonists of Van Diemen's Land'; Ground-thrush or Australian Ground-thrush, for its terrestrial habits (like most thrushes), recorded by Morris (1898); Spotted Ground-thrush, Speckled or Scaley Thrush, for the scalloped plumage (see species name) – Scaly Thrush and the next name come from the time when the species was regarded as part of the *Z. dauma* complex, see previous species; White's Thrush, for the famous English naturalist and curate Gilbert White – the name was applied by John Latham during the period when he saw no need for the newfangled Latin name fad; Large-billed Ground-thrush, from Gould's name (*Oreocincla macrorhynca*) (which he later dropped) for the Tasmanian birds; King Thrush, cited by Leach (1911) but of unknown origin; Broadbent Ground-thrush, from race *cuneata*, formerly described as a species (*Geocincla*) *cuneata* by Charles De Vis of the Queensland Museum, from a specimen collected in north-east Queensland by museum collector Kendall Broadbent. Other names sometimes cited – e.g. Golden, Tiger, Small-billed Mountain Thrush or Ground-thrush – refer to races of *Z. dauma* outside of Australia, and have not been used in Australia to any significant degree.

Zoothera lunulata (Latham, 1802) [zoh-o-THEH-ruh loo-nʊ-LAH-tuh]: 'crescent animal-hunter', see genus name, and from Latin *lunulatus*, crescent-shaped (*luna*, moon, with *-ulus*, diminutive ending), for the bird's markings (see 'Other names').

Common Blackbird (introduced breeding resident)

See family introduction. There are three other black thrushes referred to as 'blackbirds', all found in India or the Himalayas; although this species is undoubtedly commoner in Europe, it isn't at all common where the other blackbirds are found.

Other names: English or European Blackbird; Merle, a traditional English name, though more a poetic than a commonly used name, based on the species name – perhaps it had some usage here in earlier days of the bird's introduction.

Turdus merula Linnaeus, 1758 [TOOR-dʊs ME-rʊ-luh]: 'blackbird thrush', see genus name, and from Latin *merula*, blackbird (cf. French *merle*).

Island Thrush (Christmas Island; extinct on Heard and Norfolk islands)

See family introduction. There are a huge number of races, all (other than a couple in Papua New Guinea) on islands to the north of Australia.

Other names: Christmas Island Ground-thrush; other names cited refer to non-Australian races.

Turdus poliocephalus Latham, 1802 [TOOR-dʊs po-li-o-SE-fa-lʊs]: 'grey-headed thrush', see genus name, and from Greek *polios*, grey, and *kephalē*, head.

Eyebrowed Thrush (vagrant)

See family introduction. The white eyebrows are unusual in the genus, though not unique.

Other names: Eye-browed Thrush.

Turdus obscurus (Gmelin JF, 1789) [TOOR-dʊs ob-SCOO-rʊs]: 'dusky thrush', see genus name and from Latin *obscurus*, dark or dusky.

Song Thrush (introduced breeding resident)

See family introduction. Arose in England in the 17th century and soon became dominant – sounds like an implied snub to the choral prowess of other thrushes.

Other names: English or European Thrush or Song Thrush; Mavis or Throstle, both Old English names (Throstle pre-dated Thrush, and was used alongside it to modern times) – perhaps the names were used in Australia in the first years after its introduction.

Turdus philomelos Brehm CL, 1831 [TOOR-dʊs fi-lo-ME-los]: 'nightingale thrush', see genus name, and from Greek *philo-*, loving, and *melos*, song or musical phrase, and the myth of Philomela, who was turned into a nightingale after some particularly ghastly family experiences (see *Hydroprogne* under Laridae).

MUSCICAPIDAE (PASSERIFORMES): Old World flycatchers

Muscicapidae Fleming 1822 [mʊs-ki-KA-pi-dee]: the Flycatcher family, see genus name *Muscicapa*.

See Turdidae for 'thrush'; rock thrushes are not now normally regarded as true thrushes. They are generally associated with mountainous rocky habitats.

'Wheatear' is of English dialect origin, appearing as whiteeres in the 17th century, then altering to the more familiar-sounding wheatears, then 'singularised' to wheatear. The original meaning was 'white arse' (for a white rump) at a time when 'arse' was an acceptable word. The bird referred to was the Northern Wheatear (*Oenanthe oenanthe*).

'Flycatcher', with the obvious meaning, was introduced to English by the 'father of English natural history' John Ray in 1678, a translation of the Latin *muscicapa* as previously used (French *gobe-mouche*, fly-eater, is also a late arrival, dated as a bird-name in print only from 1611). Thomas Pennant, the influential 18th-century Welsh naturalist, applied it to the familiar Spotted Flycatcher (*Muscicapa grisola*) and from there it was 'used in a general and very vague way for a great many small birds from all parts of the world, which have the habit of catching flies on the wing' (Newton 1896).

The genera

Muscicapa Brisson, 1760 [mʊs-ki-KA-puh]: 'flycatcher', from Latin *musca*, fly, and *capeo*, catch or seize.

Cyanoptila Blyth, 1847 [si-a-no-TI-luh]: 'blue feathered [bird]', from Greek *kuanos/kuaneos*, a word that covers many shades of blue – some (like lapis lazuli or cornflower) very apt in the case of this lovely bird. Also Greek *ptilon*, feather or down.

Larvivora, Hodgson, 1837 [lar-VI-vo-ruh]: 'larva-eater', from Modern Latin *larva*, caterpillar (cf. Latin *larva*, mask), and Latin *-vorus*, swallowing, eating greedily, devouring).

Ficedula Brisson, 1760 [fi-SE-dʊ-luh]: 'fig-pecker', from Latin *ficedula*, fig-pecker. According to Pliny (77–79 AD), the *ficedula* changed in the winter in shape, colour and name, becoming what he called *melacoryphus*, the Blackcap *Sylvia atricapilla*. Wondrous bird!

Monticola Boie, 1822 [mon-TI-ko-luh]: 'mountain dweller', from Latin *mons/montis*, mountain, and *-cola*, dweller, inhabitant.

Oenanthe Vieillot, 1816 [oy-NAN-theh]: 'wheatear', from Greek *oinanthē*, a vine-flower; it is thought that the bird so named by Aristotle (c. 330 BC) and described as a summer visitor may have arrived when the vines were flowering. Pliny (77–79 AD) gave a similar account and Belon (1555), reading these two descriptions, identified their bird as a wheatear (Arnott 2007).

The species

Grey-streaked Flycatcher (vagrant)

See family introduction. From the grey-streaked breast (and sort of from the species name).

Muscicapa griseisticta (Swinhoe, 1861) [mʊs-ki-KA-puh gree-se-i-STIK-tuh]: 'grey-spotted fly-catcher', see genus name, and from Mediaeval Latin *griseus*, grey, and Greek *stictos*, spotted or dappled.

Dark-sided Flycatcher (vagrant)

See family introduction. For the dark flanks, which distinguish this species from otherwise very similar pale-flanked relatives.

Muscicapa sibirica Gmelin JF, 1789 [mʊs-ki-KA-puh si-BI-ri-kuh]: 'Siberian fly-catcher', see genus name, and from the placename. Said by Gmelin (1789) to live near Lake Baikal and in eastern Siberia.

Asian Brown Flycatcher (vagrant to Ashmore Reef)

See family introduction. Well, it's certainly Asian and brown, but so are many other related small flycatchers.

Muscicapa dauurica Pallas, 1811 [mʊs-ki-KA-puh dow-Ʊ-ri-kuh]: 'Daurian flycatcher', see genus name, and from Dauria or Transbaikal, a mountain region to the east of Lake Baikal in the south of the Russian Federation and part of the huge region of Siberia. 'Dauuria' and 'Dahuria' are alternative spellings.

Blue-and-White Flycatcher (vagrant)

See family introduction. For the male only, which is bright blue above and white (and black) below.

Cyanoptila cyanomelana Temminck, 1829 [si-a-no-TI-luh si-a-no-MEL-a-nuh]: 'blue-black blue-feather', see genus name, and from Greek *kuanos*/*kuaneos* (again) and *melas*, black, for the darker breast.

Siberian Blue Robin (vagrant)

See family introduction. It's slaty blue and breeds across southern Siberia; several genera of thrushes are called robins (including the familiar American Robin *Turdus migratorius).*

Larvivora cyane (Pallas, 1776) [lar-VI-vo-ruh SY-a-neh]: 'blue larva-eater', see genus name, and from Latin *cyaneus*, dark blue or sea-blue – the word Pliny (77–79 AD) used to describe the colour of the kingfisher.

Narcissus Flycatcher (vagrant to Barrow Island)

See family introduction. From the species name.

Ficedula narcissina (Temminck, 1836) [fi-SEH-dʊ-luh nar-si-SEE-nuh]: 'daffodil-yellow fig-pecker', see genus name, and from Latin *narcissus*, Greek *narkissos* (originally probably the Pheasant's Eye Narcissus, *Narcissus poeticus* or *Narcissus scrotinus*). The word is connected with Greek *narkē*, numbness or sleep, and the narcissus was believed to have narcotic properties – but really it just makes you sick. In this case, the narcissus in question is the good old yellow daffodil, which is truly the colour of the bird (well, of its underparts at least). The Greek mythological connection – poor Narcissus wasting away while admiring his reflection in a pool, then in some versions plunging a knife into his breast – is that the Pheasant's Eye Narcissus, with its orange centre, was supposed to have sprung from that blood.

Mugimaki Flycatcher (vagrant)

See family introduction. From the species name and Japanese name for the bird.

Ficedula mugimaki (Temminck, 1836) [fi-SEH-dʊ-luh mʊ-gi-MUH-ki]: 'mugimaki fig-pecker', see genus and common names.

Blue Rock Thrush (vagrant)

See family introduction. The male only is powder blue with rufous breast and belly.

Monticola solitarius (Linnaeus, 1758) [mon-TI-ko-luh so-li-TAH-ri-ʊs]: 'solitary mountain-dweller', see genus name, and from Latin *solitarius*, solitary or separate, reflecting the fact that the bird is rarely seen in groups.

Isabelline Wheatear (vagrant)

See family introduction. From the species name.

Oenanthe isabellina (Temminck, 1829) [oy-NAN-theh i-sa-bel-LEE-nuh]: 'pale yellow wheatear', see genus name, and from French *isabelle*, pale yellow. This colour name is said (apocryphal is the word that springs to mind) to be from Queen Isabella I of Castile and was adopted because, when she besieged Granada in 1491, she swore not to change her underwear (*chemise*, at least) till the town was taken. The siege began in the European spring and lasted until the end of the year ... Oh dear.

DICAEIDAE (PASSERIFORMES): flowerpeckers

Dicaeidae [di-SEH-i-dee]: the Tiny Indian family, from the genus name *Dicaeum* (see genus name).

'Flower-pecker' was introduced by Thomas Jerdon in the *Birds of India* (Jerdon 1862) for members of the genus *Dicaeum* and associated genera (mostly not now recognised). All members are primarily flower- or fruit-eaters (notably associating with mistletoes).

The genus

Dicaeum Cuvier, 1816 [di-SEH-ʊm]: 'little Indian bird'. Cuvier got the name of what he called 'a very little bird from India' from Aelian (c. 200 AD) who said in *De Natura Animalium*, 'A certain race of small birds nests in the high and steep mountains, whose size reaches [the size of] a partridge's egg, of the colour of sandarach [red lead, i.e. orange-coloured]. The Indians are used to call it '*dikairon*' in their language, and the Greeks '*dikaion*'.' So far so good, but he went on to say that the bird buries its dung, which is deadly to people. Hence the assertion that Greek *dikaion* actually 'probably refers to the scarab beetle' (HANZAB 1990–2006, vol. 7b). Arnott (2007) completed the story, pointing out that this account (originally from Ctesias) conflates 'in bizarre fashion': (a) two tiny Indian birds with patches of orange plumage, including the Orange-bellied Flowerpecker, *Dicaeum triganostoma*, (b) Indian dung beetles and (c) plants like *Cannabis sativa* and *indica*, whose extract might have been moulded into balls capable of being confused with dung balls! What can we say?

The species

Red-capped Flowerpecker (Torres Strait Islands)

See family introduction. The descriptor is true, but not uniquely so (other species include Flame-crowned, Crimson-crowned and Scarlet-headed Flowerpecker).

Dicaeum geelvinkianum Meyer AB, 1874 [di-SEH-ʊm gehl-vin-kee-AH-num]: 'Geelvink little Indian bird', see genus name, and for Geelvink Bay in West Papua, now called Cenderwasih Bay. Geelvink Bay was named for a 17th-century ship of the Dutch East India Company; the ship's name apparently came from Joan Geelvinck, a Dutch merchant and politician.

Mistletoebird (breeding resident)

The name did not derive in Australia – the English Missel-bird (and Missel or Mistle Thrush) are corruptions of it. 'Our' Mistletoebird and the thrushes are completely unrelated and dissimilar, and as late as 1898 Morris was unaware of the word; in his comprehensive *Dictionary of Austral English*, he used Flower-pecker. The first usage we can find is immediately after that (Hall 1899). It is quite likely that it was in familiar, if not written, usage well before that, because the bird's total reliance on fruiting mistletoe is obvious even to a non-specialist; the name may have arisen independently of the English usage, but was more likely influenced by it.

Other names: Swallow Warbler, used by Latham (1801), from Shaw's species name; Crimson-throated Manakin (Lewin 1808) – Gould (1848) somewhat mysteriously cited this as Crimson-throated Honeysucker, and we are uncertain of its actual origin; Swallow Dicaeum, used by Gould (1848), adopting Cuvier's genus name; Australian Flower Swallow; Australian or Mistletoe Flowerpecker or Flower Pecker.

Dicaeum hirundinaceum (Shaw, 1792) [di-SEH-ʊm hi-rʊn-di-NAH-se-um]: 'swallow-like little Indian bird', see genus name, and from Latin *hirundo*, swallow, and *-aceus*, similar to. The male bird's red, white and glossy blue-black colouring was seen as similar to that of the swallow.

NECTARINIIDAE (PASSERIFORMES): sunbirds

Nectariniidae Vigors, 1825 [nek-ta-ri-NEE-i-dee]: the Nectar bird family, from genus name *Nectarinia.*

'Sunbird' has been known since the early 19th century; Newton (1896) cited Swainson as claiming that they are 'so called by the natives of Asia in allusion to their splendid and shining plumage', though it is unclear why such 'natives' would have been naming them in English. Newton put it a little more diplomatically, pointing out that Swainson 'gives no hint as to the nation or language wherein the name originated'.

The genus

Cinnyris Cuvier, 1816 [si-NI-ris]: 'very tiny unknown Greek bird', as Cuvier says. The name is mentioned by Hesychius, but he did not specify what it might be either. Arnott (2007) suggests that Greek *kinnyris* might translate either as a name for any little nestling (plaintive mewer), or perhaps for a wagtail (tail-mover).

The species

Olive-backed Sunbird (breeding resident)

See family introduction. Not a very satisfactory name in terms of distinguishing the species from other sunbirds, but it is a very variable species found over a huge range from northern Australia to southern China (the type specimen is from the Philippines). The back is about the only characteristic common to all races.

Other names: Australian Sun-bird, used by Gould (1869), who thought it an endemic species. Yellow-bellied or Yellow-breasted Sunbird, the name by which it was long known in Australia – Schodde and Mason (1999) spearheaded the move to return to the broader species name, on the basis that not all races outside Australia have yellow undersides; Allied Sunbird, cited by CSIRO (1969) but we can find no other use of it; Yellow Sunbird; Yellow-crested Sunbird, cited by HANZAB (1990–2006, vol. 7b), but this makes no apparent sense and seems to be the only usage – the implication is that it derives from a mis-hearing of 'yellow-breasted'.

Cinnyris jugularis (Linnaeus, 1766) [si-NI-ris yʊ-gʊ-LAH-ris]: 'tiny throated unknown bird', see genus name, and from Latin *iugulis*, throat or neck, referring to the iridescent blue-black–violet bib.

PASSERIDAE (PASSERIFORMES): Old World sparrows

Passeridae Rafinesque, 1815 [pas-SE-ri-dee]: the Sparrow family, see genus name *Passer.* 'Sparrow' can be traced back to Old English *sparwa* or *spearwa*, with the modern form appearing in the 14th century. Lockwood (1984) reported that its use was long generalised to any small bird, although if so it is unclear when it became limited in England to species of *Passer* (elsewhere, notably in North America, it became used for unrelated though apparently similar species, especially of family Passerellidae).

The genus

Passer Brisson, 1760 [PAS-sehr]: 'sparrow', from Latin *passer,* meaning the same. Thinking back to the sparrow/ostrich misunderstanding (see Ostrich, under Struthionidae), we note that Latin *passer marinus* was not a sea sparrow, but an ostrich (apparently so-called because it had to be brought by sea).

The species

House Sparrow (introduced breeding resident)

Associated with human dwellings for millennia.

Other names: English or European Sparrow; Spoggy, Sproggy, Spriggy, Sprog, and so on – all direct from English dialects.

Passer domesticus (Linnaeus, 1758) [PAS-sehr do-MES-ti-kʊs]: 'house sparrow', see genus name, and from Latin *domesticus*, belonging to the house, domestic.

Eurasian Tree Sparrow (introduced breeding resident)

Perhaps a reflection of being less associated with houses rather than especially fond of trees; found right across Europe and Asia.

Other names: European or Eastern Tree Sparrow; Mountain Sparrow, from the species name.

Passer montanus (Linnaeus, 1758) [PAS-sehr mon-TAH-nʊs]: 'mountain sparrow', see genus name, and from Latin *montanus*, belonging to the mountains.

ESTRILDIDAE (PASSERIFORMES): grass-finches

Estrildidae Bonaparte, 1850 [es-TRIL-di-dee]: the Waxbill family, from the genus *Estrilda*, waxbills (not occurring in Australia), named by Swainson (1827) from Linnaeus' *Loxia astrild*, the Common Waxbill. By 1836 Swainson had started spelling it 'Estrelda' but the name was sharply criticised by Reichenbach, who 'corrected' it in 1845 to *Astrilda* but kept on complaining: 'the English have mutilated the word' (1849), 'it is neither Latin nor English nor anything else' and so on (1862), and reminding us that he had corrected it. As is the way of these things, *Estrilda*, as the first published name, held anyway. Linnaeus' original 'astrild' was from the name Astrild, now archaic, but coined in the 17th century and much used by Swedish poets of the time as a name for Cupid, god of love. It is thought to be a compound of *ast*, an old word for love, and *eld*, meaning fire. Linnaeus' choice is supposed by some to be a reference to the Waxbill's red bill and facial marking, but unless Cupid had a red nose it makes more sense to think of the name as meaning fire or flame of love! The word Astrild survives in the common names of the waxbills in many European languages.

'Finch' can be traced back to Old English *finc*, but the origins of that word are debated. The *Oxford Dictionary* noted possible associations with a Greek word for a young bird, or various words denoting colour in a range of Indo-European languages. Lockwood (1984) maintained that it is onomatopoeic, referring originally to the Chaffinch (see Fringillidae), but this is speculative and not especially evocative of the Chaffinch's rapid slurring call, though it could refer to its alarm call.

'Firetail' is now applied only to the three species of *Stagonopleura*, all with brilliant red rumps, though until recently a couple of other red-rumped species, notably Red-browed Finch, were also so named. Gould recorded that the 'Colonists of Van Diemen's Land' used the name Fire-tail for Beautiful Firetail; he adapted it to Fire-tailed Finch. By 1898 Morris, in his *Dictionary of Austral English*, reported that the name was used in Victoria for Red-browed Finch and in Tasmania for the Beautiful Firetail. Professional ornithologists were much tardier in adopting the word; it first appears formally (for the same three species for which it is now used) in the RAOU's second *Official Checklist* (RAOU 1926).

It seems to be agreed that 'mannikin' is from the Dutch *manneken*, diminutive of a man ('little man'); the question of why it was applied to birds in general, and to the genus *Lonchura* of grass-finches in particular, is not so clear. Although these birds are not to be confused with the entirely unrelated South American manakins, colourful tyrant-flycatchers of the family Pipridae, they have a common name origin and indeed the application seems to have been first to the South Americans. Newton, in his 1896 *Dictionary of Birds*, commented that manakin was 'applied to certain small birds, a name apparently introduced into English ... in or ~1743' (by George Edwards in *Natural History of Birds*); we note, however, that Newton didn't clarify the reason for it either. By Newton's time the only general ornithological use was apparently for the Pipridae. It is perhaps significant that in 1969 the CSIRO *Index of Australian Bird Names* used mannikin only for Black-headed Mannikin (*Lonchura atricapilla* or *L. malacca*), an Asian species briefly established in small areas of New South Wales; shortly afterwards the RAOU's checklist committee recommended the use of mannikin for the Australian *Lonchura* species, as was 'widely done ... in Africa ... and the Far East' (RAOU 1978). How that state of affairs transpired seems likely to remain as obscure as the original import of the word.

'Munia' is another name used in Asia (and secondarily in Africa) for many *Lonchura* (and other) finches. Newton reported that it derived in India – we might suppose from Hindi, though he didn't specify – for 'several kinds of seed-eating birds ... to quote Mr Hodgson ... who not only Anglified [sic] the word but Latinized it, making it the title of a genus to which he applied three species'. The genus name no longer exists, but the (Anglified) vernacular version survives.

'Parrotfinch' is applied to the 10 species of *Erythrura* of South-East Asia and the adjacent Pacific, plus northern Australia, for their brilliant combinations of colours. Its usage in Australia is quite recent.

The genera

Emblema Gould, 1842 [em-BLEH-muh]: 'mosaic bird', from Latin *emblema*, inlaid or raised ornaments, or mosaic work, possibly for the white spots on the flanks, but Gould was at pains to mention the 'mingled appearance of black and red' on the throat, caused by the black base of the 'deep vermilion red' throat feathers.

Stagonopleura Reichenbach, 1850 [sta-go-no-PLOO-ruh]: 'spotty flanks', from Greek *stagōn*, drop or spot, and *pleura*, side or flank.

Neochmia Gray GR, 1849 [ne-OK-mi-uh]: 'new bird', from Greek *neokhmos*, new, *neokhmia/ neokhmōsis*, innovation. It was Hombron and Jacquinot who found the bird during their journey to the South Pole. Although Hombron felt it should be in the genus *Erythura*, he did admit that the unusually short length of the bird's legs might so set it aside from others in the genus that it would need a new genus, *Neochmia*, a name that Gray adopted.

Poephila Gould, 1842 [poh-E-fil-uh]: 'grass-lover', from Greek *poa* (the genus name of some 500 grass species across the world) (also *poiē*, *poia*), and *-philos*, lover of.

Taeniopygia Reichenbach, 1862 [teh-ni-o-PI-gi-uh] 'banded tail', from Greek *tainia*, a band or fillet, and *pugē*, rump or tail.

Erythrura Swainson, 1837 [e-ri-THROO-ruh]: 'red-tail', from Greek *eruthros*, red, and *oura*, tail.

Lonchura Sykes, 1832 [lon-KOO-ruh]: 'spear-tail', from Greek *lonkhē*, spear-head, and *oura*, tail.

Heteromunia Mathews, 1913 [he-te-ro-MOO-ni-uh]: 'different from the munias', from Greek *heteros*, different, and genus name *Munia*, see family introduction.

The species

Painted Finch (breeding resident)

See family introduction. Used by Gould (1848) from his species name.

Other names: Painted Firetail; Emblema or Emblema Finch, from the genus name. Mountain Finch, a reference to its preference for rocky inland gorges, but it seems that the references in CSIRO (1969) and HANZAB (1990–2006, vol. 7b) are misleading, because the only original usage we can find is a comment by Frederick Whitlock in an article on the Kimberley (Whitlock 1927); he noted that 'the Painted Finch ... loves water in the vicinity of rocky ranges. It might well have been named the Mountain Finch'. This strongly implies to us that it was not in fact so named, and there is little other evidence of its use.

Emblema pictum Gould, 1842 [em-BLEH-muh PIK-tum]: 'painted mosaic bird', see genus name, and from Latin *pictus*, painted or embroidered. (For those who are worrying about this, *emblema* is a neuter noun, not feminine, hence the neuter adjectival ending. This has been corrected since Gould's version, which was indeed *picta* – a kind of correction which, unlike some, is permitted.)

Beautiful Firetail (breeding resident)

See family introduction. Directly from the species name; it's indisputable, but not unique.

Other names: Fire-tail, 'Colonists of Van Diemen's Land' (Gould 1848); Firetail or Firetailed Finch; Tasmanian Finch, mostly encountered in the cage-bird literature, but one example of its appearance in the ornithological literature is from A.G. Campbell (1903a) – writing of the Dandenongs, Victoria, he referred to 'a Tasmanian Finch (*Zonaeginthus bellus*), rare on the mainland'. Perhaps he was only being descriptive.

Stagonopleura bella (Latham 1802) [sta-go-no-PLOO-ruh BEL-luh]: 'handsome spotty-flanks', see genus name, and from Latin *bellus*, pretty, handsome, agreeable.

Red-eared Firetail (breeding resident)

See family introduction. For the red patch behind the eye.

Other names: Red-eared Finch (Gould 1848); Native Sparrow, 'Colonists of Swan River' (Gould 1848); Zebra Finch, for the barred flanks (but see that species); Casuarina Finch, for a common habitat – reported to be a name used in the Pemberton area (Tarr 1948).

Stagonopleura oculata (Quoy & Gaimard, 1830) [sta-go-no-PLOO-ruh o-kʊ-LAH-tuh]: 'spotty-flanks with conspicuous eyes', see genus name, and from Latin *oculatus*, with eyes, in this case referring to the spots on the belly.

Diamond Firetail (breeding resident)

See family introduction. For the spangled flanks and sides; first used in the RAOU's second *Official Checklist* (RAOU 1926).

Other names: Spotted Finch (Lewin 1808); Spotted Grossbeak (Lewin 1822); Spotted-sided Grossbeak (Latham 1823); Spotted-sided Finch (Gould 1848) – this was the name in near-universal use until suddenly supplanted in 1926; Diamond-Sparrow, recorded by Leach (1911) as an alternative name – we may suppose that this was a popular name, and that it influenced the RAOU committee, especially given

Beautiful Firetail *Stagonopleura bella,* Double-barred Finch *Taeniopygia bichenovii* race *annulosa,* Double-barred Finch *T. bichenovii* nominate race, Plum-headed Finch *Neochmia modesta* and Red-eared Firetail *Stagonopleura oculata*

that Leach convened that committee; Diamond Finch or Diamond Java Sparrow – see also Zebra Finch, and its usage here is equally unclear.

Stagonopleura guttata (Shaw, 1796) [sta-go-no-PLOO-ruh gʊ-TAH-tuh]: 'dotted spotty-flanks', see genus name, and from Latin *gutta*, a droplet. Not quite as evocative as the common name 'diamond', but it's certainly spotted.

Red-browed Finch (breeding resident)

See family introduction. For the adult's startlingly bright red brow, from well behind the eye to the equally red bill.

Other names: Temporal Finch, by Latham for his species name (though as ever with Latham it is unclear which came first); Red-eyebrowed Finch (Gould 1848); Red-Bill, 'of the Colonists' (Gould 1848); Red Head, Reddie, Redtail; Red-browed Firetail, see family introduction; Lesser Red-browed Finch, for race *minor* of Cape York Peninsula; White-cheeked Redbrow or Whitecheek, also for race *minor*; Sydney Waxbill, widely used by cage-bird fanciers since at least the early 20th century (Philipps 1902), but not much outside that community; Australian or Australian Grey Waxbill, also an aviculturalist's name.

Neochmia temporalis (Latham 1802) [ne-OK-mi-uh tem-po-RAH-lis]: 'new bird with temples', see genus name, and from Latin *temporalis*, to do with the temples (on the head, not places of worship, and as in temporal lobes of the brain), for the extended bright red eyebrow.

Crimson Finch (breeding resident)

See family introduction. For the male's richly coloured undersides; introduced by Gould (1848).

Other names: Red Finch, reported by Gould (1848) to be used by the 'Residents of Port Essington'; Blood Finch; Australian Firefinch, a reference to the somewhat similar-looking African firefinches, of genus *Lagonostica*; Cape York, Pale or White-bellied Crimson Finch, all for Cape York race *evangelinae*, which is not noticeably pale, but, in addition to having a white belly, is slightly less intensely coloured than the nominate race; Pheasant Finch, of obscure origin though perhaps referring to the colour or tail length. The only reference we can find outside HANZAB (1990–2006, vol. 7b) is to 'Rockhampton Pheasant Finch', in an 11 November 1863 *Sydney Morning Herald* report of the New South Wales Acclimatisation Society; however, there is nothing to suggest that it actually refers to this species.

Neochmia phaeton (Hombron & Jacquinot, 1841) [ne-OK-mi-uh FEH-tohn]: 'fiery new bird like a tropicbird', see genus name, and from Greek *phaethōn*, meaning shining, but also the name of the sun god Helios/Phoebus' son, who set the Earth on fire when recklessly driving the chariot of the sun across the heavens. It was also used (from the 18th century) as the name of a light horse-drawn carriage. Why this name? HANZAB (1990–2006, vol. 7b) suggested it is simply for the crimson colour and Hombron and Jacquinot (1841) didn't explain; however, their description of the bird (*Fringilla phaeton*) said, 'tail adorned with two long bright crimson feathers, more than a third longer than the other [feathers]'. Their observation of the long tapered red tail could certainly have brought to mind the Tropicbirds (genus *Phaethon*) with their long tail streamers.

Star Finch (breeding resident)

See family introduction. For the white spangles on cheeks, breast and flanks. The name seems to have appeared out of the blue in the RAOU's second *Official Checklist* (RAOU 1926).

Other names: Red-tailed Finch, used by Gould (1848), directly from his species name – it was the name of choice until the end of the 19th century; Rufous-tailed or Ruficauda Finch; Red-faced Finch, a descriptive name, which gained near-universal use in the literature from Hall (1899) until 1926; Red-faced Firetail.

Neochmia ruficauda (Gould, 1837) [ne-OK-mi-uh roo-fi-COW-duh]: 'red-tailed new bird', see genus name, and from Latin *rufus*, red, and *cauda*, tail.

Plum-headed Finch (breeding resident)

See family introduction. For the male's purple-plush forehead; it seems to have arisen spontaneously during the 19th century – Gould (1848, 1865) seemed unaware of it, but it was first-choice name in Morris' 1898 *Dictionary of Austral English*.

Other names: Plain-coloured Finch, used by Gould (1848), reflecting his species name; Plum-capped Finch or Plumhead; Modest Grass Finch, from the species name; Cherry Finch, for the same reason as Plum-headed, but primarily a cage-bird fanciers' name; Diadem Finch, presumably another reference to the rich crown – it is rare in print, but an example is in a natural history column in the Melbourne *Argus* on 15 July 1913, 'the Plum-head Finch, sometimes known as the Diadem Finch'.

Neochmia modesta (Gould, 1837) [ne-OK-mi-uh mo-DES-tuh]: 'plain new bird', see genus name, and from Latin *modestus*, meaning plain, moderate, sober. Not sure which, if any, of these really apply – the bird has quite noticeable stripes across breast and belly, as well as its purple diadem! It is, however, undoubtedly plainer than others of its genus.

Masked Finch (breeding resident)

See family introduction. As used by Gould (1848), directly from his species name.

Other names: Masked Grass Finch; White-eared Finch, Grass Finch or Masked Finch, all for race *leucotis* of Cape York Peninsula, and reflecting that name, which was applied by Gould as a separate species *Poephila leucotis*.

Poephila personata Gould, 1842 [poh-E-fil-uh pehr-so-NAH-tuh]: 'masked grass-lover', see genus name, and from Latin *persona*, a mask (for the bird's dark face), a word perhaps more familiar to us as an aspect of human behaviour (thanks to psychiatrist Carl Jung): 'adopting another persona' (i.e. wearing a different mask).

Long-tailed Finch (breeding resident)

See family introduction. For its most obvious characteristic (see also species name).

Other names: Long-tailed Grass Finch, used by Gould (1848) – see also genus name; Black-heart Finch or Blackheart, for the (vaguely heart-shaped) large black breast patch; Orange-billed or Red-billed Finch or Grass Finch, for race *hecki* from the Northern Territory and Queensland (the nominate race from Western Australia has a yellow bill). Heck's Finch, also for race *hecki*, described as a full species *Poephila hecki* by Oskar Heinroth in 1901, following a collecting expedition on behalf of the Berlin Zoo to the south Pacific, in the course of which he presumably popped in to Darwin; Heinroth was at the time working voluntarily for the zoo, of which Ludwig Heck was the director. (In 1904, perhaps in gratitude for having his very own finch, Heck promoted Heinroth to curator – a paid position that he created for Heinroth.)

Poephila acuticauda (Gould, 1840) [poh-E-fil-uh a-koo-ti-KOW-duh]: 'pointy-tailed grass-lover', see genus name, and from Latin *acutus* (pointy, like the acute angle), and *cauda*, tail.

Black-throated Finch (breeding resident)

See family introduction. A descriptive name that does nothing to distinguish it from its two closest relatives, especially the Long-tailed Finch. It's not clear when it arose (Gould, who scientifically named the bird, didn't use it) but it was recorded by Morris in 1898 as first-choice name. What seems certain is that it was in response to confusion arising from Gould's name for this species – Banded Grass Finch – and the then widely used name for Double-barred Finch, which was Banded Finch.

Other names: Banded Grass Finch, per Gould (1848) from his species name; Blackthroat; Diggle's Finch, for Cape York race *atropygialis*, for the polymath Silvester Diggles – prominent musician, ornithologist and entomologist, a founder of the Queensland Philharmonic Society and the Queensland Museum – who named the subspecies in 1876; Black-rumped Finch or Grass Finch, Black-tailed Finch, all for *atropygialis*, directly from the name; Parson Finch (sometimes offered as Parson's Finch), primarily a cage-bird enthusiasts' name though that was not necessarily its origin, and apparently deriving from the black bib; Chocolate Parson, for northern race *nigrotecta* (not widely recognised), which is darker brown.

Poephila cincta (Gould, 1837) [poh-E-fil-uh SINK-tuh]: 'girdled grass-lover', see genus name, and from Latin *cingo/cinctus*, encircle (-d) or girdle (-d). Gould put it this way: 'with a black band (*fascia*) encircling the lower part of the body'.

Zebra Finch (breeding resident)

See family introduction. For the conspicuously barred tail, though its origin has seemed surprisingly recent, given the great familiarity of the name. In the bird literature, Leach (1911) used it as an alternative name, and in 1931 Neville Cayley used it as first-choice name. However, we can trace it back to 1867 (in the *South Australian Register*, 14 December), though this use, and subsequent ones, all seem to refer to cagebirds. The first application we can find for wild birds is in Margery (1895): he uses both Zebra Finch and Chestnut-eared Finch, lumping them with two pardalote species as 'diamond birds'. It is unclear whether the name was of folk origin, largely oral, or if it derived from the cagebird industry and spread. Either way, it seems inconceivable that a bird so abundant around dams and homesteads throughout inland Australia would not have had names applied to it and it seems unlikely that country people all used Chestnut-eared Finch. Perhaps the real surprise is that so few alternative names are recorded.

Other names: Diamond Bird, see common name, though we know of no other such usage; Chestnut-eared Finch, as used by Gould (1848) from his species name (*Amadina castanotis*), though it appears infrequently. The current situation is that our Zebra Finch is seen as a subspecies of a species also found in the Lesser Sundas, though this is far from resolved – *Handbook of the Birds of the World*, for instance, separates them out, with Australian birds being *T. castanotis*, Australian Zebra Finch. Java Sparrow, cited by both CSIRO (1969) and HANZAB (1990–2006, vol. 7b), but it is not clear how or where this presumed confusion arose – other usages of the application for this species are not evident. Pimp, similarly obscure and hard to find examples of, it may refer to the little tooting call; Waxbill, cited by HANZAB (1990–2006, vol. 7b), but the only example we can find is from the *Australian National*

Dictionary (in the *Sydney Bulletin* of 1940) – 'watched a mob of waxbills (zebra finches)' – where it seems to be used more as a group name than a species name; Zebbie; hundreds and thousands, an oral usage we've encountered in central Australia.

Taeniopygia guttata (Vieillot, 1817) [teh-ni-o-PI-gi-uh gʊ-TAH-tuh]: 'spotted banded-tail', see genus name, and from Latin *gutta*, a droplet, for the white spots on the male bird's bright chestnut flanks.

Double-barred Finch (breeding resident)

See family introduction. For the pair of conspicuous black bars across throat and belly.

Other names: Bicheno's Finch, the name used by Gould from the species name; Banded Finch, long a preferred name; Double-bar; Ringed or Black-ringed Finch; Owl-faced Finch, a surprisingly descriptive name; Double-barrelled or Double-bearded Finch, surely both mistranscriptions or mishearings; Black-rumped Finch (Gould 1848) or Black-rumped Double-bar, for northern Australian race *annulosa*, whose rump differs from the white-rumped nominate race – Gould originally designated it as a separate species *Amadina annulosa* and it was still recognised as such well into the 20th century (RAOU 1926).

Taeniopygia bichenovii (Vigors & Horsfield, 1827) [teh-ni-o-PI-gi-uh bi-she-NOH-vi-ee]: 'Bicheno's banded-tail', see genus name, and for James Ebenezer Bicheno, who at the time of naming was secretary of the London Linnean Society and a Fellow of the Royal Society (so a colleague of Vigors and Horsfield); he was primarily a botanist but had some interest in birds. He was later appointed Colonial Secretary of Van Diemen's Land, where he died.

Blue-faced Parrotfinch (breeding resident)

See family introduction. It has indeed a striking and beautiful blue face, but not uniquely in the genus.

Other names: Blue-faced Finch; Tri-coloured Finch or Parrot-Finch, from the species name – though confusingly the name is also used for *E. tricolor* from Indonesia, which might seem to have an equal claim to it. Green-backed or Australian Green-backed Finch, a name formerly used for non-Australian birds, applied when the bird was first found in Australia (Kershaw, 1918).

Erythrura trichroa (Kittlitz, 1833) [e-ri-THROO-ruh tri-KRO-uh]: 'three-coloured red-tail', see genus name, and from Greek *tri-*, three-, and *khroa*, colour. The three colours are the red of the tail, the blue face and the green body and wings. Kittlitz (1833) described these as 'rusty blood-red', 'ultra-marine blue' and 'a beautiful parrot-green'.

Gouldian Finch (breeding resident)

See family introduction. For Elizabeth Gould, a highly talented artist, married to John Gould until her sadly early death at the age of 37. John Gould usually contained his feelings as per the Victorian era, but on this occasion he went to what were probably the limits that the times allowed: 'it is in fact beyond the power of my pen to describe or my pencil to portray anything like the splendour of the changeable hues of the lilac band which crosses the breast of this little gem ... It is therefore with feelings of no ordinary nature that I have ventured to dedicate this new and lovely little bird to the memory of her, who in addition to being a most affectionate wife, for several years laboured so hard and so zealously assisted me with her pencil in my various works, but who, after having made a circuit of the globe with me, and braved many dangers with a courage only equalled by her virtues, and while cheerfully engaged in illustrating the present work, was by the Divine will of her Maker suddenly called from this to a brighter and better world; and I feel assured that in dedicating this bird to the memory of Mrs Gould, I shall have the full sanction of all who were personally acquainted with her, as well as those who only knew her by her delicate works as an artist' (Gould 1848).

Other names: Gouldian Grass Finch; Painted or Rainbow Finch; Purple-breasted or Purple-chested Finch; Scarlet-headed Finch.

Erythrura gouldiae (Gould, 1844) [e-ri-THROO-ruh GOOL-di-eh]: 'Mrs Gould's red-tail', see genus name (though, strangely, the tail of this particular one is not red or even brown), and for Elizabeth Gould (see common name). The '*ae*' tells us, even if we didn't know, that it was named after Elizabeth, not John. If it had been after John it would have been *gouldii*. So why doesn't the species name here have the same ending as the genus? Well, when a person's name is used as a species name, there are two possibilities: it can become an adjective, in which case it must agree with the genus name. The species name would then have been e.g. *gouldiana*, and we couldn't have said which Gould was being commemorated. Or, as in this case, it can take the form of a noun in the genitive (or possessive) case (i.e. 'of Gould'), and must then follow the gender of the person (*-i* being the masculine ending, *ae* the feminine).

Scaly-breasted Munia (introduced breeding resident)

See family introduction. Perhaps a reference to the species name, certainly to the same character.

Other names: Nutmeg Finch or Mannikin (the name used in Australia); we are fairly confident that the reference is to the colour – 'nutmeg' has often been used in this way. Thanks to Alfred Newton's excellent *Dictionary of Birds* we can be more confident of the timing of its origin: 'the dealer's name in common use for *Munia punctulata* … but apparently of somewhat recent origin' (Newton 1896). Spice Finch, presumably by association with nutmeg, though it has been suggested to us that there may be a reference to the Spice Islands (Maluku or Moluccas), which fall within its huge range; Rice Bird, for its great fondness for rice paddies in its Asian homelands; Spotted Munia, from the species name and a former genus name (see also family introduction); Cowry Bird, explained by Newton (1896) as being the price of a caged bird in the East Indies – and who are we to say it isn't true? Other names are known, but not in significant use in Australia.

Lonchura punctulata (Linnaeus, 1758) [lon-KOO-ruh pʊnk-tʊ-LAH-tuh]: 'tiny-spotted spear-tail', see genus name, and from Latin *punctus*, a spot or point, with diminutive ending *-ulus*. The breast and belly of this bird are actually white with black scales, but can look like a dark background peppered with little white spots. Linnaeus (1758) hedged his bets, saying it was spotted with white and black!

Pale-headed Munia (vagrant)

See family introduction. For the distinctively white head.

Lonchura pallida (Wallace, 1864) [lon-KOO-ruh PAL-li-duh]: 'pale spear-tail', see genus name, and from Latin *pallidus*, pale or colourless.

Yellow-rumped Mannikin (breeding resident)

See family introduction. Used by Gould, from his species name.

Other names: Yellow-rumped Finch, Gould's name (1848); Yellow-tailed Finch; Yellow-rumped Munia (it was still in genus *Munia* into the 20th century); White-headed Finch, though it's more pale grey-fawn.

Lonchura flaviprymna (Gould, 1845) [lon-KOO-ruh flah-vi-PRIM-nuh]: 'yellow-ended spear-tail', see genus name, and from Latin *flavus*, golden or reddish yellow (from *flagro*, burn), and Greek *prumnos*, hindmost part – the rump and tail are both ochre yellow.

Chestnut-breasted Mannikin (breeding resident)

See family introduction. By Gould, from his species name.

Other names: Chestnut or Chestnut-breasted Finch; Barley-bird or Barley Sparrow, from its fondness for crops (though barley doesn't feature strongly in its range); Dun-coloured Finch – must have been a juvenile! Chestnut-breasted Grass Finch or Munia (it was still in genus *Munia* into the 20th century); Chestnut or Rockhampton Sparrow; Bullfinch, for its vague stout-billed resemblance to the European bird.

Lonchura castaneothorax (Gould, 1837) [lon-KOO-ruh kas-ta-ne-o-THOR-aks]: 'chestnut-breasted spear-tail', see genus name, and from Latin *castanea*, chestnut or chestnut-tree, hence chestnut-coloured, and *thorax*, chest.

Java Sparrow (introduced to Christmas Island)

For its home range, though it's not clear why 'sparrow' was used instead of mannikin or munia, beyond the fact that it was common and obvious to non-ornithologists.

Lonchura oryzivora (Linnaeus, 1758) [lon-KOO-ruh o-ree-ZI-vo-ruh]: 'rice-eating spear-tail', see genus name, and from Latin *oryza*, rice, and the suffix *-vorus*, eating, as in 'carnivore'.

Pictorella Mannikin (breeding resident)

See family introduction. A distortion of the species name (presumably via Pectoral Finch). It was presumably at least in oral usage – not that the bird would have been familiar in the wild to many people – but seems to first appear in print in RAOU (1926).

Other names: White-breasted Finch, the name used by Gould (1848) by way of expanding on his species name; it remained the apparently universal name of choice until that somewhat perplexing dictum of the 1926 RAOU *Official Checklist* – unfortunately the committee saw no need to explain its decisions. White-breasted Munia; Pectoral Finch or Munia, directly from the species name, though it seems to occur nearly only in the realm of cagebird enthusiasts; Pictorella Finch or Munia.

Heteromunia pectoralis (Gould, 1841) [he-te-ro-MOO-ni-uh pek-to-RAH-lis]: 'different from the munias (in spite of its common names!), and with a chest', see genus name, and from Latin *pectus/pectoris*, the breast, (as well as evoking the Old French word *pectorale* meaning a breastplate), for the pretty scalloped bib, more marked in the female bird.

MOTACILLIDAE (PASSERIFORMES): pipits and wagtails

Motacillidae Horsfield, 1821 [mo-ta-SI-li-dee]: the Wagtail family, see genus name *Motacilla*.

'Pipit' is onomatopoeic in origin but was recorded for the first time only in the late 18th century; it was reportedly used by bird-catchers around London (Lockwood 1984). At that stage no distinction was made between larks and pipits, but when the genus *Anthus* was proposed soon afterwards, recognising the differences, the newly recorded name was conscripted for the genus. However, *pipi* (and variants) is known since the 13th century in French.

'Wagtail' was coined for the obvious characteristic, and dates from the early 16th century. It replaces an older word, 'wagstart', which meant the same thing, 'start' meaning a tail and having its origins in old Germanic languages.

The genera

Dendronanthus Blyth, 1844 [den-dro-NAN-thus]: 'tree-pipit', from Greek *dendron*, tree, and Bechstein's genus *Anthus*, see family introduction.

Motacilla Linnaeus, 1758 [mo-ta-SIL-luh]: 'little mover and shaker', Latin *motacilla*, probably the White Wagtail. Varro (116–27 BC) described the bird as 'always moving its tail' but the name is not to be translated as 'moving tail', as is sometimes thought; it is probably a diminutive from *moto*, to keep moving about.

Anthus Bechstein, 1805 [AN-thʊs]: 'small bird', from Latin *anthus*, a small bird believed to be the Yellow Wagtail, *Motacilla flava*, perhaps because of the Greek word *anthos*, a flower but also meaning brilliant (of gold or bright colours). Pliny (77–79 AD) mentioned the bird but gave no details that could identify it (unless you count it neighing like a horse, or producing a precious stone in its crop!).

The species

Forest Wagtail (vagrant)

See family introduction. Although it is also found in more open habitats, it is widespread in forests; the name is probably intended to distinguish it from the similar and co-existing open country White Wagtail.

Dendronanthus indicus Gmelin JF, 1789 [den-dro-NAN-thus IN-di-kus]: 'Indian tree-pipit', see genus name, and from Latin *indicus*, a generic term here for the bird's Asian origins – Gmelin (1789) wrote that Sonnerat called it the *Bergeronnette grise des Indes*, Latham the Indian Wagtail. The bird breeds in East Asia and migrates to tropical Asia, including India and Sri Lanka, in winter.

Eastern Yellow Wagtail (non-breeding summer migrant)

Until recently Yellow Wagtail (*M. flava*) was recognised as one polytypic species – all taxa have yellow undersides, and some have yellow heads as well. Other wagtails, notably Grey Wagtail (see profile) are also yellow beneath, but generally not as extensively as Yellow. Some of the subspecies have been raised to species status; this species breeds in eastern Siberia (and a short way into adjacent North America) and winters in South-East Asia as far south as northern Australia.

Other names: Yellow Wagtail, see common name; Green-headed Yellow Wagtail, for race *taivana*, recorded in Australia, which at the time was regarded as a full species.

Motacilla tschutschensis Gmelin JF, 1789 [mo-ta-SIL-uh chʊt-CHEN-sis]: 'Chukhotskiy mover and shaker', see genus name, and for its habitat 'on the shores of Tschutschi', as Gmelin (1789) put it (i.e. the Chukchi Sea, north of the Bering Strait) and hence the Chukhotskiy peninsula, on the western side of the strait plus the suffix *-ensis*, usually indicating the place of origin of the type specimen.

Citrine Wagtail (vagrant)

From the species name.

Motacilla citreola Pallas, 1776 [mo-ta-SIL-uh sit-re-OH-luh]: 'lemon-coloured mover and shaker', see genus name, and from Greek *kitrea/kitron*, the citron tree and its fruit (*Citrus medica*), and Latin *citrus or citreus*, which Pliny stated referred to an African tree with fragrant wood (possibly the cedar *Cedrus atlantica*) but also to the citron tree. From the second (and Mediaeval Latin *citrinus*) comes our own word 'citrus', French *citron* meaning lemon (and related words). The word was applied at some stage to lemon colour, as in the citrine, a form of yellow quartz regarded as a semi-precious stone.

Grey Wagtail (non-breeding summer migrant)

From the species name.

Motacilla cinerea Tunstall, 1771 [mo-ta-SIL-uh si-NE-re-uh]: 'ashy-grey mover and shaker', see genus name, and from Latin *cinereus*, like ashes, for its head and upper parts.

White Wagtail (vagrant or uncommon regular visitor)

From the species name (which is about as unhelpful as such names get).

Motacilla alba Linnaeus, 1758 [mo-ta-SIL-uh AL-buh]: 'white mover and shaker', see genus name, and from Latin *albus*, dull white. It **is** white, apart from its grey back and wings, big black cap and bib (even in non-breeding plumage), and some of its tail. Hmmm.

Australian Pipit (breeding resident)

As currently recognised, the species is found only in Australia – and New Guinea!

Other names: Pipit; Richard's Pipit, from the relatively recent times when it was included in *A. richardi*, which has a huge Asian range; Indian Pipit, likewise; Australasian Pipit, from when it was included in New Zealand Pipit *A. novaeseelandiae*; Groundlark or Australian Groundlark; Grundy, cited by CSIRO (1969) but not otherwise evident, though we suggest with some confidence that it is an oral form derived from Groundlark.

Anthus australis Vieillot, 1818 [AN-thʊs ost-RAH-lis]: 'small southern bird', see genus name, and from Latin *australis*, southern.

Pechora Pipit (vagrant)

For the Pechora River in western Siberia; this is close to the western limits of its breeding range and may indicate where western ornithologists first encountered it, though the type specimen was not collected there.

Anthus gustavi Swinhoe, 1863 [AN-thʊs gʊs-TAH-vee]: 'Gustaaf's small bird', see genus name, and for Gustaaf (or Gustav or Gustavus) Schlegel (1840–1903), a Dutch ornithologist, who Swinhoe (1863) said 'was the first to procure the bird at Amoy' (Xiamen, China).

Red-throated Pipit (vagrant)

For the adult male's breeding plumage.

Anthus cervinus (Pallas, 1811) [AN-thʊs sehr-VEE-nʊs]: 'stag-coloured small bird', see genus name, and from Latin *cervinus*, pertaining to a stag or deer, unclear from Pallas, but probably for what he called the rusty colour of the throat.

FRINGILLIDAE (PASSERIFORMES): 'true' finches

Fringillidae Leach, 1820 [frin-GIL-li-dee]: the Finch family, see genus *Fringilla*. For a discussion of 'finch', see Estrildidae.

The genera

Fringilla Linnaeus, 1758 [frin-GIL-luh]: 'finch', from Latin *fringilla*, a small bird, probably a chaffinch.

Chloris Cuvier, 1800 [KLO-ris]: 'greenfinch', from Greek *khlōris*, greenfinch, and *khlōros*, greenish-yellow.

Acanthis Borkhausen, 1797 [a-KAN-this]: 'small bird of thorn-bushes', from Greek *akanthis* or *akanthos*, probably a finch of some kind, named for its fondness for thistles – Arnott (2007) suggested a linnet or greenfinch perhaps, on grounds of Aristotle's (c. 330 BC) account of the bird; and Latin *acanthis*, a little bird of a dark-green colour, that lives in the thorn bushes.

Carduelis Brisson, 1760 [kar-dʊ-EH-lis]: 'goldfinch', from Latin *carduelis*, goldfinch.

The species

Common Chaffinch (vagrant – from introduced New Zealand birds – to Lord Howe and Norfolk islands)

Originally Chaff Finch, for its habit of sifting through barnyard straw for seeds. (There is no uncommon chaffinch species in England; the reference may be to contrast with the congeneric Brambling *F. montifringilla*.)

Fringilla coelebs Linnaeus, 1758 [frin-GIL-luh KOY-lebz]: 'unmarried chaffinch', see genus name, and from Latin *caelebs*, unmarried (either sex). Linnaeus (1758) claimed that 'the female alone migrates via Belgium to Italy', so he presumably regarded both the members of a pair as single.

Common Greenfinch (introduced breeding resident)

For the pale olive male; the name goes back at least to the 15th century.

Other names: European Greenfinch, because its natural range is pan-European; Greenie; Green Linnet, though its use in Australia is unclear, because Linnets (*Carduelis cannabina*) have not been introduced – however, the name doubtless came from England, and Greenfinches have often been placed with Linnets in genus *Carduelis*.

Chloris chloris (Linnaeus, 1758) [KLO-ris KLO-ris]: 'greenfinch-greenfinch', see genus name.

Common Redpoll (introduced; breeding resident of Macquarie Island, vagrant – from New Zealand – to Lord Howe Island)

For the red forehead of all male redpolls (the taxonomy is debated: up to five species have been recognised, though some would see just one).

Acanthis flammea (Linnaeus, 1758) [a-KAN-this FLAM-me-uh]: 'flaming thorn-bush bird', see genus name, and from Latin *flammeus*, flaming or fiery.

European Goldfinch (introduced breeding resident)

For the yellow wing bars, especially striking in flight; the name goes back to at least the 16th century. Found virtually throughout Europe.

Other names: Eurasian Goldfinch (its range continues deep into central Asia); Thistle-finch, for its extreme fondness for seeding thistle-heads; Goldie.

Carduelis carduelis (Linnaeus, 1758) [kar-dʊ-EHL-is kar-dʊ-EHL-is]: 'goldfinch-goldfinch', see genus name.

EMBERIZIDAE (PASSERIFORMES): buntings and American sparrows

Emberizidae Vigors, 1825 [em-be-RI-zi-dee]: the Bunting family, see genus name *Emberiza*.

The genus

Emberiza Linnaeus, 1758 [em-be-REE-zuh]: 'bunting', from Old German *embritz*, bunting.

The species

Yellowhammer (vagrant to Lord Howe Island from introduced New Zealand populations)

From an Old English name for the bird, *hamer* or *ammer*, the origin of which has been lost, to which yellow was later appended, for its yellow head.

Emberiza citrinella Linnaeus, 1758 [em-be-REE-zuh si-tri-NEL-luh]: 'little lemon-coloured bunting', see genus name, and from *citrinella*, which is what Mediaeval Italians called a yellow finch (Aldrovandus 1600). Ultimately from Greek *kitrea/kitron*, the citron tree and its fruit (*Citrus medica*), and Latin *citrus or citreus*, which Pliny (77–79 AD) said referred to an African tree with fragrant wood (possibly the cedar *Cedrus atlantica*) but also to the citron tree. From the second (and Mediaeval Latin *citrinus*) comes our own word 'citrus', French *citron* meaning lemon (and related words).

References

'A Bush Naturalist' (1879) Muscicapidae – Flycatchers. *Brisbane Courier Mail*, 11 October 1879, <http://trove.nla.gov.au/ndp/del/article/885543>.

'A. stentoreus' (2004) Why 'Regent' Honeyeater? *Canberra Bird Notes* **29** (1).

Aelian [Claudius Aelianus] (c. 200 AD) *De Natura Animalium*. Jacobs, Jena, Germany.

Aldrovandus [Aldrovandi, Ulisse] (1600) *Ornithologiae Tomus Alter, cum indice copiosissimo variarum linguarum*. Bellagamba, Bologna, Italy.

Aldrovandus (1603) *Ornithologiae Tomus Tertius ac Postremus, cum indice copiosissimo variarum linguarum*. Bernia, Bologna, Italy.

Alström P, Cibois A, Irestedt M, Zuccon D, Gelang M, Fjeldså J, *et al.* (2018) Comprehensive molecular phylogeny of the grassbirds and allies (Locustellidae) reveals extensive non-monophyly of traditional genera, and a proposal for a new classification. *Molecular Phylogenetics and Evolution* **127**, 367–375. doi:10.1016/j.ympev.2018.03.029

ANBG (2019) William Swainson. Australian National Botanic Gardens, Canberra, <http://www.anbg.gov.au/biography/swainson-william.html>.

Anonymous (1911) Obituary notice. *Emu* **11**, 62. [Kendall Broadbent]

Argus (1927) Nature notes and queries; rare visitors. [Melbourne] *Argus*, 1 April 1927, <http://trove.nla.gov.au/ndp/del/article/3846879>.

Aristophanes (414 BC) *The Birds*. Tufts University website, Medford MA, USA, <http://www.perseus.tufts.edu/hopper/text?doc=Perseus%3Atext%3A1999.01.0025%3Acard%3D227>.

Aristotle (c. 330 BC) *History of Animals*. Translated by D'Arcy Wentworth Thompson. <http://ebooks.adelaide.edu.au/a/aristotle/history/complete.html>.

Arnott WG (2007) *Birds in the Ancient World from A to Z*. Routledge, London, UK.

Ashby E (1910) Description of a New Acanthiza. *Emu* **9**, 137–138. doi:10.1071/MU909137

Azara F (1805) *Apuntamientos para la Historia Natural de los Páxaros del Paragüay*. Tomo 3. Manuela Ibarra, Madrid, Spain.

Backhouse J (1843) *A Narrative of a Visit to the Australian Colonies*. London, UK.

Baker S (1945) *The Australian Language*. Angus and Robertson, Sydney.

Banks J (1770) *The Endeavour Journal of Sir Joseph Banks*. London, UK.

Banks RC, Browning MR (1995) Comments on the status of revived old names for some North American birds. *The Auk* **112**, 633–648.

Belon P (1555) *L'Histoire de la Nature des Oiseaux, avec leurs descriptions, et naïfs portraicts retirez du naturel*. Cavellat, Paris, France.

Bennett G (1834) *Wanderings in New South Wales, Batavia, Pedir Coast, Singapore, and China, being the Journal of a Naturalist in those Countries during 1832, 1833, and 1834*. London, UK.

Beolens B, Watkins M (2003) *Whose bird? Men and Women Commemorated in the Common Names of Birds*. Christopher Helm, London, UK.

Beruldsen G (1980) *A Field Guide to Nests and Eggs of Australian Birds*. Rigby, Adelaide.

Billberg GJ (1828) *Synopsis faunae Scandinaviae* **1** (2), 166.

Blakers M, Davies SJJF, Reilly PN (1984) *Atlas of Australian Birds*. Royal Australasian Ornithologists Union, Melbourne University Press, Melbourne.

Blyth E (1843) Report for December Meeting 1842. *Journal of the Asiatic Society of Bengal* **12**(2), 943.

Blyth E (1864) Report from the curator in Proceedings of the Asiatic Society of Bengal for the months of August, September and October 1863. *Journal of the Asiatic Society of Bengal* **32**, 451–461.

Boddaert P (1783) *Table des Planches Enluminéez d'Histoire Naturelle de M. D'Aubenton: Avec les denominations de M.M. de Buffon, Brisson, Edwards, Linnaeus et Latham, precedé d'une Notice des Principaux Ouvrages Zoologiques enluminés*. Utrecht, Netherlands.

Boie F (1822) Ueber Classification, in sonderheit der Europäischen Vögel. *Isis von Oken* **1822**, 563–564.

Boles W (1983) A taxonomic revision of the brown thornbill *Acathiza pusilla* Shaw (1790) with description of a new sub-species. *Emu* **83**, 51–58. doi:10.1071/MU9830051

Bonaparte CL (1830) *Annali di storia naturale*. Tomo 4. Marsigli, Bologna, Spain.

Bonaparte CL (1850) *Conspectus Generum Avium*. De Breuk, Leiden, Netherlands.

Bonaparte CL (1854) Coup d'oeil sur les pigeons (deuxième partie). *Comptes Rendus Hebdomadaires des Séances de l'Académie des Sciences* **39**, 1072–1078.

Bonaparte CL (1855) Coup d'oeil sur les pigeons (quatrième partie). *Comptes Rendus Hebdomadaires des Séances de l'Académie des Sciences* **40**, 15–24.

Bosc LAG (1792) Ardea gularis. *Actes de la Société d'Histoire Naturelle de Paris* **1**, 4.

BOU (1915) *A List of British Birds.* British Ornithologists' Union, London, UK.
Breton W (1833) *Excursions in New South Wales, Western Australia and Van Diemen's Land.* London.
Brisson MJ (1760) *Ornithologie ou Méthode contenant la Division des Oiseaux en Ordres, Sections, Genres, Especes et Leurs Variétés.* Tomes 1, 3 and 6. Bauche, Paris, France.
Brooker MG, Ridpath MG, Estbergs AJ, Bywater J, Hart DS, Jones MS (1979) Bird observations on the north-west Nullarbor Plains and neighbouring regions, 1967–1978. *Emu* **79**, 176–190. doi:10.1071/MU9790176
Buffon GLL (1765–1783) *Planches Enluminées d'Histoire Naturelle.* 10 volumes. Paris, France.
Buffon GLL (1770) *Histoire Naturelle des Oiseaux. Volume 1.* Imprimerie Royale, Paris, France.
Buffon GLL, Montbeillard G (1779) *Histoire Naturelle des Oiseaux. Volume 6.* Imprimerie Royale, Paris.
Cabanis J-L (1850) *Verzeichniss der Ornithologischen Sammlung des Oberamtmann Ferdinand Heine. Volumes 1–2. Museum Heineanum.* Halberstadt, Berlin, Germany.
Caius J (1570) *De rariorum Animalium atque Stirpium Historia.* Liber 1, <http://www.gutenberg.org/files/27655/27655-h/27655-h.htm>.
Camden W (1607) *Britannia*, Philemon Holland, UK, <http://www.philological.bham.ac.uk/cambrit/>.
Campbell AJ (1897) Some Australian Birds. *The Australasian (Melbourne)*, 20 November, p. 30.
Campbell A (1901) *Nests and Eggs of Australian Birds.* 2 parts. Pawson and Brailsford, Sheffield, UK.
Campbell AJ (1902) Description of a New *Microeca* from Northern Australia. *Emu* **2**, 85. doi:10.1080/01584197.1902.11791889
Campbell AJ (1903a) Stray feathers: some November notes. *Emu* **2**, 176–178. doi:10.1071/MU902172i
Campbell AJ (1903b) Description of a new *Acanthiza*. *Emu* **2**, 202–203. doi:10.1071/MU902202
Carter T (1921) Remarks and notes on some Western Australian birds. *Emu* **21**, 54–58. doi:10.1071/MU921054
Catesby M, Edwards G (1731) *The Natural History of Carolina, Florida and the Bahamas. Volume 1.* C. Marsh, London, UK.
Cayley N (1931) *What Bird is That?* Angus and Robertson, Sydney.
Chisholm A (1922) *Mateship with Birds.* Whitcombe and Tombs, Melbourne.
Chisholm EC (1929) Birds of the East Bogan District, County of Flinders, New South Wales. *Emu* **29**, 143–151. doi:10.1071/MU929143
Chisholm A (1950) Further notes on vocal mimicry. *Emu* **49**, 232–234. doi:10.1071/MU949232
Chisholm A (1960a) *The Romance of the Lyrebird.* Angus and Robertson, Sydney.
Chisholm A (1960b) Remarks on robins. *Emu* **60**, 221–236. doi:10.1071/MU960221
Chisholm AH (1972) 'Elsey, Joseph Ravenscroft (1834–1857)'. *Australian Dictionary of Biography*, National Centre of Biography, Australian National University, Canberra, <http://adb.anu.edu.au/biography/elsey-joseph-ravenscroft-3481/text5331>.
Christidis L, Boles W (2008) *Systematics and Taxonomy of Australian Birds.* CSIRO Publishing, Melbourne.
Cicero MT (45 BC) *De Natura Deorum 2* (translated by O. Plasberg, 1917). Teubner, Leipzig, Germany.
Clark A (1910) A new name for *Psephotus multicolor*. *The Auk* **27**, 80.
Cocker M, Mabey R (2005) *Birds Britannica.* Chatto & Windus, London, UK.
Cohn M (1934) Stray feathers: Koolides. *Emu* **34**, 140–141. doi:10.1071/MU934134k
Collett R (1898) On some pigeons and parrots from north and north-west Australia. *Proceedings of the Zoological Society of London* **1898**, 353–357.
Collins D (1798) *An Account of the English Colony of New South Wales. Volume 1.* London, UK.
Collins D (1802) *An Account of the English Colony of New South Wales. Volume 2.* London, UK.
Condon HT (1951) Notes on the birds of South Australia: occurrence, distribution and taxonomy. *South Australian Ornithologist* **20**, 26–68.
Coues E (1882) *The Coues Checklist of North American birds.* 2nd edn. Estes and Lauriat, Boston MA, USA.
Covington MA (2010) *Latin Pronunciation Demystified.* Website, <http://www.ai.uga.edu/mc/latinpro.pdf>.
CSIRO (1969) *An Index of Australian Bird Names. Division of Wildlife Research Technical Paper No. 20.* Commonwealth Scientific and Industrial Research Organisation, Canberra.
Ctesias (c. 400 BCa) *Indica* Extracts in Photius (9th century AD), <https://www.livius.org/sources/content/ctesias-overview-of-the-works/photius-excerpt-of-ctesias-indica/>.
Ctesias (c. 400 BCb) *Persica* Extracts in Photius (9th century AD), <https://www.livius.org/sources/content/ctesias-overview-of-the-works/photius-excerpt-of-ctesias-persica/>.
Cupper J, Cupper L (1981) *Hawks in Focus.* Jaclin Enterprises, Mildura.

Curtis S (2011) 'Devilbird' nr Narooma (NSW). Birding-Aus website, <http://birding-aus.org/?p= 6771>.

Cuvier G (1817) *Le règne animal distribué d'après son organisation.* Tome 1. Deterville, Paris. D'Ombrain AF (1934). The white-browed wood-swallow. *Emu* **33**, 292–297. doi:10.1071/MU933292

Dabb G (2005) *Chthonicoline Reflections.* Birding-Aus Archives, <http://bioacoustics.cse.unsw.edu.au/archives/html/canberrabirds/2005-11/msg00054.html>.

Dabb G (2008) *The Triumph of Dr Horsfield.* Birding-Aus Archives, <http://bioacoustics.cse.unsw.edu.au/archives/html/canberrabirds/2008-02/msg00181.html>.

Dampier W (1703) *A New Voyage Round the World.* London, UK.

de Freycinet L (1824) *Voyage Autour du Monde Exécuté sur l'Uranie et la Physicienne.* 17 Volumes. Paris, France.

del Hoyo J, Elliott A, Sargatal J, Christie DA, de Juana E (Eds) (2019) *Handbook of the Birds of the World Alive. Lynx Edicions,* Barcelona, Spain, <http://www.hbw.com/>.

de Sahagún B (1829) [c. 1575] *Historia General de las Cosas de Nueva España. Volume 3.* Alejandro Valdés, Mexico.

De Vis CW (1883) Description of two new birds of Queensland. *Proceedings of the Linnean Society of New South Wales* **7**, 561–563. doi:10.5962/bhl.part.22765

De Vis CW (1895) Description of a flycatcher, presumably new. *Proceedings of the Linnean Society of New South Wales* **20**(2), 171. doi:10.5962/bhl.part.24339

Debus S (1998) *The Birds of Prey of Australia: A Field Guide.* Oxford University Press, Melbourne.

Dumont C (1824) Mérion, *Dictionnaire des Sciences Naturelles, dans Lequel on Traite Méthodiquement des Differens Êtres de la Nature. Volume 30.* FG Levrault, Strasbourg, France.

Dunn O, Kelley JA (Eds) (1991) *The Diario of Christopher Columbus' First Voyage to America (1492–1493) abstracted by Fray Bartolomé de las Casas.* University of Oklahoma Press, Norman OK, USA.

Edwards G (1747) *Natural History of Birds.* Part 2. College of Physicians, London, UK.

Encyclopaedia of New Zealand (1966) Buller, Sir Walter Lawry, NZC, KCMG, FRS. Ministry for Culture and Heritage, Wellington, New Zealand, <http://www.teara.govt.nz/en/1966/buller-sir-walter-lawry-nzc-kcmg-frs/1>.

Eyton TC (1836) *A Catalogue of British Birds* and *A History of the Rarer British Birds.* Longman, Rees, Orme, Brown, Green, and Longman, London, UK.

Falla RA (1976) Notes on the gadfly petrels *Pterodroma externa* & *P. e. cervicalis. Notornis* **23**, 320–322.

Field B (Ed.) (1825) *Geographical Memoirs on New South Wales.* John Murray, London, UK.

Fletcher JA (1924) Birds of the steppes. *Emu* **24**, 107–117. doi:10.1071/MU924107

Flower SS (1925) Contributions to our knowledge of the duration of life in vertebrate animals. IV. Birds. *Proceedings of the Zoological Society of London* **95**(4), 1365–1422. doi:10.1111/j.1469-7998.1925.tb07441.x

Ford HW (1918) Birds about the Tanjil River and Ranges, Victoria, 1916–17. *Emu* **17**, 221–223. doi:10.1071/MU917221

Ford J (1983) Speciation in the ground-thrush complex *Zoothera dauma* in Australia. *Emu* **83**, 141–151. doi:10.1071/MU9830141

Ford JR, Parker SA (1974) Distribution and taxonomy of some birds from south-western Queensland. *Emu* **74**, 177–194. doi:10.1071/MU974177

Forster JR (1781) *Indische Zoologie oder Systematische Beschreibungen Seltener und Unbekannter Thiere aus Indien.* Gebauer, Halle, Germany.

Forster JR (1844) *Descriptiones Animalium Quae in Itinere ad Maris Australis Terras per Annos 1772 1773 et 1774 suscepto.* Berlin, Germany.

Frith HJ (1967) *Waterfowl in Australia.* Angus and Robertson, Sydney.

Gaza T (1513) *Habentur hoc Volumine haec Theodoro Gaza interprete, Aristotelis De Generatione Animalium.* Book V. Venice, Italy.

Gesner C (1555) *Historiae Animalium Liber III qui est de Avium natura.* Zurich, Switzerland.

Gloger C (1827) Etwas über einige ornithologische Gattungsbenennungen. *Notizen aus dem Gebiete der Natur- und Heilkunde* **16,** 279.

Gmelin JF (1788) *Caroli A. Linné, Systema Naturae per Regna Tria Naturae, Secundum Classes, Ordines, Genera, Species, cum Characteribus, Differentiis, Synonymis, Locis.* Edn 13, Volume I. Beer, Leipzig, Germany.

Gmelin JF (1789) *Caroli A. Linné, Systema Naturae per Regna Tria Naturae, Secundum Classes, Ordines, Genera, Species, cum Characteribus, Differentiis, Synonymis, Locis.* Edn 13, Volume I, Part 2. Beer, Leipzig, Germany.

Goodwin D (1970) *Pigeons and Doves of the World*. British Museum (Natural History), London, UK.
Gould J (1836a) Characters of some new birds in the Society's collection, including two new genera, *Paradoxornis* and *Actinodura*. *Proceedings of the Zoological Society of London* **4**, 17–19.
Gould J (1836b) A small collection of birds from Swan River. *Proceedings of the Zoological Society of London* **4**, 85–86.
Gould J (1837) *The Birds of Europe. Volume 2*. J Gould, London, UK.
Gould J (1838a) *Synopsis of the Birds of Australia; Companion to Gould's Handbook*. Thorn and Greenwell, Brisbane.
Gould J (1838b) Characters of a large number of new species of Australian birds. *Proceedings of the Zoological Society of London*, **5**, 138–157 [1837].
Gould J (1840) Presentation of fifty new species of Australian birds. *Proceedings of the Zoological Society of London*, **8**, 113, 147–151, 159–165, 169–179. [1841]
Gould J (1842) Some additional ornithological novelties from Australia. *Proceedings of the Zoological Society of London* **10**, 17–21.
Gould J (1843) Thirty new species of Australian birds. *Proceedings of the Zoological Society of London, [1842]* **10**, 135–140.
Gould J (1844) A new species of bird from Western Australia. *Proceedings of the Zoological Society of London* **12**, 1–2.
Gould J (1845) Meeting of July 22. *Proceedings of the Zoological Society of London* **13**, 80.
Gould J (1847) Descriptions of new species of Australian birds. *Proceedings of the Zoological Society of London* **15**, 31–35.
Gould J (1848) *The Birds of Australia. Volumes 1–7*. London, UK.
Gould J (1850) Descriptions of new birds. *Proceedings of the Zoological Society of London* **18**, 91–95.
Gould J (1851) Descriptions of a new species of *Ptilotis* and a new species of *Eopsaltria*. *Proceedings of the Zoological Society of London* **19**, 285. doi:10.1111/j.1096-3642.1851.tb01181.x
Gould J (1855) *The Birds of Australia. Supplement*. Part 2. London, UK.
Gould J (1857) On several new species of birds from various parts of the world. *Proceedings of the Zoological Society of London* **25**, 220–224. doi:10.1111/j.1096-3642.1857.tb01230.x
Gould J (1865) *Handbook to the Birds of Australia*. 2 volumes. J Gould, London, UK.
Gould J (1866) Additions to the list of the avifauna of Australia, with descriptions of three new species. *Proceedings of the Zoological Society of London* **1866**, 217–218.
Gould J (1869) *The Birds of Australia. Supplement* [originally issued in parts, 1851–69]. London, UK.
Gould J (1872) On two new species of birds. *Annals and Magazine of Natural History. Series 4* **10**, 114.
Gould J (1875) Descriptions of three new species of Australian birds. *Proceedings of the Zoological Society of London* **1875**, 314–315.
Grant M, Hazel J (2002) *Who's Who in Classical Mythology*. Routledge, London, UK.
Graves R (1996) *Greek Myths*. Folio Society, London, UK.
Gray JE (1830) *Illustrations of Indian Zoology.Volume 1*. Treutel, London, UK.
Gray GR (1849) *The Genera of Birds*. Longman, Brown, Green and Longman, London, UK.
Gray GR (1858) A list of the birds, with descriptions of new species obtained by Mr Alfred R. Wallace in the Aru and Ké Islands. *Proceedings of the Zoological Society of London* **26**, 169–198. doi:10.1111/j.1469-7998.1858.tb06363.x
Gray GR (1860) List of birds collected by Mr Wallace at the Molucca Islands, with descriptions of new species etc. *Proceedings of the Zoological Society of London* **28**, 366.
Gray GR, Wetmore A (1841) *A List of the Genera of Birds: with their Synonyma and an Indication of the Typical Species of each Genus. [With Appendix (1842)]*. Printer Richard and John E Taylor, London, UK.
Hall R (1899) *A Key to the Birds of Australia and Tasmania with their Geographical Distribution in Australia*. Melville, Mullen and Slade, Melbourne, and Dulau, London, UK.
Hall R (1907) *The Useful Birds of Southern Australia*. Lothian, Sydney.
HANZAB (1990–2006) *Handbook of Australian, New Zealand and Antarctic Birds*. 7 Volumes. Royal Australasian Ornithologists Union and Birds Australia, Melbourne.
Harcourt EV (1851) A sketch of Madeira: containing information for the traveller or invalid visitor. John Murray, London, UK.
Hartert E (1892) Coraciae, of the familes Cypselidae, Caprimulgidae, Podargidae, and Steatornithindae. *Catalogue of the Picariae in the Collection of the British Museum*. British Museum (Natural History), London.

Hay J, MacLagan M, Elizabeth G (2008) *New Zealand English.* Edinburgh University Press.
Heine F, Reichenow A (1890) *Nomenclator Musei Heineani Ornithologici; Verzeichniss der Vogel-Sammlung des Königlichen Oberamtmanns Ferdinand Heine.* R. Friedländer & Sohn, Berlin, Germany.
Henderson J (1832) *Observations of the Colonies of New South Wales and Van Diemen's Land.* Calcutta, India.
Hernandez F (1651) *Rerum medicarum Novae Hispaniae Thesaurus, seu, Plantarum Animalium Mineralium Mexicanorum Historia.* Rome, Italy.
Herodotus (c. 435 BC) *The Histories.* Tufts University website, Medford MA, USA, <http://www.perseus.tufts.edu/hopper/text?doc=Perseus:abo:tlg,0016,001:4:175>.
Hill HE (1902) Some notes from the Geelong and Otway Districts. *Emu* **2**, 161–167. doi:10.1071/MU902161
Hill HE (1903) Some notes from Brookton, W.A. *Emu* **3**, 104–107. doi:10.1071/MU903104
Hince B (2000) *The Antarctic Dictionary: A Complete Guide to Antarctic English.* CSIRO Publishing, Melbourne.
Hindwood K (1932) The ground parrot. *Emu* **32**(4), 241–246. doi:10.1071/MU932241
Hindwood K (1949a) A note on William Swainson. *Emu* **49**(3), 208–210. doi:10.1071/MU949208
Hindwood K (1949b) *Pardalotus xanthopygus*: a competition in 'christening'. *Emu* **49**(3), 205–208. doi:10.1071/MU949205
Hodgson BH (1836) The ornithology of Nepal. *Asiatick Researches* **19**, 143–192.
Hodgson BH (1837) On some new species of the Edolian and Ceblepyrine sub-families of the Laniidae of Nepal. *India Review and Journal of Foreign Science and the Arts* **1**, 324–329.
Hombron JB, Jacquinot H (1841) Descriptions de plusieurs oiseaux nouveaux ou peu connus, provenant de l'expédition autour du monde faite sur les corvettes *l'Astrolabe* et *la Zélée. Annales des sciences naturelles series 2* **16**, 312–320.
Homer (c. 700 BCa) *The Odyssey* (trans. EV Rieu, 1946). Penguin Books, London, UK.
Homer (c. 700 BCb) *The Iliad* (translated by EV Rieu, 1950). Penguin Books, London, UK.
Horsfield T (1821) Systematic arrangements and description of birds from the island of Java. *Transactions of the Linnean Society of London* **13**, 133–200. doi:10.1111/j.1095-8339.1821.tb00061.x
Horsfield T (1824). *Zoological Researches in Java and the Neighbouring Islands.* Kingsbury, Parbury & Allen, London, UK.
Howitt R (1845) *Australia: Historical, Descriptive, and Statistic; With an Account of a Four Years' Residence in that Colony; Notes of a Voyage Round the World; Australian Poems, &c.* Longman, Brown, Green and Longmans, London, UK.
Hunter J (1793) *An Historical Journal of the Transactions at Port Jackson and Norfolk Island.* London.
Illustrated London News (1851) *The Illustrated London Reading Book.* Illustrated London News, London, UK.
Imber MJ, Tennyson AJD (2001) A new petrel species (Procellariidae) from the south-west Pacific. *Emu* **101**, 123–127. doi:10.1071/MU00067
Iredale T (1933) Thomas Skottowe, naturalist. *Emu* **33**, 273–278. doi:10.1071/MU933273
Iredale T (1939) The eclipse plumage of the elfin wren (*Ryania melanocephala*). *Emu* **39**, 39–40. doi:10.1071/MU939039
Iredale T (1946) A new Australian parrot. *Emu* **46**(1), 1–2. doi:10.1071/MU946001
Iredale T (1956) *Birds of New Guinea.* Georgian House, Melbourne.
Jackson S (1908) In the Barron River Valley, North Queensland. *Emu* **8**, 233–283. doi:10.1071/MU908233
Jacquinot H, Pucheran J (1853) Mammifères et oiseaux. In *Voyage au Pôle Sud et dans l'Océanie sur les Corvettes L'Astrolabe et La Zélée Tome Troisième: Zoologie.* (Eds JB Hombron and H Jacquinot) p. 69. Gide et J Baudry, Paris, France.
Jardine F, Jardine A (1867) *Narrative of the Overland Expedition of the Messrs. Jardine, from Rockhampton to Cape York, Northern Queensland.* JW Buxton, Brisbane.
Jardine W (1831) Descriptions of new or little-known species of birds. *Edinburgh Journal of Natural and Geographic Sciences* **3**, 209–212.
Jardine W (1834) *Naturalist's Library: Ornithology. Volume IV.* Lizars, Stirling and Kenney, Edinburgh, UK.
Jardine W, Selby PJ (1826–1835) *Illustrations of Ornithology. Volumes 1 and 2.* WH Lizars, Edinburgh, UK.

Jerdon T (1862) *Birds of India. Volume 1.* Calcutta, India.
Jerrard CHH (1931) Vocal powers of the yellow oriole. *Emu* **31**, 42–43. doi:10.1071/MU931042
Jobling JA (2010) *Dictionary of Scientific Bird Names.* Christopher Helm, London, UK.
Jobling JA (2018) *Key to Scientific Names in Ornithology.* In *Handbook of the Birds of the World Alive.* (Eds J del Hoyo, A Elliott, J Sargatal, DA Christie and E de Juana). Lynx Edicions, Barcelon, Spain, <https://www.hbw.com>.
Kear J (2005) *Ducks, Geese and Swans.* Oxford University Press, Oxford, UK.
Kerr R (1792) *The Animal Kingdom, or Zoological System, of the Celebrated Sir Charles Linnæus.* Strahan, Cadell and Creech, Edinburgh, UK.
Kershaw JA (1918) Australian green-backed finch (*Erythrura trichroa magillivrayi*). *Emu* **18**, 1. doi:10.1071/MU918001
King PP (1827) *Narrative of a Survey of the Intertropical and Western Coasts of Australia Performed between the Years 1818 and 1822. Volume 2.* John Murray, London, UK.
Kittlitz H (1833) Über einige noch unbeschriebene Vögel von der Insel Luzon, den Carolinen und den Marianen. *Mémoires Présentés à l'Académie des Sciences de Saint Pétersbourg par divers Savants et lus dans ses Assemblées* **2**, 1–10.
Kloot T (1986) Why I became interested in natural history. *The La Trobe Journal* **38**, 33–34.
Kuhl H (1820a) *Beiträge zur Zoologie und vergleichenden Anatomie.* Frankfurt am Main, Germany.
Kuhl H (1820b) *Conspectus Psittacorum: cum Specierum Definitionibus, Novarum Descriptionibus, Synonymis et Circa Patriam Singularum Naturalem Adversariis, Adjecto Indice Museorum, ubi earum Artificiosae Exuviae Servantur: cum Tabulis III.* Bonn, Germany.
Latham J (1785) *General Synopsis of Birds. Volume 3 Part 2.* Leigh & Sotheby, London, UK.
Latham J (1790) *Index Ornithologicus, sive Systema Ornithologiae: Complectens Avium Divisionem in Classes, Ordines, Genera, Species, Ipsarumque Varietates: Adjectis Synonymis, Locis, Descriptionibus, etc. Volume 1.* Leigh & Sotheby, London, UK.
Latham J (1801) *General Synopsis of Birds. Supplement 2.* Leigh & Sotheby, London, UK.
Latham J (1802) *Supplementum Indicis Ornithologici sive Systematis Ornithologiae.* Leigh & Sotheby, London, UK.
Latham J (1821) *A General History of Birds. Volume 1.* Jacob & Johnson,Winchester, UK.
Latham J (1823) *A General History of Birds. Volume 2.* Jacob & Johnson,Winchester, UK.
Lavery HJ, Seton D (1967) *Cisticola juncidis* (Rafinesque) in north-east Queensland. *Emu* **67**, 125–132. doi:10.1071/MU967125
Leach J (1911) *An Australian Bird Book.* Whitcombe and Tombs, Melbourne.
Lear E (1832) *Illustrations of the Family of the Psittacidae, or Parrots.* E Lear, London, UK.
Leem K, Gunnerus JE (1787) *Beskrivelse over Finmarkens Lapper: deres Tungemaal, Levemaade og forrige Afgudsdyrkelse.* Copenhagen, Denmark.
Leichhardt L (1847) *Journal of an Overland Expedition in Australia, 1844–46.* T and W Boone, London, UK.
Le Maout E (1853) *Histoire Naturelle des Oiseaux: Suivant la Classification de M. Isidore Geoffroy-Saint-Hilaire.* L. Curmer, Paris, France.
Le Soeuf D (1899) On the habits of the mound-building birds of Australia. *The Ibis* **5**, 9–19.
Lesson RP (1825) Distribution géographique de quelques oiseaux marins, observés dans le voyage autour du monde de la corvette *La Coquille. Annales des Sciences Naturelles* **6**, 88–103.
Lesson RP (1827) Nouveau genre d'oiseau. *Bulletin des Sciences Naturelles et de Géologie. Deuxième Section du Bulletin Universel des Sciences et de l'Industrie* **11**, 443–444.
Lesson RP (1828) *Manuel d'ornithologie, ou Description des Genres et des Principales Espèces d'oiseaux.* Roret, Paris, France.
Lesson RP (1831) *Traité d'Ornithologie ou Table Méthodique des ordres, sous-ordres, familles, tribus, genres, sous-genres et races d'oiseaux. Volumes 1 and 2.* FG Levrault, Paris, France.
Lesson RP (1837) Histoire naturelle. In *Journal de la Navigation autour du Globe de la Frégate la Thétis et de la Corvette L'Espérance. (H Bougainville) Volume 2.* Rignoux, Paris, France.
Lesson RP (1839a) *Voyage Autour du Monde Entrepris par Ordre du Gouvernement sur la corvette La Coquille. Volumes 1 and 2.* Pourrat Frères, Paris, France.
Lesson RP (1839b) Liste d'oiseaux nouveaux de la collection du docteur Abeillé de Bordeaux. *Revue Zoologique par la Société Cuvierienne* **2**, 167.
Lesson RP, Garnot P (1826) Description d'une nouvelle espèce de Cassican (*Barita Keraudrenii*). *Bulletin des Sciences naturelles et de Géologie (Paris)* 8(1) 110–111.

Lesson RP, Garnot P (1827). *Voyage autour du Monde exécuté par order du Roi, sur la Corvette de Sa Majeste, la Coquille, pendant les années 1822, 1823, 1824 et 1825*. Vol. 1, Zoologie. Livr. 5. Arthus Bertrand, Paris, France.

Lewin JW (1805) *Prodromus Entomology. Natural History of Lepidopterous Insects of New South Wales, Collected, Engraved and Faithfully Painted after Nature*. Thomas Lewin, London, UK.

Lewin J (1808) *Birds of New Holland with their Natural History*. London, UK.

Lewin J (1813) *Birds of New South Wales with their Natural History*. G Howe, Sydney.

Lewin J (1822) *Natural History of the Birds of New South Wales*. JH Bohte, London, UK.

Lewis CT, Short C (1879) *A Latin Dictionary*. Tufts University website, Medford MA, USA, http://www.perseus.tufts.edu/hopper/text?doc= Perseus%3Atext%3A1999.04.0059>.

Liddell HG, Scott R (1940) *A Greek-English Lexicon*. Tufts University website, Medford MA, USA, http://www.perseus.tufts.edu/hopper/text?doc=Perseus:text:1999.04.0057>.

Lindsay WM (1921) *The Corpus, Épinal, Erfurt and Leyden Glossaries*. Philological Society, London, UK.

Linnaeus C (1746) *Fauna Svecica Sistens Animalia Sveciæ Regni: Quadrupedia, Aves, Amphibia, Pisces, Insecta, Vermes*. Conrad & Georg Jacob Wishoff, Leiden, Netherlands.

Linnaeus C (1758) *Systema Naturae per Regna Tria Naturae, Secundum Classes, Ordines, Genera, Species, cum Characteribus, Differentiis, Synonymis, Locis*. Editio 10, Tomus I. Laurentius Salvius, Stockholm, Sweden.

Linnaeus C (1767) *Mantissa Plantarum. Generum editionis VI et Specierum editionis II. Appendix: Regni Animalis, Aves*. Laurentius Salvius, Stockholm, Sweden.

Linnaeus C (1771) *Mantissa Plantarum Altera. Generum Editio VI et Specierum Edition II*. Laurentius Salvius, Stockholm, Sweden.

Littler F (1903) Notes on some birds peculiar to Tasmania. *Emu* **3**, 23–30. doi:10.1071/MU903023

Lockwood WB (1984) *The Oxford Dictionary of British Bird Names*. Oxford University Press, Oxford, UK.

Lord C (1927) The 'button-grass' parrot. *Emu* **27**, 42–43. doi:10.1071/MU927042

Lucas AHS, Le Soeuf WHD (1911) *The Birds of Australia*. Whitcombe and Tombs, Melbourne.

Lyons CM (1901) Some notes on the birds of Lake Eyre District. *Emu* **1**(3), 133–138. doi:10.1071/MU901133

Marcgrave G (1648) *Historiæ rerum naturalium Brasiliæ*. Ioannes de Laet, Antwerp, Belgium.

Margery AT (1895) *Aboriginal Water-Quest*. Proceedings of the Royal Geographical Society of Australasia, South Australian Branch. Adelaide.

Marsh JH (1948) *No Pathway Here*. Timmins, Cape Town, South Africa.

Mathews G (1909) Additions to the 'Handlist of the Birds of Australasia'. *Emu* **9**, 92. doi:10.1071/MU909092a

Mathews G (1910a) Additions to the 'Handlist of the Birds of Australasia'. *Emu* **10**, 57. doi:10.1071/MU910057

Mathews G (1910b) On some necessary alterations in the nomenclature of birds. *Novitates Zoologicae* **17**, 492–503. doi:10.5962/bhl.part.13695

Mathews G (1910*c*) In the address by British Ornithologists' Club Chairman P. Sclater. *Bulletin of the British Ornithologists' Club* **27**, 16.

Mathews G (1912a) A reference-list to the birds of Australia. *Novitates Zoologicae* **18**(3), 171–446. doi:10.5962/bhl.part.1694

Mathews G (1912b) Australian Black-browed Mollymawk. *The Birds of Australia* **2**, 267–272.

Mathews G (1912c) Additions and corrections to my reference list. *Austral Avian Record* **1**(2), 25–52.

Mathews G (1913) *A List of the Birds of Australia*. Witherby and Co., London, UK.

Mathews G (1946) *A Working List of Australian Birds*. Shepherd and Newman, Sydney.

McDonald JD (1973) *Birds of Australia*. Reed, Sydney.

McGilp JN (1934) Stray feathers: Koolides. *Emu* **34**, 140. doi:10.1071/MU934134j

McGilp JN (1935) Birds of the Musgrave Ranges. *Emu* **34**, 163–176. doi:10.1071/MU934163

Mearns B, Mearns R (1988) *Biographies for Birdwatchers: The Lives of those Commemorated in Western Palaearctic Bird Names*. Academic Press, London, UK.

Mearns B, Mearns R (1992) *Audubon to Xántus: The Lives of those Commemorated in North American Bird Names*. Academic Press, London, UK.

Mellor JW (1912) Description of a new grasswren. *Emu* **12**(3), 166–167.

Menkhorst P, Rogers D, Clarke R, David J, Marsack P, Franklin K (2017) *The Australian Bird Guide*. CSIRO Publishing, Melbourne.

von Middendorff ATh (1848) *Reise in den Äussersten Norden und Osten Sibiriens Während der Jahre 1843 und 1844*. Buchdruckerei der Kaiserlichen Akademie der Wissenschaften, St Petersburg, Russia.
Mitchell T (1838) *Three Expeditions into the Interior of Eastern Australia*. London, UK.
Morcombe M (2000) *Field Guide to Australian Birds*. Steve Parish Publishing, Brisbane.
Morgan AM (1926) *Cinclosoma castanotum clarum* (Chestnut-backed Ground-bird). *South Australian Ornithologist* **8**(5), 138.
Morris E (1898) *Dictionary of Austral English*. London, UK.
Mouchard A (2011) Historias zoologicas. Website, <http://historiaszoologicas.blogspot.com.au/2011_07_01_ archive.html>.
Müller PLS (1776) *Des Ritters Carl von Linné vollständiges Natursystem Supplements und Register-Band*. Gabriel Nicolaus Raspe, Nuremberg, Germany.
Müller S (1839–44) *Land- en Volkenkunde*. In *Verhandelingen over de Natuurlijke Geschiedenis der Nederlandsche Overzeesche Bezittingen*. (Ed. CJ Temminck) p. 177. J.A.Susanna, Leiden, Netherlands.
Murphy RC (1917) A New Albatross from the West Coast of South America. *Bulletin of the American Museum of Natural History* **37**, 861.
Murphy RC (1930) Birds Collected During the Whitney South Sea Expedition XI. *American Museum Novitates* **419**, 4–6.
Murphy RC (1936) *Oceanic Birds of South America*. Macmillan, New York, USA.
Nash R (1926) *The Conquest of Brazil*. Harcourt Brace, New York, USA.
Nash D (2014) *Berigora: a word that clawed on – from where?* Endangered Languages and Cultures website, <http://www.paradisec.org.au/blog/2014/01/berigora-a-word-that-clawed-on-from-where/>.
National Library of New Zealand – (1966) Walter Lawry Buller. National Library of New Zealand website, Wellington, <http://www.natlib.govt.nz/collections/online-exhibitions/early-explorers-and-collectors/walter-buller>.
Newton A (Ed.) (1882) *Scopoli's Ornithological Papers from his Deliciae Florae et Faunae Insubricae*. Willughby Society, London, UK.
Newton A (1896) *A Dictionary of Birds*. Adam and Charles Black, London, UK.
Nicholls B (1924) A Trip to Mungeranie, Central Australia. *Emu* **24**, 45–59. doi:10.1071/MU924045
North AJ (1901) Description of a new species of the genus *Malurus*. *Victorian Naturalist* **18**, 29–30.
North AJ (1902) On three apparently undescribed species of Australian birds. *Victorian Naturalist* **19**, 102–104.
North A (1904) *Nests and Eggs of Birds found Breeding in Australia and Tasmania*. Australian Museum, Sydney.
North A (1909) Notes on the nesting-site of *Gerygone personata*, Gould. *Records of the Australian Museum* 7(3), 186–188. doi:10.3853/j.0067-1975.7.1909.961
North A (1910) Description of a new genus and species of honeyeater from Western Australia. *Victorian Naturalist* **26**, 138–139.
North A (1916) The birds of Coolabah and Brewarrina, north-western New South Wales. *Records of the Australian Museum* **11**(6), 121–162. doi:10.3853/j.0067-1975.11.1916.913
Norton A (1907) Mental Development in Animals. *Proceedings of the Royal Society of Queensland* **9**, 41–52.
Oates EW (1889) *The Fauna of British India. Birds. Volume 1*. Taylor & Francis, London, UK.
Ogilvie-Grant WR (1909) *Malurus bernieri* sp. nov. *Bulletin of the British Ornithologists' Club* **23**, 72.
Olsen P (1995) *Australian Birds of Prey*. University of New South Wales Press, Sydney.
Olsen P (2007) *Glimpses of Paradise: The Quest for the Beautiful Parrakeet*. National Library of Australia, Canberra.
Olsen P, Joseph L (2011) *Stray Feathers: Reflections on the Structure, Behaviour and Evolution of Birds*. CSIRO Publishing, Melbourne.
Olson SL (1998) Comment on the proposed suppression of all prior usages of generic and specific names of birds (Aves) by John Gould and others conventionally accepted as published in the Proceedings of the Zoological Society of London (1). *Bulletin of Zoological Nomenclature* **55**(3), 176–181. doi:10.5962/bhl.part.180
Olson SL (2000) A new genus for the Kerguelen Petrel. *Bulletin of the British Ornithologists' Club* **120**, 61.
Olson SL, Warheit KL (1988) A new genus for *Sula abbotti*. *Bulletin of the British Ornithologists' Club* **108**, 9–12.
Ovid (c. 8 AD) *Metamorphoses*. Republished 1916 in Loeb Classical Library, Harvard University Press, Cambridge MA, USA.

Ovid (c. 16 BC) *Amores*. Book 2, VI. <http://www.sacred-texts.com/cla/ovid/lboo/index.htm>.
Parker S (1973) The identity of *Microeca brunneicauda*, Campbell 1902. *Emu* **73**, 23–25. doi:10.1071/MU973019c
Parker S (1977) Distribution and occurrence in South Australia of owls of the genus *Tyto*. *South Australian Ornithologist* **27**, 207–215.
Parker S (1982) Notes on *Amytornis striatus merrotsyi* Mellor, a subspecies of the striated grasswren inhabiting the Flinders Ra. *South Australian Ornithologist* **29**, 13–16.
Parker S (1984) The identity of *Sericornis tyrannula* De Vis. *Emu* **84**, 108–110. doi:10.1071/MU9840108
Peale T (1848) *Mammalia and Ornithology: United States Exploring Expedition. Volume VIII*. C Sherman, Philadelphia PA, USA.
Pennant T (1776) *British Zoology: Birds*. London, UK.
Penhallurick J (2012) World Bird Info website, <http://worldbirdinfo.net/default.aspx>.
Penhallurick J, Wink M (2004) Analysis of the taxonomy and nomenclature of the Procellariiformes based on complete nucleotide sequences of the mitochondrial cytochrome *b* gene. *Emu* **104**, 125–147. doi:10.1071/MU01060
Peterson AP (2011) Zoonomen website, <www.zoonomen.net>.
Philipps R (1902) The Australian waxbill. *Avicultural Magazine* **8**(12), 209.
Pizzey G (1980) *Field Guide to the Birds of Australia*. 1st edn. Collins, Sydney.
Pizzey G, Knight F (2003) *Field Guide to the Birds of Australia*. 7th edn. Harper Collins, Sydney.
Pizzey G, Knight F (2007) *Field Guide to the Birds of Australia*. 8th edn. Harper Collins, Sydney.
Pliny (77–79 AD) *Natural History*. Book X, <http://penelope.uchicago.edu/Thayer/E/Roman/ Texts/Pliny_the_Elder/home.html>[English]; <http://penelope.uchicago.edu/Thayer/L/Roman/Texts/Pliny_the_Elder/10*.html>[Latin].
Polaris (1993) *Il Vocabolario Etimologico della Lingua Italiana di Pianigiani*. Dizionario Etimologico Online, <http://www.etimo.it>.
Poley S (2012) *Scientific Bird Names Explained*. UK.rec.birdwatching newsgroup website. <http://sbpoley.home.xs4all. nl/ukrb/scientific_names.html>.
Pollard J (1930) Whisper-song. *Emu* **30**, 62–63. doi:10.1071/MU930062
Purnell H (1914) *Gregory M. Mathews with the nest of* Sphenura broadbenti *at Point Addis, November 22, 1914*. National Library of Australia, Canberra, <http://nla.gov.au/nla.pic-vn3793566>.
Quémada B (1988) *Trésor de la Langue Française*. Volume 13. Gallimard, Paris, France.
Quintilian (c. 95 AD) *Institutio Oratoria*. Latin Library website, http://www.thelatinlibrary.com/quintilian/quintilian. institutio1.shtml#6>.
Quoy JRC, Gaimard JP (1830) *Voyage de la Corvette Astrolabe Exécuté par Ordre du Roi Pendant les Années 1826–1827–1828–1829, sous le Commandement de M. J. Dumont d'Urville. Zoologie* Tome 1, Pt 1. J. Tastu, Paris, France.
Rabelais F (1534) *Oeuvres Complètes: Gargantua*. 1961. Société d'Édition "Les Belles Lettres", Paris, France.
Ralphe PG (1905) *Birds of the Isle of Man*. Isle of Man website, <http://www.isle-of-man.com/manxnotebook/fulltext/bd1905/index.htm>.
Ramsay EP (1874) Descriptions of five new species of birds from Queensland and the egg of *Chlamydera maculata*. *Proceedings of the Zoological Society of London* **1874**, 603.
Rand AL, Gilliard ET (1967) *Handbook of New Guinea Birds*. London, UK.
Ranzani C (1822) *Elementi di Zoologia. Tomo Terzo contenente la Storia Naturale degli Uccelli*. Annesio Nobile, Bologna.
RAOU (1913) *The Official Checklist of the Birds of Australia*. Royal Australasian Ornithologists Union. Supplement to *Emu* **13**.
RAOU (1926) *The Official Checklist of the Birds of Australia*. Royal Australasian Ornithologists Union, Wolstenholme, Sydney.
RAOU (1978) *Recommended English Names for Australian Birds*. Royal Australasian Ornithologists Union. Supplement to *Emu* **77**.
Ray J (1713) *Synopsis Methodica Avium & Piscium: Opus Posthumum, Quod Vivus Recensuit & Perfecit Ipse Insignissimus Author: in Quo Multas Species, in Ipsius Ornithologia & Ichthyologia Desideratas, Adjecit: Methodumque Suam Piscium Naturæ Magis Conv*. Gulielmi Innys, London, UK.
Reichenbach L (1849) *Ein Beitrag zur Naturgeschichte Australiens Band 2*. Expedition der vollständigste Naturgeschichte, Dresden and Leipzig, Germany.
Reichenbach L (1852) *Avium Systema Naturale. Das Natürliche System der Vögel*. Expedition der vollständigste Naturgeschichte, Dresden and Leipzig.

Reichenbach L (1862) *Die Singvögel als Fortsetzung der vollständigsten Naturgeschichte.* Expedition der vollständigste Naturgeschichte, Dresden and Leipzig, Germany.

Richardson J, Kirby W, Swainson W (1831) *Fauna Boreali-Americana or the Zoology of the Northern Parts of British America.* Volume 2. John Murray, London, UK.

Richmond CW (1902) List of generic terms proposed for birds during the years 1890 to 1900, inclusive. *Proceedings of the United States National Museum* **24**, 663–730. doi:10.5479/si.00963801.1267.663

Richmond CW (1917) Generic names applied to birds during the years 1906 to 1915, inclusive. *Proceedings of the United States National Museum* **53**, 565–636. doi:10.5479/si.00963801.2221.565

Roberts S (1934) *The brown honeyeater* (Gliciphila indistincta). *Emu* **34**, 33–35. doi:10.1071/MU934033

Robin L (2001) The Flight of the Emu; a hundred years of Australian ornithology 1901–2001. Melbourne University Press, Melbourne.

Rocamora G, Yeatman-Berthelot D (2018) Crow-billed Drongo (*Dicrurus annectens*). In *Handbook of the Birds of the World Alive.* (Eds J del Hoyo, A Elliott, J Sargatal, DA Christie and E Juana). Lynx Edicions, Barcelona, Spain, <https://www.hbw.com/node/60576>.

Ross J (1819) *A Voyage of Discovery: Made under the Orders of the Admiralty, in His Majesty's Ships Isabella and Alexander for the Purpose of Exploring Baffin Bay and Inquiring into the Probability of a North-West Passage. Appendix 2: Zoological Memoranda.* John Murray, London, UK.

Rothschild W (1893) *Diomedea immutabilis* sp.nov. *Bulletin of the British Ornithologists' Club* **9**, 48.

Rothschild W (1903) *Thalassogeron carteri* n.sp. *Bulletin of the British Ornithologists' Club* **14**, 6.

Rothschild W (1907) *Extinct Birds: an Attempt to Unite in one Volume a Short Account of those Birds which have Become Extinct in Historical Times: that is, within the Last Six or Seven Hundred Years: to Which are Added a Few which Still Exist, but are on the Verge of Extinction.* Hutchinson & Company, London, UK.

Rowland P (2008) *Bowerbirds.* CSIRO Publishing, Melbourne.

Rowley I (1974) *Bird Life.* Collins, Sydney.

Sabine J (1819) An account of a new species of gull lately discovered off the west coast of Greenland. *Transactions of the Linnean Society of London* **12**, 520. doi:10.1111/j.1095-8339.1817.tb00244.x

Salvadori T (1878) Uccelli Papuani. *Annali del Museo civico di storia naturale di Genova* 1 (12), 335. Salvin O (1888). Critical notes on the Procellariidae. *The Ibis* **30**(3), 357.

Salvadori T (1891) Description of two new species of Parrots of the genus *Cyanoramphus* in the British Museum. *The Annals and Magazine of Natural History. Zoology, Botany, and Geology. Series 6* **7**, 68.

Saunders DA (1979) Distribution and Taxonomy of the White-tailed and Yellow-tailed Black-Cockatoos *Calyptorhynchus* spp. *Emu* **79**, 215–227.

Schodde R, Bock WJ (1998) On the proposed suppression of all prior usages of generic and specific names of birds (Aves) by John Gould and others conventionally accepted as published in the Proceedings of the Zoological Society of London (2). *Bulletin of Zoological Nomenclature* **55**(3), 181–185.

Schodde R, Bock WJ (2009) *Aplonis* Gould, 1836 (Aves: STURNIDAE): proposed conservation of spelling. *Bulletin of Zoological Nomenclature* **66**(1), 56–63. doi:10.21805/bzn.v66i1.a7

Schodde R, Mason I (1997) *Zoological Catalogue of Australia.* CSIRO Publishing, Melbourne.

Schodde R, Mason I (1999) *The Directory of Australian Birds.* CSIRO Publishing, Melbourne.

Schodde R, Dickinson EC, Steinheimer FD, Bock WJ (2010) The date of Latham's *Supplementum Indicis Ornithologici* 1801 or 1802? *South Australian Ornithologist* **35**(8), 232–235.

Sharpe RB (1883) Catalogue of the Passeriformes or perching birds. *Catalogue of the Birds of the British Museum* **7**, 336.

Sharpe RB (1896) Catalogue of the Limicolae. *Catalogue of the Birds of the British Museum* **24**, 307.

Shaw G (1794) *Zoology of New Holland.* London, UK.

Shaw G (1796) *Cimelia physica. Figures of Rare and Curious Quadrupeds, Birds, &c. Together with Several of the Most Elegant Plants.* T Bensley, London, UK. [2nd edn of Miller JF (1785) *Icones animalium*]

Shaw G (1819) *General Zoology, or Systematic Natural History. Volumes 1, 11 and 12, Birds.* Printed for G Kearsley, London, UK.

Siebold PF, Temminck CJ, Schlegel H (1850) *Fauna japonica, sive, Descriptio animalium, quae in itinere per Japoniam, jussu et Auspiciis, Superiorum, qui Summum in India Batava Imperium Tenent, Suscepto, Annis 1823–1830.* T4 Aves. Arnz et Socios, Leiden, Netherlands.

Slater P (1970) *A Field Guide to Australian Birds: Non-passerines.* Rigby, Adelaide.

Slater P (1974) *A Field Guide to Australian Birds: Passerines.* Rigby, Adelaide.

Sonnerat P (1776) *Voyage à la Nouvelle Guinée.* Ruault, Paris, France.

Soothill E, Whitehead P (1978) *Wildfowl, a World Guide.* Peerage Books, London, UK.

Stejneger L (1884) *Analecta Ornithologica* Second Series. *The Auk* **1**, 231–233. doi:10.2307/4067229

Stephens JF, Shaw G (1824) *General Zoology or Systematic Natural History. Volume 12, Birds*. Printed for J and A Arch, London.

Stephens JF, Shaw G (1826) *General Zoology or Systematic Natural History. Volume 13, Birds*. Printed for J and A Arch, London.

Strickland HE, Phillips J, Richardson J, Owen R, Jenyns L, Broderip WJ, *et al.* (1842) Report of a Committee appointed 'to consider the rules by which the nomenclature of Zoology may be established on a uniform and permanent basis'. John Murray for the British Association for the Advancement of Science, London.

Stroud PT (2000) *The Emperor of Nature: Charles-Lucien Bonaparte and his World*. University of Pennsylvania Press, Philadelphia PA, USA.

Sturt C (1833) *Two Expeditions into the Interior of Southern Australia during the Years 1828, 1829, 1830, 1831*. London.

Sturt C (1849) *Narrative of an Expedition into Central Australia*. London.

Sullivan AM (1911) Some mallee birds. *Emu* **11**, 114–119. doi:10.1071/MU911114

Sullivan C (1931) Notes from north-western New South Wales. *Emu* **31**, 124–135. doi:10.1071/MU931124

Sundevall C (1872) *Methodi Naturalis Avium Disponendarum Tentamen. Försök till Fogelklassens Naturenliga Uppställnung*. Samson & Wallin, London, UK.

Swainson W (1825) On the characters and natural affinities of several new birds from Australasia; including some observations on the Columbidae. *Zoological Journal* **1**, 463–484.

Swainson W (1827) On several groups and forms in ornithology, not hitherto defined. *Zoological Journal* **3**, 343–362.

Swainson W (1831a) On those birds which exhibit the typical perfection of the Family Anatidae. *Journal of the Royal Institution of Great Britain* **2**, 11–29.

Swainson W (1831b). Appendix 1. In *Fauna Boreali-Americana or the Zoology of the Northern Parts of British America. Volume 2*. (Eds J Richardson, W Kirby and W Swainson). John Murray, London, UK.

Swainson W (1837) *On the Natural History and Classification of Birds. Volume 2*. Longman, Rees, Orme, Brown, Green, and Longman, London, UK.

Swinhoe R (1863) On new and little-known birds from China. *Proceedings of the Zoological Society of London* **1863**, 87–94.

Swinhoe R (1867) Jottings on birds from my Amoy journal. *Ibis. Series 2* **3**, 386.

Sydney Morning Herald (1913) *Bird Notes*. National Library of Australia Archives, Canberra, http://trove.nla.gov.au/ndp/del/article/15460468?searchTerm=jakul&searchLimits=>.

Tarr H (1948) Stray feathers: a note on the red-eared firetail. *Emu* **48**, 161. doi:10.1071/MU948158d

Temminck CJ (1815) *Manuel d'Ornithologie ou Tableau Systématique des Oiseaux qui se Trouvent en Europe*. 1st edn. Gabriel Dufour, Paris, France.

Temminck CJ (1820) *Manuel d'ornithologie ou tableau systématique des oiseaux qui se trouvent en Europe*. 2nd edn. Part 1. Gabriel Dufour, Paris, France.

Temminck MCJ (1822) Account of some new species of birds of the genera Psittacis and Columba, in the Museum of the Linnean Society. *Transactions of the Linnean Society of London* **13**, 107–130. doi:10.1111/j.1095-8339.1821.tb00059.x

Temminck CJ (1826–27) *Nouveau Recueil de Planches Coloriées d'Oiseaux: pour servir de suite et de complément aux planches enluminées de Buffon*. FG Levrault, Paris.

Temminck CJ, Laugier de Chartrouse G (1838) *Nouveau Recueil de Planches Coloriées d'Oiseaux: pour servir de suite et de complément aux planches enluminées de Buffon. 5 Volumes*. FG Levrault, Paris, France.

Tench W (1793) *A Complete Account of the Settlement at Port Jackson*. London.

Thieberger N, McGregor W (Eds) (1994) *Macquarie Aboriginal Words*. Macquarie University Press, Sydney.

Thomas N, Berghof O (Eds) (2000) *Voyage Round the World: George Forster. Volume 1*. University of Hawaii Press, Honolulu HI, USA.

Thompson DW (1895) *Glossary of Greek Birds*. Henry Frowde, London, UK.

Troy J (1994) *The Sydney Language*. Produced with the assistance of the Australian Dictionaries Project and the Australian Institute of Aboriginal and Torres Strait Islander Studies, Canberra.

Van Gasse P (1999) *Naming of Rufous Songlark*. Birding-Aus Archives, <http://bioacoustics.cse.unsw.edu.au/birding-aus/1999–03/msg00221.html>.

Varro MT (116–27 BC) *De Lingua Latina*. Latin Library website, <http://www.thelatinlibrary.com/varro.ll5.html>.
Vieillot LJP (1816) *Analyse d'une Nouvelle Ornithologie Élémentaire*. Deterville, Paris, France.
Vieillot LJP (1816–19) *Nouveau Dictionnaire d'Histoire Naturelle*. 36 Volumes, 1816–19. Deterville, Paris.
Vigors N (1825) Descriptions of some rare, interesting, or hitherto uncharacterized subjects of Zoology. *Zoological Journal* **1**, 526–536.
Vigors N, Horsfield T (1827) A description of the Australian birds in the collection of the Linnean Society; with an attempt at arranging them according to their natural affinities. *Transactions of the Linnean Society of London* **15**(1), 170–331. doi:10.1111/j.1095-8339.1826.tb00115.x
von Middendorf ATh (1853) *Reise in den Äussersten Norden und Osten Sibiriens Während der Jahre 1843 und 1844*. Band 2, Teil 2. St Petersburg, Russia.
Wagler JG (1831) Einige Mittheilungen über Thiere Mexicos. *Isis* **24**, 510–535.
Wagler JG (1832a) Mittheilungen über einige merkwürdige Thiere. *Isis* **25**, 275–282.
Wagler JG (1832b) *Monographia Psittacorum. Abhandlungen der Königlich Bayerischen Akademie der Wissenschaften* **1**, 463–750.
Wakelin H (1968) Some notes on the birds of Norfolk Island. *Notornis* **15**(3), 156–175.
Walker RB (1952) Indian mynas on the Darling Downs. *Emu* **52**, 64–65. doi:10.1071/MU952062c
Wallace AR (1863) A list of the birds inhabiting the islands of Timor, Flores and Lombock, with descriptions of the new species. *Proceedings of the Zoological Society of London* **1863**, 489.
Watling T (1792–97) 'Soft-tailed Flycatcher', native name 'Mereangeree'. Watling Drawing no. 262. Natural History Museum, London, <http://www.nhm.ac.uk/jdsml/nature-online/first-fleet/nathist.dsml?sa=1&lastDisp=list¬es=true&beginIndex=254&>.
Wentworth WC (1819) *Statistical, Historical, and Political Description of NSW*. G and WB Whittaker, London, UK.
White HL (1911) Descriptions of two nests and eggs. *Emu* **11**(4), 249–250.
White HL (1916a) Description of new honeyeater of the genus *Ptilotis*, from North Australia. *Emu* **16**, 165.
White HL (1916b) Descriptions of new or rare eggs. *Emu* **16**, 159–164. doi:10.1071/MU916159
White J (1790) *Journal of a Voyage to New South Wales*. J Debrett, London, UK.
Whitlock F (1910) On the East Murchison. Four months' collecting trip. *Emu* **9**, 181–219 10.1071/MU909181.
Whitlock F (1922) Notes from the Nullarbor Plain. *Emu* **21**, 170–187. doi:10.1071/MU921170
Whitlock F (1927) Peculiarities in the distribution of birds in Western Australia. *Emu* **27**, 179–184. doi:10.1071/MU927179
Whittell HM (1940) Frederick Lawson Whitlock. *Emu* **39**, 279–286. doi:10.1071/MU939279
Wigan ML (1931) The Dusky Wood-Swallow. *Emu* **31**, 93–95. doi:10.1071/MU931093
Williamson C (1809) Oriental field sports. *The Sporting Magazine* **34**, 275–281.
Willughby F, Ray J (1676) *Ornithologiae Libri Tres*. Johanne Martin, London, UK.
Wilson J (1831) *Illustrations of Zoology: Being Representations of New, Rare, or Remarkable Subjects of the Animal Kingdom, Drawn and Coloured after Nature with Historical and Descriptive Details*. William Blackwood, Edinburgh.
Worthy TH (1997) A survey of historical laughing owl (*Sceloglaux albifacies*) specimens in museum collections. *Notornis* **44**(4), 241–252.
Xenophon (c. 380 BC) *Anabasis*. Tufts University website, Medford MA, USA, http://www.perseus.tufts.edu/hopper/text?doc=Perseus:abo:tlg,0032,006>.
Zeuner FE (1963) *A History of Domesticated Animals*. Harper & Row, New York, USA.
Zoological Club of the Linnean Society (1826) From January 1825 to April 1826. *Zoological Journal* **II**, 136.

Index of common names

Index of scientific names

Page numbers in **bold** type refer to illustrations